Open the door to the fascinating world of physics

Physics, the most fundamental of all natural sciences, will reveal to you the basic principles of the Universe. And while physics can seem challenging, its true beauty lies in the sheer simplicity of fundamental physical theories—theories and concepts that can alter and expand your view of the world around you. Other courses that follow will use the same principles, so it is important that you understand and are able to apply the various concepts and theories discussed in the text. **Physics for Scientists and Engineers, Sixth Edition** is your guide to this fascinating science.

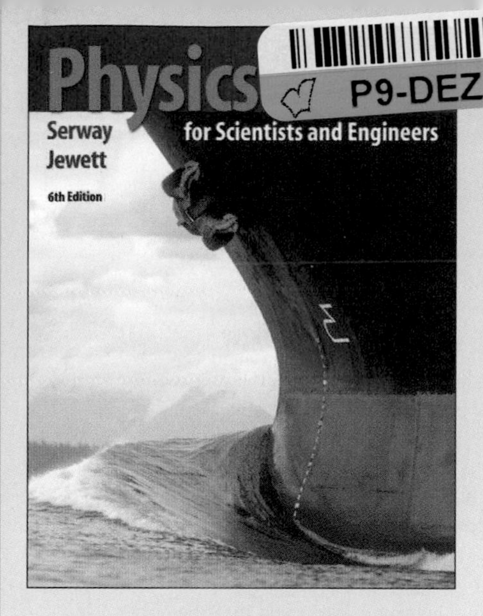

Your quick start for studying smart

Achieve success in your physics course by making the most of what **Physics for Scientists and Engineers, Sixth Edition** has to offer you. From a host of in-text features to a range of Web resources, you'll have everything you need to understand the natural forces and principles of physics:

- **Dynamic built-in study aids.** Throughout every chapter the authors have built in a wide range of examples, exercises, and illustrations that will help you understand and appreciate the laws of physics. *See pages 2 and 3 for more information.*
- **A powerful Web-based learning system.** The text is fully integrated with **PhysicsNow**, an interactive learning system that tailors itself to your needs in the course. It's like having a personal tutor available whenever you need it! *See pages 4–7 to explore* **PhysicsNow**.

Your *Quick Start for Studying Smart* begins with this special tour through the book. On the following pages you'll discover how **Physics for Scientists and Engineers, Sixth Edition** and **PhysicsNow** not only enhance your experience in this course, but help you to succeed!

Quick Start for Studying Smart!

Take charge of your success

Everything you need to succeed in your course is available to you in **Physics for Scientists and Engineers, Sixth Edition.** Authors Serway and Jewett have filled their text with learning tools and study aids that will clarify concepts and help you build a solid base of knowledge. The end result: confidence in the classroom, in your study sessions, and in your exams.

THE RIGHT APPROACH

Start out right! Early on in the text the authors outline a general problem-solving strategy that will enable you to increase your accuracy in solving problems, enhance your understanding of physical concepts, eliminate initial worry or lack of direction in approaching a problem, and organize your work. The problem-solving strategy is integrated into the *Coached Problems* found on **PhysicsNow** to reinforce this key skill. (See pages 4–7 for more information about the **PhysicsNow** Web-based and student-centered learning system.)

GENERAL PROBLEM-SOLVING STRATEGY

Conceptualize

- The first thing to do when approaching a problem is to *think about* and *understand* the situation. Study carefully any diagrams, graphs, tables, or photographs that accompany the problem. Imagine a movie, running in your mind, of what happens in the problem.
- If a diagram is not provided, you should almost always make a quick drawing of the situation. Indicate any known values, perhaps in a table or directly on your sketch.
- Now focus on what algebraic or numerical information is given in the problem. Carefully read the problem statement, looking for key phrases such as "starts from at rest" ($v_i = 0$), "stops" ($v_f = 0$), or "freely falls" ($a_y = -g = -9.80$ m/s^2).
- Now focus on the expected result of solving the problem. Exactly what is the question asking? Will the final result be numerical or algebraic? Do you know what units to expect?
- Don't forget to incorporate information from your own experiences and common sense. What should a reasonable answer look like? You wouldn't expect to calculate the speed of an automobile to be

Analyze

- Now you must analyze the problem and strive for a mathematical solution. Because you have already categorized the problem, it should not be too difficult to select relevant equations that apply to the type of situation in the problem. For example, if the problem involves a particle moving under constant acceleration, Equations 2.9 to 2.13 are relevant.
- Use algebra (and calculus, if necessary) to solve symbolically for the unknown variable in terms of what is given. Substitute in the appropriate numbers, calculate the result, and round it to the proper number of significant figures.

Finalize

- This is the most important part. Examine your numerical answer. Does it have the correct units? Does it meet your expectations from your conceptualization of the problem? What about the algebraic form of the result — before you substituted numerical values? Does it make sense? Examine the variables in the problem to see whether the answer would change in a physically meaningful way if they were drastically increased or decreased or even became zero. Looking at limiting cases to see whether they yield expected values is a very useful way to make sure that you are obtaining reasonable results.

  Think about how this problem compares with others you have done. How was it similar? In what critical ways did it differ? Why was this problem assigned? You should have learned something by doing it. Can you figure out what? If it is a new category of problem, be sure you understand it so that you can use it as a model for solving future problems in the same category.

[W]hen solving complex problems, you may need to iden[ti]fy a series of sub-problems and apply the problem-solv-

PROBLEM-SOLVING HINTS

Problem-Solving Hints help you approach homework assignments with greater confidence. General strategies and suggestions are included for solving the types of problems featured in the worked examples, end-of-chapter problems, and **PhysicsNow**. This feature helps you identify the essential steps in solving problems and increases your skills as a problem solver.

PROBLEM-SOLVING HINTS

Applying Newton's Laws

The following procedure is recommended when dealing with problems involving Newton's laws:

- Draw a simple, neat diagram of the system to help *conceptualize* the problem.
- *Categorize* the problem: if any acceleration component is zero, the particle is in equilibrium in this direction and $\Sigma F = 0$. If not, the particle is undergoing an acceleration, the problem is one of nonequilibrium in this direction, and $\Sigma F = ma$.
- *Analyze* the problem by isolating the object whose motion is being analyzed. Draw a free-body diagram for this object. For systems containing more than one object, draw *separate* free-body diagrams for each object. *Do not* include in the free-body diagram forces exerted by the object on its surroundings.
- Establish convenient coordinate axes for each object and find the components of the forces along these axes. Apply Newton's second law, $\Sigma \mathbf{F} = m\mathbf{a}$, in component form. Check your dimensions to make sure that all terms have uni[ts of force.]
- Solve the comp[...] have as many i[...] complete soluti[...]
- *Finalize* by maki[...] Also check the [...] variables. By d[...]

WORKED EXAMPLES

Reinforce your understanding of essential problem-solving techniques using a large number of realistic *Worked Examples*. In many cases, these examples serve as models for solving the end-of-chapter problems. Numerous *Worked Examples* include specific references to the general problem-solving strategy to illustrate the underlying concepts and methodology used in arriving at a correct solution. This will help you understand the logic behind the solution and the advantage of using a particular approach to solve the problem. **PhysicsNow** also features a number of worked examples to further enhance your understanding of problem solving and to give you even more practice solving problems.

Example 4.3 The Long Jump

A long-jumper (Fig. 4.12) leaves the ground at an angle of 20.0° above the horizontal and at a speed of 11.0 m/s.

(A) How far does he jump in the horizontal direction? (Assume his motion is equivalent to that of a particle.)

Solution We *conceptualize* the motion of the long-jumper as equivalent to that of a simple projectile such as the ball in Example 4.2, and *categorize* this problem as a projectile motion problem. Because the initial speed and launch angle are given, and because the final height is the same as the initial height, we further categorize this problem as satisfying the conditions for which Equations 4.13 and 4.14 can be used. This is the most direct way to *analyze* this problem [...] describi[...] the gen[...]

provides a graphical representation of the flight of the long-jumper. As before, we set our origin of coordinates at the takeoff point and label the peak as Ⓐ and the landing point as Ⓑ. The horizontal motion is described by Equation 4.11:

$$x_f = x_B = (v_i \cos \theta_i) t_B = (11.0 \text{ m/s})(\cos 20.0°) t_B$$

The value of x_B can be found if the time of landing t_B is known. We can find t_B by remembering that $a_y = -g$ and by using the y part of Equation 4.8a. We also note that at the top of the jump the vertical component of velocity v_{yA} is zero:

$$v_{yf} = v_{yA} = v_i \sin \theta_i - g t_A$$

Figure 4.12 (Example 4.3) Mike Powell, current holder of the world long jump record of 8.95 m.

Mike Powell/Allsport/Getty Images

This is the time at which the long-jumper is at the *top* of the jump. Because of the symmetry of the vertical motion, another 0.384 s passes before the jumper returns to the ground. Therefore, the time at which the jumper lands is $t_B = 2t_A = 0.768$ s. Substituting this value into the above expression for x_f gives

$$x_f = x_B = (11.0 \text{ m/s})(\cos 20.0°)(0.768 \text{ s}) = \boxed{7.94 \text{ m}}$$

This is a reasonable distance for a world-class athlete.

(B) What is the maximum height reached?

Solution We find the maximum height reached by using Equation 4.12:

$$y_{max} = y_A = (v_i \sin \theta_i) t_A - \tfrac{1}{2} g t_A^2$$
$$= (11.0 \text{ m/s})(\sin 20.0°)(0.384 \text{ s})$$
$$-\tfrac{1}{2}(9.80 \text{ m/s}^2)(0.384 \text{ s})^2 = \boxed{0.722 \text{ m}}$$

To *finalize* this problem, find the answers to parts (a) and (b) using Equations 4.13 and 4.14. The results should agree. Treating the long-jumper as a particle is an oversimplification. Nevertheless, the values obtained are consistent with experience in sports. We learn that we can model a complicated system such as a long-jumper as a particle and still obtain results that are reasonable.

$n = -1$, and we can write the acceleration expression as

where k is a dimensionless constant of proportionality. Knowing the dimensions of a, r, and v, we see that the dimensional equation must be

$$\frac{L}{T^2} = L^n\left(\frac{L}{T}\right)^m = \frac{L^{n+m}}{T^m}$$

$$a = kr^{-1}v^2 = k\frac{v^2}{r}$$

When we discuss uniform circular motion later, we shall see that $k = 1$ if a consistent set of units is used. The constant k would not equal 1 if, for example, v were in km/h and you wanted a in m/s^2.

▲ PITFALL PREVENTION

1.5 Always Include Units

When performing calculations, include the units for every quantity and carry the units through the entire calculation. Avoid the temptation to drop the units early and then attach the expected units once you have an answer. By including the units in every step, you can detect errors if the units for the answer turn out to be incorrect.

1.5 Conversion of Units

Sometimes it is necessary to convert units from one measurement system to another, or to convert within a system, for example, from kilometers to meters. Equalities between SI and U.S. customary units of length are as follows:

1 mile = 1 609 m = 1.609 km 1 ft = 0.304 8 m = 30.48 cm

1 m = 39.37 in. = 3.281 ft 1 in. = 0.025 4 m = 2.54 cm (exactly)

A more complete list of conversion factors can be found in Appendix A.

Units can be treated as algebraic quantities that can cancel each other. For example, suppose we wish to convert 15.0 in. to centimeters. Because 1 in. is defined as exactly 2.54 cm, we find that

PITFALL PREVENTIONS

An easy way to make sure you avoid common mistakes—while studying *and* while taking exams! These helpful notes, found in the margins, allow you to correct misconceptions before applying them to new material.

Example 4.5 That's Quite an Arm!

A stone is thrown from the top of a building upward at an angle of 30.0° to the horizontal with an initial speed of 20.0 m/s, as shown in Figure 4.14. If the height of the building is 45.0 m,

(A) how long before the stone hits the ground?

Solution We *conceptualize* the problem by studying Figure 4.14, in which we have indicated the various parameters. By now, it should be natural to *categorize* this as a projectile motion problem.

To *analyze* the problem, let us once again separate motion into two components. The initial x and y components of the stone's velocity are

$$v_{xi} = v_i \cos\theta_i = (20.0\text{ m/s})\cos 30.0° = 17.3\text{ m/s}$$

$$v_{yi} = v_i \sin\theta_i = (20.0\text{ m/s})\sin 30.0° = 10.0\text{ m/s}$$

To find t, we can use $y_f = y_i + v_{yi}t + \frac{1}{2}a_y t^2$ (Eq. 4.9a) with $y_i = 0$, $y_f = -45.0$ m, $a_y = -g$, and $v_{yi} = 10.0$ m/s (there is a negative sign on the numerical value of y_f because we have chosen the top of the building as the origin):

$$-45.0\text{ m} = (10.0\text{ m/s})t - \tfrac{1}{2}(9.80\text{ m/s}^2)t^2$$

Solving the quadratic equation for t gives, for the positive root, $t = 4.22$ s. To *finalize* this part, think: Does the negative root have any physical meaning?

(B) What is the speed of the stone just before it strikes the ground?

Solution We can use Equation 4.8a, $v_{yf} = v_{yi} + a_y t$, with $t = 4.22$ s to obtain the y component of the velocity just before the stone strikes the ground:

$$v_{yf} = 10.0\text{ m/s} - (9.80\text{ m/s}^2)(4.22\text{ s}) = -31.4\text{ m/s}$$

Because $v_{xf} = v_{xi} = 17.3$ m/s, the required speed is

$$v_f = \sqrt{v_{xf}^2 + v_{yf}^2} = \sqrt{(17.3)^2 + (-31.4)^2}\text{ m/s} = 35.9\text{ m/s}$$

What If? What if a horizontal wind is blowing in the same direction as the ball is thrown and it causes the ball to have a horizontal acceleration component $a_x = 0.500$ m/s^2. Which part of this example, (a) or (b), will have a different answer?

Answer Recall that the motions in the x and y directions are independent. Thus, the horizontal wind cannot affect the vertical motion. The vertical motion determines the time of the projectile in the air, so the answer to (a) does not change. The wind will cause the horizontal velocity component to increase with time, so that the final speed will change in part (b).

We can find the new final horizontal velocity component by using Equation 4.8a:

$$v_{xf} = v_{xi} + a_x t = 17.3\text{ m/s} + (0.500\text{ m/s}^2)(4.22\text{ s}) = 19.4\text{ m/s}$$

and the new final speed:

$$v_f = \sqrt{v_{xf}^2 + v_{yf}^2} = \sqrt{(19.4)^2 + (-31.4)^2}\text{ m/s} = 36.9\text{ m/s}$$

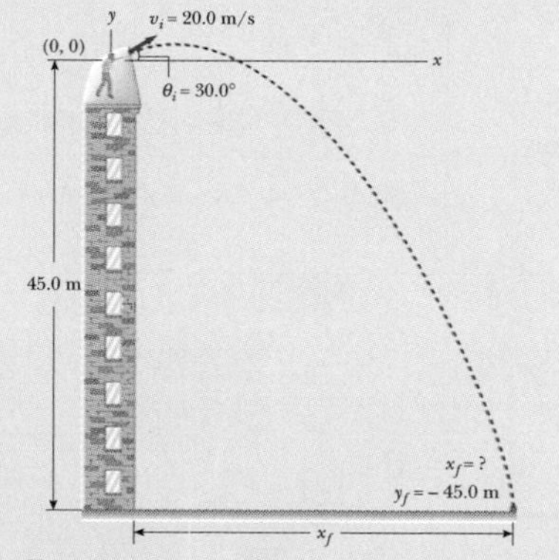

Figure 4.14 (Example 4.5) A stone is thrown from the top of a building.

...nt of ... final

...mple link at http://www.pse6.com.

WHAT IF?

What if you could get even more out of the text's worked examples? You can with this feature. The authors change some of the data or assumptions in about one-third of the *Worked Examples* and then explore the consequences. This allows you to apply concepts and problem-solving skills to new situations as well as test the final result to see if it's realistic.

Quick Quiz 5.2 An object experiences no acceleration. Which of the following *cannot* be true for the object? (a) A single force acts on the object. (b) No forces act on the object. (c) Forces act on the object, but the forces cancel.

Quick Quiz 5.3 An object experiences a net force and exhibits an acceleration in response. Which of the following statements is *always* true? (a) The object moves in the direction of the force. (b) The ... velocity. (c) The acceleration is in the same ... the object increases.

Quick Quiz 5.4 You push an objec... with a constant force for a time interval Δ... object. You repeat the experiment, but with ... terval is now required to reach the same fina... (e) $\Delta t/4$.

Answers to Quick Quizzes

5.1 (d). Choice (a) is true. Newton's first law tells us that motion requires no force: an object in motion continues to move at constant velocity in the absence of external forces. Choice (b) is also true. A stationary object can have several forces acting on it, but if the vector sum of all these external forces is zero, there is no net force and the object remains stationary.

5.2 (a). If a single force acts, this force constitutes the net force and there is an acceleration according to Newton's second law.

5.3 (c). Newton's second law relates only the force and the acceleration. Direction of motion is part of an object's *velocity*, and force determines the direction of acceleration, not that of velocity.

5.4 (d). With twice the force, the object will experience twice the acceleration. Because the force is constant, the acceler-

QUICK QUIZZES

Test your understanding of the concepts before investigating the worked examples. Found throughout each chapter, these quizzes give you an opportunity to test your conceptual understanding while also making the material more interactive. Comprehensive answers are found at the end of each chapter. A supplementary set of *Quick Quizzes* is available on the text's Web site, accessible through **www.pse6.com**.

"You do not know anything until you have practiced."

R. P. Feynman, Nobel Laureate in Physics

What do you need to learn now?

Take charge of your learning with **PhysicsNow**, a powerful, interactive study tool that will help you manage and maximize your study time. This collection of dynamic technology resources will assess your unique study needs, giving you an individualized learning plan that will enhance your conceptual and computational skills. Designed to maximize your time investment, **PhysicsNow** helps you succeed by focusing your study time on the concepts you need to master. And best of all, **PhysicsNow** is FREE with every new copy of this text!

INTERACTIVE LEARNING BEGINS WITH THIS BOOK

PhysicsNow directly links with **Physics for Scientists and Engineers, Sixth Edition**. **PhysicsNow** and this text were built in concert to enhance each other and provide you with a seamless, integrated learning system. As you work through the chapters you'll see notes that direct you to the media-enhanced activities on **PhysicsNow (www.pse6.com).** This precise page-by-page integration means you'll spend less time flipping through pages and navigating Web sites for useful exercises. It is far easier to understand physics if you see it in action, and **PhysicsNow** enables you to *become* a part of the action!

Take a practice test for this chapter by clicking on the Practice Test link at http://www.pse6.com.

SUMMARY

Scalar quantities are those that have only magnitude and no associated direction. **Vector quantities** have both magnitude and direction and obey the laws of vector addition. The magnitude of a vector is *always* a positive number.

When two or more vectors are added together, all of them must have the same units and all of them must be the same type of quantity. We can add two vectors **A** and **B** graphically. In this method (Fig. 3.6), the resultant vector **R** = **A** + **B** runs from the tail of **A** to the tip of **B**.

A second method of adding vectors involves **components** of the vectors. The x component A_x of the vector **A** is equal to the projection of **A** along the x axis of a coordinate system, as shown in Figure 3.13, where $A_x = A\cos\theta$. The y component A_y of **A** is the projection of **A** along the y axis, where $A_y = A\sin\theta$. Be sure you can determine which trigonometric functions you should use in all situations, especially when θ is defined as something other than the counterclockwise angle from the positive x axis.

If a vector **A** has an x component A_x and a y component A_y, the vector can be expressed in unit – vector form as $\mathbf{A} = A_x\hat{\mathbf{i}} + A_y\hat{\mathbf{j}}$. In this is notation, $\hat{\mathbf{i}}$ is a unit vector pointing in the positive x direction, and $\hat{\mathbf{j}}$ is a unit vector pointing in the positive y direction. Because $\hat{\mathbf{i}}$ and $\hat{\mathbf{j}}$ are unit vectors, $|\hat{\mathbf{i}}| = |\hat{\mathbf{j}}| = 1$.

We can find the resultant of two or more vectors by resolving all vectors into their x and y components, adding their resultant x and y components, and then using the Pythagorean theorem to find the magnitude of the resultant vector. We can find the angle that the resultant vector makes with respect to the x axis by using a suitable trigonometric function.

QUESTIONS

1. Two vectors have unequal magnitudes. Can their sum be zero? Explain.
2. Can the magnitude of a particle's displacement be greater

4. Which of the following are vectors and which are not: force, temperature, the volume of water in a can, the ratings of a TV show, the height of a building, the velocity of ... the Universe?

... plane. For what orientations of **A** ... ents be negative? For what orienta... s have opposite signs?

Example 4.5 That's Quite an Arm!

A stone is thrown from the top of a building upward at an angle of 30.0° to the horizontal with an initial speed of 20.0 m/s, as shown in Figure 4.14. If the height of the building is 45.0 m,

(A) how long before the stone hits the ground?

Solution We *conceptualize* the problem by studying Figure 4.14, in which we have indicated the various parameters. By now, it should be natural to *categorize* this as a projectile motion problem.

To *analyze* the problem, let us once again separate motion into two components. The initial x and y components of the stone's velocity are

$$v_{xi} = v_i\cos\theta_i = (20.0\text{ m/s})\cos 30.0° = 17.3\text{ m/s}$$

$$v_{yi} = v_i\sin\theta_i = (20.0\text{ m/s})\sin 30.0° = 10.0\text{ m/s}$$

To find t, we can use $y_f = y_i + v_{yi}t + \frac{1}{2}a_y t^2$(Eq. 4.9a) with $y_i = 0$, $y_f = -45.0$ m, $a_y = -g$, and $v_{yi} = 10.0$ m/s (there is a negative sign on the numerical value of y_f because we have chosen the top of the building as the origin):

$$-45.0\text{ m} = (10.0\text{ m/s})t - \tfrac{1}{2}(9.80\text{ m/s}^2)t^2$$

Solving the quadratic equation for t gives, for the positive root, $t = $ 4.22 s. To *finalize* this part, think: Does the negative root have any physical meaning?

(B) What is the speed of the stone just before it strikes the ground?

Solution We can use Equation 4.8a, $v_{yf} = v_{yi} + a_y t$, with $t = 4.22$ s to obtain the y component of the velocity just before the stone strikes the ground:

$$v_{yf} = 10.0\text{ m/s} - (9.80\text{ m/s}^2)(4.22\text{ s}) = -31.4\text{ m/s}$$

Because $v_{xf} = v_{xi} = 17.3$ m/s, the required speed is

$$v_f = \sqrt{v_{xf}^2 + v_{yf}^2} = \sqrt{(17.3)^2 + (-31.4)^2}\text{ m/s} = 35.9\text{ m/s}$$

To *finalize* this part, is it reasonable that the y component of the final velocity is negative? Is it reasonable that the final speed is larger than the initial speed of 20.0 m/s?

What If? What if a horizontal wind is blowing in the same direction as the ball is thrown and it causes the ball to have a horizontal acceleration component $a_x = 0.500$ m/s^2. Which part of this example, (a) or (b), will have a different answer?

Answer Recall that the motions in the x and y directions are independent. Thus, the horizontal wind cannot affect the vertical motion. The vertical motion determines the time of the projectile in the air, so the answer to (a) does not change. The wind will cause the horizontal velocity component to increase with time, so that the final speed will change in part (b).

We can find the new final horizontal velocity component by using Equation 4.8a:

$$v_{xf} = v_{xi} + a_x t = 17.3\text{ m/s} + (0.500\text{ m/s}^2)(4.22\text{ s}) = 19.4\text{ m/s}$$

and the new final speed:

$$v_f = \sqrt{v_{xf}^2 + v_{yf}^2} = \sqrt{(19.4)^2 + (-31.4)^2}\text{ m/s} = 36.9\text{ m/s}$$

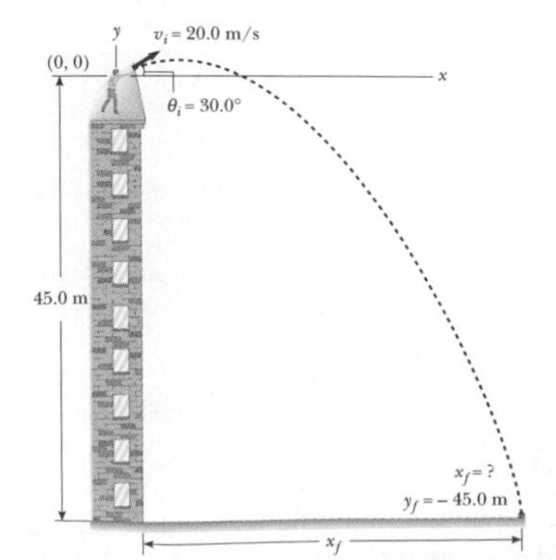

Figure 4.14 (Example 4.5) A stone is thrown from the top of a building.

Investigate this situation at the Interactive Worked Example link at http://www.pse6.com.

Quick Start for Studying Smart!

GO ONLINE AT www.pse6.com

Log on to **PhysicsNow** at **www.pse6.com** by using the free pincode packaged with this text.* You'll immediately notice the system's easy-to-use, browser-based format. Getting to where you need to go is as easy as a click of the mouse. The **PhysicsNow** system is made up of three interrelated parts:

- **How Much Do I Know?**
- **What Do I Need to Learn?**
- **What Have I Learned?**

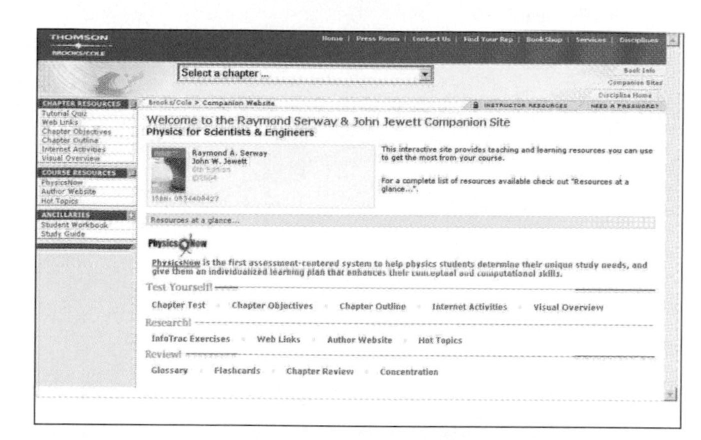

These three interrelated elements work together, but are distinct enough to allow you the freedom to explore only those assets that meet your personal needs. You can use **PhysicsNow** like a traditional Web site, accessing all assets of a particular chapter and exploring on your own. The best way to maximize the system and *your* time is to start by taking the *Pre-Test*.

* Free PIN codes are only available with new copies of **Physics for Scientists and Engineers, Sixth Edition.**

HOW MUCH DO I KNOW?

The Pre-Test is the first step in creating your *Personalized Learning Plan*. Each *Pre-Test* is based on the end-of-chapter homework problems and includes approximately 15 questions.

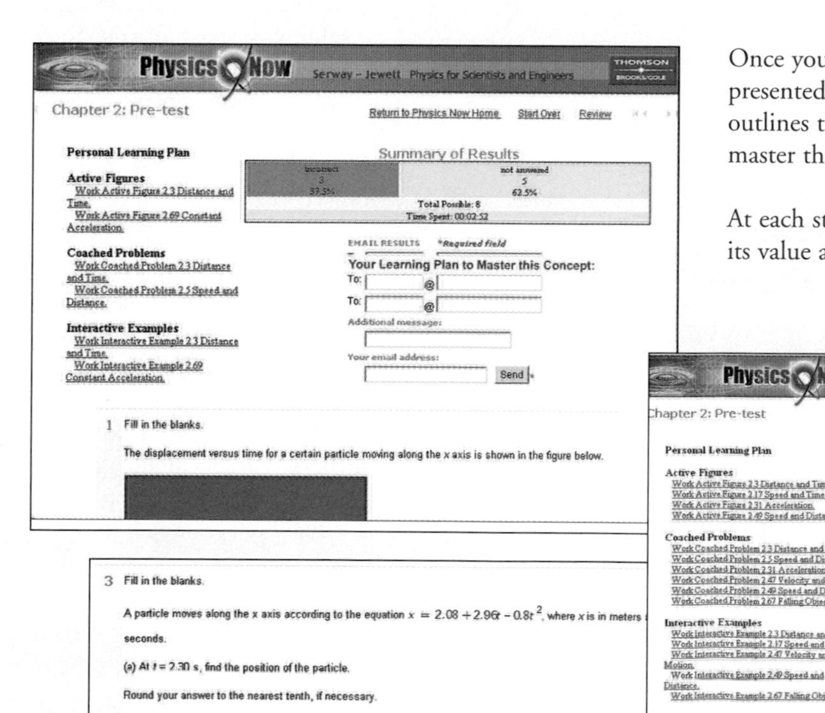

Once you've completed the *Pre-Test* you'll be presented with a detailed *Learning Plan* that outlines the elements you need to review to master the chapter's most essential concepts.

At each stage, the text is referenced to reinforce its value as a learning tool.

3 Fill in the blanks.

A particle moves along the x axis according to the equation $x = 2.08 + 2.96t - 0.8t^2$, where x is in meters and t is in seconds.

(a) At t = 2.30 s, find the position of the particle.

Round your answer to the nearest tenth, if necessary.

2 m

(b) What is its velocity?

3 m/s

(c) What is its acceleration?

4 m/s^2

status: incorrect (0.0)
correct 4.7
−0.72
−1.6
your answer: 2
3
4

Personal Learning Plan
Work Active Figure 2.17 Speed and Time.
Work Interactive Example 2.17 Speed and Time.

Turn the page to view problems from a sample *Personalized Learning Plan*.

WHAT DO I NEED TO LEARN?

Once you've completed the *Pre-Test* you're ready to work the problems in your *Personalized Learning Plan*—problems that will help you master concepts essential to your success in this course.

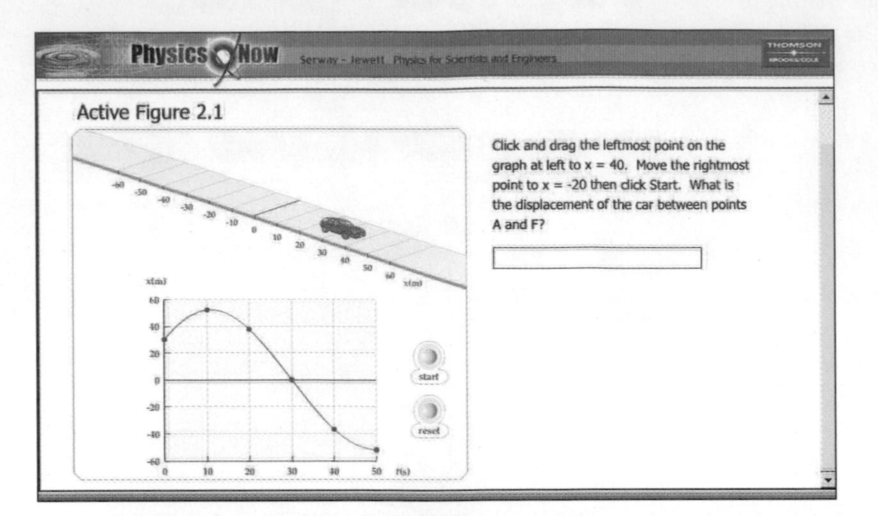

More than 200 *Active Figures* are taken from the text and animated to help you visualize physics in action. Each figure is paired with a question to help you focus on physics at work, and a brief quiz ensures that you understand the concept played out in the animations.

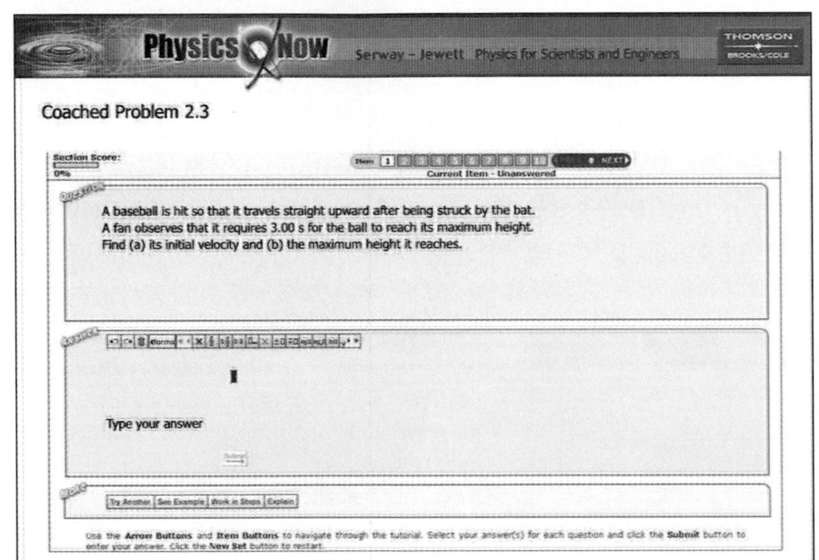

You'll continue to master the concepts though *Coached Problems.* These engaging problems reinforce the lessons in the text by taking a step-by-step approach to problem-solving methodology. Each *Coached Problem* gives you the option of working a question and receiving feedback, or seeing a solution worked for you. You'll find approximately five *Coached Problems* per chapter.

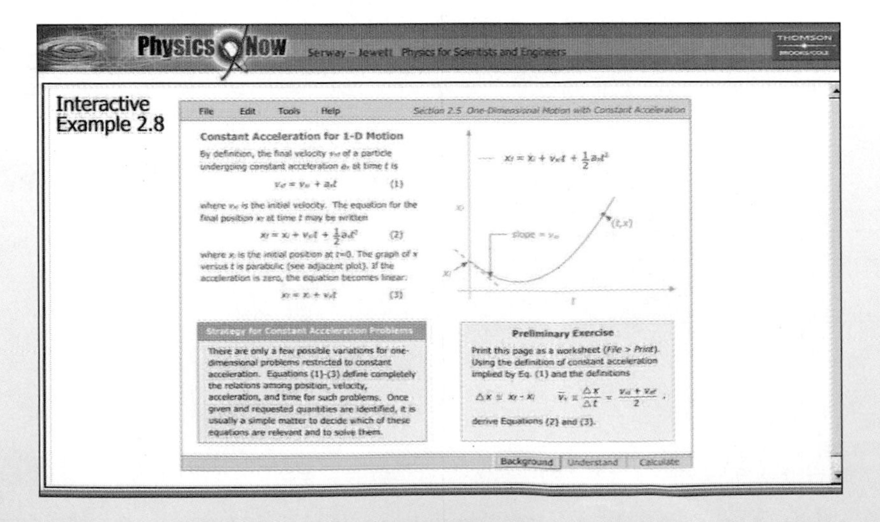

You'll strengthen your problem-solving and visualization skills by working through the *Interactive Examples.* Each step in the examples uses the authors' problem-solving methodology that is introduced in the text (see page 2 of this Visual Preface). You'll find *Interactive Examples* for each chapter of the text.

WHAT HAVE I LEARNED?

After working through the problems highlighted in your personal *Learning Plan* you'll move on to a *Chapter Quiz.* These multiple-choice quizzes present you with questions that are similar to those you might find in an exam. You can even e-mail your quiz results to your instructor.

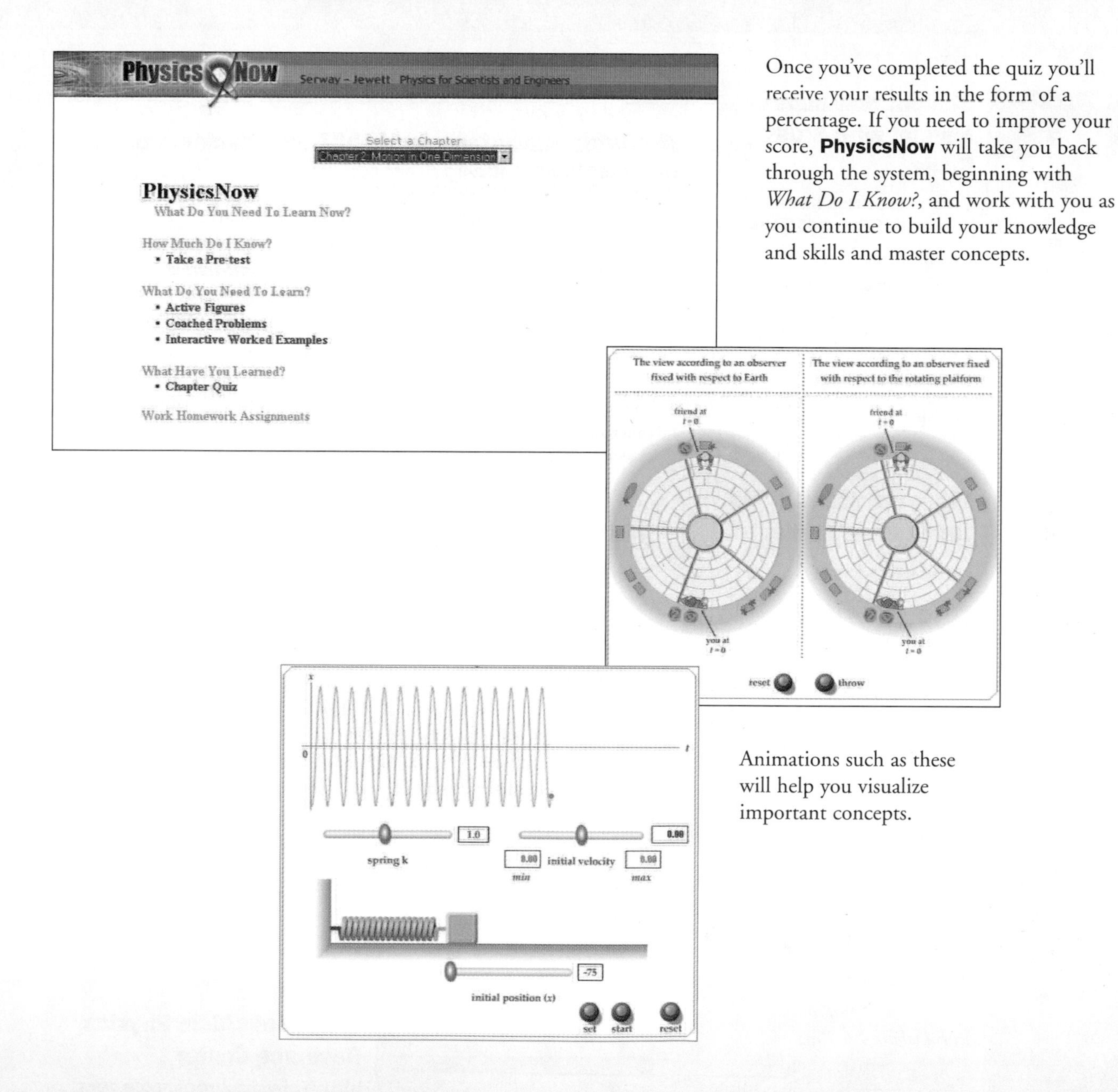

Once you've completed the quiz you'll receive your results in the form of a percentage. If you need to improve your score, **PhysicsNow** will take you back through the system, beginning with *What Do I Know?*, and work with you as you continue to build your knowledge and skills and master concepts.

Animations such as these will help you visualize important concepts.

Chart your own course for success . . .

Log on to **www.pse6.com** to take advantage of **PhysicsNow!**

Make the most of the course and your time with these exclusive study tools

Enrich your experience outside of the classroom with a host of resources designed to help you excel in the course. To purchase any of these supplements, contact your campus bookstore or visit our online BookStore at www.brookscole.com.

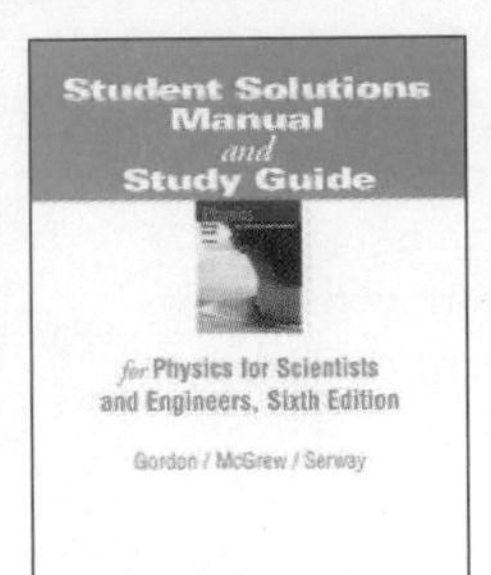

Student Solutions Manual with Study Guide

Volume I ISBN: 0-534-40855-9
Volume II ISBN: 0-534-40856-7

by John R. Gordon, Ralph McGrew, and Raymond Serway

This two-volume manual features detailed solutions to 20% of the end-of-chapter problems from the text. The manual also features a list of important equations, concepts, and answers to selected end-of-chapter questions.

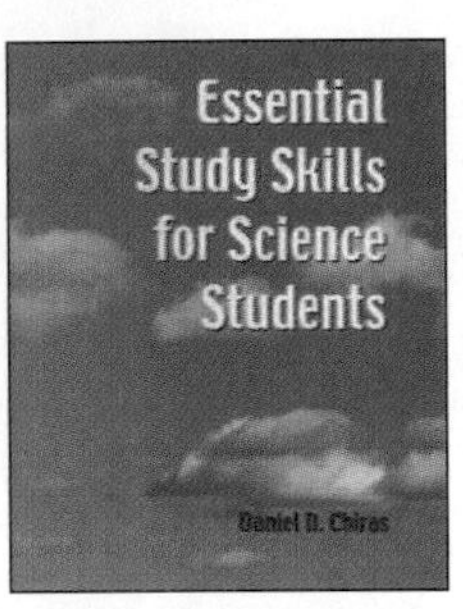

Essential Study Skills for Science Students

ISBN: 0-534-37595-2

by Daniel D. Chiras

Written specifically for science students, this book discusses how to develop good study habits, sharpen memory, learn more quickly, get the most out of lectures, prepare for tests, produce excellent term papers, and improve critical-thinking skills.

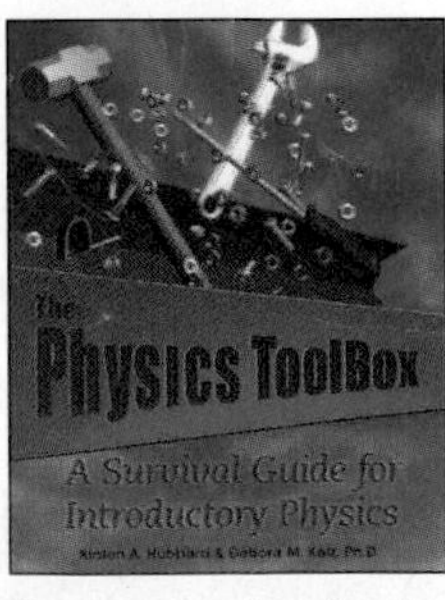

The Physics Toolbox: A Survival Guide for Introductory Physics

ISBN: 0-03-034652-5

by Kirsten A. Hubbard and Debora M. Katz

This "paperback mentor" gives you the material critical for success in physics, including an introduction to the nature of physics and science, a look at what to expect and how to succeed, a verbal overview of the concepts you'll encounter, and an extensive review of the math you'll need to solve the problems.

WebTUTOR™ Advantage

WebTutor™ Advantage on WebCT and Blackboard

WebCT ISBN: 0-534-40859-1
Blackboard ISBN: 0-534-40950-4

WebTutor Advantage offers real-time access to a full array of study tools, including chapter outlines, summaries, learning objectives, glossary flashcards (with audio), practice quizzes, **InfoTrac® College Edition** exercises, and Web links. **WebTutor Advantage** also provides robust communication tools, such as a course calendar, asynchronous discussion, real-time chat, a whiteboard, and an integrated e-mail system. Also new to **WebTutor Advantage** is access to *NewsEdge*, an online news service that brings the latest news to the **WebTutor Advantage** site daily.

Contact your instructor for more information.

Additional Web resources . . .

available FREE to you with each new copy of this text:

InfoTrac® College Edition

When you purchase a new copy of **Physics for Scientists and Engineers, Sixth Edition**, you automatically receive a FREE four-month subscription to **InfoTrac College Edition!** Newly improved, this extensive online library opens the door to the full text (not just abstracts) of countless articles from thousands of publications including *American Scientist, Physical Review, Science, Science Weekly*, and more! Use the passcode included with the new copy of this text and log on to **www.infotrac-college.com** to explore the wealth of resources available to you—24 hours a day and seven days a week!

Available only to college and university students. Journals subject to change.

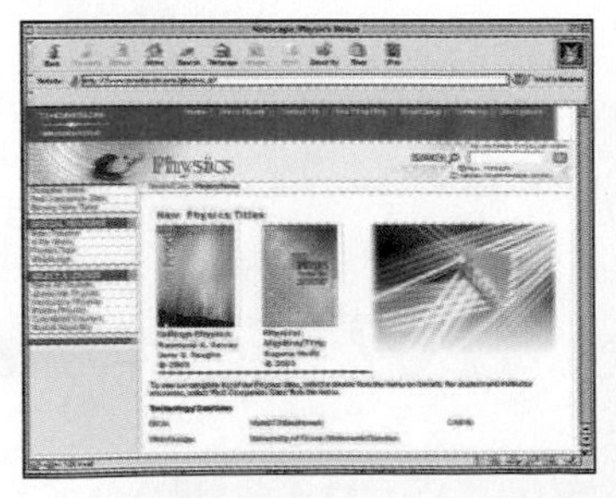

The Brooks/Cole Physics Resource Center

http://physics.brookscole.com

Here you'll find even more opportunities to hone your skills and expand your knowledge. **The Brooks/Cole Physics Resource Center** is filled with helpful content that will engage you while you master the material. You'll find additional online quizzes, Web links, animations, and *NewEdge*—an online news service that brings the latest news to this site daily.

We dedicate this book to the courageous astronauts who died on the space shuttle Columbia *on February 1, 2003. The women and men of the international team lost their lives not in a contest between countries or a struggle for necessities but in advancing one of humankind's noblest creations—science.*

6th Edition

PHYSICS

for Scientists and Engineers

Chapters 15 – 22

2

Raymond A. Serway

John W. Jewett, Jr.

California State Polytechnic University–Pomona

Australia • Canada • Mexico • Singapore • Spain
United Kingdom • United States

Editor-in-Chief: Michelle Julet
Publisher: David Harris
Physics Editor: Chris Hall
Development Editor: Susan Dust Pashos
Assistant Editor: Rebecca Heider, Alyssa White
Editorial Assistant: Seth Dobrin, Jessica Howard
Technology Project Manager: Sam Subity
Marketing Manager: Kelley McAllister
Marketing Assistant: Sandra Perin
Advertising Project Manager: Stacey Purviance
Project Manager, Editorial Production: Teri Hyde
Print/Media Buyer: Barbara Britton
Permissions Editor: Joohee Lee
Production Service: Sparkpoint Communications, a division of J. B. Woolsey Associates
Text Designer: Lisa Devenish
Photo Researcher: Terri Wright
Copy Editor: Andrew Potter
Illustrator: Rolin Graphics
Cover Designer: Lisa Devenish
Cover Image: Water Displaced by Oil Tanker, © Stuart Westmorland/CORBIS
Compositor: Progressive Information Technologies
Cover Printer: Quebecor World, Versailles
Printer: Quebecor World, Versailles

Printed in the United States of America
1 2 3 4 5 6 7 07 06 05 04 03

For more information about our products, contact us at:
Thomson Learning Academic Resource Center
1-800-423-0563

For permission to use material from this text, contact us by:
Phone: 1-800-730-2214
Fax: 1-800-730-2215
Web: http://www.thomsonrights.com

Library of Congress Control Number: 2003100126

PHYSICS FOR SCIENTISTS AND ENGINEERS, Sixth Edition

0-534-40848-6, Volume 1 (Chapters 1–14)

0-534-40849-4, Volume 2 (Chapters 15–22)

0-534-40850-8, Volume 3 (Chapters 23–34)

0-534-40853-2, Volume 4 (Chapters 35–39)

Brooks/Cole—Thomson Learning
10 Davis Drive
Belmont, CA 94002
USA

Asia
Thomson Learning
5 Shenton Way #01-01
UIC Building
Singapore 068808

Australia
Nelson Thomson Learning
102 Dodds Street
South Melbourne, Victoria 3205
Australia

Canada
Nelson Thomson Learning
1120 Birchmount Road
Toronto, Ontario M1K 5G4
Canada

Europe/Middle East/Africa
Thomson Learning
High Holborn House
50/51 Bedford Row
London WC1R 4LR
United Kingdom

Latin America
Thomson Learning
Seneca, 53
Colonia Polanco
11560 Mexico D.F.
Mexico

Spain
Paraninfo Thomson Learning
Calle/Magallanes, 25
28015 Madrid, Spain

Contents Overview

PART 1 Mechanics 1

1 Physics and Measurement *2*
2 Motion in One Dimension *23*
3 Vectors *58*
4 Motion in Two Dimensions *77*
5 The Laws of Motion *111*
6 Circular Motion and Other Applications of Newton's Laws *150*
7 Energy and Energy Transfer *181*
8 Potential Energy *217*
9 Linear Momentum and Collisions *251*
10 Rotation of a Rigid Object About a Fixed Axis *292*
11 Angular Momentum *336*
12 Static Equilibrium and Elasticity *362*
13 Universal Gravitation *389*
14 Fluid Mechanics *420*

PART 2 Oscillations and Mechanical Waves 451

15 Oscillatory Motion *452*
16 Wave Motion *486*
17 Sound Waves *512*
18 Superposition and Standing Waves *543*

PART 3 Thermodynamics 579

19 Temperature *580*
20 Heat and the First Law of Thermodynamics *604*
21 The Kinetic Theory of Gases *640*
22 Heat Engines, Entropy, and the Second Law of Thermodynamics *667*

PART 4 Electricity and Magnetism 705

23 Electric Fields *706*
24 Gauss's Law *739*
25 Electric Potential *762*
26 Capacitance and Dielectrics *795*
27 Current and Resistance *831*
28 Direct Current Circuits *858*
29 Magnetic Fields *894*
30 Sources of the Magnetic Field *926*
31 Faraday's Law *967*
32 Inductance *1003*
33 Alternating Current Circuits *1033*
34 Electromagnetic Waves *1066*

PART 5 Light and Optics 1093

35 The Nature of Light and the Laws of Geometric Optics *1094*
36 Image Formation *1126*
37 Interference of Light Waves *1176*
38 Diffraction Patterns and Polarization *1205*

PART 6 Modern Physics 1243

39 Relativity *1244*

Appendices *A.1*
Answers to Odd-Numbered Problems *A.37*
Index *I.1*

Steve Niedorf/Getty Images

Table of Contents

PART 1 **Mechanics** 1

Chapter 1 Physics and Measurement *2*
1.1 Standards of Length, Mass, and Time *4*
1.2 Matter and Model Building *7*
1.3 Density and Atomic Mass *9*
1.4 Dimensional Analysis *10*
1.5 Conversion of Units *12*
1.6 Estimates and Order-of-Magnitude Calculations *13*
1.7 Significant Figures *15*

Chapter 2 Motion in One Dimension *23*
2.1 Position, Velocity, and Speed *24*
2.2 Instantaneous Velocity and Speed *28*
2.3 Acceleration *31*
2.4 Motion Diagrams *34*
2.5 One-Dimensional Motion with Constant Acceleration *36*
2.6 Freely Falling Objects *40*
2.7 Kinematic Equations Derived from Calculus *44*
General Problem-Solving Strategy *47*

Chapter 3 Vectors *58*
3.1 Coordinate Systems *59*
3.2 Vector and Scalar Quantities *60*
3.3 Some Properties of Vectors *61*
3.4 Components of a Vector and Unit Vectors *65*

elektraVision/Index Stock Imagery

Chapter 4 Motion in Two Dimensions *77*
4.1 The Position, Velocity, and Acceleration Vectors *78*
4.2 Two-Dimensional Motion with Constant Acceleration *80*
4.3 Projectile Motion *83*
4.4 Uniform Circular Motion *91*
4.5 Tangential and Radial Acceleration *94*
4.6 Relative Velocity and Relative Acceleration *96*

Chapter 5 The Laws of Motion *111*
5.1 The Concept of Force *112*
5.2 Newton's First Law and Inertial Frames *114*
5.3 Mass *116*
5.4 Newton's Second Law *116*
5.5 The Gravitational Force and Weight *119*
5.6 Newton's Third Law *120*
5.7 Some Applications of Newton's Laws *122*
5.8 Forces of Friction *131*

Chapter 6 Circular Motion and Other Applications of Newton's Laws *150*
6.1 Newton's Second Law Applied to Uniform Circular Motion *151*
6.2 Nonuniform Circular Motion *157*
6.3 Motion in Accelerated Frames *159*
6.4 Motion in the Presence of Resistive Forces *162*
6.5 Numerical Modeling in Particle Dynamics *167*

Chapter 7 Energy and Energy Transfer *181*
7.1 Systems and Environments *182*
7.2 Work Done by a Constant Force *183*
7.3 The Scalar Product of Two Vectors *186*
7.4 Work Done by a Varying Force *188*
7.5 Kinetic Energy and the Work–Kinetic Energy Theorem *193*
7.6 The Nonisolated System—Conservation of Energy *196*
7.7 Situations Involving Kinetic Friction *199*
7.8 Power *203*
7.9 Energy and the Automobile *205*

Chapter 8 Potential Energy *217*
8.1 Potential Energy of a System *218*
8.2 The Isolated System—Conservation of Mechanical Energy *220*
8.3 Conservative and Nonconservative Forces *228*
8.4 Changes in Mechanical Energy for Nonconservative Forces *229*
8.5 Relationship Between Conservative Forces and Potential Energy *234*
8.6 Energy Diagrams and Equilibrium of a System *236*

Courtesy Tourism Malaysia

Chapter 9 **Linear Momentum and Collisions** *251*

9.1 Linear Momentum and Its Conservation *252*
9.2 Impulse and Momentum *256*
9.3 Collisions in One Dimension *260*
9.4 Two-Dimensional Collisions *267*
9.5 The Center of Mass *270*
9.6 Motion of a System of Particles *274*
9.7 Rocket Propulsion *277*

Chapter 10 **Rotation of a Rigid Object About a Fixed Axis** *292*

10.1 Angular Position, Velocity, and Acceleration *293*
10.2 Rotational Kinematics: Rotational Motion with Constant Angular Acceleration *296*
10.3 Angular and Linear Quantities *297*
10.4 Rotational Kinetic Energy *300*
10.5 Calculation of Moments of Inertia *302*
10.6 Torque *306*
10.7 Relationship Between Torque and Angular Acceleration *307*
10.8 Work, Power, and Energy in Rotational Motion *312*
10.9 Rolling Motion of a Rigid Object *316*

Chapter 11 **Angular Momentum** *336*

11.1 The Vector Product and Torque *337*
11.2 Angular Momentum *339*
11.3 Angular Momentum of a Rotating Rigid Object *343*
11.4 Conservation of Angular Momentum *345*
11.5 The Motion of Gyroscopes and Tops *350*
11.6 Angular Momentum as a Fundamental Quantity *351*

Chapter 12 **Static Equilibrium and Elasticity** *362*

12.1 The Conditions for Equilibrium *363*
12.2 More on the Center of Gravity *365*
12.3 Examples of Rigid Objects in Static Equilibrium *366*
12.4 Elastic Properties of Solids *373*

Chapter 13 **Universal Gravitation** *389*

13.1 Newton's Law of Universal Gravitation *390*
13.2 Measuring the Gravitational Constant *393*
13.3 Free-Fall Acceleration and the Gravitational Force *394*
13.4 Kepler's Laws and the Motion of Planets *396*
13.5 The Gravitational Field *401*
13.6 Gravitational Potential Energy *403*
13.7 Energy Considerations in Planetary and Satellite Motion *405*

Chapter 14 **Fluid Mechanics** *420*

14.1 Pressure *421*
14.2 Variation of Pressure with Depth *423*
14.3 Pressure Measurements *426*
14.4 Buoyant Forces and Archimedes's Principle *427*
14.5 Fluid Dynamics *431*
14.6 Bernoulli's Equation *433*
14.7 Other Applications of Fluid Dynamics *436*

PART 2 **Oscillations and Mechanical Waves 451**

Chapter 15 **Oscillatory Motion** *452*

15.1 Motion of an Object Attached to a Spring *453*
15.2 Mathematical Representation of Simple Harmonic Motion *454*
15.3 Energy of the Simple Harmonic Oscillator *462*
15.4 Comparing Simple Harmonic Motion with Uniform Circular Motion *465*
15.5 The Pendulum *468*
15.6 Damped Oscillations *471*
15.7 Forced Oscillations *472*

Don Bonsey/Getty Images

Chapter 16 **Wave Motion** *486*

16.1 Propagation of a Disturbance *487*
16.2 Sinusoidal Waves *491*
16.3 The Speed of Waves on Strings *496*
16.4 Reflection and Transmission *499*
16.5 Rate of Energy Transfer by Sinusoidal Waves on Strings *501*
16.6 The Linear Wave Equation *503*

Chapter 17 **Sound Waves** *512*

17.1 Speed of Sound Waves *513*
17.2 Periodic Sound Waves *515*
17.3 Intensity of Periodic Sound Waves *516*
17.4 The Doppler Effect *522*
17.5 Digital Sound Recording *528*
17.6 Motion Picture Sound *532*

Chapter 18 **Superposition and Standing Waves** *543*

18.1 Superposition and Interference *544*
18.2 Standing Waves *549*
18.3 Standing Waves in a String Fixed at Both Ends *552*
18.4 Resonance *558*
18.5 Standing Waves in Air Columns *559*
18.6 Standing Waves in Rods and Membranes *563*
18.7 Beats: Interference in Time *564*
18.8 Nonsinusoidal Wave Patterns *566*

PART 3 Thermodynamics 579

Chapter 19 **Temperature** *580*

19.1 Temperature and the Zeroth Law of Thermodynamics *581*
19.2 Thermometers and the Celsius Temperature Scale *583*
19.3 The Constant-Volume Gas Thermometer and the Absolute Temperature Scale *584*

© 1973 Kim Vandiver & Harold E. Edgerton/Courtesy of Palm Press, Inc.

Lowell Georgia/CORBIS

19.4 Thermal Expansion of Solids and Liquids *586*
19.5 Macroscopic Description of an Ideal Gas *591*

Chapter 20 **Heat and the First Law of Thermodynamics** *604*

20.1 Heat and Internal Energy *605*
20.2 Specific Heat and Calorimetry *607*
20.3 Latent Heat *611*
20.4 Work and Heat in Thermodynamic Processes *615*
20.5 The First Law of Thermodynamics *618*
20.6 Some Applications of the First Law of Thermodynamics *619*
20.7 Energy Transfer Mechanisms *623*

Chapter 21 **The Kinetic Theory of Gases** *640*

21.1 Molecular Model of an Ideal Gas *641*
21.2 Molar Specific Heat of an Ideal Gas *646*
21.3 Adiabatic Processes for an Ideal Gas *649*
21.4 The Equipartition of Energy *650*
21.5 The Boltzmann Distribution Law *654*
21.6 Distribution of Molecular Speeds *655*
21.7 Mean Free Path *658*

Chapter 22 **Heat Engines, Entropy, and the Second Law of Thermodynamics** *667*

22.1 Heat Engines and the Second Law of Thermodynamics *669*
22.2 Heat Pumps and Refrigerators *671*
22.3 Reversible and Irreversible Processes *673*
22.4 The Carnot Engine *675*
22.5 Gasoline and Diesel Engines *679*
22.6 Entropy *683*
22.7 Entropy Changes in Irreversible Processes *687*
22.8 Entropy on a Microscopic Scale *690*

PART 4 Electricity and Magnetism 705

Chapter 23 **Electric Fields** *706*

23.1 Properties of Electric Charges *707*
23.2 Charging Objects by Induction *709*

23.3 Coulomb's Law *711*
23.4 The Electric Field *715*
23.5 Electric Field of a Continuous Charge Distribution *719*
23.6 Electric Field Lines *723*
23.7 Motion of Charged Particles in a Uniform Electric Field *725*

Chapter 24 Gauss's Law *739*

24.1 Electric Flux *740*
24.2 Gauss's Law *743*
24.3 Application of Gauss's Law to Various Charge Distributions *746*
24.4 Conductors in Electrostatic Equilibrium *750*
24.5 Formal Derivation of Gauss's Law *752*

Chapter 25 Electric Potential *762*

25.1 Potential Difference and Electric Potential *763*
25.2 Potential Differences in a Uniform Electric Field *765*
25.3 Electric Potential and Potential Energy Due to Point Charges *768*
25.4 Obtaining the Value of the Electric Field from the Electric Potential *772*
25.5 Electric Potential Due to Continuous Charge Distributions *774*
25.6 Electric Potential Due to a Charged Conductor *778*
25.7 The Millikan Oil-Drop Experiment *781*
25.8 Applications of Electrostatics *782*

Chapter 26 Capacitance and Dielectrics *795*

26.1 Definition of Capacitance *796*
26.2 Calculating Capacitance *797*
26.3 Combinations of Capacitors *802*
26.4 Energy Stored in a Charged Capacitor *807*
26.5 Capacitors with Dielectrics *810*
26.6 Electric Dipole in an Electric Field *815*
26.7 An Atomic Description of Dielectrics *817*

Chapter 27 Current and Resistance *831*

27.1 Electric Current *832*
27.2 Resistance *835*
27.3 A Model for Electrical Conduction *841*
27.4 Resistance and Temperature *843*
27.5 Superconductors *844*
27.6 Electrical Power *845*

Chapter 28 Direct Current Circuits *858*

28.1 Electromotive Force *859*
28.2 Resistors in Series and Parallel *862*
28.3 Kirchhoff's Rules *869*
28.4 *RC* Circuits *873*
28.5 Electrical Meters *879*
28.6 Household Wiring and Electrical Safety *880*

Chapter 29 Magnetic Fields *894*

29.1 Magnetic Fields and Forces *896*
29.2 Magnetic Force Acting on a Current-Carrying Conductor *900*
29.3 Torque on a Current Loop in a Uniform Magnetic Field *904*
29.4 Motion of a Charged Particle in a Uniform Magnetic Field *907*
29.5 Applications Involving Charged Particles Moving in a Magnetic Field *910*
29.6 The Hall Effect *914*

Chapter 30 Sources of the Magnetic Field *926*

30.1 The Biot–Savart Law *927*
30.2 The Magnetic Force Between Two Parallel Conductors *932*
30.3 Ampère's Law *933*
30.4 The Magnetic Field of a Solenoid *938*
30.5 Magnetic Flux *940*
30.6 Gauss's Law in Magnetism *941*
30.7 Displacement Current and the General Form of Ampère's Law *942*
30.8 Magnetism in Matter *944*
30.9 The Magnetic Field of the Earth *953*

Chapter 31 Faraday's Law *967*

31.1 Faraday's Law of Induction *968*
31.2 Motional emf *973*
31.3 Lenz's Law *977*
31.4 Induced emf and Electric Fields *981*
31.5 Generators and Motors *982*
31.6 Eddy Currents *986*
31.7 Maxwell's Equations *988*

Chapter 32 Inductance *1003*

32.1 Self-Inductance *1004*
32.2 *RL* Circuits *1006*
32.3 Energy in a Magnetic Field *1011*
32.4 Mutual Inductance *1013*
32.5 Oscillations in an *LC* Circuit *1015*
32.6 The *RLC* Circuit *1020*

Richard Megna/Fundamental Photographs

Chapter 33 Alternating Current Circuits *1033*
33.1 AC Sources *1033*
33.2 Resistors in an AC Circuit *1034*
33.3 Inductors in an AC Circuit *1038*
33.4 Capacitors in an AC Circuit *1041*
33.5 The *RLC* Series Circuit *1043*
33.6 Power in an AC Circuit *1047*
33.7 Resonance in a Series *RLC* Circuit *1049*
33.8 The Transformer and Power Transmission *1052*
33.9 Rectifiers and Filters *1054*

Chapter 34 Electromagnetic Waves *1066*
34.1 Maxwell's Equations and Hertz's Discoveries *1067*
34.2 Plane Electromagnetic Waves *1069*
34.3 Energy Carried by Electromagnetic Waves *1074*
34.4 Momentum and Radiation Pressure *1076*
34.5 Production of Electromagnetic Waves by an Antenna *1079*
34.6 The Spectrum of Electromagnetic Waves *1080*

PART 5 Light and Optics 1093

Chapter 35 The Nature of Light and the Laws of Geometric Optics *1094*
35.1 The Nature of Light *1095*
35.2 Measurements of the Speed of Light *1096*
35.3 The Ray Approximation in Geometric Optics *1097*
35.4 Reflection *1098*
35.5 Refraction *1102*
35.6 Huygen's Principle *1107*
35.7 Dispersion and Prisms *1109*
35.8 Total Internal Reflection *1111*
35.9 Fermat's Principle *1114*

Chapter 36 Image Formation *1126*
36.1 Images Formed by Flat Mirrors *1127*
36.2 Images Formed by Spherical Mirrors *1131*
36.3 Images Formed by Refraction *1138*
36.4 Thin Lenses *1141*
36.5 Lens Aberrations *1152*
36.6 The Camera *1153*
36.7 The Eye *1155*
36.8 The Simple Magnifier *1159*
36.9 The Compound Microscope *1160*
36.10 The Telescope *1162*

Chapter 37 Interference of Light Waves *1176*
37.1 Conditions for Interference *1177*
37.2 Young's Double-Slit Experiment *1177*
37.3 Intensity Distribution of the Double-Slit Interference Pattern *1182*
37.4 Phasor Addition of Waves *1184*
37.5 Change of Phase Due to Reflection *1188*
37.6 Interference in Thin Films *1189*
37.7 The Michelson Interferometer *1194*

Chapter 38 Diffraction Patterns and Polarization *1205*
38.1 Introduction to Diffraction Patterns *1206*
38.2 Diffraction Patterns from Narrow Slits *1207*
38.3 Resolution of Single-Slit and Circular Apertures *1214*
38.4 The Diffraction Grating *1217*
38.5 Diffraction of X-Rays by Crystals *1224*
38.6 Polarization of Light Waves *1225*

PART 6 Modern Physics 1243

Chapter 39 Relativity *1244*
39.1 The Principle of Galilean Relativity *1246*
39.2 The Michelson–Morley Experiment *1248*
39.3 Einstein's Principle of Relativity *1250*
39.4 Consequences of the Special Theory of Relativity *1251*
39.5 The Lorentz Transformation Equations *1262*
39.6 The Lorentz Velocity Transformation Equations *1264*
39.7 Relativistic Linear Momentum and the Relativistic Form of Newton's Laws *1267*
39.8 Relativistic Energy *1268*
39.9 Mass and Energy *1272*
39.10 The General Theory of Relativity *1273*

Appendix A Tables *A.1*
Table A.1 Conversion Factors *A.1*
Table A.2 Symbols, Dimensions, and Units of Physical Quantities *A.2*
Table A.3 Table of Atomic Masses *A.4*

Appendix B Mathematics Review *A.14*
B.1 Scientific Notation *A.14*
B.2 Algebra *A.15*
B.3 Geometry *A.20*
B.4 Trigonometry *A.21*
B.5 Series Expansions *A.23*
B.6 Differential Calculus *A.23*
B.7 Integral Calculus *A.25*
B.8 Propagation of Uncertainty *A.28*

Appendix C Periodic Table of the Elements *A.30*

Appendix D SI Units *A.32*

Appendix E Nobel Prizes *A.33*

Answers to Odd-Numbered Problems *A.37*

Index *I.1*

About the Authors

Raymond A. Serway received his doctorate at Illinois Institute of Technology and is Professor Emeritus at James Madison University. Dr. Serway began his teaching career at Clarkson University, where he conducted research and taught from 1967 to 1980. His second academic appointment was at James Madison University as Professor of Physics and Head of the Physics Department from 1980 to 1986. He remained at James Madison University until his retirement in 1997. He was the recipient of the Madison Scholar Award at James Madison University in 1990, the Distinguished Teaching Award at Clarkson University in 1977, and the Alumni Achievement Award from Utica College in 1985. As Guest Scientist at the IBM Research Laboratory in Zurich, Switzerland, he worked with K. Alex Müller, 1987 Nobel Prize recipient. Dr. Serway also held research appointments at Rome Air Development Center from 1961 to 1963, at IIT Research Institute from 1963 to 1967, and as a visiting scientist at Argonne National Laboratory, where he collaborated with his mentor and friend, Sam Marshall. In addition to earlier editions of this textbook, Dr. Serway is the co-author of the high-school textbook *Physics* with Jerry Faughn, published by Holt, Rinehart, & Winston and co-author of the third edition of *Principles of Physics* with John Jewett, the sixth edition of *College Physics* with Jerry Faughn, and the second edition of *Modern Physics* with Clem Moses and Curt Moyer. In addition, Dr. Serway has published more than 40 research papers in the field of condensed matter physics and has given more than 70 presentations at professional meetings. Dr. Serway and his wife Elizabeth enjoy traveling, golfing, gardening, and spending quality time with their four children and five grandchildren.

John W. Jewett, Jr. earned his doctorate at Ohio State University, specializing in optical and magnetic properties of condensed matter. Dr. Jewett began his academic career at Richard Stockton College of New Jersey, where he taught from 1974 to 1984. He is currently Professor of Physics at California State Polytechnic University, Pomona. Throughout his teaching career, Dr. Jewett has been active in promoting science education. In addition to receiving four National Science Foundation grants, he helped found and direct the Southern California Area Modern Physics Institute (SCAMPI). He also directed Science IMPACT (Institute for Modern Pedagogy and Creative Teaching), which works with teachers and schools to develop effective science curricula. Dr. Jewett's honors include the Stockton Merit Award at Richard Stockton College, the Outstanding Professor Award at California State Polytechnic University for 1991–1992, and the Excellence in Undergraduate Physics Teaching Award from the American Association of Physics Teachers (AAPT) in 1998. He has given over 80 presentations at professional meetings, including presentations at international conferences in China and Japan. In addition to his work on this textbook, he is co-author of the third edition of *Principles of Physics* with Ray Serway and author of *The World of Physics . . . Mysteries, Magic, and Myth.* Dr. Jewett enjoys playing piano, traveling, and collecting antiques that can be used as demonstration apparatus in physics lectures, as well as spending time with his wife Lisa and their children and grandchildren.

Preface

Mark Cooper/Corbis Stock Market

In writing this sixth edition of *Physics for Scientists and Engineers,* we continue our ongoing efforts to improve the clarity of presentation and we again include new pedagogical features that help support the learning and teaching processes. Drawing on positive feedback from users of the fifth edition and reviewers' suggestions, we have refined the text in order to better meet the needs of students and teachers. We have for the first time integrated a powerful collection of media resources into many of the illustrations, examples, and end-of-chapter problems in the text. These resources compose the Web-based learning system ***PhysicsNow*** and are flagged by the media icon . Further details are described below.

This textbook is intended for a course in introductory physics for students majoring in science or engineering. The entire contents of the text in its extended version could be covered in a three-semester course, but it is possible to use the material in shorter sequences with the omission of selected chapters and sections. The mathematical background of the student taking this course should ideally include one semester of calculus. If that is not possible, the student should be enrolled in a concurrent course in introductory calculus.

Objectives

This introductory physics textbook has two main objectives: to provide the student with a clear and logical presentation of the basic concepts and principles of physics, and to strengthen an understanding of the concepts and principles through a broad range of interesting applications to the real world. To meet these objectives, we have placed emphasis on sound physical arguments and problem-solving methodology. At the same time, we have attempted to motivate the student through practical examples that demonstrate the role of physics in other disciplines, including engineering, chemistry, and medicine.

Changes in the Sixth Edition

A large number of changes and improvements have been made in preparing the sixth edition of this text. Some of the new features are based on our experiences and on current trends in science education. Other changes have been incorporated in response to comments and suggestions offered by users of the fifth edition and by reviewers of the manuscript. The following represent the major changes in the sixth edition:

Active Figures Many diagrams from the text have been animated to form **Active Figures,** part of the ***PhysicsNow*** integrated Web-based learning system. By visualizing phenomena and processes that cannot be fully represented on a static page, students greatly increase their conceptual understanding. **Active Figures** are identified with the media icon . An addition to the figure caption in **blue type** describes briefly the nature and contents of the animation.

Interactive Worked Examples Approximately 76 of the worked examples in the text have been identified as interactive, labeled with the media icon . As part of the ***PhysicsNow*** Web-based learning system, students can engage in an extension of the problem solved in the example. This often includes elements of both visualization and calculation, and may also involve prediction and intuition building. Often the interactivity is inspired by the **"What If?"** question we posed in the example text.

What If? Approximately one-third of the worked examples in the text contain this new feature. At the completion of the example solution, a **What If?** question offers a variation on the situation posed in the text of the example. For instance, this feature might explore the effects of changing the conditions of the situation, determine what happens when a quantity is taken to a particular limiting value, or question whether additional information can be determined about the problem situation. The answer to the question generally includes both a conceptual response and a mathematical response. This feature encourages students to think about the results of the example and assists in conceptual understanding of the principles. It also prepares students to encounter novel problems featured on exams. Some of the end-of-chapter problems also carry the **"What If?"** feature.

Quick Quizzes The number of Quick Quiz questions in each chapter has been increased. Quick Quizzes provide students with opportunities to test their understanding of the physical concepts presented. The questions require students to make decisions on the basis of sound reasoning, and some of them have been written to help students overcome common misconceptions. Quick Quizzes have been cast in an objective format, including multiple choice, true–false, and ranking. Answers to all Quick Quiz questions are found at the end of each chapter. Additional Quick Quizzes that can be used in classroom teaching are available on the instructor's companion Web site. Many instructors choose to use such questions in a "peer instruction" teaching style, but they can be used in standard quiz format as well.

Pitfall Preventions These new features are placed in the margins of the text and address common student misconceptions and situations in which students often follow unproductive paths. Over 200 Pitfall Preventions are provided to help students avoid common mistakes and misunderstandings.

General Problem-Solving Strategy A general strategy to be followed by the student is outlined at the end of Chapter 2 and provides students with a structured process for solving problems. In Chapters 3 through 5, the strategy is employed explicitly in every example so that students learn how it is applied. In the remaining chapters, the strategy appears explicitly in one example per chapter so that students are encouraged throughout the course to follow the procedure.

Line-by-Line Revision The entire text has been carefully edited to improve clarity of presentation and precision of language. We hope that the result is a book that is both accurate and enjoyable to read.

Problems A substantial revision of the end-of-chapter problems was made in an effort to improve their variety and interest, while maintaining their clarity and quality. Approximately 17% of the problems (about 550) are new. All problems have been carefully edited. Solutions to approximately 20% of the end-of-chapter problems are included in the *Student Solutions Manual and Study Guide.* These problems are identified by boxes around their numbers. A smaller subset of solutions, identified by the media icon www, are available on the World Wide Web **(http://www.pse6.com)** as coached solutions with hints. Targeted feedback is provided for students whose instructors adopt *Physics for Scientists and Engineers,* sixth edition. See the next section for a complete description of other features of the problem set.

Content Changes The content and organization of the textbook is essentially the same as that of the fifth edition. An exception is that Chapter 13 (Oscillatory Motion) in the fifth edition has been moved to the Chapter 15 position in the sixth edition, in order to form a cohesive four-chapter Part 2 on oscillations and waves. Many sections in various chapters have been streamlined, deleted, or combined with other sections to allow for a more balanced presentation. The chapters on Modern Physics, Chapters 39–46, have been extensively rewritten to provide more up-to-date material as well as modern applications. A more detailed list of content changes can be found on the instructor's companion Web site.

Content

The material in this book covers fundamental topics in classical physics and provides an introduction to modern physics. The book is divided into six parts. Part 1 (Chapters 1 to 14) deals with the fundamentals of Newtonian mechanics and the physics of fluids, Part 2 (Chapters 15 to 18) covers oscillations, mechanical waves, and sound, Part 3 (Chapters 19 to 22) addresses heat and thermodynamics, Part 4 (Chapters 23 to 34) treats electricity and magnetism, Part 5 (Chapters 35 to 38) covers light and optics, and Part 6 (Chapters 39 to 46) deals with relativity and modern physics. Each part opener includes an overview of the subject matter covered in that part, as well as some historical perspectives.

Text Features

Most instructors would agree that the textbook selected for a course should be the student's primary guide for understanding and learning the subject matter. Furthermore, the textbook should be easily accessible and should be styled and written to facilitate instruction and learning. With these points in mind, we have included many pedagogical features in the textbook that are intended to enhance its usefulness to both students and instructors. These features are as follows:

Style To facilitate rapid comprehension, we have attempted to write the book in a style that is clear, logical, and engaging. We have chosen a writing style that is somewhat informal and relaxed so that students will find the text appealing and enjoyable to read. New terms are carefully defined, and we have avoided the use of jargon.

Previews All chapters begin with a brief preview that includes a discussion of the chapter's objectives and content.

Important Statements and Equations Most important statements and definitions are set in **boldface** type or are highlighted with a background screen for added emphasis and ease of review. Similarly, important equations are highlighted with a background screen to facilitate location.

Bruce Ayers/Getty Images

Problem-Solving Hints In several chapters, we have included general strategies for solving the types of problems featured both in the examples and in the end-of-chapter problems. This feature helps students to identify necessary steps in problem solving and to eliminate any uncertainty they might have. Problem-solving strategies are highlighted with a light red background screen for emphasis and ease of location.

Marginal Notes Comments and notes appearing in blue type in the margin can be used to locate important statements, equations, and concepts in the text.

Pedagogical Use of Color Readers should consult the **pedagogical color chart** (second page inside the front cover) for a listing of the color-coded symbols used in the text diagrams, Web-based **Active Figures,** and diagrams within **Interactive Worked Examples.** This system is followed consistently whenever possible, with slight variations made necessary by the complexity of physical situations depicted in Part 4.

Mathematical Level We have introduced calculus gradually, keeping in mind that students often take introductory courses in calculus and physics concurrently. Most steps are shown when basic equations are developed, and reference is often made to mathematical appendices at the end of the textbook. Vector products are introduced later in the text, where they are needed in physical applications. The dot product is introduced in Chapter 7, which addresses energy and energy transfer; the cross product is introduced in Chapter 11, which deals with angular momentum.

Worked Examples A large number of worked examples of varying difficulty are presented to promote students' understanding of concepts. In many cases, the examples serve as models for solving the end-of-chapter problems. Because of the increased emphasis on understanding physical concepts, many examples are conceptual in nature and are labeled as such. The examples are set off in boxes, and the answers to examples with numerical solutions are highlighted with a background screen. We have already mentioned that a number of examples are designated as interactive and are part of the ***PhysicsNow*** Web-based learning system.

Questions Questions of a conceptual nature requiring verbal or written responses are provided at the end of each chapter. Over 1 000 questions are included in this edition. Some questions provide the student with a means of self-testing the concepts presented in the chapter. Others could serve as a basis for initiating classroom discussions. Answers to selected questions are included in the *Student Solutions Manual and Study Guide,* and answers to all questions are found in the *Instructor's Solutions Manual.*

Significant Figures Significant figures in both worked examples and end-of-chapter problems have been handled with care. Most numerical examples are worked out to either two or three significant figures, depending on the precision of the data provided. End-of-chapter problems regularly state data and answers to three-digit precision.

Problems An extensive set of problems is included at the end of each chapter; in all, over 3 000 problems are given throughout the text. Answers to odd-numbered problems are provided at the end of the book in a section whose pages have colored edges for ease of location. For the convenience of both the student and the instructor, about two thirds of the problems are keyed to specific sections of the chapter. The remaining problems, labeled "Additional Problems," are not keyed to specific sections.

Usually, the problems within a given section are presented so that the straightforward problems (those with black problem numbers) appear first. For ease of identification, the numbers of intermediate-level problems are printed in blue, and those of challenging problems are printed in magenta.

- **Review Problems** Many chapters include review problems requiring the student to combine concepts covered in the chapter with those discussed in previous chapters. These problems reflect the cohesive nature of the principles in the text and verify that physics is not a scattered set of ideas. When facing real-world issues such as global warming or nuclear weapons, it may be necessary to call on ideas in physics from several parts of a textbook such as this one.
- **Paired Problems** To allow focused practice in solving problems stated in symbolic terms, some end-of-chapter numerical problems are paired with the same problems in symbolic form. Paired problems are identified by a common light red background screen.
- **Computer- and Calculator-Based Problems** Many chapters include one or more problems whose solution requires the use of a computer or graphing calculator. Computer modeling of physical phenomena enables students to obtain graphical representations of variables and to perform numerical analyses.
- **Coached Problems with Hints** These have been described above as part of the ***PhysicsNow*** Web-based learning system. These problems are identified by the media icon www and targeted feedback is provided to students of instructors adopting the sixth edition.

Units The international system of units (SI) is used throughout the text. The U.S. customary system of units is used only to a limited extent in the chapters on mechanics, heat, and thermodynamics.

Summaries Each chapter contains a summary that reviews the important concepts and equations discussed in that chapter. A marginal note in **blue type** next to each chapter summary directs students to a practice test (Post-Test) for the chapter.

Appendices and Endpapers Several appendices are provided at the end of the textbook. Most of the appendix material represents a review of mathematical concepts and techniques used in the text, including scientific notation, algebra, geometry, trigonometry, differential calculus, and integral calculus. Reference to these appendices is made

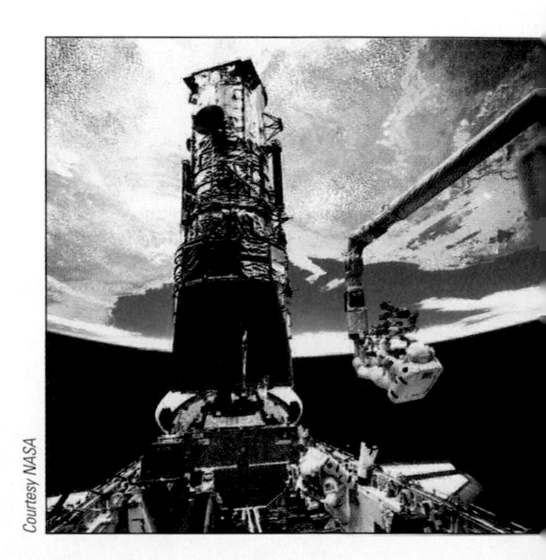
Courtesy NASA

throughout the text. Most mathematical review sections in the appendices include worked examples and exercises with answers. In addition to the mathematical reviews, the appendices contain tables of physical data, conversion factors, atomic masses, and the SI units of physical quantities, as well as a periodic table of the elements. Other useful information, including fundamental constants and physical data, planetary data, a list of standard prefixes, mathematical symbols, the Greek alphabet, and standard abbreviations of units of measure, appears on the endpapers.

Student Ancillaries

Student Solutions Manual and Study Guide by John R. Gordon, Ralph McGrew, and Raymond Serway. This two-volume manual features detailed solutions to 20% of the end-of-chapter problems from the text. The manual also features a list of important equations, concepts, and notes from key sections of the text, in addition to answers to selected end-of-chapter questions. Volume 1 contains Chapters 1 through 22 and Volume 2 contains Chapters 23 through 46.

WebTutor™ on WebCT and Blackboard **WebTutor** offers students real-time access to a full array of study tools, including chapter outlines, summaries, learning objectives, glossary flashcards (with audio), practice quizzes, **InfoTrac® College Edition** exercises, and Web links.

InfoTrac® College Edition Adopters and their students automatically receive a four-month subscription to **InfoTrac® College Edition** with every new copy of this book. Newly improved, this extensive online library opens the door to the full text (not just abstracts) of countless articles from thousands of publications including *American Scientist, Physical Review, Science, Science Weekly,* and more! Available only to college and university students. Journals subject to change.

The Brooks/Cole Physics Resource Center You will find additional online quizzes, Web links and animations at **http://physics.brookscole.com.**

Ancillaries for Instructors

The first four ancillaries below are available to qualified adopters. Please consult your local sales representative for details.

Instructor's Solutions Manual by Ralph McGrew and James A. Currie. This two-volume manual contains complete worked solutions to all of the end-of-chapter problems in the textbook as well as answers to even-numbered problems. The solutions to problems new to the sixth edition are marked for easy identification by the instructor. New to this edition are complete answers to the conceptual questions in the main text. Volume 1 contains Chapters 1 through 22 and Volume 2 contains Chapters 23 through 46.

Printed Test Bank by Edward Adelson. This two-volume test bank contains approximately 2 300 multiple-choice questions. These questions are also available in electronic format with complete answers and solutions in the Brooks/Cole Assessment test program. Volume 1 contains Chapters 1 through 22 and Volume 2 contains Chapters 23 through 46.

Multimedia Manager This easy-to-use multimedia lecture tool allows you to quickly assemble art and database files with notes to create fluid lectures. The CD-ROM set (Volume 1, Chapters 1–22; Volume 2, Chapters 23–46) includes a database of animations, video clips, and digital art from the text as well as electronic files of the *Instructor's Solutions Manual and Test Bank.* The simple interface makes it easy for you to incorporate graphics, digital video, animations, and audio clips into your lectures.

Transparency Acetates Each volume contains approximately 100 acetates featuring art from the text. Volume 1 contains Chapters 1 through 22 and Volume 2 contains Chapters 23 through 46.

George Semple

Brooks/Cole Assessment With a balance of efficiency, high performance, simplicity and versatility, **Brooks/Cole Assessment (BCA)** gives you the power to transform the learning and teaching experience. **BCA** is fully integrated testing, tutorial, and course management software accessible by instructors and students anytime, anywhere. Delivered for FREE in a browser-based format without the need for any proprietary software or plug-ins, **BCA** uses correct scientific notation to provide the drill of basic skills that students need, enabling the instructor to focus more time in higher-level learning activities (i.e., concepts and applications). Students can have unlimited practice in questions and problems, building their own confidence and skills. Results flow automatically to a grade book for tracking so that instructors will be better able to assess student understanding of the material, even prior to class or an actual test.

WebTutor™ on WebCT and Blackboard With **WebTutor's** text-specific, preformatted content and total flexibility, instructors can easily create and manage their own personal Web site. **WebTutor's** course management tool gives instructors the ability to provide virtual office hours, post syllabi, set up threaded discussions, track student progress with the quizzing material, and much more. **WebTutor** also provides robust communication tools, such as a course calendar, asynchronous discussion, real-time chat, a whiteboard, and an integrated e-mail system.

Additional Options for Online Homework For detailed information and demonstrations, contact your Thomson • Brooks/Cole representative or visit the following:

- WebAssign: A Web-based Homework System
  http://www.webassign.net or contact WebAssign at *webassign@ncsu.edu*
- Homework Service
  http://hw.ph.utexas.edu/hw.html or contact *moore@physics.utexas.edu*
- CAPA: A Computer-Assisted Personalized Approach
  http://capa4.lite.msu.edu/homepage/

Instructor's Companion Web Site Consult the instructor's site at *http://www.pse6.com* for additional Quick Quiz questions, a detailed list of content changes since the fifth edition, a problem correlation guide, images from the text, and sample PowerPoint lectures. Instructors adopting the sixth edition of *Physics for Scientists and Engineers* may download these materials after securing the appropriate password from their local Thomson • Brooks/Cole sales representative.

Teaching Options

The topics in this textbook are presented in the following sequence: classical mechanics, oscillations and mechanical waves, and heat and thermodynamics followed by electricity and magnetism, electromagnetic waves, optics, relativity, and modern physics. This presentation represents a traditional sequence, with the subject of mechanical waves being presented before electricity and magnetism. Some instructors may prefer to cover this material after completing electricity and magnetism (i.e., after Chapter 34). The chapter on relativity is placed near the end of the text because this topic often is treated as an introduction to the era of "modern physics." If time permits, instructors may choose to cover Chapter 39 after completing Chapter 13, as it concludes the material on Newtonian mechanics.

For those instructors teaching a two-semester sequence, some sections and chapters could be deleted without any loss of continuity. The following sections can be considered optional for this purpose:

2.7 Kinematic Equations Derived from Calculus
4.6 Relative Velocity and Relative Acceleration
6.3 Motion in Accelerated Frames
6.4 Motion in the Presence of Resistive Forces
6.5 Numerical Modeling in Particle Dynamics
7.9 Energy and the Automobile

8.6	Energy Diagrams and Equilibrium of a System
9.7	Rocket Propulsion
11.5	The Motion of Gyroscopes and Tops
11.6	Angular Momentum as a Fundamental Quantity
14.7	Other Applications of Fluid Dynamics
15.6	Damped Oscillations
15.7	Forced Oscillations
17.5	Digital Sound Recording
17.6	Motion Picture Sound
18.6	Standing Waves in Rods and Membranes
18.8	Nonsinusoidal Wave Patterns
21.7	Mean Free Path
22.8	Entropy on a Microscopic Scale
24.5	Formal Derivation of Gauss's Law
25.7	The Millikan Oil-Drop Experiment
25.8	Applications of Electrostatics
26.7	An Atomic Description of Dielectrics
27.5	Superconductors
28.5	Electrical Meters
28.6	Household Wiring and Electrical Safety
29.5	Applications Involving Charged Particles Moving in a Magnetic Field
29.6	The Hall Effect
30.8	Magnetism in Matter
30.9	The Magnetic Field of the Earth
31.6	Eddy Currents
33.9	Rectifiers and Filters
34.5	Production of Electromagnetic Waves by an Antenna
35.9	Fermat's Principle
36.5	Lens Aberrations
36.6	The Camera
36.7	The Eye
36.8	The Simple Magnifier
36.9	The Compound Microscope
36.10	The Telescope
38.5	Diffraction of X-Rays by Crystals
39.10	The General Theory of Relativity

Topham Picturepoint/The Image Works

Acknowledgments

The sixth edition of this textbook was prepared with the guidance and assistance of many professors who reviewed selections of the manuscript, the pre-revision text, or both. We wish to acknowledge the following scholars and express our sincere appreciation for their suggestions, criticisms, and encouragement:

Edward Adelson, *Ohio State University*
Michael R. Cohen, *Shippensburg University*
Jerry D. Cook, *Eastern Kentucky University*
J. William Dawicke, *Milwaukee School of Engineering*
N. John DiNardo, *Drexel University*
Andrew Duffy, *Boston University*
Robert J. Endorf, *University of Cincinnati*
F. Paul Esposito, *University of Cincinnati*
Joe L. Ferguson, *Mississippi State University*
Perry Ganas, *California State University, Los Angeles*
John C. Hardy, *Texas A&M University*
Michael Hayes, *University of Pretoria (South Africa)*
John T. Ho, *The State University of New York, Buffalo*
Joseph W. Howard, *Salisbury University*
Robert Hunt, *Johnson County Community College*
Walter S. Jaronski, *Radford University*
Sangyong Jeon, *McGill University, Quebec*
Stan Jones, *University of Alabama*
L. R. Jordan, *Palm Beach Community College*
Teruki Kamon, *Texas A & M University*
Louis E. Keiner, *Coastal Carolina University*
Mario Klarič, *Midlands Technical College*
Laird Kramer, *Florida International University*
Edwin H. Lo, *American University*
James G. McLean, *The State University of New York, Geneseo*
Richard E. Miers, *Indiana University–Purdue University, Fort Wayne*
Oscar Romulo Ochoa, *The College of New Jersey*

Paul S. Ormsby, *Moraine Valley Community College*
Didarul I. Qadir, *Central Michigan University*
Judith D. Redling, *New Jersey Institute of Technology*
Richard W. Robinett, *Pennsylvania State University*
Om P. Rustgi, *SUNY College at Buffalo*
Mesgun Sebhatu, *Winthrop University*
Natalia Semushkina, *Shippensburg University*
Daniel Stump, *Michigan State University*
Uwe C. Täuber, *Virginia Polytechnic Institute*
Perry A. Tompkins, *Samford University*
Doug Welch, *McMaster University, Ontario*
Augden Windelborn, *Northern Illinois University*
Jerzy M. Wrobel, *University of Missouri, Kansas City*
Jianshi Wu, *Fayetteville State University*
Michael Zincani, *University of Dallas*

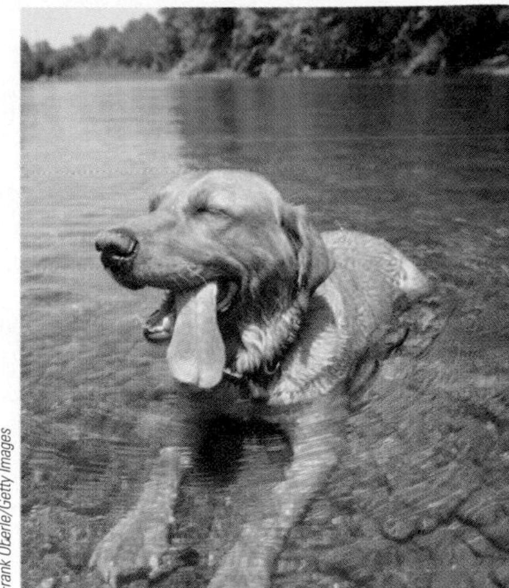
Frank Oberle/Getty Images

This title was carefully checked for accuracy by Michael Kotlarchyk *(Rochester Institute of Technology)*, Chris Vuille *(Embry-Riddle Aeronautical University)*, Laurencin Dunbar *(St. Louis Community College)*, William Dawicke *(Milwaukee School of Engineering)*, Ioan Kosztin *(University of Missouri)*, Tom Barrett *(Ohio State University)*, Z. M. Stadnik *(University of Ottawa)*, Ronald E. Jodoin *(Rochester Institute of Technology)*, Brian A. Raue *(Florida International University)*, Peter Moeck *(Portland State University)*, and Grant Hart *(Brigham Young University)*. We thank them for their diligent efforts under schedule pressure!

We are grateful to Ralph McGrew for organizing the end-of-chapter problems, writing many new problems, and his excellent suggestions for improving the content of the textbook. Problems new to this edition were written by Edward Adelson, Ronald Bieniek, Michael Browne, Andrew Duffy, Robert Forsythe, Perry Ganas, Michael Hones, John Jewett, Boris Korsunsky, Edwin Lo, Ralph McGrew, Raymond Serway, and Jerzy Wrobel, with the help of Bennett Simpson and JoAnne Maniago. Students Alexander Coto, Karl Payne, and Eric Peterman made corrections to problems taken from previous editions, as did teachers David Aspnes, Robert Beichner, Joseph Biegen, Tom Devlin, Vasili Haralambous, Frank Hayes, Erika Hermon, Ken Menningen, Henry Nebel, and Charles Teague. We are grateful to authors John R. Gordon and Ralph McGrew and compositor Michael Rudmin for preparing the *Student Solutions Manual and Study Guide*. Authors Ralph McGrew and James Currie and compositor Mary Toscano have prepared an excellent *Instructor's Solutions Manual*, and we thank them. Edward Adelson has carefully edited and improved the Test Bank for the sixth edition. Kurt Vandervoort prepared extra Quick Quiz questions for the instructor's companion Web site.

Special thanks and recognition go to the professional staff at the Brooks/Cole Publishing Company—in particular Susan Pashos, Rebecca Heider and Alyssa White (who managed the ancillary program and so much more), Jessica Howard, Seth Dobrin, Peter McGahey, Teri Hyde, Michelle Julet, David Harris, and Chris Hall—for their fine work during the development and production of this textbook. We are most appreciative of Sam Subity's masterful management of the ***PhysicsNow*** media program. Kelley McAllister is our energetic Marketing Manager, and Stacey Purviance coordinates our marketing communications. We recognize the skilled production service provided by the staff at Sparkpoint Communications, the excellent artwork produced by Rolin Graphics, and the dedicated photo research efforts of Terri Wright.

Finally, we are deeply indebted to our wives and children for their love, support, and long-term sacrifices.

Raymond A. Serway
Leesburg, Virginia

John W. Jewett, Jr.
Pomona, California

To the Student

It is appropriate to offer some words of advice that should be of benefit to you, the student. Before doing so, we assume that you have read the Preface, which describes the various features of the text that will help you through the course.

How to Study

Very often instructors are asked, "How should I study physics and prepare for examinations?" There is no simple answer to this question, but we would like to offer some suggestions that are based on our own experiences in learning and teaching over the years.

First and foremost, maintain a positive attitude toward the subject matter, keeping in mind that physics is the most fundamental of all natural sciences. Other science courses that follow will use the same physical principles, so it is important that you understand and are able to apply the various concepts and theories discussed in the text.

Concepts and Principles

It is essential that you understand the basic concepts and principles before attempting to solve assigned problems. You can best accomplish this goal by carefully reading the textbook before you attend your lecture on the covered material. When reading the text, you should jot down those points that are not clear to you. We've purposely left wide margins in the text to give you space for making notes. Also be sure to make a diligent attempt at answering the questions in the Quick Quizzes as you come to them in your reading. We have worked hard to prepare questions that help you judge for yourself how well you understand the material. Study carefully the **What If?** features that appear with many of the worked examples. These will help you to extend your understanding beyond the simple act of arriving at a numerical result. The Pitfall Preventions will also help guide you away from common misunderstandings about physics. During class, take careful notes and ask questions about those ideas that are unclear to you. Keep in mind that few people are able to absorb the full meaning of scientific material after only one reading. Several readings of the text and your notes may be necessary. Your lectures and laboratory work supplement reading of the textbook and should clarify some of the more difficult material. You should minimize your memorization of material. Successful memorization of passages from the text, equations, and derivations does not necessarily indicate that you understand the material. Your understanding of the material will be enhanced through a combination of efficient study habits, discussions with other students and with instructors, and your ability to solve the problems presented in the textbook. Ask questions whenever you feel clarification of a concept is necessary.

Study Schedule

It is important that you set up a regular study schedule, preferably a daily one. Make sure that you read the syllabus for the course and adhere to the schedule set by your instructor. The lectures will make much more sense if you read the corresponding text material before attending them. As a general rule, you should devote about two hours of study time for every hour you are in class. If you are having trouble with the course, seek the advice of the instructor or other students who have taken the course. You may find it necessary to seek further instruction from experienced students. Very often, instructors offer review sessions in addition to regular class periods. It is important that you avoid the practice of delaying study until a day or two before an exam. More often than not, this approach has disastrous results. Rather than undertake an all-night study session, briefly review the basic concepts and equations, and get a good night's rest. If you feel you need additional help in understanding the concepts, in preparing for exams, or in problem solving, we suggest that you acquire a copy of the *Student Solutions Manual and Study Guide* that accompanies this textbook; this manual should be available at your college bookstore.

George Semple

Use the Features

You should make full use of the various features of the text discussed in the Preface. For example, marginal notes are useful for locating and describing important equations and concepts, and **boldfaced** type indicates important statements and definitions. Many useful tables are contained in the Appendices, but most are incorporated in the text where they are most often referenced. Appendix B is a convenient review of mathematical techniques.

Answers to odd-numbered problems are given at the end of the textbook, answers to Quick Quizzes are located at the end of each chapter, and answers to selected end-of-chapter questions are provided in the *Student Solutions Manual and Study Guide.* Problem-Solving Strategies and Hints are included in selected chapters throughout the text and give you additional information about how you should solve problems. The Table of Contents provides an overview of the entire text, while the Index enables you to locate specific material quickly. Footnotes sometimes are used to supplement the text or to cite other references on the subject discussed.

After reading a chapter, you should be able to define any new quantities introduced in that chapter and to discuss the principles and assumptions that were used to arrive at certain key relations. The chapter summaries and the review sections of the *Student Solutions Manual and Study Guide* should help you in this regard. In some cases, it may be necessary for you to refer to the index of the text to locate certain topics. You should be able to associate with each physical quantity the correct symbol used to represent that quantity and the unit in which the quantity is specified. Furthermore, you should be able to express each important equation in a concise and accurate prose statement.

Problem Solving

R. P. Feynman, Nobel laureate in physics, once said, "You do not know anything until you have practiced." In keeping with this statement, we strongly advise that you develop the skills necessary to solve a wide range of problems. Your ability to solve problems will be one of the main tests of your knowledge of physics, and therefore you should try to solve as many problems as possible. It is essential that you understand basic concepts and principles before attempting to solve problems. It is good practice to try to find alternate solutions to the same problem. For example, you can solve problems in mechanics using Newton's laws, but very often an alternative method that draws on energy considerations is more direct. You should not deceive yourself into thinking that you understand a problem merely because you have seen it solved in class. You must be able to solve the problem and similar problems on your own.

The approach to solving problems should be carefully planned. A systematic plan is especially important when a problem involves several concepts. First, read the problem several times until you are confident you understand what is being asked. Look for any key words that will help you interpret the problem and perhaps allow you to make certain assumptions. Your ability to interpret a question properly is an integral part of problem solving. Second, you should acquire the habit of writing down the information given in a problem and those quantities that need to be found; for example, you might construct a table listing both the quantities given and the quantities to be found. This procedure is sometimes used in the worked examples of the textbook. Finally, after you have decided on the method you feel is appropriate for a given problem, proceed with your solution. Specific problem-solving strategies (Hints) of this type are included in the text and are highlighted with a light red screen. We have also developed a General Problem-Solving Strategy to help guide you through complex problems. If you follow the steps of this procedure *(Conceptualize, Categorize, Analyze, Finalize)*, you will not only find it easier to come up with a solution, but you will also gain more from your efforts. This Strategy is located at the end of Chapter 2 (page 47) and is used in all worked examples in Chapters 3 through 5 so that you can learn how to apply it. In the remaining chapters, the Strategy is used in one example per chapter as a reminder of its usefulness.

Often, students fail to recognize the limitations of certain equations or physical laws in a particular situation. It is very important that you understand and remember the assumptions that underlie a particular theory or formalism. For example, certain equations in kinematics apply only to a particle moving with constant acceleration. These equations are not valid for describing motion whose acceleration is not constant, such as the motion of an object connected to a spring or the motion of an object through a fluid.

© Phil Degginger/Stone/Getty

Experiments

Physics is a science based on experimental observations. In view of this fact, we recommend that you try to supplement the text by performing various types of "hands-on" experiments, either at home or in the laboratory. These can be used to test ideas and models discussed in class or in the textbook. For example, the common Slinky™ toy is excellent for studying traveling waves; a ball swinging on the end of a long string can be used to investigate pendulum motion; various masses attached to the end of a vertical spring or rubber band can be used to determine their elastic nature; an old pair of Polaroid sunglasses and some discarded lenses and a magnifying glass are the components of various experiments in optics; and an approximate measure of the free-fall acceleration can be determined simply by measuring with a stopwatch the time it takes for a ball to drop from a known height. The list of such experiments is endless. When physical models are not available, be imaginative and try to develop models of your own.

New Media

We strongly encourage you to use the ***PhysicsNow*** Web-based learning system that accompanies this textbook. It is far easier to understand physics if you see it in action, and these new materials will enable you to become a part of that action. ***PhysicsNow*** media described in the Preface are accessed at the URL *http://www.pse6.com,* and feature a three-step learning process consisting of a Pre-Test, a personalized learning plan, and a Post-Test.

In addition to other elements, ***PhysicsNow*** includes the following Active Figures and Interactive Worked Examples:

Chapter 2
Active Figures 2.1, 2.3, 2.9, 2.10, 2.11, and 2.13
Examples 2.8 and 2.12

Chapter 3
Active Figures 3.2, 3.3, 3.6, and 3.16
Example 3.5

Chapter 4
Active Figures 4.5, 4.7, and 4.11
Examples 4.4, 4.5, and 4.18

Chapter 5
Active Figure 5.16
Examples 5.9, 5.10, 5.12, and 5.14

Chapter 6
Active Figures 6.2, 6.8, 6.12, and 6.15
Examples 6.4, 6.5, and 6.7

Chapter 7
Active Figure 7.10
Examples 7.9 and 7.11

Chapter 8
Active Figures 8.3, 8.4, and 8.16
Examples 8.2 and 8.4

Chapter 9
Active Figures 9.8, 9.9, 9.13, 9.16, and 9.17
Examples 9.1, 9.5, and 9.8

Chapter 10
Active Figures 10.4, 10.14, and 10.30
Examples 10.12, 10.13, and 10.14

Chapter 11
Active Figures 11.1, 11.3, and 11.4
Examples 11.6 and 11.10

Chapter 12
Active Figures 12.14, 12.16, 12.17, and 12.18
Examples 12.3 and 12.4

Chapter 13
Active Figures 13.1, 13.5, and 13.7
Example 13.1

Chapter 14
Active Figures 14.9 and 14.10
Examples 14.2 and 14.10

Chapter 15
Active Figures 15.1, 15.2, 15.7, 15.9, 15.10, 15.11, 15.14, 15.17, and 15.22

Chapter 16
Active Figures 16.4, 16.7, 16.8, 16.10, 16.14, 16.15, and 16.17
Example 16.5

Chapter 17
Active Figures 17.2, 17.8, and 17.9
Examples 17.1 and 17.6

Chapter 18
Active Figures 18.1, 18.2, 18.4, 18.9, 18.10, 18.22, and 18.25
Examples 18.4 and 18.5

Chapter 19
Active Figures 19.8 and 19.12
Example 19.7

Chapter 20
Active Figures 20.4, 20.5, and 20.7
Example 20.10

Chapter 21
Active Figures 21.2, 21.4, 21.11, and 21.12
Example 21.4

Chapter 22
Active Figures 22.2, 22.5, 22.10, 22.11, 22.12, 22.13, and 22.19
Example 22.10

Chapter 23
Active Figures 23.7, 23.13, 23.24, and 23.26
Examples 23.3 and 23.11

Chapter 24
Active Figures 24.4 and 24.9
Examples 24.5 and 24.10

Chapter 25
Active Figures 25.10 and 25.27
Examples 25.2 and 25.3

Chapter 26
Active Figures 26.4, 26.9, and 26.10
Examples 26.4 and 26.5

Chapter 27
Active Figures 27.9 and 27.13
Examples 27.3 and 27.8

Chapter 28
Active Figures 28.2, 28.4, 28.6, 28.19, 28.21, 28.27, and 28.29
Examples 28.1, 28.5, and 28.9

Chapter 29
Active Figures 29.1, 29.14, 29.18, 29.19, 29.23, and 29.24
Example 29.7

Chapter 30
Active Figures 30.8, 30.9, and 30.21
Examples 30.1, 30.3, and 30.8

Chapter 31
Active Figures 31.1, 31.2, 31.10, 31.21, 31.23, and 31.26
Examples 31.4 and 31.5

Chapter 32
Active Figures 32.3, 32.4, 32.6, 32.7, 32.17, 32.18, 32.21, and 32.23
Examples 32.3 and 32.7

Chapter 33
Active Figures 33.2, 33.3, 33.6, 33.7, 33.9, 33.10, 33.13, 33.15, 33.19, 33.25, and 33.26
Examples 33.5 and 33.7

Chapter 34
Active Figure 34.3
Examples 34.1 and 34.4

Chapter 35
Active Figures 35.4, 35.6, 35.10, 35.11, 35.23, and 35.26
Examples 35.2 and 35.6

Chapter 36
Active Figures 36.2, 36.15, 36.20, 36.28, 36.44, and 36.45
Examples 36.4, 36.5, 36.9, 36.10, and 36.12

Chapter 37
Active Figures 37.2, 37.11, 37.13, 37.22
Examples 37.1 and 37.4

Chapter 38
Active Figures 38.4, 38.11, 38.17, 38.18, 38.30, and 38.31
Examples 38.1 and 38.7

Chapter 39
Active Figures 39.4, 39.6, and 39.11
Examples 39.4 and 39.9

An Invitation to Physics

It is our sincere hope that you too will find physics an exciting and enjoyable experience and that you will profit from this experience, regardless of your chosen profession. Welcome to the exciting world of physics!

> *The scientist does not study nature because it is useful; he studies it because he delights in it, and he delights in it because it is beautiful. If nature were not beautiful, it would not be worth knowing, and if nature were not worth knowing, life would not be worth living.*
>
> **—Henri Poincaré**

Oscillations and Mechanical Waves

We begin this new part of the text by studying a special type of motion called *periodic* motion. This is a *repeating* motion of an object in which the object continues to return to a given position after a fixed time interval. Familiar objects that exhibit periodic motion include a pendulum and a beach ball floating on the waves at a beach. The back and forth movements of such an object are called *oscillations.* We will focus our attention on a special case of periodic motion called *simple harmonic motion.* We shall find that all periodic motions can be modeled as combinations of simple harmonic motions. Thus, simple harmonic motion forms a basic building block for more complicated periodic motion.

Simple harmonic motion also forms the basis for our understanding of *mechanical waves.* Sound waves, seismic waves, waves on stretched strings, and water waves are all produced by some source of oscillation. As a sound wave travels through the air, elements of the air oscillate back and forth; as a water wave travels across a pond, elements of the water oscillate up and down and backward and forward. In general, as waves travel through any medium, the elements of the medium move in repetitive cycles. Therefore, the motion of the elements of the medium bears a strong resemblance to the periodic motion of an oscillating pendulum or an object attached to a spring.

To explain many other phenomena in nature, we must understand the concepts of oscillations and waves. For instance, although skyscrapers and bridges appear to be rigid, they actually oscillate, a fact that the architects and engineers who design and build them must take into account. To understand how radio and television work, we must understand the origin and nature of electromagnetic waves and how they propagate through space. Finally, much of what scientists have learned about atomic structure has come from information carried by waves. Therefore, we must first study oscillations and waves if we are to understand the concepts and theories of atomic physics. ■

◀ *Drops of water fall from a leaf into a pond. The disturbance caused by the falling water causes the water surface to oscillate. These oscillations are associated with waves moving away from the point at which the water fell. In Part 2 of the text, we will explore the principles related to oscillations and waves. (Don Bonsey/Getty Images)*

Chapter 15

Oscillatory Motion

CHAPTER OUTLINE

15.1 Motion of an Object Attached to a Spring

15.2 Mathematical Representation of Simple Harmonic Motion

15.3 Energy of the Simple Harmonic Oscillator

15.4 Comparing Simple Harmonic Motion with Uniform Circular Motion

15.5 The Pendulum

15.6 Damped Oscillations

15.7 Forced Oscillations

▲ *In the Bay of Fundy, Nova Scotia, the tides undergo oscillations with very large amplitudes, such that boats often end up sitting on dry ground for part of the day. In this chapter, we will investigate the physics of oscillatory motion. (www.comstock.com)*

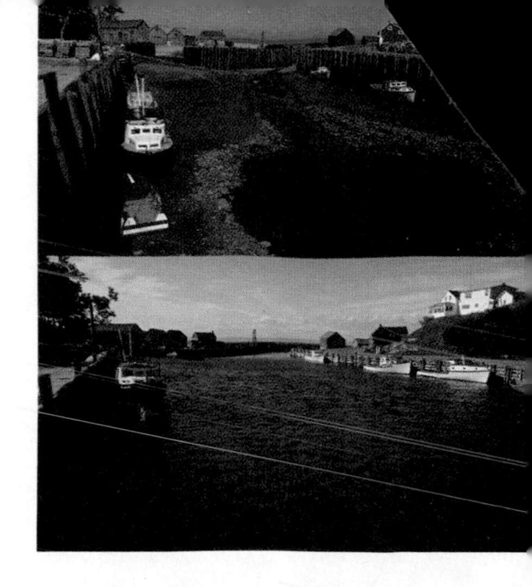

Periodic motion is motion of an object that regularly repeats—the object returns to a given position after a fixed time interval. With a little thought, we can identify several types of periodic motion in everyday life. Your car returns to the driveway each afternoon. You return to the dinner table each night to eat. A bumped chandelier swings back and forth, returning to the same position at a regular rate. The Earth returns to the same position in its orbit around the Sun each year, resulting in the variation among the four seasons. The Moon returns to the same relationship with the Earth and the Sun, resulting in a full Moon approximately once a month.

In addition to these everyday examples, numerous other systems exhibit periodic motion. For example, the molecules in a solid oscillate about their equilibrium positions; electromagnetic waves, such as light waves, radar, and radio waves, are characterized by oscillating electric and magnetic field vectors; and in alternating-current electrical circuits, voltage, current, and electric charge vary periodically with time.

A special kind of periodic motion occurs in mechanical systems when the force acting on an object is proportional to the position of the object relative to some equilibrium position. If this force is always directed toward the equilibrium position, the motion is called *simple harmonic motion,* which is the primary focus of this chapter.

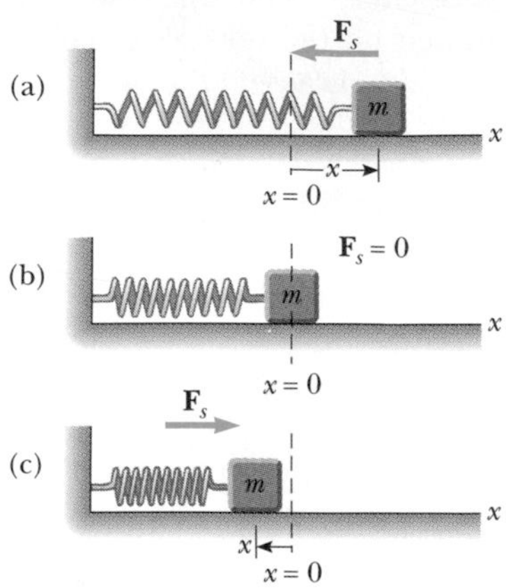

Active Figure 15.1 A block attached to a spring moving on a frictionless surface. (a) When the block is displaced to the right of equilibrium ($x > 0$), the force exerted by the spring acts to the left. (b) When the block is at its equilibrium position ($x = 0$), the force exerted by the spring is zero. (c) When the block is displaced to the left of equilibrium ($x < 0$), the force exerted by the spring acts to the right.

At the Active Figures link, at http://www.pse6.com, you can choose the spring constant and the initial position and velocities of the block to see the resulting simple harmonic motion.

15.1 Motion of an Object Attached to a Spring

As a model for simple harmonic motion, consider a block of mass m attached to the end of a spring, with the block free to move on a horizontal, frictionless surface (Fig. 15.1). When the spring is neither stretched nor compressed, the block is at the position called the **equilibrium position** of the system, which we identify as $x = 0$. We know from experience that such a system oscillates back and forth if disturbed from its equilibrium position.

We can understand the motion in Figure 15.1 qualitatively by first recalling that when the block is displaced to a position x, the spring exerts on the block a force that is proportional to the position and given by **Hooke's law** (see Section 7.4):

Hooke's law

$$F_s = -kx \qquad (15.1)$$

We call this a **restoring force** because it is always directed toward the equilibrium position and therefore *opposite* the displacement from equilibrium. That is, when the block is displaced to the right of $x = 0$ in Figure 15.1, then the position is positive and the restoring force is directed to the left. When the block is displaced to the left of $x = 0$, then the position is negative and the restoring force is directed to the right.

Applying Newton's second law $\Sigma F_x = ma_x$ to the motion of the block, with Equation 15.1 providing the net force in the x direction, we obtain

$$-kx = ma_x$$

$$a_x = -\frac{k}{m}x \qquad (15.2)$$

▲ PITFALL PREVENTION

15.1 The Orientation of the Spring

Figure 15.1 shows a *horizontal* spring, with an attached block sliding on a frictionless surface. Another possibility is a block hanging from a *vertical* spring. All of the results that we discuss for the horizontal spring will be the same for the vertical spring, except that when the block is placed on the vertical spring, its weight will cause the spring to extend. If the resting position of the block is defined as $x = 0$, the results of this chapter will apply to this vertical system also.

That is, the acceleration is proportional to the position of the block, and its direction is opposite the direction of the displacement from equilibrium. Systems that behave in this way are said to exhibit **simple harmonic motion. An object moves with simple harmonic motion whenever its acceleration is proportional to its position and is oppositely directed to the displacement from equilibrium.**

If the block in Figure 15.1 is displaced to a position $x = A$ and released from rest, its *initial* acceleration is $-kA/m$. When the block passes through the equilibrium position $x = 0$, its acceleration is zero. At this instant, its speed is a maximum because the acceleration changes sign. The block then continues to travel to the left of equilibrium with a positive acceleration and finally reaches $x = -A$, at which time its acceleration is $+kA/m$ and its speed is again zero, as discussed in Sections 7.4 and 8.6. The block completes a full cycle of its motion by returning to the original position, again passing through $x = 0$ with maximum speed. Thus, we see that the block oscillates between the turning points $x = \pm A$. In the absence of friction, because the force exerted by the spring is conservative, this idealized motion will continue forever. Real systems are generally subject to friction, so they do not oscillate forever. We explore the details of the situation with friction in Section 15.6.

As Pitfall Prevention 15.1 points out, the principles that we develop in this chapter are also valid for an object hanging from a vertical spring, as long as we recognize that the weight of the object will stretch the spring to a new equilibrium position $x = 0$. To prove this statement, let x_s represent the total extension of the spring from its equilibrium position *without* the hanging object. Then, $x_s = -(mg/k) + x$, where $-(mg/k)$ is the extension of the spring due to the weight of the hanging object and x is the instantaneous extension of the spring due to the simple harmonic motion. The magnitude of the net force on the object is then $F_s - F_g = -k(-(mg/k) + x) - mg = -kx$. The net force on the object is the same as that on a block connected to a horizontal spring as in Equation 15.1, so the same simple harmonic motion results.

Quick Quiz 15.1 A block on the end of a spring is pulled to position $x = A$ and released. In one full cycle of its motion, through what total distance does it travel? (a) $A/2$ (b) A (c) $2A$ (d) $4A$

15.2 Mathematical Representation of Simple Harmonic Motion

Let us now develop a mathematical representation of the motion we described in the preceding section. We model the block as a particle subject to the force in Equation 15.1. We will generally choose x as the axis along which the oscillation occurs; hence, we will drop the subscript-x notation in this discussion. Recall that, by definition, $a = dv/dt = d^2x/dt^2$, and so we can express Equation 15.2 as

$$\frac{d^2x}{dt^2} = -\frac{k}{m}x \tag{15.3}$$

15.2 A Nonconstant Acceleration

Notice that the acceleration of the particle in simple harmonic motion is not constant. Equation 15.3 shows that it varies with position x. Thus, we *cannot* apply the kinematic equations of Chapter 2 in this situation.

If we denote the ratio k/m with the symbol ω^2 (we choose ω^2 rather than ω in order to make the solution that we develop below simpler in form), then

$$\omega^2 = \frac{k}{m} \tag{15.4}$$

and Equation 15.3 can be written in the form

$$\frac{d^2x}{dt^2} = -\omega^2 x \tag{15.5}$$

What we now require is a mathematical solution to Equation 15.5—that is, a function $x(t)$ that satisfies this second-order differential equation. This is a mathematical representation of the position of the particle as a function of time. We seek a function $x(t)$ whose second derivative is the same as the original function with a negative sign and multiplied by ω^2. The trigonometric functions sine and cosine exhibit this behavior, so we can build a solution around one or both of these. The following cosine function is a solution to the differential equation:

$$x(t) = A\cos(\omega t + \phi) \tag{15.6}$$

Position versus time for an object in simple harmonic motion

where A, ω, and ϕ are constants. To see explicitly that this equation satisfies Equation 15.5, note that

$$\frac{dx}{dt} = A\frac{d}{dt}\cos(\omega t + \phi) = -\omega A\sin(\omega t + \phi) \tag{15.7}$$

$$\frac{d^2x}{dt^2} = -\omega A\frac{d}{dt}\sin(\omega t + \phi) = -\omega^2 A\cos(\omega t + \phi) \tag{15.8}$$

Comparing Equations 15.6 and 15.8, we see that $d^2x/dt^2 = -\omega^2 x$ and Equation 15.5 is satisfied.

The parameters A, ω, and ϕ are constants of the motion. In order to give physical significance to these constants, it is convenient to form a graphical representation of the motion by plotting x as a function of t, as in Figure 15.2a. First, note that A, called the **amplitude** of the motion, is simply **the maximum value of the position of the particle in either the positive or negative x direction.** The constant ω is called the **angular frequency,** and has units of rad/s.[1] It is a measure of how rapidly the oscillations are occurring—the more oscillations per unit time, the higher is the value of ω. From Equation 15.4, the angular frequency is

$$\omega = \sqrt{\frac{k}{m}} \tag{15.9}$$

The constant angle ϕ is called the **phase constant** (or initial phase angle) and, along with the amplitude A, is determined uniquely by the position and velocity of the particle at $t = 0$. If the particle is at its maximum position $x = A$ at $t = 0$, the phase constant is $\phi = 0$ and the graphical representation of the motion is shown in Figure 15.2b. The quantity $(\omega t + \phi)$ is called the **phase** of the motion. Note that the function $x(t)$ is periodic and its value is the same each time ωt increases by 2π radians.

Equations 15.1, 15.5, and 15.6 form the basis of the mathematical representation of simple harmonic motion. If we are analyzing a situation and find that the force on a particle is of the mathematical form of Equation 15.1, we know that the motion will be that of a simple harmonic oscillator and that the position of the particle is described by Equation 15.6. If we analyze a system and find that it is described by a differential equation of the form of Equation 15.5, the motion will be that of a simple harmonic oscillator. If we analyze a situation and find that the position of a particle is described by Equation 15.6, we know the particle is undergoing simple harmonic motion.

PITFALL PREVENTION

15.3 Where's the Triangle?

Equation 15.6 includes a trigonometric function, a *mathematical function* that can be used whether it refers to a triangle or not. In this case, the cosine function happens to have the correct behavior for representing the position of a particle in simple harmonic motion.

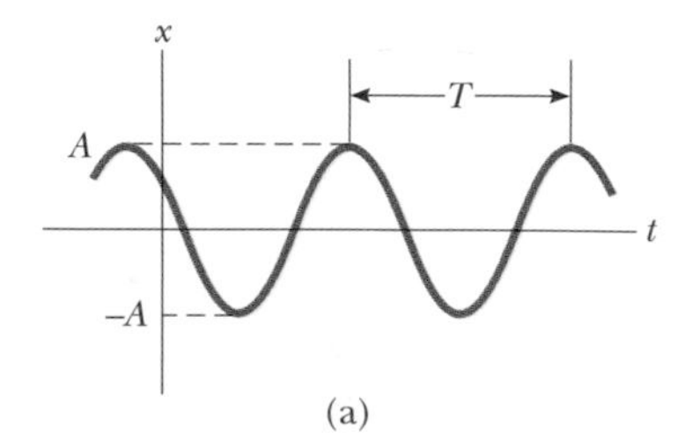

(a)

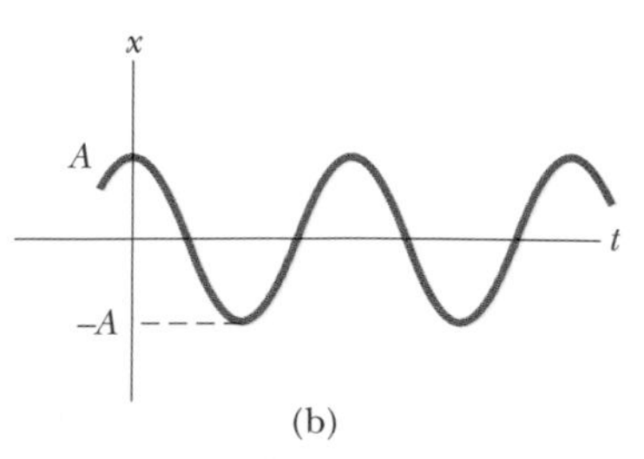

(b)

Active Figure 15.2 (a) An x-vs.-t graph for an object undergoing simple harmonic motion. The amplitude of the motion is A, the period (page 456) is T, and the phase constant is ϕ. (b) The x-vs.-t graph in the special case in which $x = A$ at $t = 0$ and hence $\phi = 0$.

At the Active Figures link at http://www.pse6.com, you can adjust the graphical representation and see the resulting simple harmonic motion of the block in Figure 15.1.

[1] We have seen many examples in earlier chapters in which we evaluate a trigonometric function of an angle. The argument of a trigonometric function, such as sine or cosine, *must* be a pure number. The radian is a pure number because it is a ratio of lengths. Angles in degrees are pure numbers simply because the degree is a completely artificial "unit"—it is not related to measurements of lengths. The notion of requiring a pure number for a trigonometric function is important in Equation 15.6, where the angle is expressed in terms of other measurements. Thus, ω *must* be expressed in rad/s (and not, for example, in revolutions per second) if t is expressed in seconds. Furthermore, other types of functions such as logarithms and exponential functions require arguments that are pure numbers.

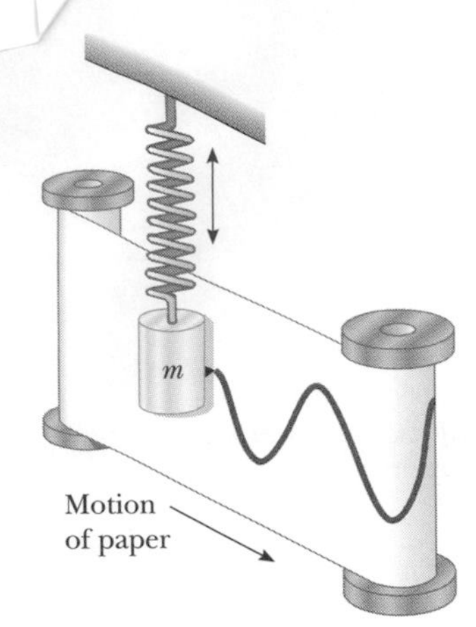

Figure 15.3 An experimental apparatus for demonstrating simple harmonic motion. A pen attached to the oscillating object traces out a sinusoidal pattern on the moving chart paper.

An experimental arrangement that exhibits simple harmonic motion is illustrated in Figure 15.3. An object oscillating vertically on a spring has a pen attached to it. While the object is oscillating, a sheet of paper is moved perpendicular to the direction of motion of the spring, and the pen traces out the cosine curve in Equation 15.6.

Quick Quiz 15.2 Consider a graphical representation (Fig. 15.4) of simple harmonic motion, as described mathematically in Equation 15.6. When the object is at point Ⓐ on the graph, its (a) position and velocity are both positive (b) position and velocity are both negative (c) position is positive and its velocity is zero (d) position is negative and its velocity is zero (e) position is positive and its velocity is negative (f) position is negative and its velocity is positive.

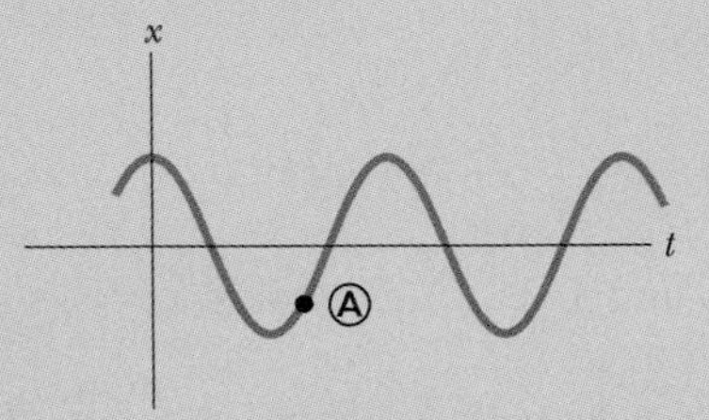

Figure 15.4 (Quick Quiz 15.2) An x-t graph for an object undergoing simple harmonic motion. At a particular time, the object's position is indicated by Ⓐ in the graph.

Quick Quiz 15.3 Figure 15.5 shows two curves representing objects undergoing simple harmonic motion. The correct description of these two motions is that the simple harmonic motion of object B is (a) of larger angular frequency and larger amplitude than that of object A (b) of larger angular frequency and smaller amplitude than that of object A (c) of smaller angular frequency and larger amplitude than that of object A (d) of smaller angular frequency and smaller amplitude than that of object A.

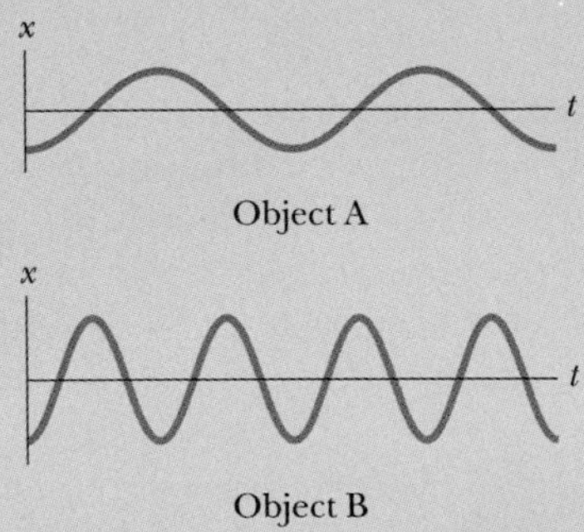

Figure 15.5 (Quick Quiz 15.3) Two x-t graphs for objects undergoing simple harmonic motion. The amplitudes and frequencies are different for the two objects.

Let us investigate further the mathematical description of simple harmonic motion. The **period** T of the motion is the time interval required for the particle to go through one full cycle of its motion (Fig. 15.2a). That is, the values of x and v for the particle at time t equal the values of x and v at time $t + T$. We can relate the period to the angular frequency by using the fact that the phase increases by 2π radians in a time interval of T:

$$[\omega(t + T) + \phi] - (\omega t + \phi) = 2\pi$$

Simplifying this expression, we see that $\omega T = 2\pi$, or

$$T = \frac{2\pi}{\omega} \tag{15.10}$$

The inverse of the period is called the **frequency** f of the motion. Whereas the period is the time interval per oscillation, the frequency represents the **number of oscillations that the particle undergoes per unit time interval:**

$$f = \frac{1}{T} = \frac{\omega}{2\pi} \tag{15.11}$$

The units of f are cycles per second, or **hertz** (Hz). Rearranging Equation 15.11 gives

$$\omega = 2\pi f = \frac{2\pi}{T} \tag{15.12}$$

We can use Equations 15.9, 15.10, and 15.11 to express the period and frequency of the motion for the particle–spring system in terms of the characteristics m and k of the system as

Period

$$T = \frac{2\pi}{\omega} = 2\pi\sqrt{\frac{m}{k}} \tag{15.13}$$

Frequency

$$f = \frac{1}{T} = \frac{1}{2\pi}\sqrt{\frac{k}{m}} \tag{15.14}$$

That is, the period and frequency depend *only* on the mass of the particle and the force constant of the spring, and *not* on the parameters of the motion, such as A or ϕ. As we might expect, the frequency is larger for a stiffer spring (larger value of k) and decreases with increasing mass of the particle.

We can obtain the velocity and acceleration[2] of a particle undergoing simple harmonic motion from Equations 15.7 and 15.8:

Velocity of an object in simple harmonic motion

$$v = \frac{dx}{dt} = -\omega A\sin(\omega t + \phi) \tag{15.15}$$

Acceleration of an object in simple harmonic motion

$$a = \frac{d^2x}{dt^2} = -\omega^2 A\cos(\omega t + \phi) \tag{15.16}$$

From Equation 15.15 we see that, because the sine and cosine functions oscillate between ± 1, the extreme values of the velocity v are $\pm\,\omega A$. Likewise, Equation 15.16 tells us that the extreme values of the acceleration a are $\pm\,\omega^2 A$. Therefore, the *maximum* values of the magnitudes of the velocity and acceleration are

Maximum magnitudes of speed and acceleration in simple harmonic motion

$$v_{\max} = \omega A = \sqrt{\frac{k}{m}}\,A \tag{15.17}$$

$$a_{\max} = \omega^2 A = \frac{k}{m}\,A \tag{15.18}$$

Figure 15.6a plots position versus time for an arbitrary value of the phase constant. The associated velocity–time and acceleration–time curves are illustrated in Figures 15.6b and 15.6c. They show that the phase of the velocity differs from the phase of the position by $\pi/2$ rad, or 90°. That is, when x is a maximum or a minimum, the velocity is zero. Likewise, when x is zero, the speed is a maximum. Furthermore, note that the

▲ PITFALL PREVENTION

15.4 Two Kinds of Frequency

We identify two kinds of frequency for a simple harmonic oscillator—f, called simply the *frequency*, is measured in hertz, and ω, the *angular frequency*, is measured in radians per second. Be sure that you are clear about which frequency is being discussed or requested in a given problem. Equations 15.11 and 15.12 show the relationship between the two frequencies.

[2] Because the motion of a simple harmonic oscillator takes place in one dimension, we will denote velocity as v and acceleration as a, with the direction indicated by a positive or negative sign, as in Chapter 2.

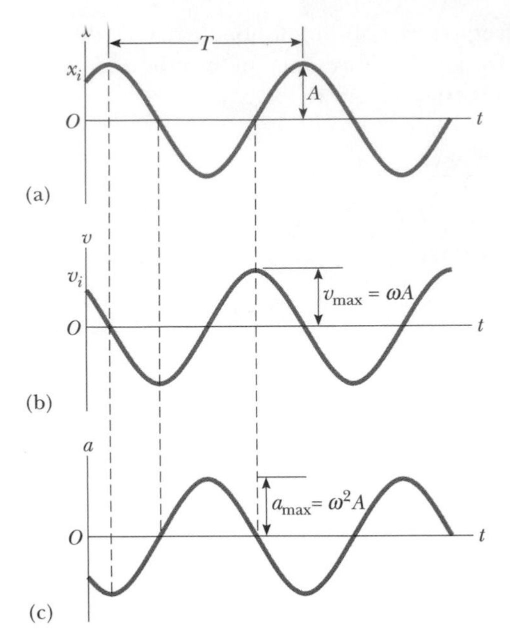

Figure 15.6 Graphical representation of simple harmonic motion. (a) Position versus time. (b) Velocity versus time. (c) Acceleration versus time. Note that at any specified time the velocity is 90° out of phase with the position and the acceleration is 180° out of phase with the position.

phase of the acceleration differs from the phase of the position by π radians, or 180°. For example, when x is a maximum, a has a maximum magnitude in the opposite direction.

Quick Quiz 15.4 Consider a graphical representation (Fig. 15.4) of simple harmonic motion, as described mathematically in Equation 15.6. When the object is at position Ⓐ on the graph, its (a) velocity and acceleration are both positive (b) velocity and acceleration are both negative (c) velocity is positive and its acceleration is zero (d) velocity is negative and its acceleration is zero (e) velocity is positive and its acceleration is negative (f) velocity is negative and its acceleration is positive.

Quick Quiz 15.5 An object of mass m is hung from a spring and set into oscillation. The period of the oscillation is measured and recorded as T. The object of mass m is removed and replaced with an object of mass $2m$. When this object is set into oscillation, the period of the motion is (a) $2T$ (b) $\sqrt{2}T$ (c) T (d) $T/\sqrt{2}$ (e) $T/2$.

Equation 15.6 describes simple harmonic motion of a particle in general. Let us now see how to evaluate the constants of the motion. The angular frequency ω is evaluated using Equation 15.9. The constants A and ϕ are evaluated from the initial conditions, that is, the state of the oscillator at $t = 0$.

Suppose we initiate the motion by pulling the particle from equilibrium by a distance A and releasing it from rest at $t = 0$, as in Figure 15.7. We must then require that

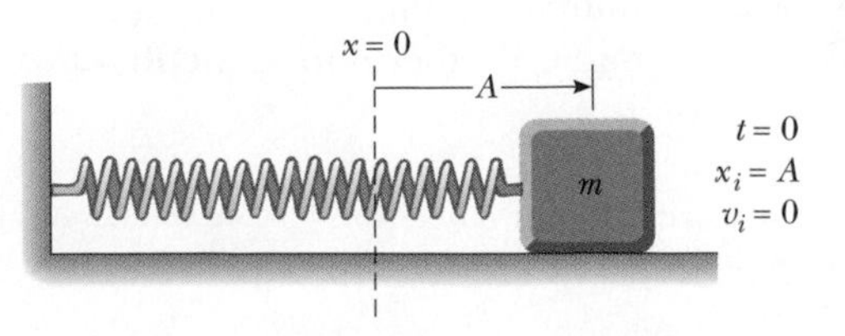

Active Figure 15.7 A block–spring system that begins its motion from rest with the block at $x = A$ at $t = 0$. In this case, $\phi = 0$ and thus $x = A \cos \omega t$.

At the Active Figures link at http://www.pse6.com, you can compare the oscillations of two blocks starting from different initial positions to see that the frequency is independent of the amplitude.

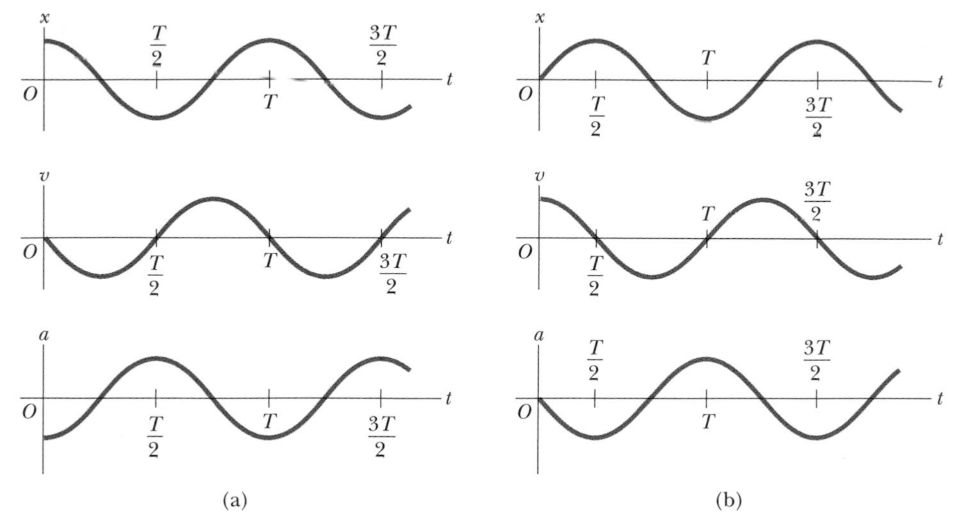

Figure 15.8 (a) Position, velocity, and acceleration versus time for a block undergoing simple harmonic motion under the initial conditions that at $t = 0$, $x(0) = A$ and $v(0) = 0$. (b) Position, velocity, and acceleration versus time for a block undergoing simple harmonic motion under the initial conditions that at $t = 0$, $x(0) = 0$ and $v(0) = v_i$.

our solutions for $x(t)$ and $v(t)$ (Eqs. 15.6 and 15.15) obey the initial conditions that $x(0) = A$ and $v(0) = 0$:

$$x(0) = A\cos\phi = A$$

$$v(0) = -\omega A\sin\phi = 0$$

These conditions are met if we choose $\phi = 0$, giving $x = A\cos\omega t$ as our solution. To check this solution, note that it satisfies the condition that $x(0) = A$, because $\cos 0 = 1$.

The position, velocity, and acceleration versus time are plotted in Figure 15.8a for this special case. The acceleration reaches extreme values of $\mp\omega^2 A$ when the position has extreme values of $\pm A$. Furthermore, the velocity has extreme values of $\pm\omega A$, which both occur at $x = 0$. Hence, the quantitative solution agrees with our qualitative description of this system.

Let us consider another possibility. Suppose that the system is oscillating and we define $t = 0$ as the instant that the particle passes through the unstretched position of the spring while moving to the right (Fig. 15.9). In this case we must require that our solutions for $x(t)$ and $v(t)$ obey the initial conditions that $x(0) = 0$ and $v(0) = v_i$:

$$x(0) = A\cos\phi = 0$$

$$v(0) = -\omega A\sin\phi = v_i$$

The first of these conditions tells us that $\phi = \pm\pi/2$. With these choices for ϕ, the second condition tells us that $A = \mp v_i/\omega$. Because the initial velocity is positive and the amplitude must be positive, we must have $\phi = -\pi/2$. Hence, the solution is given by

$$x = \frac{v_i}{\omega}\cos\left(\omega t - \frac{\pi}{2}\right)$$

The graphs of position, velocity, and acceleration versus time for this choice of $t = 0$ are shown in Figure 15.8b. Note that these curves are the same as those in Figure 15.8a, but shifted to the right by one fourth of a cycle. This is described mathematically by the phase constant $\phi = -\pi/2$, which is one fourth of a full cycle of 2π.

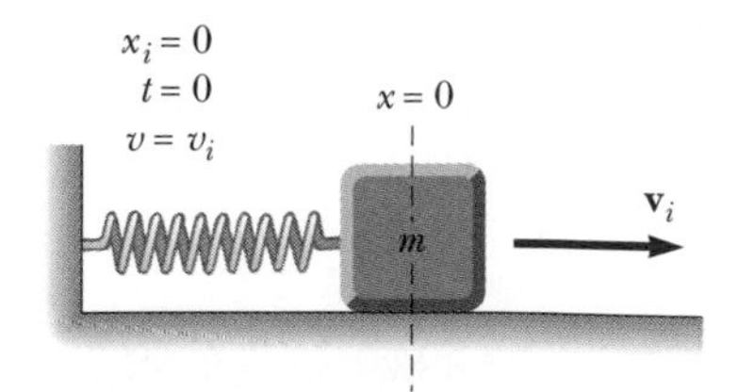

Active Figure 15.9 The block–spring system is undergoing oscillation, and $t = 0$ is defined at an instant when the block passes through the equilibrium position $x = 0$ and is moving to the right with speed v_i.

At the Active Figures link at http://www.pse6.com, you can compare the oscillations of two blocks with different velocities at t = 0 to see that the frequency is independent of the amplitude.

Example 15.1 An Oscillating Object

An object oscillates with simple harmonic motion along the x axis. Its position varies with time according to the equation

$$x = (4.00\text{ m})\cos\left(\pi t + \frac{\pi}{4}\right)$$

where t is in seconds and the angles in the parentheses are in radians.

(A) Determine the amplitude, frequency, and period of the motion.

Solution By comparing this equation with Equation 15.6, $x = A\cos(\omega t + \phi)$, we see that $A =$ 4.00 m and $\omega = \pi$ rad/s. Therefore, $f = \omega/2\pi = \pi/2\pi =$ 0.500 Hz and $T = 1/f =$ 2.00 s.

(B) Calculate the velocity and acceleration of the object at any time t.

Solution Differentiating x to find v, and v to find a, we obtain

$$v = \frac{dx}{dt} = -(4.00\text{ m/s})\sin\left(\pi t + \frac{\pi}{4}\right)\frac{d}{dt}(\pi t)$$

$$= -(4.00\pi\text{ m/s})\sin\left(\pi t + \frac{\pi}{4}\right)$$

$$a = \frac{dv}{dt} = -(4.00\pi\text{ m/s})\cos\left(\pi t + \frac{\pi}{4}\right)\frac{d}{dt}(\pi t)$$

$$= -(4.00\pi^2\text{ m/s}^2)\cos\left(\pi t + \frac{\pi}{4}\right)$$

(C) Using the results of part (B), determine the position, velocity, and acceleration of the object at $t = 1.00$ s.

Solution Noting that the angles in the trigonometric functions are in radians, we obtain, at $t = 1.00$ s,

$$x = (4.00\text{ m})\cos\left(\pi + \frac{\pi}{4}\right) = (4.00\text{ m})\cos\left(\frac{5\pi}{4}\right)$$

$$= (4.00\text{ m})(-0.707) = -2.83\text{ m}$$

$$v = -(4.00\pi\text{ m/s})\sin\left(\frac{5\pi}{4}\right)$$

$$= -(4.00\pi\text{ m/s})(-0.707) = 8.89\text{ m/s}$$

$$a = -(4.00\pi^2\text{ m/s}^2)\cos\left(\frac{5\pi}{4}\right)$$

$$= -(4.00\pi^2\text{ m/s}^2)(-0.707) = 27.9\text{ m/s}^2$$

(D) Determine the maximum speed and maximum acceleration of the object.

Solution In the general expressions for v and a found in part (B), we use the fact that the maximum values of the sine and cosine functions are unity. Therefore, v varies between $\pm 4.00\pi$ m/s, and a varies between $\pm 4.00\pi^2$ m/s^2. Thus,

$$v_{\max} = 4.00\pi\text{ m/s} = 12.6\text{ m/s}$$

$$a_{\max} = 4.00\pi^2\text{ m/s}^2 = 39.5\text{ m/s}^2$$

We obtain the same results using the relations $v_{\max} = \omega A$ and $a_{\max} = \omega^2 A$, where $A = 4.00$ m and $\omega = \pi$ rad/s.

(E) Find the displacement of the object between $t = 0$ and $t = 1.00$ s.

Solution The position at $t = 0$ is

$$x_i = (4.00\text{ m})\cos\left(0 + \frac{\pi}{4}\right) = (4.00\text{ m})(0.707) = 2.83\text{ m}$$

In part (C), we found that the position at $t = 1.00$ s is -2.83 m; therefore, the displacement between $t = 0$ and $t = 1.00$ s is

$$\Delta x = x_f - x_i = -2.83\text{ m} - 2.83\text{ m} = -5.66\text{ m}$$

Because the object's velocity changes sign during the first second, the magnitude of Δx is not the same as the distance traveled in the first second. (By the time the first second is over, the object has been through the point $x = -2.83$ m once, traveled to $x = -4.00$ m, and come back to $x = -2.83$ m.)

Example 15.2 Watch Out for Potholes!

A car with a mass of 1 300 kg is constructed so that its frame is supported by four springs. Each spring has a force constant of 20 000 N/m. If two people riding in the car have a combined mass of 160 kg, find the frequency of vibration of the car after it is driven over a pothole in the road.

Solution To conceptualize this problem, think about your experiences with automobiles. When you sit in a car, it moves downward a small distance because your weight is compressing the springs further. If you push down on the front bumper and release, the front of the car oscillates a couple of times. We can model the car as being supported by a single spring and categorize this as an oscillation problem based on our simple spring model. To analyze the problem, we first need to consider the effective spring constant of the four springs combined. For a given extension x of the springs, the combined force on the car is the sum of the forces from the individual springs:

$$F_{\text{total}} = \sum(-kx) = -\left(\sum k\right)x$$

where x has been factored from the sum because it is the

same for all four springs. We see that the effective spring constant for the combined springs is the sum of the individual spring constants:

$$k_{\text{eff}} = \sum k = 4 \times 20\,000 \text{ N/m} = 80\,000 \text{ N/m}$$

Hence, the frequency of vibration is, from Equation 15.14,

$$f = \frac{1}{2\pi}\sqrt{\frac{k_{\text{eff}}}{m}} = \frac{1}{2\pi}\sqrt{\frac{80\,000 \text{ N/m}}{1\,460 \text{ kg}}} = 1.18 \text{ Hz}$$

To finalize the problem, note that the mass we used here is that of the car plus the people, because this is the total mass that is oscillating. Also note that we have explored only up-and-down motion of the car. If an oscillation is established in which the car rocks back and forth such that the front end goes up when the back end goes down, the frequency will be different.

What If? **Suppose the two people exit the car on the side of the road. One of them pushes downward on the car and releases it so that it oscillates vertically. Is the frequency of the oscillation the same as the value we just calculated?**

Answer The suspension system of the car is the same, but the mass that is oscillating is smaller—it no longer includes the mass of the two people. Thus, the frequency should be higher. Let us calculate the new frequency:

$$f = \frac{1}{2\pi}\sqrt{\frac{k_{\text{eff}}}{m}} = \frac{1}{2\pi}\sqrt{\frac{80\,000 \text{ N/m}}{1\,300 \text{ kg}}} = 1.25 \text{ Hz}$$

As we predicted conceptually, the frequency is a bit higher.

Example 15.3 A Block–Spring System

A 200-g block connected to a light spring for which the force constant is 5.00 N/m is free to oscillate on a horizontal, frictionless surface. The block is displaced 5.00 cm from equilibrium and released from rest, as in Figure 15.7.

(A) Find the period of its motion.

Solution From Equations 15.9 and 15.10, we know that the angular frequency of a block–spring system is

$$\omega = \sqrt{\frac{k}{m}} = \sqrt{\frac{5.00 \text{ N/m}}{200 \times 10^{-3} \text{ kg}}} = 5.00 \text{ rad/s}$$

and the period is

$$T = \frac{2\pi}{\omega} = \frac{2\pi}{5.00 \text{ rad/s}} = 1.26 \text{ s}$$

(B) Determine the maximum speed of the block.

Solution We use Equation 15.17:

$$v_{\max} = \omega A = (5.00 \text{ rad/s})(5.00 \times 10^{-2} \text{ m}) = 0.250 \text{ m/s}$$

(C) What is the maximum acceleration of the block?

Solution We use Equation 15.18:

$$a_{\max} = \omega^2 A = (5.00 \text{ rad/s})^2(5.00 \times 10^{-2} \text{ m}) = 1.25 \text{ m/s}^2$$

(D) Express the position, speed, and acceleration as functions of time.

Solution We find the phase constant from the initial condition that $x = A$ at $t = 0$:

$$x(0) = A\cos\phi = A$$

which tells us that $\phi = 0$. Thus, our solution is $x = A\cos\omega t$. Using this expression and the results from (A), (B), and (C), we find that

$$x = A\cos\omega t = (0.050\,0 \text{ m})\cos 5.00t$$

$$v = \omega A\sin\omega t = -(0.250 \text{ m/s})\sin 5.00t$$

$$a = -\omega^2 A\cos\omega t = -(1.25 \text{ m/s}^2)\cos 5.00t$$

What If? **What if the block is released from the same initial position, $x_i = 5.00$ cm, but with an initial velocity of $v_i = -0.100$ m/s? Which parts of the solution change and what are the new answers for those that do change?**

Answers Part (A) does not change—the period is independent of how the oscillator is set into motion. Parts (B), (C), and (D) will change. We begin by considering position and velocity expressions for the initial conditions:

$$(1) \qquad x(0) = A\cos\phi = x_i$$

$$(2) \qquad v(0) = -\omega A\sin\phi = v_i$$

Dividing Equation (2) by Equation (1) gives us the phase constant:

$$\frac{-\omega A\sin\phi}{A\cos\phi} = \frac{v_i}{x_i}$$

$$\tan\phi = -\frac{v_i}{\omega x_i} = -\frac{-0.100 \text{ m}}{(5.00 \text{ rad/s})(0.050\,0 \text{ m})} = 0.400$$

$$\phi = 0.12\pi$$

Now, Equation (1) allows us to find A:

$$A = \frac{x_i}{\cos\phi} = \frac{0.050\,0 \text{ m}}{\cos(0.12\pi)} = 0.053\,9 \text{ m}$$

The new maximum speed is

$$v_{\max} = \omega A = (5.00 \text{ rad/s})(5.39 \times 10^{-2} \text{ m}) = 0.269 \text{ m/s}$$

The new magnitude of the maximum acceleration is

$$a_{\max} = \omega^2 A = (5.00 \text{ rad/s})^2(5.39 \times 10^{-2} \text{ m}) = 1.35 \text{ m/s}^2$$

The new expressions for position, velocity, and acceleration are

$$x = (0.053\,9 \text{ m})\cos(5.00t + 0.12\pi)$$

$$v = -(0.269 \text{ m/s})\sin(5.00t + 0.12\pi)$$

$$a = -(1.35 \text{ m/s}^2)\cos(5.00t + 0.12\pi)$$

As we saw in Chapters 7 and 8, many problems are easier to solve with an energy approach rather than one based on variables of motion. This particular **What If?** is easier to solve from an energy approach. Therefore, in the next section we shall investigate the energy of the simple harmonic oscillator.

15.3 Energy of the Simple Harmonic Oscillator

Let us examine the mechanical energy of the block–spring system illustrated in Figure 15.1. Because the surface is frictionless, we expect the total mechanical energy of the system to be constant, as was shown in Chapter 8. We assume a massless spring, so the kinetic energy of the system corresponds only to that of the block. We can use Equation 15.15 to express the kinetic energy of the block as

Kinetic energy of a simple harmonic oscillator

$$K = \tfrac{1}{2}mv^2 = \tfrac{1}{2}m\omega^2 A^2 \sin^2(\omega t + \phi) \tag{15.19}$$

The elastic potential energy stored in the spring for any elongation x is given by $\frac{1}{2}kx^2$ (see Eq. 8.11). Using Equation 15.6, we obtain

Potential energy of a simple harmonic oscillator

$$U = \tfrac{1}{2}kx^2 = \tfrac{1}{2}kA^2 \cos^2(\omega t + \phi) \tag{15.20}$$

We see that K and U are *always* positive quantities. Because $\omega^2 = k/m$, we can express the total mechanical energy of the simple harmonic oscillator as

$$E = K + U = \tfrac{1}{2}kA^2[\sin^2(\omega t + \phi) + \cos^2(\omega t + \phi)]$$

From the identity $\sin^2 \theta + \cos^2 \theta = 1$, we see that the quantity in square brackets is unity. Therefore, this equation reduces to

Total energy of a simple harmonic oscillator

$$E = \tfrac{1}{2}kA^2 \tag{15.21}$$

That is, **the total mechanical energy of a simple harmonic oscillator is a constant of the motion and is proportional to the square of the amplitude.** Note that U is small when K is large, and vice versa, because the sum must be constant. In fact, the total mechanical energy is equal to the maximum potential energy stored in the spring when $x = \pm A$ because $v = 0$ at these points and thus there is no kinetic energy. At the equilibrium position, where $U = 0$ because $x = 0$, the total energy, all in the form of kinetic energy, is again $\frac{1}{2}kA^2$. That is,

$$E = \tfrac{1}{2}mv_{\text{max}}^2 = \tfrac{1}{2}m\omega^2 A^2 = \tfrac{1}{2}m\frac{k}{m}A^2 = \tfrac{1}{2}kA^2 \qquad (\text{at } x = 0)$$

Plots of the kinetic and potential energies versus time appear in Figure 15.10a, where we have taken $\phi = 0$. As already mentioned, both K and U are always positive, and at all times their sum is a constant equal to $\frac{1}{2}kA^2$, the total energy of the system. The variations of K and U with the position x of the block are plotted in Figure 15.10b.

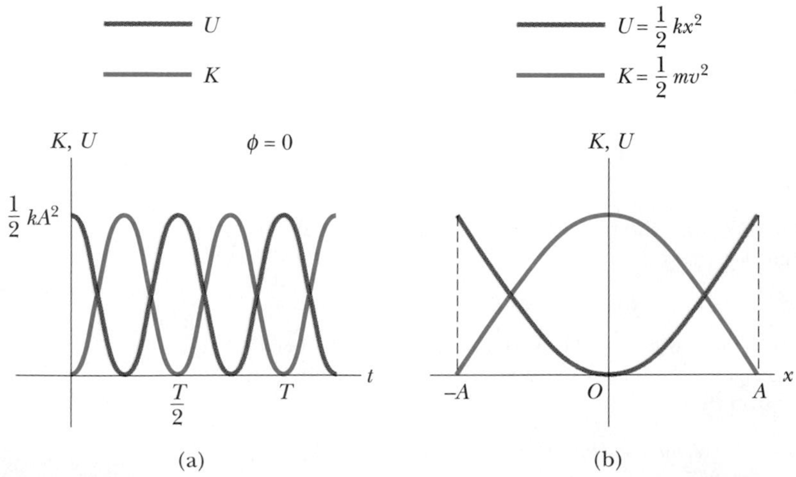

Active Figure 15.10 (a) Kinetic energy and potential energy versus time for a simple harmonic oscillator with $\phi = 0$. (b) Kinetic energy and potential energy versus position for a simple harmonic oscillator. In either plot, note that $K + U =$ constant.

At the Active Figures link at* http://www.pse6.com, *you can compare the physical oscillation of a block with energy graphs in this figure as well as with energy bar graphs.

Energy is continuously being transformed between potential energy stored in the spring and kinetic energy of the block.

Figure 15.11 illustrates the position, velocity, acceleration, kinetic energy, and potential energy of the block–spring system for one full period of the motion. Most of the ideas discussed so far are incorporated in this important figure. Study it carefully.

Finally, we can use the principle of conservation of energy to obtain the velocity for an arbitrary position by expressing the total energy at some arbitrary position x as

$$E = K + U = \tfrac{1}{2}mv^2 + \tfrac{1}{2}kx^2 = \tfrac{1}{2}kA^2$$

$$v = \pm\sqrt{\frac{k}{m}(A^2 - x^2)} = \pm\omega\sqrt{A^2 - x^2} \quad (15.22)$$

Velocity as a function of position for a simple harmonic oscillator

When we check Equation 15.22 to see whether it agrees with known cases, we find that it verifies the fact that the speed is a maximum at $x = 0$ and is zero at the turning points $x = \pm A$.

You may wonder why we are spending so much time studying simple harmonic oscillators. We do so because they are good models of a wide variety of physical phenomena. For example, recall the Lennard–Jones potential discussed in Example 8.11. This complicated function describes the forces holding atoms together. Figure 15.12a shows that, for small displacements from the equilibrium position, the potential energy curve

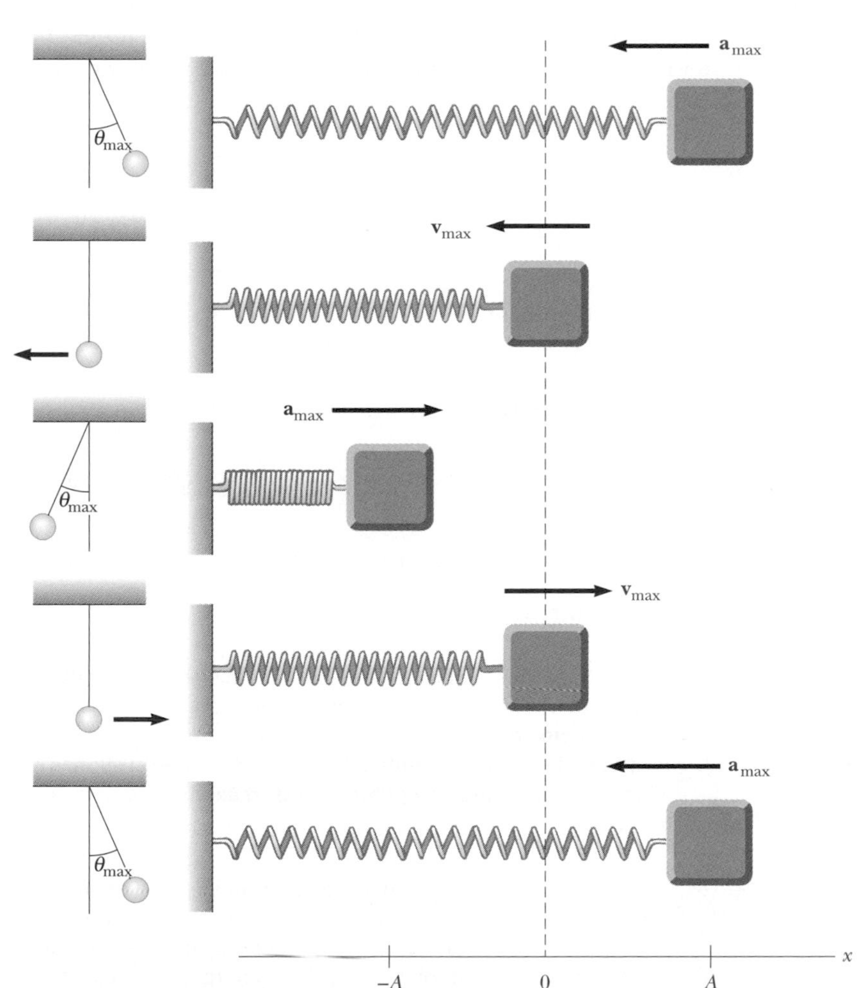

t	x	v	a	K	U
0	A	0	$-\omega^2 A$	0	$\frac{1}{2}kA^2$
$T/4$	0	$-\omega A$	0	$\frac{1}{2}kA^2$	0
$T/2$	$-A$	0	$\omega^2 A$	0	$\frac{1}{2}kA^2$
$3T/4$	0	ωA	0	$\frac{1}{2}kA^2$	0
T	A	0	$-\omega^2 A$	0	$\frac{1}{2}kA^2$

Active Figure 15.11 Simple harmonic motion for a block–spring system and its analogy to the motion of a simple pendulum (Section 15.5). The parameters in the table at the right refer to the block–spring system, assuming that at $t = 0$, $x = A$ so that $x = A\cos\omega t$.

At the Active Figures link at* http://www.pse6.com, *you can set the initial position of the block and see the block–spring system and the analogous pendulum in motion.

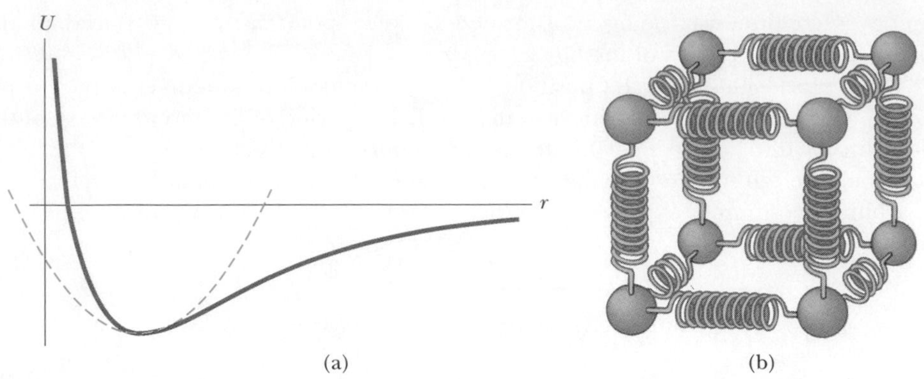

Figure 15.12 (a) If the atoms in a molecule do not move too far from their equilibrium positions, a graph of potential energy versus separation distance between atoms is similar to the graph of potential energy versus position for a simple harmonic oscillator (blue curve). (b) The forces between atoms in a solid can be modeled by imagining springs between neighboring atoms.

for this function approximates a parabola, which represents the potential energy function for a simple harmonic oscillator. Thus, we can model the complex atomic binding forces as being due to tiny springs, as depicted in Figure 15.12b.

The ideas presented in this chapter apply not only to block–spring systems and atoms, but also to a wide range of situations that include bungee jumping, tuning in a television station, and viewing the light emitted by a laser. You will see more examples of simple harmonic oscillators as you work through this book.

Example 15.4 Oscillations on a Horizontal Surface

A 0.500-kg cart connected to a light spring for which the force constant is 20.0 N/m oscillates on a horizontal, frictionless air track.

(A) Calculate the total energy of the system and the maximum speed of the cart if the amplitude of the motion is 3.00 cm.

Solution Using Equation 15.21, we obtain

$$E = \tfrac{1}{2}kA^2 = \tfrac{1}{2}(20.0\ \text{N/m})(3.00 \times 10^{-2}\ \text{m})^2$$
$$= \boxed{9.00 \times 10^{-3}\ \text{J}}$$

When the cart is located at $x = 0$, we know that $U = 0$ and $E = \tfrac{1}{2}mv_{\text{max}}^2$; therefore,

$$\tfrac{1}{2}mv_{\text{max}}^2 = 9.00 \times 10^{-3}\ \text{J}$$

$$v_{\text{max}} = \sqrt{\frac{2(9.00 \times 10^{-3}\ \text{J})}{0.500\ \text{kg}}} = \boxed{0.190\ \text{m/s}}$$

(B) What is the velocity of the cart when the position is 2.00 cm?

Solution We can apply Equation 15.22 directly:

$$v = \pm\sqrt{\frac{k}{m}(A^2 - x^2)}$$
$$= \pm\sqrt{\frac{20.0\ \text{N/m}}{0.500\ \text{kg}}[(0.0300\ \text{m})^2 - (0.0200\ \text{m})^2]}$$
$$= \boxed{\pm 0.141\ \text{m/s}}$$

The positive and negative signs indicate that the cart could be moving to either the right or the left at this instant.

(C) Compute the kinetic and potential energies of the system when the position is 2.00 cm.

Solution Using the result of (B), we find that

$$K = \tfrac{1}{2}mv^2 = \tfrac{1}{2}(0.500\ \text{kg})(0.141\ \text{m/s})^2 = \boxed{5.00 \times 10^{-3}\ \text{J}}$$

$$U = \tfrac{1}{2}kx^2 = \tfrac{1}{2}(20.0\ \text{N/m})(0.0200\ \text{m})^2 = \boxed{4.00 \times 10^{-3}\ \text{J}}$$

Note that $K + U = E$.

What If? The motion of the cart in this example could have been initiated by releasing the cart from rest at $x = 3.00$ cm. What if the cart were released from the same position, but with an initial velocity of $v = -0.100$ m/s? What are the new amplitude and maximum speed of the cart?

Answer This is the same type of question as we asked at the end of Example 15.3, but here we apply an energy approach. First let us calculate the total energy of the system at $t = 0$, which consists of both kinetic energy and potential energy:

$$E = \tfrac{1}{2}mv^2 + \tfrac{1}{2}kx^2$$
$$= \tfrac{1}{2}(0.500\ \text{kg})(-0.100\ \text{m/s})^2 + \tfrac{1}{2}(20.0\ \text{N/m})(0.030\,0\ \text{m})^2$$
$$= 1.15 \times 10^{-2}\ \text{J}$$

To find the new amplitude, we equate this total energy to the potential energy when the cart is at the end point of the motion:

$$E = \tfrac{1}{2}kA^2$$

$$A = \sqrt{\frac{2E}{k}} = \sqrt{\frac{2(1.15 \times 10^{-2}\text{ J})}{20.0\text{ N/m}}} = 0.033\,9\text{ m}$$

Note that this is larger than the previous amplitude of 0.030 0 m. To find the new maximum speed, we equate this total energy to the kinetic energy when the cart is at the equilibrium position:

$$E = \tfrac{1}{2}mv_{\text{max}}^2$$

$$v_{\text{max}} = \sqrt{\frac{2E}{m}} = \sqrt{\frac{2(1.15 \times 10^{-2}\text{ J})}{0.500\text{ kg}}} = 0.214\text{ m/s}$$

This is larger than the value found in part (a) as expected because the cart has an initial velocity at $t = 0$.

15.4 Comparing Simple Harmonic Motion with Uniform Circular Motion

Some common devices in our everyday life exhibit a relationship between oscillatory motion and circular motion. For example, the pistons in an automobile engine (Figure 15.13a) go up and down—oscillatory motion—yet the net result of this motion is circular motion of the wheels. In an old-fashioned locomotive (Figure 15.13b), the drive shaft goes back and forth in oscillatory motion, causing a circular motion of the wheels. In this section, we explore this interesting relationship between these two types of motion. We shall use this relationship again when we study electromagnetism and when we explore optics.

Figure 15.14 is an overhead view of an experimental arrangement that shows this relationship. A ball is attached to the rim of a turntable of radius A, which is illuminated from the side by a lamp. The ball casts a shadow on a screen. We find that **as the turntable rotates with constant angular speed, the shadow of the ball moves back and forth in simple harmonic motion.**

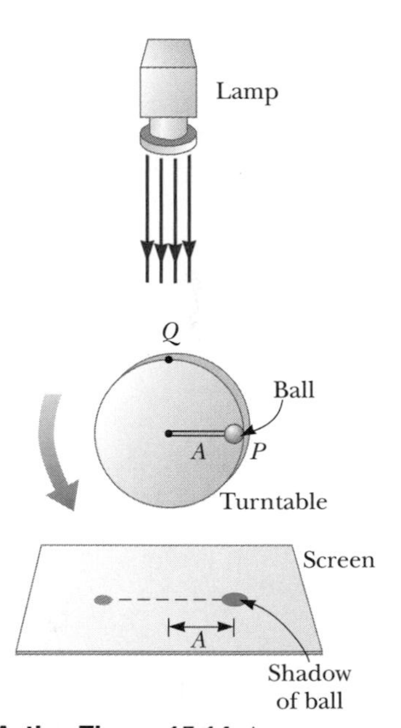

Active Figure 15.14 An experimental setup for demonstrating the connection between simple harmonic motion and uniform circular motion. As the ball rotates on the turntable with constant angular speed, its shadow on the screen moves back and forth in simple harmonic motion.

At the Active Figures link at http://www.pse6.com, you can adjust the frequency and radial position of the ball and see the resulting simple harmonic motion of the shadow.

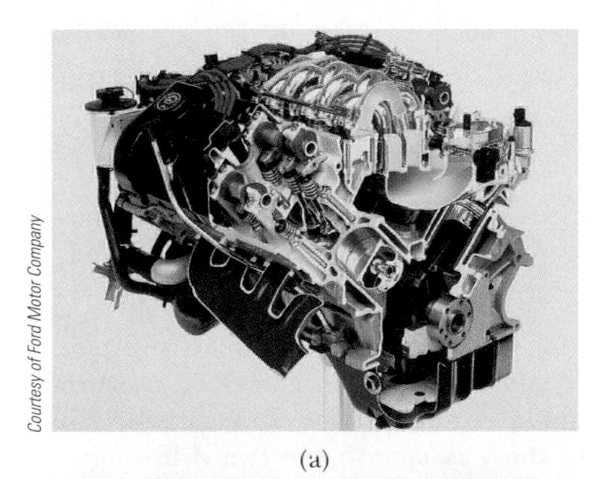

Courtesy of Ford Motor Company

(a)

© Link / Visuals Unlimited

(b)

Figure 15.13 (a) The pistons of an automobile engine move in periodic motion along a single dimension. This photograph shows a cutaway view of two of these pistons. This motion is converted to circular motion of the crankshaft, at the lower right, and ultimately of the wheels of the automobile. (b) The back-and-forth motion of pistons (in the curved housing at the left) in an old-fashioned locomotive is converted to circular motion of the wheels.

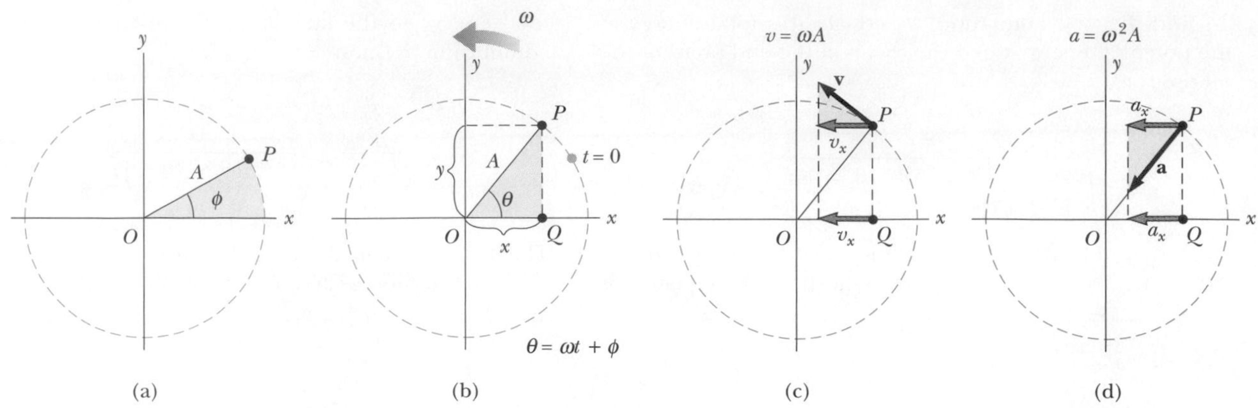

Figure 15.15 Relationship between the uniform circular motion of a point P and the simple harmonic motion of a point Q. A particle at P moves in a circle of radius A with constant angular speed ω. (a) A reference circle showing the position of P at $t = 0$. (b) The x coordinates of points P and Q are equal and vary in time according to the expression $x = A\cos(\omega t + \phi)$. (c) The x component of the velocity of P equals the velocity of Q. (d) The x component of the acceleration of P equals the acceleration of Q.

Consider a particle located at point P on the circumference of a circle of radius A, as in Figure 15.15a, with the line OP making an angle ϕ with the x axis at $t = 0$. We call this circle a *reference circle* for comparing simple harmonic motion with uniform circular motion, and we take the position of P at $t = 0$ as our reference position. If the particle moves along the circle with constant angular speed ω until OP makes an angle θ with the x axis, as in Figure 15.15b, then at some time $t > 0$, the angle between OP and the x axis is $\theta = \omega t + \phi$. As the particle moves along the circle, the projection of P on the x axis, labeled point Q, moves back and forth along the x axis between the limits $x = \pm A$.

Note that points P and Q always have the same x coordinate. From the right triangle OPQ, we see that this x coordinate is

$$x(t) = A\cos(\omega t + \phi) \qquad (15.23)$$

This expression is the same as Equation 15.6 and shows that the point Q moves with simple harmonic motion along the x axis. Therefore, we conclude that

simple harmonic motion along a straight line can be represented by the projection of uniform circular motion along a diameter of a reference circle.

We can make a similar argument by noting from Figure 15.15b that the projection of P along the y axis also exhibits simple harmonic motion. Therefore, **uniform circular motion can be considered a combination of two simple harmonic motions, one along the x axis and one along the y axis, with the two differing in phase by 90°.**

This geometric interpretation shows that the time interval for one complete revolution of the point P on the reference circle is equal to the period of motion T for simple harmonic motion between $x = \pm A$. That is, the angular speed ω of P is the same as the angular frequency ω of simple harmonic motion along the x axis. (This is why we use the same symbol.) The phase constant ϕ for simple harmonic motion corresponds to the initial angle that OP makes with the x axis. The radius A of the reference circle equals the amplitude of the simple harmonic motion.

Because the relationship between linear and angular speed for circular motion is $v = r\omega$ (see Eq. 10.10), the particle moving on the reference circle of radius A has a velocity of magnitude ωA. From the geometry in Figure 15.15c, we see that the x component of this velocity is $-\omega A\sin(\omega t + \phi)$. By definition, point Q has a velocity given by dx/dt. Differentiating Equation 15.23 with respect to time, we find that the velocity of Q is the same as the x component of the velocity of P.

The acceleration of P on the reference circle is directed radially inward toward O and has a magnitude $v^2/A = \omega^2 A$. From the geometry in Figure 15.15d, we see that the x component of this acceleration is $-\omega^2 A \cos(\omega t + \phi)$. This value is also the acceleration of the projected point Q along the x axis, as you can verify by taking the second derivative of Equation 15.23.

Quick Quiz 15.6 Figure 15.16 shows the position of an object in uniform circular motion at $t = 0$. A light shines from above and projects a shadow of the object on a screen below the circular motion. The correct values for the *amplitude* and *phase constant* (relative to an x axis to the right) of the simple harmonic motion of the shadow are (a) 0.50 m and 0 (b) 1.00 m and 0 (c) 0.50 m and π (d) 1.00 m and π.

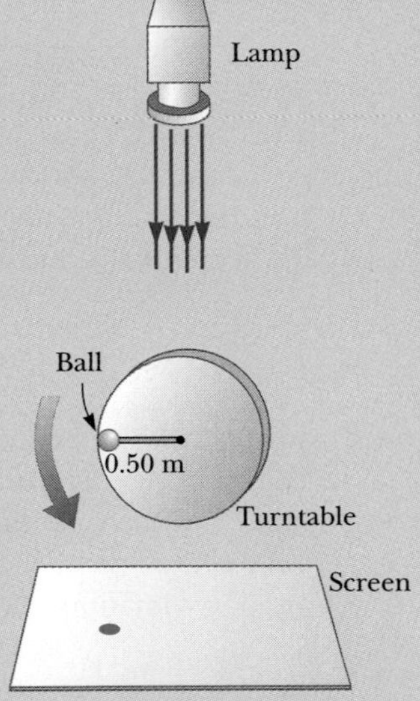

Figure 15.16 (Quick Quiz 15.6) An object moves in circular motion, casting a shadow on the screen below. Its position at an instant of time is shown.

Example 15.5 Circular Motion with Constant Angular Speed

A particle rotates counterclockwise in a circle of radius 3.00 m with a constant angular speed of 8.00 rad/s. At $t = 0$, the particle has an x coordinate of 2.00 m and is moving to the right.

(A) Determine the x coordinate as a function of time.

Solution Because the amplitude of the particle's motion equals the radius of the circle and $\omega = 8.00$ rad/s, we have

$$x = A\cos(\omega t + \phi) = (3.00\text{ m})\cos(8.00t + \phi)$$

We can evaluate ϕ by using the initial condition that $x = 2.00$ m at $t = 0$:

$$2.00\text{ m} = (3.00\text{ m})\cos(0 + \phi)$$

$$\phi = \cos^{-1}\left(\frac{2.00\text{ m}}{3.00\text{ m}}\right)$$

If we were to take our answer as $\phi = 48.2° = 0.841$ rad, then the coordinate $x = (3.00\text{ m})\cos(8.00t + 0.841)$ would be decreasing at time $t = 0$ (that is, moving to the left). Because our particle is first moving to the right, we must choose $\phi = -0.841$ rad. The x coordinate as a function of time is then

$$x = (3.00\text{ m})\cos(8.00t - 0.841)$$

Note that the angle ϕ in the cosine function must be in radians.

(B) Find the x components of the particle's velocity and acceleration at any time t.

Solution

$$v_x = \frac{dx}{dt} = (-3.00\text{ m})(8.00\text{ rad/s})\sin(8.00t - 0.841)$$

$$= -(24.0\text{ m/s})\sin(8.00t - 0.841)$$

$$a_x = \frac{dv}{dt} = (-24.0\text{ m/s})(8.00\text{ rad/s})\cos(8.00t - 0.841)$$

$$= -(192\text{ m/s}^2)\cos(8.00t - 0.841)$$

From these results, we conclude that $v_{max} = 24.0$ m/s and that $a_{max} = 192$ m/s^2.

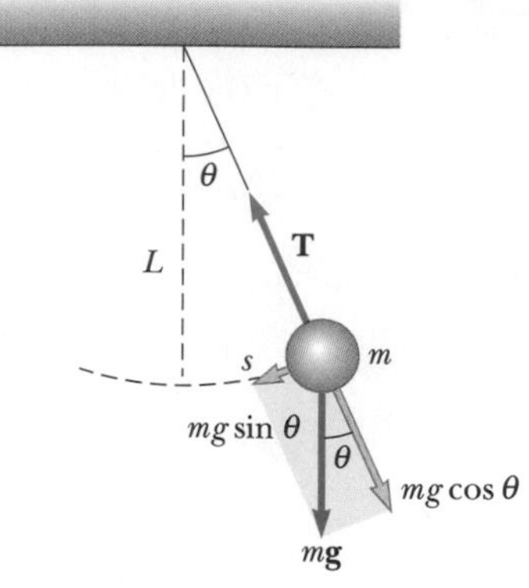

Active Figure 15.17 When θ is small, a simple pendulum oscillates in simple harmonic motion about the equilibrium position $\theta = 0$. The restoring force is $-mg\sin\theta$, the component of the gravitational force tangent to the arc.

At the Active Figures link at http://www.pse6.com, you can adjust the mass of the bob, the length of the string, and the initial angle and see the resulting oscillation of the pendulum.

15.5 The Pendulum

The **simple pendulum** is another mechanical system that exhibits periodic motion. It consists of a particle-like bob of mass m suspended by a light string of length L that is fixed at the upper end, as shown in Figure 15.17. The motion occurs in the vertical plane and is driven by the gravitational force. We shall show that, provided the angle θ is small (less than about 10°), the motion is very close to that of a simple harmonic oscillator.

The forces acting on the bob are the force $\mathbf{T}$ exerted by the string and the gravitational force $m\mathbf{g}$. The tangential component $mg\sin\theta$ of the gravitational force always acts toward $\theta = 0$, opposite the displacement of the bob from the lowest position. Therefore, the tangential component is a restoring force, and we can apply Newton's second law for motion in the tangential direction:

$$F_t = -mg\sin\theta = m\frac{d^2s}{dt^2}$$

where s is the bob's position measured along the arc and the negative sign indicates that the tangential force acts toward the equilibrium (vertical) position. Because $s = L\theta$ (Eq. 10.1a) and L is constant, this equation reduces to

$$\frac{d^2\theta}{dt^2} = -\frac{g}{L}\sin\theta$$

Considering θ as the position, let us compare this equation to Equation 15.3—does it have the same mathematical form? The right side is proportional to $\sin\theta$ rather than to θ; hence, we would not expect simple harmonic motion because this expression is not of the form of Equation 15.3. However, if we assume that θ is *small*, we can use the approximation $\sin\theta \approx \theta$; thus, in this approximation, the equation of motion for the simple pendulum becomes

▲ PITFALL PREVENTION

15.5 Not True Simple Harmonic Motion

Remember that the pendulum *does not* exhibit true simple harmonic motion for *any* angle. If the angle is less than about 10°, the motion is close to and can be *modeled* as simple harmonic.

$$\frac{d^2\theta}{dt^2} = -\frac{g}{L}\theta \quad \text{(for small values of } \theta\text{)} \qquad (15.24)$$

Now we have an expression that has the same form as Equation 15.3, and we conclude that the motion for small amplitudes of oscillation is simple harmonic motion. Therefore, the function θ can be written as $\theta = \theta_{max}\cos(\omega t + \phi)$, where θ_{max} is the *maximum angular position* and the angular frequency ω is

Angular frequency for a simple pendulum

$$\omega = \sqrt{\frac{g}{L}} \qquad (15.25)$$

The period of the motion is

Period of a simple pendulum

$$T = \frac{2\pi}{\omega} = 2\pi\sqrt{\frac{L}{g}} \qquad (15.26)$$

In other words, **the period and frequency of a simple pendulum depend only on the length of the string and the acceleration due to gravity.** Because the period is independent of the mass, we conclude that all simple pendula that are of equal length and are at the same location (so that g is constant) oscillate with the same period. The analogy between the motion of a simple pendulum and that of a block–spring system is illustrated in Figure 15.11.

The simple pendulum can be used as a timekeeper because its period depends only on its length and the local value of g. It is also a convenient device for making precise measurements of the free-fall acceleration. Such measurements are important because variations in local values of g can provide information on the location of oil and of other valuable underground resources.

Quick Quiz 15.7 A grandfather clock depends on the period of a pendulum to keep correct time. Suppose a grandfather clock is calibrated correctly and then a mischievous child slides the bob of the pendulum downward on the oscillating rod. Does the grandfather clock run (a) slow (b) fast (c) correctly?

Quick Quiz 15.8 Suppose a grandfather clock is calibrated correctly at sea level and is then taken to the top of a very tall mountain. Does the grandfather clock run (a) slow (b) fast (c) correctly?

Example 15.6 A Connection Between Length and Time

Christian Huygens (1629–1695), the greatest clockmaker in history, suggested that an international unit of length could be defined as the length of a simple pendulum having a period of exactly 1 s. How much shorter would our length unit be had his suggestion been followed?

Solution Solving Equation 15.26 for the length gives

$$L = \frac{T^2 g}{4\pi^2} = \frac{(1.00 \text{ s})^2(9.80 \text{ m/s}^2)}{4\pi^2} = 0.248 \text{ m}$$

Thus, the meter's length would be slightly less than one fourth of its current length. Note that the number of significant digits depends only on how precisely we know g because the time has been defined to be exactly 1 s.

What If? **What if Huygens had been born on another planet? What would the value for *g* have to be on that planet such that the meter based on Huygens's pendulum would have the same value as our meter?**

Answer We solve Equation 15.26 for g:

$$g = \frac{4\pi^2 L}{T^2} = \frac{4\pi^2(1.00 \text{ m})}{(1.00 \text{ s})^2} = 4\pi^2 \text{ m/s}^2 = 39.5 \text{ m/s}^2$$

No planet in our solar system has an acceleration due to gravity that is this large.

Physical Pendulum

Suppose you balance a wire coat hanger so that the hook is supported by your extended index finger. When you give the hanger a small angular displacement (with your other hand) and then release it, it oscillates. If a hanging object oscillates about a fixed axis that does not pass through its center of mass and the object cannot be approximated as a point mass, we cannot treat the system as a simple pendulum. In this case the system is called a **physical pendulum.**

Consider a rigid object pivoted at a point O that is a distance d from the center of mass (Fig. 15.18). The gravitational force provides a torque about an axis through O, and the magnitude of that torque is $mgd \sin\theta$, where θ is as shown in Figure 15.18. Using the rotational form of Newton's second law, $\Sigma\tau = I\alpha$, where I is the moment of inertia about the axis through O, we obtain

$$-mgd \sin\theta = I\frac{d^2\theta}{dt^2}$$

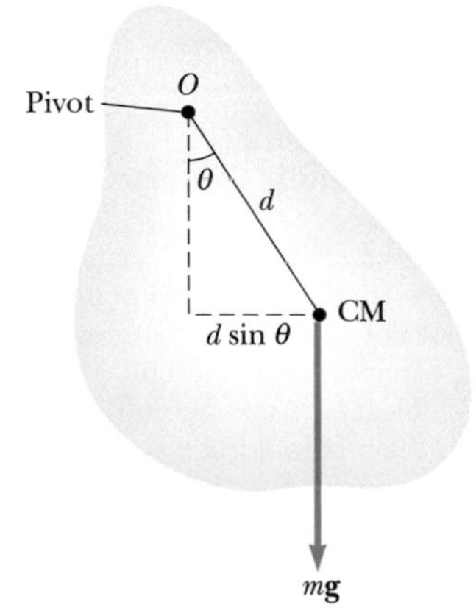

Figure 15.18 A physical pendulum pivoted at O.

The negative sign indicates that the torque about O tends to decrease θ. That is, the gravitational force produces a restoring torque. If we again assume that θ is small, the approximation $\sin\theta \approx \theta$ is valid, and the equation of motion reduces to

$$\frac{d^2\theta}{dt^2} = -\left(\frac{mgd}{I}\right)\theta = -\omega^2\theta \qquad (15.27)$$

Because this equation is of the same form as Equation 15.3, the motion is simple harmonic motion. That is, the solution of Equation 15.27 is $\theta = \theta_{\text{max}}\cos(\omega t + \phi)$, where θ_{max} is the maximum angular position and

$$\omega = \sqrt{\frac{mgd}{I}}$$

The period is

Period of a physical pendulum

$$T = \frac{2\pi}{\omega} = 2\pi\sqrt{\frac{I}{mgd}} \qquad (15.28)$$

One can use this result to measure the moment of inertia of a flat rigid object. If the location of the center of mass—and hence the value of d—is known, the moment of inertia can be obtained by measuring the period. Finally, note that Equation 15.28 reduces to the period of a simple pendulum (Eq. 15.26) when $I = md^2$—that is, when all the mass is concentrated at the center of mass.

Example 15.7 A Swinging Rod

A uniform rod of mass M and length L is pivoted about one end and oscillates in a vertical plane (Fig. 15.19). Find the period of oscillation if the amplitude of the motion is small.

Solution In Chapter 10 we found that the moment of inertia of a uniform rod about an axis through one end is $\frac{1}{3}ML^2$. The distance d from the pivot to the center of mass is $L/2$. Substituting these quantities into Equation 15.28 gives

$$T = 2\pi\sqrt{\frac{\frac{1}{3}ML^2}{Mg(L/2)}} = 2\pi\sqrt{\frac{2L}{3g}}$$

Comment In one of the Moon landings, an astronaut walking on the Moon's surface had a belt hanging from his space suit, and the belt oscillated as a physical pendulum. A scientist on the Earth observed this motion on television and used it to estimate the free-fall acceleration on the Moon. How did the scientist make this calculation?

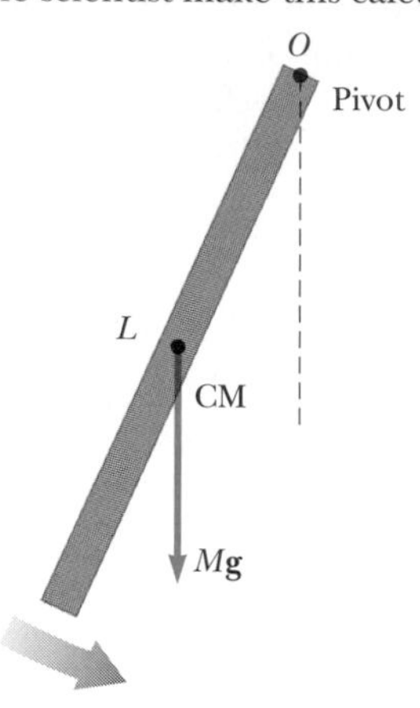

Figure 15.19 A rigid rod oscillating about a pivot through one end is a physical pendulum with $d = L/2$ and, from Table 10.2, $I = \frac{1}{3}ML^2$.

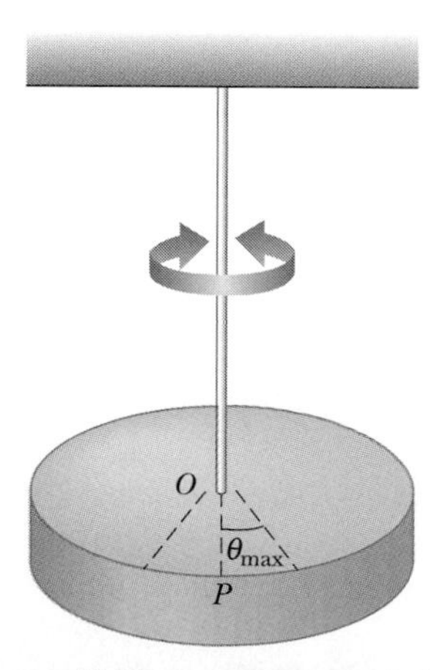

Figure 15.20 A torsional pendulum consists of a rigid object suspended by a wire attached to a rigid support. The object oscillates about the line OP with an amplitude θ_{max}.

Torsional Pendulum

Figure 15.20 shows a rigid object suspended by a wire attached at the top to a fixed support. When the object is twisted through some angle θ, the twisted wire exerts on the object a restoring torque that is proportional to the angular position. That is,

$$\tau = -\kappa\theta$$

where κ (kappa) is called the *torsion constant* of the support wire. The value of κ can be obtained by applying a known torque to twist the wire through a measurable angle θ. Applying Newton's second law for rotational motion, we find

$$\tau = -\kappa\theta = I\frac{d^2\theta}{dt^2}$$

$$\frac{d^2\theta}{dt^2} = -\frac{\kappa}{I}\theta \qquad (15.29)$$

Again, this is the equation of motion for a simple harmonic oscillator, with $\omega = \sqrt{\kappa/I}$ and a period

Period of a torsional pendulum

$$T = 2\pi\sqrt{\frac{I}{\kappa}} \qquad (15.30)$$

This system is called a *torsional pendulum*. There is no small-angle restriction in this situation as long as the elastic limit of the wire is not exceeded.

15.6 Damped Oscillations

The oscillatory motions we have considered so far have been for ideal systems—that is, systems that oscillate indefinitely under the action of only one force—a linear restoring force. In many real systems, nonconservative forces, such as friction, retard the motion. Consequently, the mechanical energy of the system diminishes in time, and the motion is said to be *damped*. Figure 15.21 depicts one such system: an object attached to a spring and submersed in a viscous liquid.

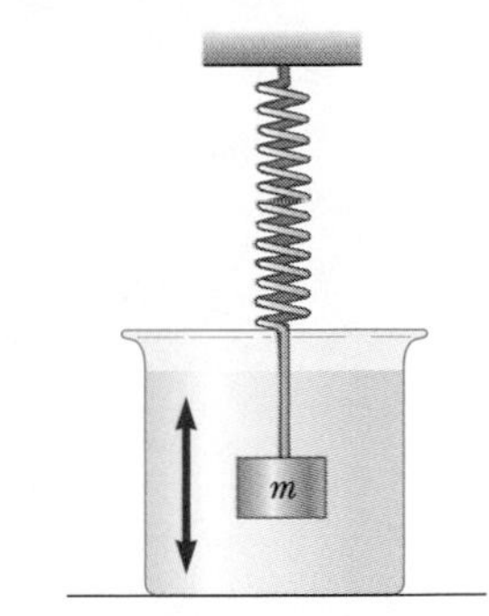

Figure 15.21 One example of a damped oscillator is an object attached to a spring and submersed in a viscous liquid.

One common type of retarding force is the one discussed in Section 6.4, where the force is proportional to the speed of the moving object and acts in the direction opposite the motion. This retarding force is often observed when an object moves through air, for instance. Because the retarding force can be expressed as $\mathbf{R} = -b\mathbf{v}$ (where b is a constant called the *damping coefficient*) and the restoring force of the system is $-kx$, we can write Newton's second law as

$$\sum F_x = -kx - bv_x = ma_x$$

$$-kx - b\frac{dx}{dt} = m\frac{d^2x}{dt^2} \quad (15.31)$$

The solution of this equation requires mathematics that may not be familiar to you; we simply state it here without proof. When the retarding force is small compared with the maximum restoring force—that is, when b is small—the solution to Equation 15.31 is

$$x = Ae^{-\frac{b}{2m}t}\cos(\omega t + \phi) \quad (15.32)$$

where the angular frequency of oscillation is

$$\omega = \sqrt{\frac{k}{m} - \left(\frac{b}{2m}\right)^2} \quad (15.33)$$

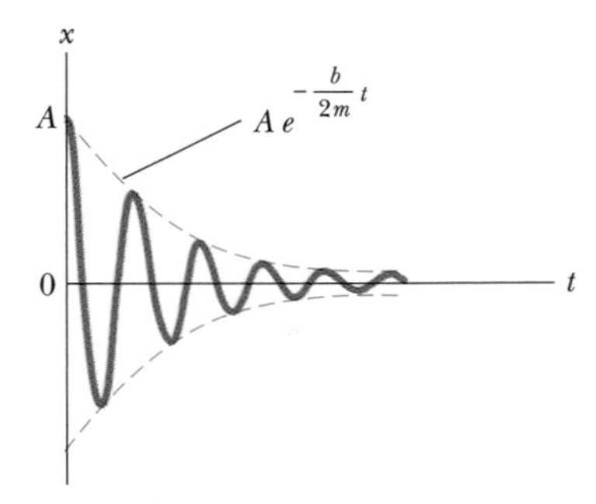

Active Figure 15.22 Graph of position versus time for a damped oscillator. Note the decrease in amplitude with time.

At the Active Figures link at http://www.pse6.com, you can adjust the spring constant, the mass of the object, and the damping constant and see the resulting damped oscillation of the object.

This result can be verified by substituting Equation 15.32 into Equation 15.31.

Figure 15.22 shows the position as a function of time for an object oscillating in the presence of a retarding force. We see that **when the retarding force is small, the oscillatory character of the motion is preserved but the amplitude decreases in time, with the result that the motion ultimately ceases.** Any system that behaves in this way is known as a **damped oscillator.** The dashed blue lines in Figure 15.22, which define the *envelope* of the oscillatory curve, represent the exponential factor in Equation 15.32. This envelope shows that **the amplitude decays exponentially with time.** For motion with a given spring constant and object mass, the oscillations dampen more rapidly as the maximum value of the retarding force approaches the maximum value of the restoring force.

It is convenient to express the angular frequency (Eq. 15.33) of a damped oscillator in the form

$$\omega = \sqrt{\omega_0^2 - \left(\frac{b}{2m}\right)^2}$$

where $\omega_0 = \sqrt{k/m}$ represents the angular frequency in the absence of a retarding force (the undamped oscillator) and is called the **natural frequency** of the system.

When the magnitude of the maximum retarding force $R_{max} = bv_{max} < kA$, the system is said to be **underdamped.** The resulting motion is represented by the blue curve in Figure 15.23. As the value of b increases, the amplitude of the oscillations decreases more and more rapidly. When b reaches a critical value b_c such that $b_c/2m = \omega_0$, the system does not oscillate and is said to be **critically damped.** In this case the system, once released from rest at some nonequilibrium position, approaches but does not pass through the equilibrium position. The graph of position versus time for this case is the red curve in Figure 15.23.

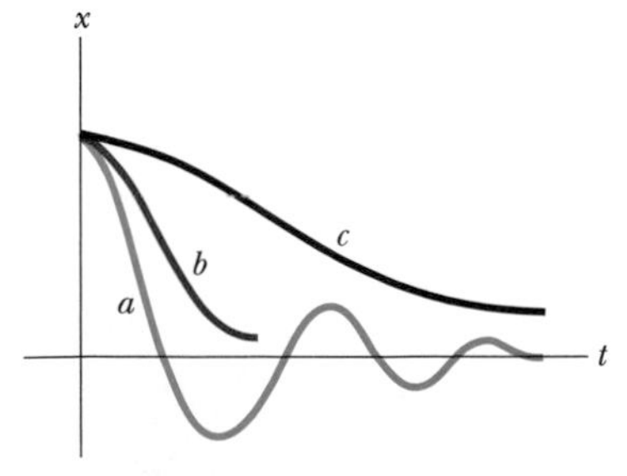

Figure 15.23 Graphs of position versus time for (a) an underdamped oscillator, (b) a critically damped oscillator, and (c) an overdamped oscillator.

If the medium is so viscous that the retarding force is greater than the restoring force—that is, if $R_{max} = bv_{max} > kA$ and $b/2m > \omega_0$—the system is **overdamped.** Again, the displaced system, when free to move, does not oscillate but simply returns to its equilibrium position. As the damping increases, the time interval required for the system to approach equilibrium also increases, as indicated by the black curve in Figure 15.23. For critically damped and overdamped systems, there is no angular frequency ω and the solution in Equation 15.32 is not valid.

Whenever friction is present in a system, whether the system is overdamped or underdamped, the energy of the oscillator eventually falls to zero. The lost mechanical energy is transformed into internal energy in the object and the retarding medium.

Quick Quiz 15.9 An automotive suspension system consists of a combination of springs and shock absorbers, as shown in Figure 15.24. If you were an automotive engineer, would you design a suspension system that was (a) underdamped (b) critically damped (c) overdamped?

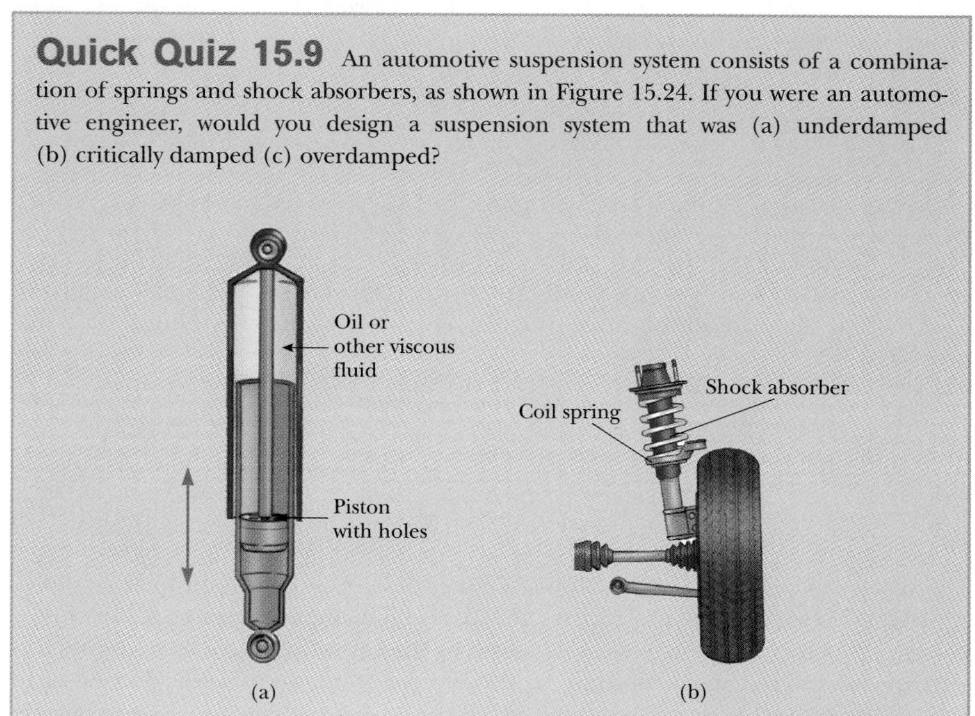

Figure 15.24 (a) A shock absorber consists of a piston oscillating in a chamber filled with oil. As the piston oscillates, the oil is squeezed through holes between the piston and the chamber, causing a damping of the piston's oscillations. (b) One type of automotive suspension system, in which a shock absorber is placed inside a coil spring at each wheel.

15.7 Forced Oscillations

We have seen that the mechanical energy of a damped oscillator decreases in time as a result of the resistive force. It is possible to compensate for this energy decrease by applying an external force that does positive work on the system. At any instant, energy can be transferred into the system by an applied force that acts in the direction of motion of the oscillator. For example, a child on a swing can be kept in motion by appropriately timed "pushes." The amplitude of motion remains constant if the energy input per cycle of motion exactly equals the decrease in mechanical energy in each cycle that results from resistive forces.

A common example of a forced oscillator is a damped oscillator driven by an external force that varies periodically, such as $F(t) = F_0 \sin \omega t$, where ω is the angular frequency of the driving force and F_0 is a constant. In general, the frequency ω of the driving force is variable while the natural frequency ω_0 of the oscillator is fixed by the values of k and m. Newton's second law in this situation gives

$$\sum F = ma \longrightarrow F_0 \sin \omega t - b\frac{dx}{dt} - kx = m\frac{d^2x}{dt^2} \qquad (15.34)$$

Again, the solution of this equation is rather lengthy and will not be presented. After the driving force on an initially stationary object begins to act, the amplitude of the oscillation will increase. After a sufficiently long period of time, when the energy input per cycle from the driving force equals the amount of mechanical energy transformed to internal energy for each cycle, a steady-state condition is reached in which the oscillations proceed with constant amplitude. In this situation, Equation 15.34 has the solution

$$x = A\cos(\omega t + \phi) \qquad (15.35)$$

where

$$A = \frac{F_0/m}{\sqrt{(\omega^2 - \omega_0{}^2)^2 + \left(\frac{b\omega}{m}\right)^2}} \qquad (15.36)$$

Amplitude of a driven oscillator

and where $\omega_0 = \sqrt{k/m}$ is the natural frequency of the undamped oscillator ($b = 0$).

Equations 15.35 and 15.36 show that the forced oscillator vibrates at the frequency of the driving force and that the amplitude of the oscillator is constant for a given driving force because it is being driven in steady-state by an external force. For small damping, the amplitude is large when the frequency of the driving force is near the natural frequency of oscillation, or when $\omega \approx \omega_0$. The dramatic increase in amplitude near the natural frequency is called **resonance,** and the natural frequency ω_0 is also called the **resonance frequency** of the system.

The reason for large-amplitude oscillations at the resonance frequency is that energy is being transferred to the system under the most favorable conditions. We can better understand this by taking the first time derivative of x in Equation 15.35, which gives an expression for the velocity of the oscillator. We find that v is proportional to $\sin(\omega t + \phi)$, which is the same trigonometric function as that describing the driving force. Thus, the applied force $\mathbf{F}$ is in phase with the velocity. The rate at which work is done on the oscillator by $\mathbf{F}$ equals the dot product $\mathbf{F}\cdot\mathbf{v}$; this rate is the power delivered to the oscillator. Because the product $\mathbf{F}\cdot\mathbf{v}$ is a maximum when $\mathbf{F}$ and $\mathbf{v}$ are in phase, we conclude that **at resonance the applied force is in phase with the velocity and the power transferred to the oscillator is a maximum.**

Figure 15.25 is a graph of amplitude as a function of frequency for a forced oscillator with and without damping. Note that the amplitude increases with decreasing damping ($b \rightarrow 0$) and that the resonance curve broadens as the damping increases. Under steady-state conditions and at any driving frequency, the energy transferred into the system equals the energy lost because of the damping force; hence, the average total energy of the oscillator remains constant. In the absence of a damping force ($b = 0$), we see from Equation 15.36 that the steady-state amplitude approaches infinity as ω approaches ω_0. In other words, if there are no losses in the system and if we continue to drive an initially motionless oscillator with a periodic force that is in phase with the velocity, the amplitude of motion builds without limit (see the brown curve in Fig. 15.25). This limitless building does not occur in practice because some damping is always present in reality.

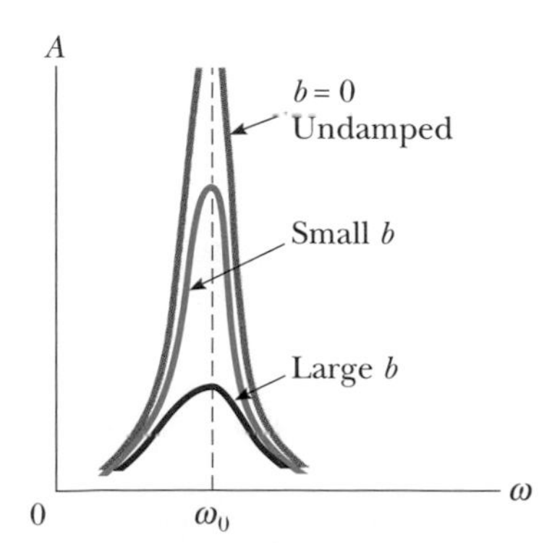

Figure 15.25 Graph of amplitude versus frequency for a damped oscillator when a periodic driving force is present. When the frequency ω of the driving force equals the natural frequency ω_0 of the oscillator, resonance occurs. Note that the shape of the resonance curve depends on the size of the damping coefficient b.

Later in this book we shall see that resonance appears in other areas of physics. For example, certain electric circuits have natural frequencies. A bridge has natural frequencies that can be set into resonance by an appropriate driving force. A dramatic example of such resonance occurred in 1940, when the Tacoma Narrows Bridge in the state of Washington was destroyed by resonant vibrations. Although the winds were not particularly strong on that occasion, the "flapping" of the wind across the roadway (think of the "flapping" of a flag in a strong wind) provided a periodic driving force whose frequency matched that of the bridge. The resulting oscillations of the bridge caused it to ultimately collapse (Fig. 15.26) because the bridge design had inadequate built-in safety features.

(a)

(b)

Figure 15.26 (a) In 1940 turbulent winds set up torsional vibrations in the Tacoma Narrows Bridge, causing it to oscillate at a frequency near one of the natural frequencies of the bridge structure. (b) Once established, this resonance condition led to the bridge's collapse.

Many other examples of resonant vibrations can be cited. A resonant vibration that you may have experienced is the "singing" of telephone wires in the wind. Machines often break if one vibrating part is in resonance with some other moving part. Soldiers marching in cadence across a bridge have been known to set up resonant vibrations in the structure and thereby cause it to collapse. Whenever any real physical system is driven near its resonance frequency, you can expect oscillations of very large amplitudes.

SUMMARY

***Take a practice test for this chapter by clicking on the Practice Test link at* http://www.pse6.com.**

When the acceleration of an object is proportional to its position and is in the direction opposite the displacement from equilibrium, the object moves with simple harmonic motion. The position x of a simple harmonic oscillator varies periodically in time according to the expression

$$x(t) = A\cos(\omega t + \phi) \qquad (15.6)$$

where A is the **amplitude** of the motion, ω is the **angular frequency,** and ϕ is the **phase constant.** The value of ϕ depends on the initial position and initial velocity of the oscillator.

The time interval T needed for one complete oscillation is defined as the **period** of the motion:

$$T = \frac{2\pi}{\omega} \qquad (15.10)$$

A block–spring system moves in simple harmonic motion on a frictionless surface with a period

$$T = \frac{2\pi}{\omega} = 2\pi\sqrt{\frac{m}{k}} \qquad (15.13)$$

The inverse of the period is the **frequency** of the motion, which equals the number of oscillations per second.

The velocity and acceleration of a simple harmonic oscillator are

$$v = \frac{dx}{dt} = -\omega A\sin(\omega t + \phi) \qquad (15.15)$$

$$a = \frac{d^2x}{dt^2} = -\omega^2 A\cos(\omega t + \phi) \qquad (15.16)$$

$$v = \pm\omega\sqrt{A^2 - x^2} \qquad (15.22)$$

Thus, the maximum speed is ωA, and the maximum acceleration is $\omega^2 A$. The speed is zero when the oscillator is at its turning points $x = \pm A$ and is a maximum when the oscillator is at the equilibrium position $x = 0$. The magnitude of the acceleration is a maximum at the turning points and zero at the equilibrium position.

The kinetic energy and potential energy for a simple harmonic oscillator vary with time and are given by

$$K = \frac{1}{2}mv^2 = \frac{1}{2}m\omega^2A^2\sin^2(\omega t + \phi) \qquad (15.19)$$

$$U = \frac{1}{2}kx^2 = \frac{1}{2}kA^2\cos^2(\omega t + \phi) \qquad (15.20)$$

The total energy of a simple harmonic oscillator is a constant of the motion and is given by

$$E = \frac{1}{2}kA^2 \qquad (15.21)$$

The potential energy of the oscillator is a maximum when the oscillator is at its turning points and is zero when the oscillator is at the equilibrium position. The kinetic energy is zero at the turning points and a maximum at the equilibrium position.

A **simple pendulum** of length L moves in simple harmonic motion for small angular displacements from the vertical. Its period is

$$T = 2\pi\sqrt{\frac{L}{g}} \qquad (15.26)$$

For small angular displacements from the vertical, a **physical pendulum** moves in simple harmonic motion about a pivot that does not go through the center of mass. The period of this motion is

$$T = 2\pi\sqrt{\frac{I}{mgd}} \qquad (15.28)$$

where I is the moment of inertia about an axis through the pivot and d is the distance from the pivot to the center of mass.

If an oscillator experiences a damping force $\mathbf{R} = -b\mathbf{v}$, its position for small damping is described by

$$x = Ae^{-\frac{b}{2m}t}\cos(\omega t + \phi) \qquad (15.32)$$

where

$$\omega = \sqrt{\frac{k}{m} - \left(\frac{b}{2m}\right)^2} \qquad (15.33)$$

If an oscillator is subject to a sinusoidal driving force $F(t) = F_0 \sin \omega t$, it exhibits **resonance,** in which the amplitude is largest when the driving frequency matches the natural frequency of the oscillator.

QUESTIONS

1. Is a bouncing ball an example of simple harmonic motion? Is the daily movement of a student from home to school and back simple harmonic motion? Why or why not?

2. If the coordinate of a particle varies as $x = -A\cos\omega t$, what is the phase constant in Equation 15.6? At what position is the particle at $t = 0$?

3. Does the displacement of an oscillating particle between $t = 0$ and a later time t necessarily equal the position of the particle at time t? Explain.

4. Determine whether or not the following quantities can be in the same direction for a simple harmonic oscillator: (a) position and velocity, (b) velocity and acceleration, (c) position and acceleration.

5. Can the amplitude A and phase constant ϕ be determined for an oscillator if only the position is specified at $t = 0$? Explain.

6. Describe qualitatively the motion of a block–spring system when the mass of the spring is not neglected.

7. A block is hung on a spring, and the frequency f of the oscillation of the system is measured. The block, a second identical block, and the spring are carried in the Space Shuttle to space. The two blocks are attached to the ends of the spring, and the system is taken out into space on a space walk. The spring is extended, and the system is released to oscillate while floating in space. What is the frequency of oscillation for this system, in terms of f?

8. A block–spring system undergoes simple harmonic motion with amplitude A. Does the total energy change if the mass is doubled but the amplitude is not changed? Do the kinetic and potential energies depend on the mass? Explain.
9. The equations listed in Table 2.2 give position as a function of time, velocity as a function of time, and velocity as function of position for an object moving in a straight line with constant acceleration. The quantity v_{xi} appears in every equation. Do any of these equations apply to an object moving in a straight line with simple harmonic motion? Using a similar format, make a table of equations describing simple harmonic motion. Include equations giving acceleration as a function of time and acceleration as a function of position. State the equations in such a form that they apply equally to a block–spring system, to a pendulum, and to other vibrating systems. What quantity appears in every equation?
10. What happens to the period of a simple pendulum if the pendulum's length is doubled? What happens to the period if the mass of the suspended bob is doubled?
11. A simple pendulum is suspended from the ceiling of a stationary elevator, and the period is determined. Describe the changes, if any, in the period when the elevator (a) accelerates upward, (b) accelerates downward, and (c) moves with constant velocity.
12. Imagine that a pendulum is hanging from the ceiling of a car. As the car coasts freely down a hill, is the equilibrium position of the pendulum vertical? Does the period of oscillation differ from that in a stationary car?
13. A simple pendulum undergoes simple harmonic motion when θ is small. Is the motion periodic when θ is large? How does the period of motion change as θ increases?
14. If a grandfather clock were running slow, how could we adjust the length of the pendulum to correct the time?
15. Will damped oscillations occur for any values of b and k? Explain.
16. Is it possible to have damped oscillations when a system is at resonance? Explain.
17. At resonance, what does the phase constant ϕ equal in Equation 15.35? (*Suggestion:* Compare this equation with the expression for the driving force, which must be in phase with the velocity at resonance.)
18. You stand on the end of a diving board and bounce to set it into oscillation. You find a maximum response, in terms of the amplitude of oscillation of the end of the board, when you bounce at frequency f. You now move to the middle of the board and repeat the experiment. Is the resonance frequency for forced oscillations at this point higher, lower, or the same as f? Why?
19. Some parachutes have holes in them to allow air to move smoothly through the chute. Without the holes, the air gathered under the chute as the parachutist falls is sometimes released from under the edges of the chute alternately and periodically from one side and then the other. Why might this periodic release of air cause a problem?
20. You are looking at a small tree. You do not notice any breeze, and most of the leaves on the tree are motionless. However, one leaf is fluttering back and forth wildly. After you wait for a while, that leaf stops moving and you notice a different leaf moving much more than all the others. Explain what could cause the large motion of one particular leaf.
21. A pendulum bob is made with a sphere filled with water. What would happen to the frequency of vibration of this pendulum if there were a hole in the sphere that allowed the water to leak out slowly?

PROBLEMS

1, 2, 3 = straightforward, intermediate, challenging ☐ = full solution available in the *Student Solutions Manual and Study Guide*

= coached solution with hints available at http://www.pse6.com = computer useful in solving problem

= paired numerical and symbolic problems

Note: Neglect the mass of every spring, except in problems 66 and 68.

Section 15.1 Motion of an Object Attached to a Spring

Problems 15, 16, 19, 23, 56, and 62 in Chapter 7 can also be assigned with this section.

1. A ball dropped from a height of 4.00 m makes a perfectly elastic collision with the ground. Assuming no mechanical energy is lost due to air resistance, (a) show that the ensuing motion is periodic and (b) determine the period of the motion. (c) Is the motion simple harmonic? Explain.

Section 15.2 Mathematical Representation of Simple Harmonic Motion

2. In an engine, a piston oscillates with simple harmonic motion so that its position varies according to the expression

$$x = (5.00\ \text{cm})\cos(2t + \pi/6)$$

where x is in centimeters and t is in seconds. At $t = 0$, find (a) the position of the piston, (b) its velocity, and (c) its acceleration. (d) Find the period and amplitude of the motion.

3. The position of a particle is given by the expression $x = (4.00\ \text{m})\cos(3.00\pi t + \pi)$, where x is in meters and t is in seconds. Determine (a) the frequency and period of the motion, (b) the amplitude of the motion, (c) the phase constant, and (d) the position of the particle at $t = 0.250$ s.

4. (a) A hanging spring stretches by 35.0 cm when an object of mass 450 g is hung on it at rest. In this situation, we define its position as $x = 0$. The object is pulled down an additional 18.0 cm and released from rest to oscillate without friction. What is its position x at a time 84.4 s later? (b) **What If?** A hanging spring stretches by 35.5 cm when an object of mass 440 g is hung on it at rest. We define this new position as $x = 0$. This object is also pulled down an additional 18.0 cm and released from rest to oscillate without friction. Find its position 84.4 s later. (c) Why are the answers to (a) and (b) different by such a large percentage when the data are so similar? Does this circumstance reveal a fundamental difficulty in calculating the future? (d) Find the distance traveled by the vibrating object in part (a). (e) Find the distance traveled by the object in part (b).

5. A particle moving along the x axis in simple harmonic motion starts from its equilibrium position, the origin, at $t = 0$ and moves to the right. The amplitude of its motion is 2.00 cm, and the frequency is 1.50 Hz. (a) Show that the position of the particle is given by

$$x = (2.00 \text{ cm})\sin(3.00\pi t)$$

Determine (b) the maximum speed and the earliest time ($t > 0$) at which the particle has this speed, (c) the maximum acceleration and the earliest time ($t > 0$) at which the particle has this acceleration, and (d) the total distance traveled between $t = 0$ and $t = 1.00$ s.

6. The initial position, velocity, and acceleration of an object moving in simple harmonic motion are x_i, v_i, and a_i; the angular frequency of oscillation is ω. (a) Show that the position and velocity of the object for all time can be written as

$$x(t) = x_i \cos \omega t + \left(\frac{v_i}{\omega}\right) \sin \omega t$$

$$v(t) = -x_i\omega \sin \omega t + v_i \cos \omega t$$

(b) If the amplitude of the motion is A, show that

$$v^2 - ax = v_i^2 - a_i x_i = \omega^2 A^2$$

7. A simple harmonic oscillator takes 12.0 s to undergo five complete vibrations. Find (a) the period of its motion, (b) the frequency in hertz, and (c) the angular frequency in radians per second.

8. A vibration sensor, used in testing a washing machine, consists of a cube of aluminum 1.50 cm on edge mounted on one end of a strip of spring steel (like a hacksaw blade) that lies in a vertical plane. The mass of the strip is small compared to that of the cube, but the length of the strip is large compared to the size of the cube. The other end of the strip is clamped to the frame of the washing machine, which is not operating. A horizontal force of 1.43 N applied to the cube is required to hold it 2.75 cm away from its equilibrium position. If the cube is released, what is its frequency of vibration?

9. A 7.00-kg object is hung from the bottom end of a vertical spring fastened to an overhead beam. The object is set into vertical oscillations having a period of 2.60 s. Find the force constant of the spring.

10. A piston in a gasoline engine is in simple harmonic motion. If the extremes of its position relative to its center point are ± 5.00 cm, find the maximum velocity and acceleration of the piston when the engine is running at the rate of 3 600 rev/min.

11. A 0.500-kg object attached to a spring with a force constant of 8.00 N/m vibrates in simple harmonic motion with an amplitude of 10.0 cm. Calculate (a) the maximum value of its speed and acceleration, (b) the speed and acceleration when the object is 6.00 cm from the equilibrium position, and (c) the time interval required for the object to move from $x = 0$ to $x = 8.00$ cm.

12. A 1.00-kg glider attached to a spring with a force constant of 25.0 N/m oscillates on a horizontal, frictionless air track. At $t = 0$ the glider is released from rest at $x = -3.00$ cm. (That is, the spring is compressed by 3.00 cm.) Find (a) the period of its motion, (b) the maximum values of its speed and acceleration, and (c) the position, velocity, and acceleration as functions of time.

13. A 1.00-kg object is attached to a horizontal spring. The spring is initially stretched by 0.100 m, and the object is released from rest there. It proceeds to move without friction. The next time the speed of the object is zero is 0.500 s later. What is the maximum speed of the object?

14. A particle that hangs from a spring oscillates with an angular frequency ω. The spring is suspended from the ceiling of an elevator car and hangs motionless (relative to the elevator car) as the car descends at a constant speed v. The car then stops suddenly. (a) With what amplitude does the particle oscillate? (b) What is the equation of motion for the particle? (Choose the upward direction to be positive.)

Section 15.3 Energy of the Simple Harmonic Oscillator

15. A block of unknown mass is attached to a spring with a spring constant of 6.50 N/m and undergoes simple harmonic motion with an amplitude of 10.0 cm. When the block is halfway between its equilibrium position and the end point, its speed is measured to be 30.0 cm/s. Calculate (a) the mass of the block, (b) the period of the motion, and (c) the maximum acceleration of the block.

16. A 200-g block is attached to a horizontal spring and executes simple harmonic motion with a period of 0.250 s. If the total energy of the system is 2.00 J, find (a) the force constant of the spring and (b) the amplitude of the motion.

17. An automobile having a mass of 1 000 kg is driven into a brick wall in a safety test. The bumper behaves like a spring of force constant 5.00×10^6 N/m and compresses 3.16 cm as the car is brought to rest. What was the speed of the car before impact, assuming that no mechanical energy is lost during impact with the wall?

18. A block–spring system oscillates with an amplitude of 3.50 cm. If the spring constant is 250 N/m and the mass of the block is 0.500 kg, determine (a) the mechanical energy of the system, (b) the maximum speed of the block, and (c) the maximum acceleration.

19. A 50.0-g object connected to a spring with a force constant of 35.0 N/m oscillates on a horizontal, frictionless surface with an amplitude of 4.00 cm. Find (a) the total energy of the system and (b) the speed of the object when the position is 1.00 cm. Find (c) the kinetic energy and (d) the potential energy when the position is 3.00 cm.

20. A 2.00-kg object is attached to a spring and placed on a horizontal, smooth surface. A horizontal force of 20.0 N is required to hold the object at rest when it is pulled 0.200 m from its equilibrium position (the origin of the x axis). The object is now released from rest with an initial position of $x_i = 0.200$ m, and it subsequently undergoes simple harmonic oscillations. Find (a) the force constant of the spring, (b) the frequency of the oscillations, and (c) the maximum speed of the object. Where does this maximum speed occur? (d) Find the maximum acceleration of the object. Where does it occur? (e) Find the total energy of the oscillating system. Find (f) the speed and (g) the acceleration of the object when its position is equal to one third of the maximum value.

21. The amplitude of a system moving in simple harmonic motion is doubled. Determine the change in (a) the total energy, (b) the maximum speed, (c) the maximum acceleration, and (d) the period.

22. A 65.0-kg bungee jumper steps off a bridge with a light bungee cord tied to herself and to the bridge (Figure P15.22). The unstretched length of the cord is 11.0 m. She reaches the bottom of her motion 36.0 m below the bridge before bouncing back. Her motion can be separated into an 11.0-m free fall and a 25.0-m section of simple harmonic oscillation. (a) For what time interval is she in free fall? (b) Use the principle of conservation of energy to find the spring constant of the bungee cord. (c) What is the location of the equilibrium point where the spring force balances the gravitational force acting on the jumper? Note that this point is taken as the origin in our mathematical description of simple harmonic oscillation. (d) What is the angular frequency of the oscillation? (e) What time interval is required for the cord to stretch by 25.0 m? (f) What is the total time interval for the entire 36.0-m drop?

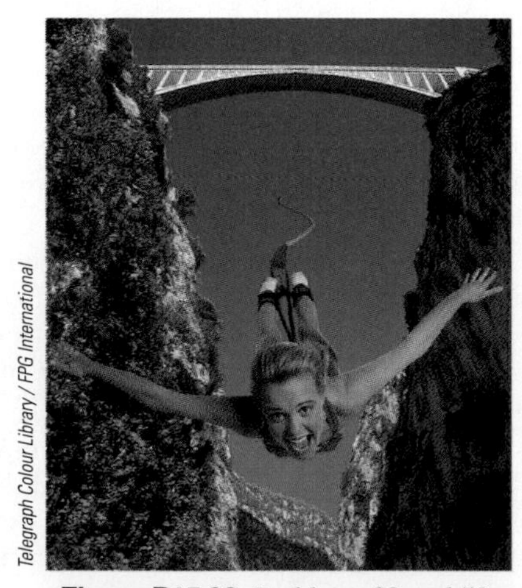
Telegraph Colour Library / FPG International

Figure P15.22 Problems 22 and 58.

23. A particle executes simple harmonic motion with an amplitude of 3.00 cm. At what position does its speed equal half its maximum speed?

24. A cart attached to a spring with constant 3.24 N/m vibrates with position given by $x = (5.00 \text{ cm}) \cos(3.60t \text{ rad/s})$. (a) During the first cycle, for $0 < t < 1.75$ s, just when is the system's potential energy changing most rapidly into kinetic energy? (b) What is the maximum rate of energy transformation?

Section 15.4 Comparing Simple Harmonic Motion with Uniform Circular Motion

25. While riding behind a car traveling at 3.00 m/s, you notice that one of the car's tires has a small hemispherical bump on its rim, as in Figure P15.25. (a) Explain why the bump, from your viewpoint behind the car, executes simple harmonic motion. (b) If the radii of the car's tires are 0.300 m, what is the bump's period of oscillation?

Figure P15.25

26. Consider the simplified single-piston engine in Figure P15.26. If the wheel rotates with constant angular speed, explain why the piston rod oscillates in simple harmonic motion.

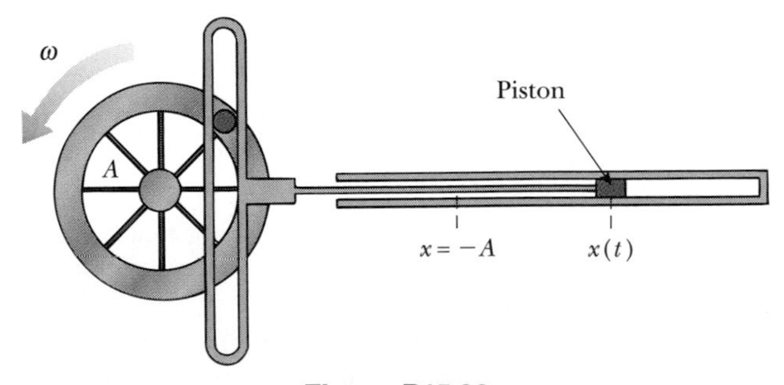

Figure P15.26

Section 15.5 The Pendulum

Problem 60 in Chapter 1 can also be assigned with this section.

27. A man enters a tall tower, needing to know its height. He notes that a long pendulum extends from the ceiling almost to the floor and that its period is 12.0 s. (a) How tall is the tower? (b) **What If?** If this pendulum is taken to the Moon, where the free-fall acceleration is 1.67 m/s^2, what is its period there?

28. A "seconds pendulum" is one that moves through its equilibrium position once each second. (The period of the pendulum is precisely 2 s.) The length of a seconds pendulum is 0.992 7 m at Tokyo, Japan and 0.994 2 m at Cambridge, England. What is the ratio of the free-fall accelerations at these two locations?

29. A rigid steel frame above a street intersection supports standard traffic lights, each of which is hinged to hang immediately below the frame. A gust of wind sets a light swinging in a vertical plane. Find the order of magnitude of its period. State the quantities you take as data and their values.

30. The angular position of a pendulum is represented by the equation $\theta = (0.320 \text{ rad})\cos \omega t$, where θ is in radians and $\omega = 4.43$ rad/s. Determine the period and length of the pendulum.

31. A simple pendulum has a mass of 0.250 kg and a length of 1.00 m. It is displaced through an angle of 15.0° and then released. What are (a) the maximum speed, (b) the maximum angular acceleration, and (c) the maximum restoring force? **What If?** Solve this problem by using the simple harmonic motion model for the motion of the pendulum, and then solve the problem more precisely by using more general principles.

32. **Review problem.** A simple pendulum is 5.00 m long. (a) What is the period of small oscillations for this pendulum if it is located in an elevator accelerating upward at 5.00 m/s^2? (b) What is its period if the elevator is accelerating downward at 5.00 m/s^2? (c) What is the period of this pendulum if it is placed in a truck that is accelerating horizontally at 5.00 m/s^2?

33. A particle of mass m slides without friction inside a hemispherical bowl of radius R. Show that, if it starts from rest with a small displacement from equilibrium, the particle moves in simple harmonic motion with an angular frequency equal to that of a simple pendulum of length R. That is, $\omega = \sqrt{g/R}$.

34. A small object is attached to the end of a string to form a simple pendulum. The period of its harmonic motion is measured for small angular displacements and three lengths, each time clocking the motion with a stopwatch for 50 oscillations. For lengths of 1.000 m, 0.750 m, and 0.500 m, total times of 99.8 s, 86.6 s, and 71.1 s are measured for 50 oscillations. (a) Determine the period of motion for each length. (b) Determine the mean value of g obtained from these three independent measurements, and compare it with the accepted value. (c) Plot T^2 versus L, and obtain a value for g from the slope of your best-fit straight-line graph. Compare this value with that obtained in part (b).

35. A physical pendulum in the form of a planar body moves in simple harmonic motion with a frequency of 0.450 Hz. If the pendulum has a mass of 2.20 kg and the pivot is located 0.350 m from the center of mass, determine the moment of inertia of the pendulum about the pivot point.

36. A very light rigid rod with a length of 0.500 m extends straight out from one end of a meter stick. The stick is suspended from a pivot at the far end of the rod and is set into oscillation. (a) Determine the period of oscillation. *Suggestion:* Use the parallel-axis theorem from Section 10.5. (b) By what percentage does the period differ from the period of a simple pendulum 1.00 m long?

37. Consider the physical pendulum of Figure 15.18. (a) If its moment of inertia about an axis passing through its center of mass and parallel to the axis passing through its pivot point is I_{CM}, show that its period is

$$T = 2\pi\sqrt{\frac{I_{CM} + md^2}{mgd}}$$

where d is the distance between the pivot point and center of mass. (b) Show that the period has a minimum value when d satisfies $md^2 = I_{CM}$.

38. A torsional pendulum is formed by taking a meter stick of mass 2.00 kg, and attaching to its center a wire. With its upper end clamped, the vertical wire supports the stick as the stick turns in a horizontal plane. If the resulting period is 3.00 minutes, what is the torsion constant for the wire?

39. A clock balance wheel (Fig. P15.39) has a period of oscillation of 0.250 s. The wheel is constructed so that its mass of 20.0 g is concentrated around a rim of radius 0.500 cm. What are (a) the wheel's moment of inertia and (b) the torsion constant of the attached spring?

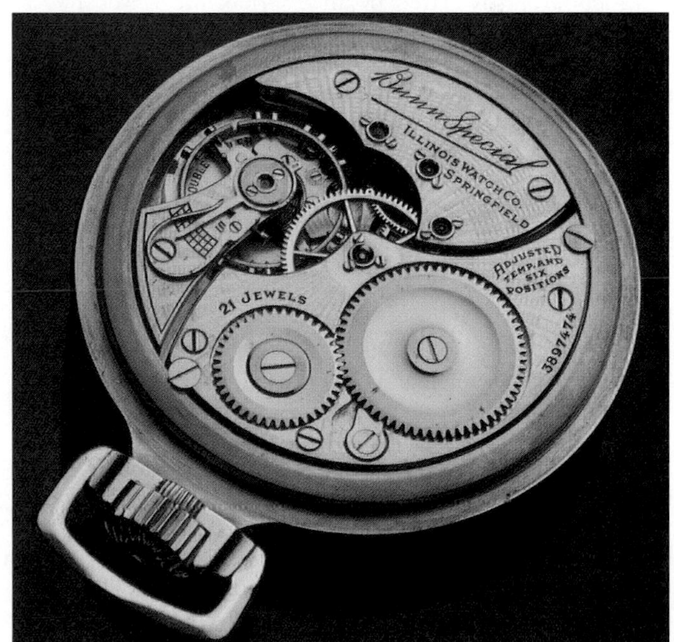

George Semple

Figure P15.39

Section 15.6 Damped Oscillations

40. Show that the time rate of change of mechanical energy for a damped, undriven oscillator is given by $dE/dt = -bv^2$ and hence is always negative. Proceed as follows: Differentiate the expression for the mechanical energy of an oscillator, $E = \frac{1}{2}mv^2 + \frac{1}{2}kx^2$, and use Equation 15.31.

41. A pendulum with a length of 1.00 m is released from an initial angle of 15.0°. After 1 000 s, its amplitude has been reduced by friction to 5.50°. What is the value of $b/2m$?

42. Show that Equation 15.32 is a solution of Equation 15.31 provided that $b^2 < 4mk$.

43. A 10.6-kg object oscillates at the end of a vertical spring that has a spring constant of 2.05×10^4 N/m. The effect of air resistance is represented by the damping coefficient $b = 3.00$ N·s/m. (a) Calculate the frequency of the damped oscillation. (b) By what percentage does the amplitude of the oscillation decrease in each cycle? (c) Find the time interval that elapses while the energy of the system drops to 5.00% of its initial value.

Section 15.7 Forced Oscillations

44. The front of her sleeper wet from teething, a baby rejoices in the day by crowing and bouncing up and down in her crib. Her mass is 12.5 kg, and the crib mattress can be modeled as a light spring with force constant 4.30 kN/m. (a) The baby soon learns to bounce with maximum amplitude and minimum effort by bending her knees at what frequency? (b) She learns to use the mattress as a trampoline—losing contact with it for part of each cycle—when her amplitude exceeds what value?

45. A 2.00-kg object attached to a spring moves without friction and is driven by an external force given by $F = (3.00\text{ N})\sin(2\pi t)$. If the force constant of the spring is 20.0 N/m, determine (a) the period and (b) the amplitude of the motion.

46. Considering an undamped, forced oscillator ($b = 0$), show that Equation 15.35 is a solution of Equation 15.34, with an amplitude given by Equation 15.36.

47. A weight of 40.0 N is suspended from a spring that has a force constant of 200 N/m. The system is undamped and is subjected to a harmonic driving force of frequency 10.0 Hz, resulting in a forced-motion amplitude of 2.00 cm. Determine the maximum value of the driving force.

48. Damping is negligible for a 0.150-kg object hanging from a light 6.30-N/m spring. A sinusoidal force with an amplitude of 1.70 N drives the system. At what frequency will the force make the object vibrate with an amplitude of 0.440 m?

49. You are a research biologist. You take your emergency pager along to a fine restaurant. You switch the small pager to vibrate instead of beep, and you put it into a side pocket of your suit coat. The arm of your chair presses the light cloth against your body at one spot. Fabric with a length of 8.21 cm hangs freely below that spot, with the pager at the bottom. A coworker urgently needs instructions and calls you from your laboratory. The motion of the pager makes the hanging part of your coat swing back and forth with remarkably large amplitude. The waiter and nearby diners notice immediately and fall silent. Your daughter pipes up and says, "Daddy, look! Your cockroaches must have gotten out again!" Find the frequency at which your pager vibrates.

50. Four people, each with a mass of 72.4 kg, are in a car with a mass of 1 130 kg. An earthquake strikes. The driver manages to pull off the road and stop, as the vertical oscillations of the ground surface make the car bounce up and down on its suspension springs. When the frequency of the shaking is 1.80 Hz, the car exhibits a maximum amplitude of vibration. The earthquake ends, and the four people leave the car as fast as they can. By what distance does the car's undamaged suspension lift the car body as the people get out?

Additional Problems

51. A small ball of mass M is attached to the end of a uniform rod of equal mass M and length L that is pivoted at the top (Fig. P15.51). (a) Determine the tensions in the rod at the pivot and at the point P when the system is stationary. (b) Calculate the period of oscillation for small displacements from equilibrium, and determine this period for $L = 2.00$ m. (*Suggestions*: Model the object at the end of the rod as a particle and use Eq. 15.28.)

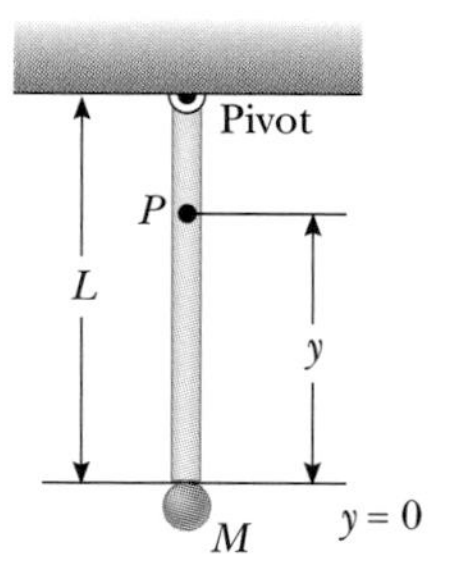

Figure P15.51

52. An object of mass $m_1 = 9.00$ kg is in equilibrium while connected to a light spring of constant $k = 100$ N/m that is fastened to a wall as shown in Figure P15.52a. A second object, $m_2 = 7.00$ kg, is slowly pushed up against m_1, compressing the spring by the amount $A = 0.200$ m, (see Figure P15.52b). The system is then released, and both objects start moving to the right on the frictionless surface. (a) When m_1 reaches the equilibrium point, m_2 loses contact with m_1 (see Fig. P15.5c) and moves to the right with speed v. Determine the value of v. (b) How far apart are the objects when the spring is fully stretched for the first time (D in Fig. P15.52d)? (*Suggestion:* First determine the period of oscillation and the amplitude of the m_1–spring system after m_2 loses contact with m_1.)

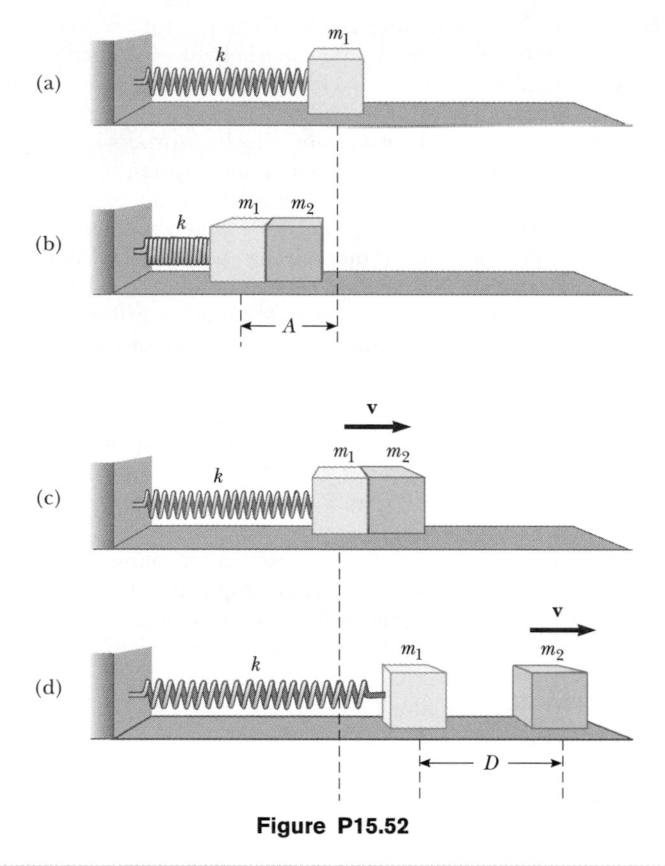

Figure P15.52

53. A large block P executes horizontal simple harmonic motion as it slides across a frictionless surface with a frequency $f = 1.50$ Hz. Block B rests on it, as shown in Figure P15.53, and the coefficient of static friction between the two is $\mu_s = 0.600$. What maximum amplitude of oscillation can the system have if block B is not to slip?

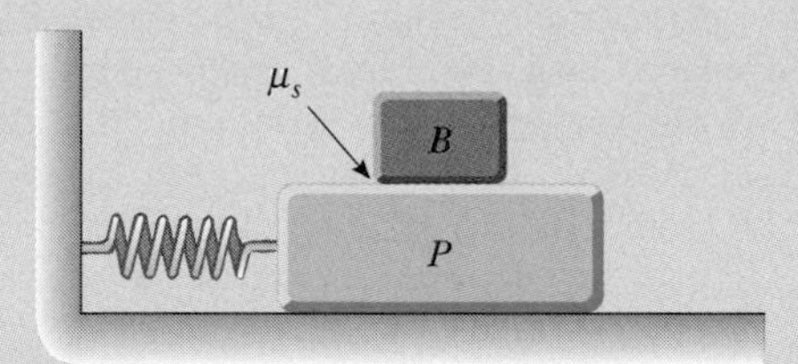

Figure P15.53 Problems 53 and 54.

54. A large block P executes horizontal simple harmonic motion as it slides across a frictionless surface with a frequency f. Block B rests on it, as shown in Figure P15.53, and the coefficient of static friction between the two is μ_s. What maximum amplitude of oscillation can the system have if the upper block is not to slip?

55. The mass of the deuterium molecule (D_2) is twice that of the hydrogen molecule (H_2). If the vibrational frequency of H_2 is 1.30×10^{14} Hz, what is the vibrational frequency of D_2? Assume that the "spring constant" of attracting forces is the same for the two molecules.

56. A solid sphere (radius = R) rolls without slipping in a cylindrical trough (radius = $5R$) as shown in Figure P15.56. Show that, for small displacements from equilibrium perpendicular to the length of the trough, the sphere executes simple harmonic motion with a period $T = 2\pi\sqrt{28R/5g}$.

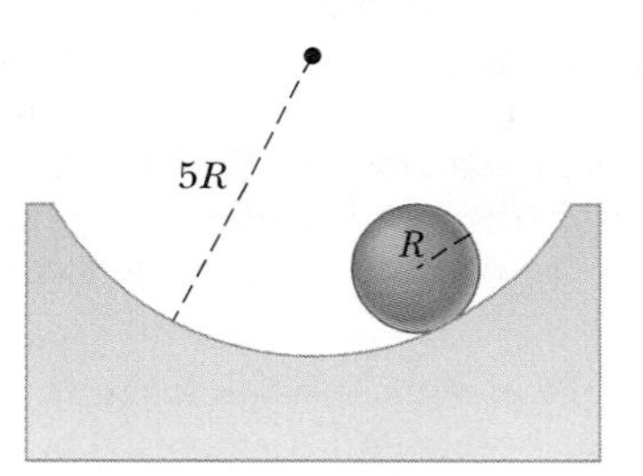

Figure P15.56

57. A light, cubical container of volume a^3 is initially filled with a liquid of mass density ρ. The cube is initially supported by a light string to form a simple pendulum of length L_i, measured from the center of mass of the filled container, where $L_i \gg a$. The liquid is allowed to flow from the bottom of the container at a constant rate (dM/dt). At any time t, the level of the fluid in the container is h and the length of the pendulum is L (measured relative to the instantaneous center of mass). (a) Sketch the apparatus and label the dimensions a, h, L_i, and L. (b) Find the time rate of change of the period as a function of time t. (c) Find the period as a function of time.

58. After a thrilling plunge, bungee-jumpers bounce freely on the bungee cord through many cycles (Fig. P15.22). After the first few cycles, the cord does not go slack. Your little brother can make a pest of himself by figuring out the mass of each person, using a proportion which you set up by solving this problem: An object of mass m is oscillating freely on a vertical spring with a period T. An object of unknown mass m' on the same spring oscillates with a period T'. Determine (a) the spring constant and (b) the unknown mass.

59. A pendulum of length L and mass M has a spring of force constant k connected to it at a distance h below its point of suspension (Fig. P15.59). Find the frequency of vibration

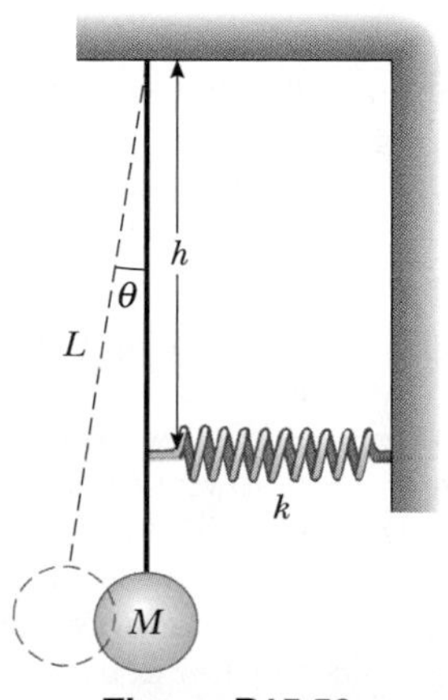

Figure P15.59

of the system for small values of the amplitude (small θ). Assume the vertical suspension of length L is rigid, but ignore its mass.

60. A particle with a mass of 0.500 kg is attached to a spring with a force constant of 50.0 N/m. At time $t = 0$ the particle has its maximum speed of 20.0 m/s and is moving to the left. (a) Determine the particle's equation of motion, specifying its position as a function of time. (b) Where in the motion is the potential energy three times the kinetic energy? (c) Find the length of a simple pendulum with the same period. (d) Find the minimum time interval required for the particle to move from $x = 0$ to $x = 1.00$ m.

61. A horizontal plank of mass m and length L is pivoted at one end. The plank's other end is supported by a spring of force constant k (Fig P15.61). The moment of inertia of the plank about the pivot is $\frac{1}{3}mL^2$. The plank is displaced by a small angle θ from its horizontal equilibrium position and released. (a) Show that it moves with simple harmonic motion with an angular frequency $\omega = \sqrt{3k/m}$. (b) Evaluate the frequency if the mass is 5.00 kg and the spring has a force constant of 100 N/m.

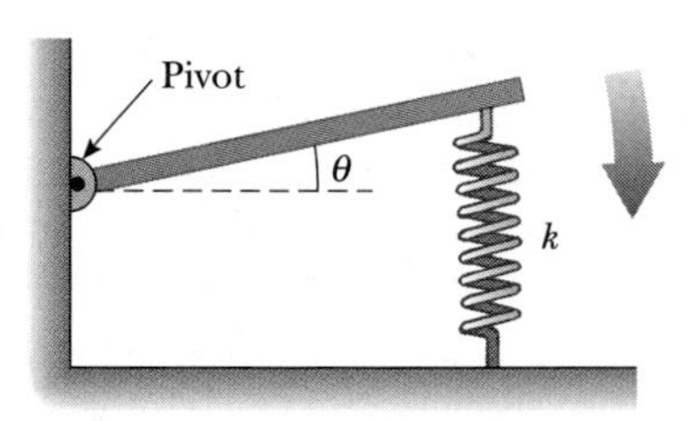

Figure P15.61

62. Review problem. A particle of mass 4.00 kg is attached to a spring with a force constant of 100 N/m. It is oscillating on a horizontal frictionless surface with an amplitude of 2.00 m. A 6.00-kg object is dropped vertically on top of the 4.00-kg object as it passes through its equilibrium point. The two objects stick together. (a) By how much does the amplitude of the vibrating system change as a result of the collision? (b) By how much does the period change? (c) By how much does the energy change? (d) Account for the change in energy.

63. A simple pendulum with a length of 2.23 m and a mass of 6.74 kg is given an initial speed of 2.06 m/s at its equilibrium position. Assume it undergoes simple harmonic motion, and determine its (a) period, (b) total energy, and (c) maximum angular displacement.

64. Review problem. One end of a light spring with force constant 100 N/m is attached to a vertical wall. A light string is tied to the other end of the horizontal spring. The string changes from horizontal to vertical as it passes over a solid pulley of diameter 4.00 cm. The pulley is free to turn on a fixed smooth axle. The vertical section of the string supports a 200-g object. The string does not slip at its contact with the pulley. Find the frequency of oscillation of the object if the mass of the pulley is (a) negligible, (b) 250 g, and (c) 750 g.

65. People who ride motorcycles and bicycles learn to look out for bumps in the road, and especially for *washboarding,* a condition in which many equally spaced ridges are worn into the road. What is so bad about washboarding? A motorcycle has several springs and shock absorbers in its suspension, but you can model it as a single spring supporting a block. You can estimate the force constant by thinking about how far the spring compresses when a big biker sits down on the seat. A motorcyclist traveling at highway speed must be particularly careful of washboard bumps that are a certain distance apart. What is the order of magnitude of their separation distance? State the quantities you take as data and the values you measure or estimate for them.

66. A block of mass M is connected to a spring of mass m and oscillates in simple harmonic motion on a horizontal, frictionless track (Fig. P15.66). The force constant of the spring is k and the equilibrium length is ℓ. Assume that all portions of the spring oscillate in phase and that the velocity of a segment dx is proportional to the distance x from the fixed end; that is, $v_x = (x/\ell)v$. Also, note that the mass of a segment of the spring is $dm = (m/\ell)\,dx$. Find (a) the kinetic energy of the system when the block has a speed v and (b) the period of oscillation.

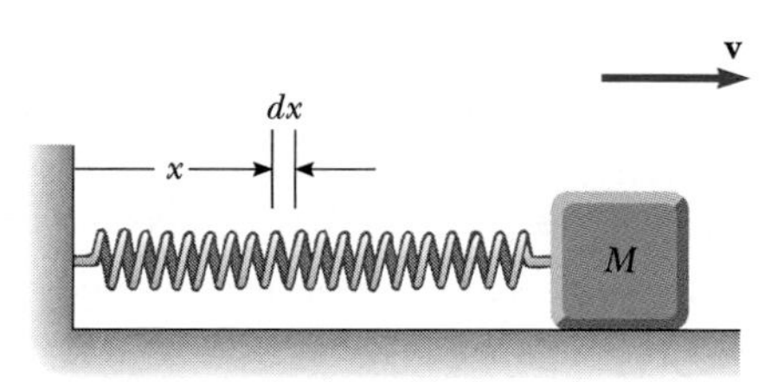

Figure P15.66

67. A ball of mass m is connected to two rubber bands of length L, each under tension T, as in Figure P15.67. The ball is displaced by a small distance y perpendicular to the length of the rubber bands. Assuming that the tension does not change, show that (a) the restoring force is $-(2T/L)y$ and (b) the system exhibits simple harmonic motion with an angular frequency $\omega = \sqrt{2T/mL}$.

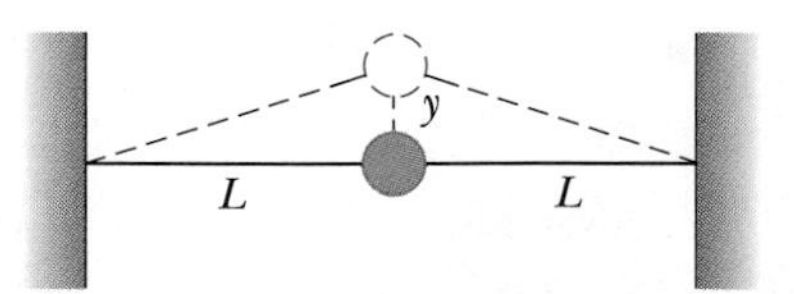

Figure P15.67

68. When a block of mass M, connected to the end of a spring of mass $m_s = 7.40$ g and force constant k, is set into simple harmonic motion, the period of its motion is

$$T = 2\pi\sqrt{\frac{M + (m_s/3)}{k}}$$

A two-part experiment is conducted with the use of blocks of various masses suspended vertically from the spring, as shown in Figure P15.68. (a) Static extensions of 17.0, 29.3, 35.3, 41.3, 47.1, and 49.3 cm are measured for M values of 20.0, 40.0, 50.0, 60.0, 70.0, and 80.0 g, respectively. Construct a graph of Mg versus x, and perform a linear least-squares fit to the data. From the slope of your graph, determine a value for k for this spring. (b) The system is now set into simple harmonic motion, and periods are measured with a stopwatch. With $M = 80.0$ g, the total time for 10 oscillations is measured to be 13.41 s. The experiment is repeated with M values of 70.0, 60.0, 50.0, 40.0, and 20.0 g, with corresponding times for 10 oscillations of 12.52, 11.67, 10.67, 9.62, and 7.03 s. Compute the experimental value for T from each of these measurements. Plot a graph of T^2 versus M, and determine a value for k from the slope of the linear least-squares fit through the data points. Compare this value of k with that obtained in part (a). (c) Obtain a value for m_s from your graph and compare it with the given value of 7.40 g.

Figure P15.68

69. A smaller disk of radius r and mass m is attached rigidly to the face of a second larger disk of radius R and mass M as shown in Figure P15.69. The center of the small disk is located at the edge of the large disk. The large disk is mounted at its center on a frictionless axle. The assembly is rotated through a small angle θ from its equilibrium position and released. (a) Show that the speed of the center of the small disk as it passes through the equilibrium position is

$$v = 2\left[\frac{Rg(1 - \cos\theta)}{(M/m) + (r/R)^2 + 2}\right]^{1/2}$$

(b) Show that the period of the motion is

$$T = 2\pi\left[\frac{(M + 2m)R^2 + mr^2}{2mgR}\right]^{1/2}$$

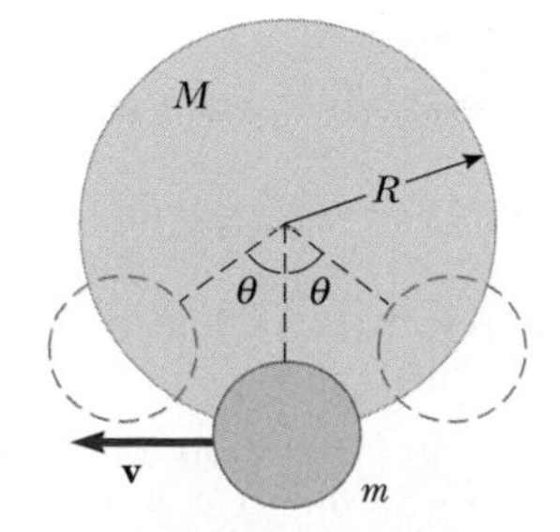

Figure P15.69

70. Consider a damped oscillator as illustrated in Figures 15.21 and 15.22. Assume the mass is 375 g, the spring constant is 100 N/m, and $b = 0.100$ N·s/m. (a) How long does it takes for the amplitude to drop to half its initial value? (b) **What If?** How long does it take for the mechanical energy to drop to half its initial value? (c) Show that, in general, the fractional rate at which the amplitude decreases in a damped harmonic oscillator is half the fractional rate at which the mechanical energy decreases.

71. A block of mass m is connected to two springs of force constants k_1 and k_2 as shown in Figures P15.71a and P15.71b. In each case, the block moves on a frictionless table after it is displaced from equilibrium and released. Show that in the two cases the block exhibits simple harmonic motion with periods

(a) $$T = 2\pi\sqrt{\frac{m(k_1 + k_2)}{k_1 k_2}}$$

(b) $$T = 2\pi\sqrt{\frac{m}{k_1 + k_2}}$$

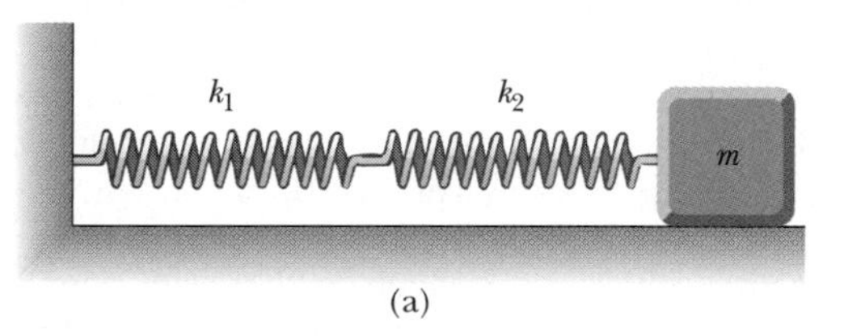

(a)

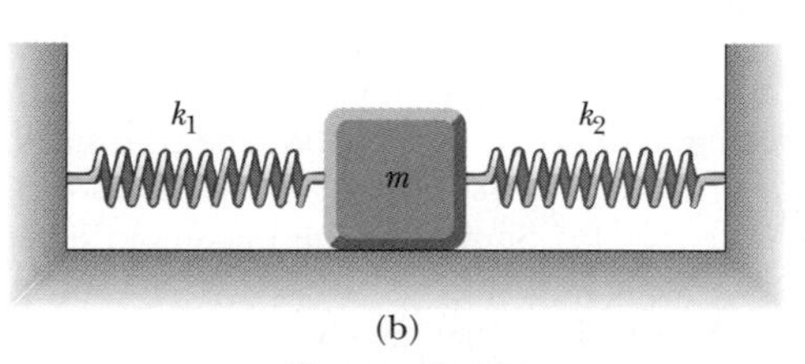

(b)

Figure P15.71

72. A lobsterman's buoy is a solid wooden cylinder of radius r and mass M. It is weighted at one end so that it floats upright in calm sea water, having density ρ. A passing shark tugs on the slack rope mooring the buoy to a lobster trap, pulling the buoy down a distance x from its equilibrium position and releasing it. Show that the buoy will execute simple harmonic motion if the resistive effects of the water are neglected, and determine the period of the oscillations.

73. Consider a bob on a light stiff rod, forming a simple pendulum of length $L = 1.20$ m. It is displaced from the vertical by an angle θ_{max} and then released. Predict the subsequent angular positions if θ_{max} is small or if it is large. Proceed as follows: Set up and carry out a numerical method to integrate the equation of motion for the simple pendulum:

$$\frac{d^2\theta}{dt^2} = -\frac{g}{L}\sin\theta$$

Take the initial conditions to be $\theta = \theta_{max}$ and $d\theta/dt = 0$ at $t = 0$. On one trial choose $\theta_{max} = 5.00°$, and on another trial take $\theta_{max} = 100°$. In each case find the position θ as a function of time. Using the same values of θ_{max}, compare your results for θ with those obtained from $\theta(t) = \theta_{max} \cos \omega t$. How does the period for the large value of θ_{max} compare with that for the small value of θ_{max}? *Note:* Using the Euler method to solve this differential equation, you may find that the amplitude tends to increase with time. The fourth-order Runge–Kutta method would be a better choice to solve the differential equation. However, if you choose Δt small enough, the solution using Euler's method can still be good.

74. Your thumb squeaks on a plate you have just washed. Your sneakers often squeak on the gym floor. Car tires squeal when you start or stop abruptly. You can make a goblet sing by wiping your moistened finger around its rim. As you slide it across the table, a Styrofoam cup may not make much sound, but it makes the surface of some water inside it dance in a complicated resonance vibration. When chalk squeaks on a blackboard, you can see that it makes a row of regularly spaced dashes. As these examples suggest, vibration commonly results when friction acts on a moving elastic object. The oscillation is not simple harmonic motion, but is called *stick-and-slip*. This problem models stick-and-slip motion.

A block of mass m is attached to a fixed support by a horizontal spring with force constant k and negligible mass (Fig. P15.74). Hooke's law describes the spring both in extension and in compression. The block sits on a long horizontal board, with which it has coefficient of static friction μ_s and a smaller coefficient of kinetic friction μ_k. The board moves to the right at constant speed v. Assume that the block spends most of its time sticking to the board and moving to the right, so that the speed v is small in comparison to the average speed the block has as it slips back toward the left. (a) Show that the maximum extension of the spring from its unstressed position is very nearly given by $\mu_s mg/k$. (b) Show that the block oscillates around an equilibrium position at which the spring is stretched by $\mu_k mg/k$. (c) Graph the block's position versus time. (d) Show that the amplitude of the block's motion is

$$A = \frac{(\mu_s - \mu_k)\, mg}{k}$$

(e) Show that the period of the block's motion is

$$T = \frac{2(\mu_s - \mu_k)\, mg}{vk} + \pi\sqrt{\frac{m}{k}}$$

(f) Evaluate the frequency of the motion if $\mu_s = 0.400$, $\mu_k = 0.250$, $m = 0.300$ kg, $k = 12.0$ N/m, and $v = 2.40$ cm/s. (g) **What If?** What happens to the frequency if the mass increases? (h) If the spring constant increases? (i) If the speed of the board increases? (j) If the coefficient of static friction increases relative to the coefficient of kinetic friction? Note that it is the excess of static over kinetic friction that is important for the vibration. "The squeaky wheel gets the grease" because even a viscous fluid cannot exert a force of static friction.

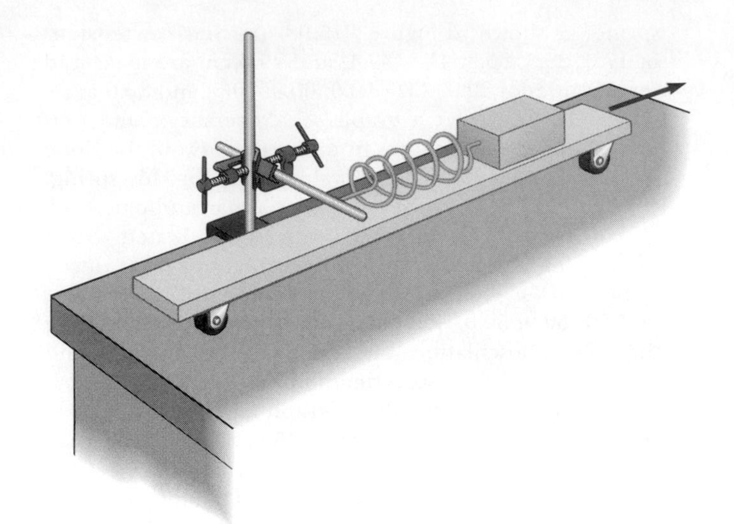

Figure P15.74

75. Review problem. Imagine that a hole is drilled through the center of the Earth to the other side. An object of mass m at a distance r from the center of the Earth is pulled toward the center of the Earth only by the mass within the sphere of radius r (the reddish region in Fig. P15.75). (a) Write Newton's law of gravitation for an object at the distance r from the center of the Earth, and show that the force on it is of Hooke's law form, $F = -kr$, where the effective force constant is $k = (4/3)\pi\rho Gm$. Here ρ is the density of the Earth, assumed uniform, and G is the gravitational constant. (b) Show that a sack of mail dropped into the hole will execute simple harmonic motion if it moves without friction. When will it arrive at the other side of the Earth?

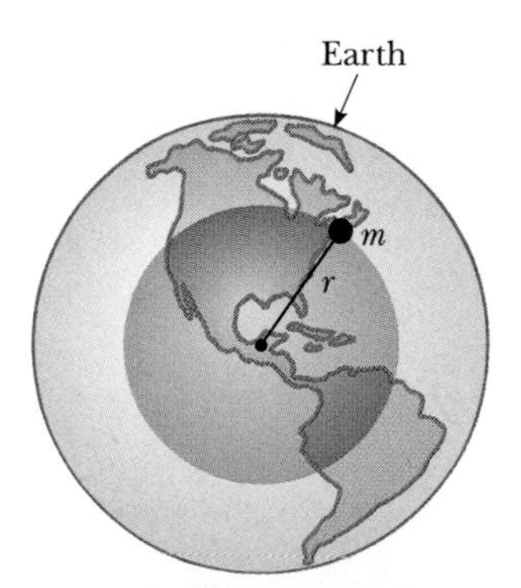

Figure P15.75

Answers to Quick Quizzes

15.1 (d). From its maximum positive position to the equilibrium position, the block travels a distance A. It then goes an equal distance past the equilibrium position to its maximum negative position. It then repeats these two motions in the reverse direction to return to its original position and complete one cycle.

15.2 (f). The object is in the region $x < 0$, so the position is negative. Because the object is moving back toward the origin in this region, the velocity is positive.

15.3 (a). The amplitude is larger because the curve for Object B shows that the displacement from the origin (the vertical axis on the graph) is larger. The frequency is larger for Object B because there are more oscillations per unit time interval.

15.4 (a). The velocity is positive, as in Quick Quiz 15.2. Because the spring is pulling the object toward equilibrium from the negative x region, the acceleration is also positive.

15.5 (b). According to Equation 15.13, the period is proportional to the square root of the mass.

15.6 (c). The amplitude of the simple harmonic motion is the same as the radius of the circular motion. The initial position of the object in its circular motion is π radians from the positive x axis.

15.7 (a). With a longer length, the period of the pendulum will increase. Thus, it will take longer to execute each swing, so that each second according to the clock will take longer than an actual second—the clock will run *slow*.

15.8 (a). At the top of the mountain, the value of g is less than that at sea level. As a result, the period of the pendulum will increase and the clock will run slow.

15.9 (a). If your goal is simply to stop the bounce from an absorbed shock as rapidly as possible, you should critically damp the suspension. Unfortunately, the stiffness of this design makes for an uncomfortable ride. If you underdamp the suspension, the ride is more comfortable but the car bounces. If you overdamp the suspension, the wheel is displaced from its equilibrium position longer than it should be. (For example, after hitting a bump, the spring stays compressed for a short time and the wheel does not quickly drop back down into contact with the road after the wheel is past the bump—a dangerous situation.) Because of all these considerations, automotive engineers usually design suspensions to be slightly underdamped. This allows the suspension to absorb a shock rapidly (minimizing the roughness of the ride) and then return to equilibrium after only one or two noticeable oscillations.

Chapter 16

Wave Motion

CHAPTER OUTLINE

16.1 Propagation of a Disturbance

16.2 Sinusoidal Waves

16.3 The Speed of Waves on Strings

16.4 Reflection and Transmission

16.5 Rate of Energy Transfer by Sinusoidal Waves on Strings

16.6 The Linear Wave Equation

▲ *The rich sound of a piano is due to waves on strings that are under tension. Many such strings can be seen in this photograph. Waves also travel on the soundboard, which is visible below the strings. In this chapter, we study the fundamental principles of wave phenomena. (Kathy Ferguson Johnson/PhotoEdit/PictureQuest)*

Most of us experienced waves as children when we dropped a pebble into a pond. At the point where the pebble hits the water's surface, waves are created. These waves move outward from the creation point in expanding circles until they reach the shore. If you were to examine carefully the motion of a beach ball floating on the disturbed water, you would see that the ball moves vertically and horizontally about its original position but does not undergo any net displacement away from or toward the point where the pebble hit the water. The small elements of water in contact with the beach ball, as well as all the other water elements on the pond's surface, behave in the same way. That is, the water *wave* moves from the point of origin to the shore, but the water is not carried with it.

The world is full of waves, the two main types being *mechanical* waves and *electromagnetic* waves. In the case of mechanical waves, some physical medium is being disturbed—in our pebble and beach ball example, elements of water are disturbed. Electromagnetic waves do not require a medium to propagate; some examples of electromagnetic waves are visible light, radio waves, television signals, and x-rays. Here, in this part of the book, we study only mechanical waves.

The wave concept is abstract. When we observe what we call a water wave, what we see is a rearrangement of the water's surface. Without the water, there would be no wave. A wave traveling on a string would not exist without the string. Sound waves could not travel from one point to another if there were no air molecules between the two points. With mechanical waves, what we interpret as a wave corresponds to the propagation of a disturbance through a medium.

Considering further the beach ball floating on the water, note that we have caused the ball to move at one point in the water by dropping a pebble at another location. The ball has gained kinetic energy from our action, so energy must have transferred from the point at which we drop the pebble to the position of the ball. This is a central feature of wave motion—*energy* is transferred over a distance, but *matter* is not.

All waves carry energy, but the amount of energy transmitted through a medium and the mechanism responsible for that transport of energy differ from case to case. For instance, the power of ocean waves during a storm is much greater than the power of sound waves generated by a single human voice.

16.1 Propagation of a Disturbance

In the introduction, we alluded to the essence of wave motion—the transfer of energy through space without the accompanying transfer of matter. In the list of energy transfer mechanisms in Chapter 7, two mechanisms depend on waves—mechanical waves and electromagnetic radiation. By contrast, in another mechanism—matter transfer—the energy transfer is accompanied by a movement of matter through space.

All mechanical waves require (1) some source of disturbance, (2) a medium that can be disturbed, and (3) some physical mechanism through which elements of the medium can influence each other. One way to demonstrate wave

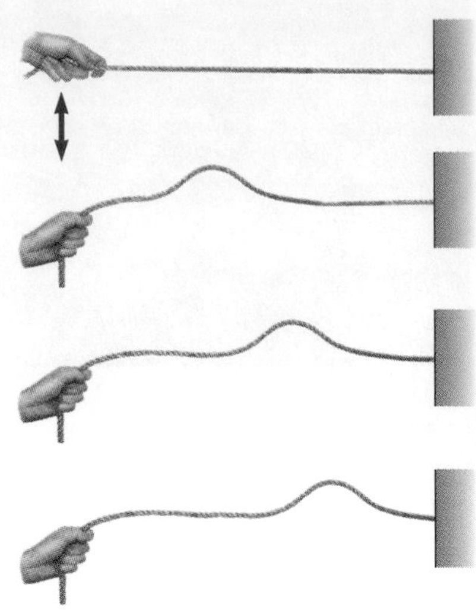

Figure 16.1 A pulse traveling down a stretched rope. The shape of the pulse is approximately unchanged as it travels along the rope.

motion is to flick one end of a long rope that is under tension and has its opposite end fixed, as shown in Figure 16.1. In this manner, a single bump (called a *pulse*) is formed and travels along the rope with a definite speed. Figure 16.1 represents four consecutive "snapshots" of the creation and propagation of the traveling pulse. The rope is the medium through which the pulse travels. The pulse has a definite height and a definite speed of propagation along the medium (the rope). As we shall see later, the properties of this particular medium that determine the speed of the disturbance are the tension in the rope and its mass per unit length. The shape of the pulse changes very little as it travels along the rope.[1]

We shall first focus our attention on a pulse traveling through a medium. Once we have explored the behavior of a pulse, we will then turn our attention to a *wave*, which is a *periodic* disturbance traveling through a medium. We created a pulse on our rope by flicking the end of the rope once, as in Figure 16.1. If we were to move the end of the rope up and down repeatedly, we would create a traveling wave, which has characteristics that a pulse does not have. We shall explore these characteristics in Section 16.2.

As the pulse in Figure 16.1 travels, each disturbed element of the rope moves in a direction *perpendicular* to the direction of propagation. Figure 16.2 illustrates this point for one particular element, labeled *P*. Note that no part of the rope ever moves in the direction of the propagation.

A traveling wave or pulse that causes the elements of the disturbed medium to move perpendicular to the direction of propagation is called a **transverse wave.**

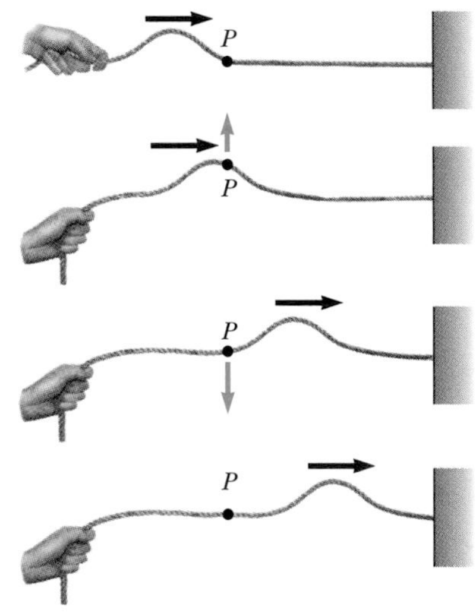

Figure 16.2 A transverse pulse traveling on a stretched rope. The direction of motion of any element *P* of the rope (blue arrows) is perpendicular to the direction of propagation (red arrows).

Compare this with another type of pulse—one moving down a long, stretched spring, as shown in Figure 16.3. The left end of the spring is pushed briefly to the right and then pulled briefly to the left. This movement creates a sudden compression of a region of the coils. The compressed region travels along the spring (to the right in Figure 16.3). The compressed region is followed by a region where the coils are extended. Notice that the direction of the displacement of the coils is *parallel* to the direction of propagation of the compressed region.

A traveling wave or pulse that causes the elements of the medium to move parallel to the direction of propagation is called a **longitudinal wave.**

Sound waves, which we shall discuss in Chapter 17, are another example of longitudinal waves. The disturbance in a sound wave is a series of high-pressure and low-pressure regions that travel through air.

Some waves in nature exhibit a combination of transverse and longitudinal displacements. Surface water waves are a good example. When a water wave travels on the surface of deep water, elements of water at the surface move in nearly circular paths, as shown in Figure 16.4. Note that the disturbance has both transverse and longitudinal

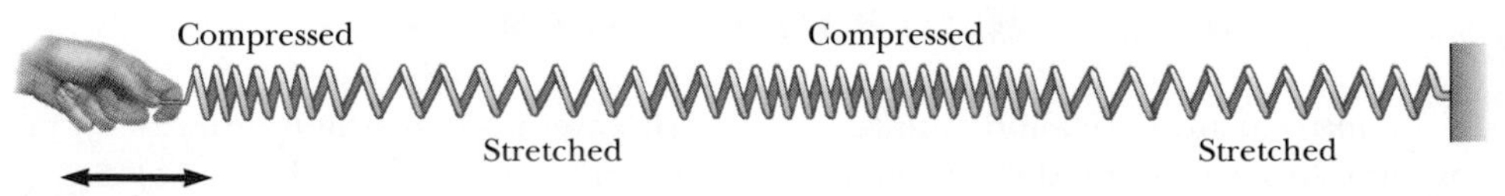

Figure 16.3 A longitudinal pulse along a stretched spring. The displacement of the coils is parallel to the direction of the propagation.

[1] In reality, the pulse changes shape and gradually spreads out during the motion. This effect is called *dispersion* and is common to many mechanical waves as well as to electromagnetic waves. We do not consider dispersion in this chapter.

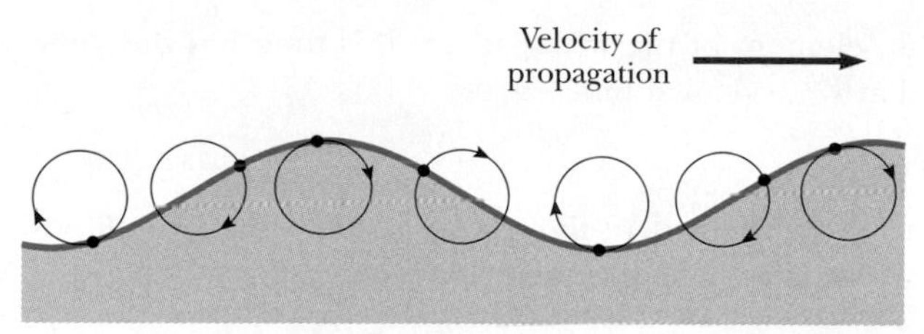

Active Figure 16.4 The motion of water elements on the surface of deep water in which a wave is propagating is a combination of transverse and longitudinal displacements, with the result that elements at the surface move in nearly circular paths. Each element is displaced both horizontally and vertically from its equilibrium position.

At the Active Figures link at* http://www.pse6.com, *you can observe the displacement of water elements at the surface of the moving waves.

components. The transverse displacements seen in Figure 16.4 represent the variations in vertical position of the water elements. The longitudinal displacement can be explained as follows: as the wave passes over the water's surface, water elements at the highest points move in the direction of propagation of the wave, whereas elements at the lowest points move in the direction opposite the propagation.

The three-dimensional waves that travel out from points under the Earth's surface along a fault at which an earthquake occurs are of both types—transverse and longitudinal. The longitudinal waves are the faster of the two, traveling at speeds in the range of 7 to 8 km/s near the surface. These are called **P waves** (with "P" standing for *primary*) because they travel faster than the transverse waves and arrive at a seismograph (a device used to detect waves due to earthquakes) first. The slower transverse waves, called **S waves** (with "S" standing for *secondary*), travel through the Earth at 4 to 5 km/s near the surface. By recording the time interval between the arrivals of these two types of waves at a seismograph, the distance from the seismograph to the point of origin of the waves can be determined. A single measurement establishes an imaginary sphere centered on the seismograph, with the radius of the sphere determined by the difference in arrival times of the P and S waves. The origin of the waves is located somewhere on that sphere. The imaginary spheres from three or more monitoring stations located far apart from each other intersect at one region of the Earth, and this region is where the earthquake occurred.

Consider a pulse traveling to the right on a long string, as shown in Figure 16.5. Figure 16.5a represents the shape and position of the pulse at time $t = 0$. At this time, the shape of the pulse, whatever it may be, can be represented by some mathematical function which we will write as $y(x, 0) = f(x)$. This function describes the transverse position y of the element of the string located at each value of x at time $t = 0$. Because the speed of the pulse is v, the pulse has traveled to the right a distance vt at the time t (Fig. 16.5b). We assume that the shape of the pulse does not change with time. Thus, at time t, the shape of the pulse is the same as it was at time $t = 0$, as in Figure 16.5a.

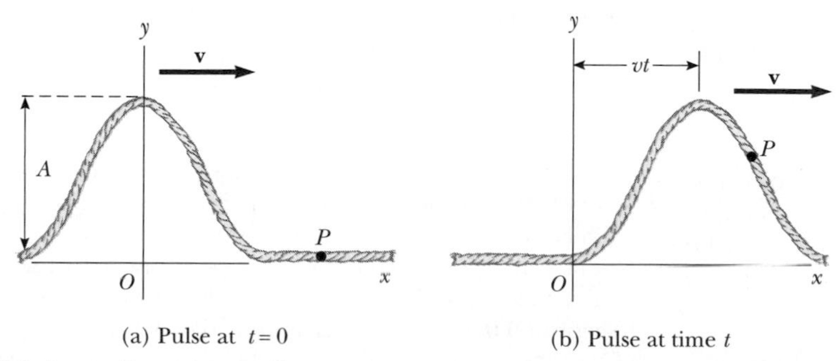

Figure 16.5 A one-dimensional pulse traveling to the right with a speed v. (a) At $t = 0$, the shape of the pulse is given by $y = f(x)$. (b) At some later time t, the shape remains unchanged and the vertical position of an element of the medium any point P is given by $y = f(x - vt)$.

Consequently, an element of the string at x at this time has the same y position as an element located at $x - vt$ had at time $t = 0$:

$$y(x, t) = y(x - vt, 0)$$

In general, then, we can represent the transverse position y for all positions and times, measured in a stationary frame with the origin at O, as

Pulse traveling to the right

$$y(x, t) = f(x - vt) \qquad (16.1)$$

Similarly, if the pulse travels to the left, the transverse positions of elements of the string are described by

Pulse traveling to the left

$$y(x, t) = f(x + vt) \qquad (16.2)$$

The function y, sometimes called the **wave function,** depends on the two variables x and t. For this reason, it is often written $y(x, t)$, which is read "y as a function of x and t."

It is important to understand the meaning of y. Consider an element of the string at point P, identified by a particular value of its x coordinate. As the pulse passes through P, the y coordinate of this element increases, reaches a maximum, and then decreases to zero. **The wave function $y(x, t)$ represents the y coordinate—the transverse position—of any element located at position x at any time t.** Furthermore, if t is fixed (as, for example, in the case of taking a snapshot of the pulse), then the wave function $y(x)$, sometimes called the **waveform,** defines a curve representing the actual geometric shape of the pulse at that time.

Quick Quiz 16.1 In a long line of people waiting to buy tickets, the first person leaves and a pulse of motion occurs as people step forward to fill the gap. As each person steps forward, the gap moves through the line. Is the propagation of this gap (a) transverse (b) longitudinal?

Quick Quiz 16.2 Consider the "wave" at a baseball game: people stand up and shout as the wave arrives at their location, and the resultant pulse moves around the stadium. Is this wave (a) transverse (b) longitudinal?

Example 16.1 A Pulse Moving to the Right

A pulse moving to the right along the x axis is represented by the wave function

$$y(x, t) = \frac{2}{(x - 3.0t)^2 + 1}$$

where x and y are measured in centimeters and t is measured in seconds. Plot the wave function at $t = 0$, $t = 1.0$ s, and $t = 2.0$ s.

Solution First, note that this function is of the form $y = f(x - vt)$. By inspection, we see that the wave speed is $v = 3.0$ cm/s. Furthermore, the maximum value of y is given by $A = 2.0$ cm. (We find the maximum value of the function representing y by letting $x - 3.0t = 0$.) The wave function expressions are

$$y(x, 0) = \frac{2}{x^2 + 1} \quad \text{at } t = 0$$

$$y(x, 1.0) = \frac{2}{(x - 3.0)^2 + 1} \quad \text{at } t = 1.0 \text{ s}$$

$$y(x, 2.0) = \frac{2}{(x - 6.0)^2 + 1} \quad \text{at } t = 2.0 \text{ s}$$

We now use these expressions to plot the wave function versus x at these times. For example, let us evaluate $y(x, 0)$ at $x = 0.50$ cm:

$$y(0.50, 0) = \frac{2}{(0.50)^2 + 1} = 1.6 \text{ cm}$$

Likewise, at $x = 1.0$ cm, $y(1.0, 0) = 1.0$ cm, and at $y = 2.0$ cm, $y(2.0, 0) = 0.40$ cm. Continuing this procedure for other values of x yields the wave function shown in Figure 16.6a. In a similar manner, we obtain the graphs of $y(x, 1.0)$ and $y(x, 2.0)$, shown in Figure 16.6b and c, respectively. These snapshots show that the pulse moves to the right without changing its shape and that it has a constant speed of 3.0 cm/s.

What If? (A) What if the wave function were

$$y(x, t) = \frac{2}{(x + 3.0t)^2 + 1}$$

How would this change the situation?

(B) What if the wave function were

$$y(x, t) = \frac{4}{(x - 3.0t)^2 + 1}$$

How would this change the situation?

Answer (A) The new feature in this expression is the plus sign in the denominator rather than the minus sign. This results in a pulse with the same shape as that in Figure 16.6, but moving to the left as time progresses.

(B) The new feature here is the numerator of 4 rather than 2. This results in a pulse moving to the right, but with twice the height of that in Figure 16.6.

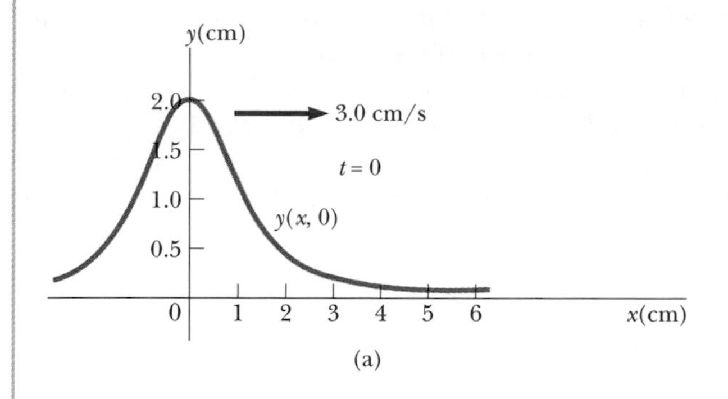

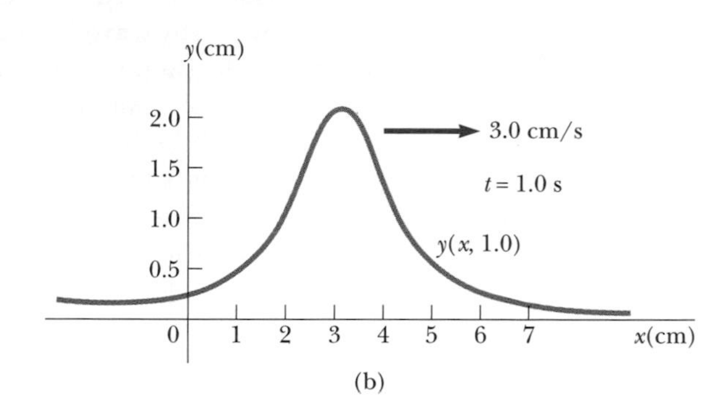

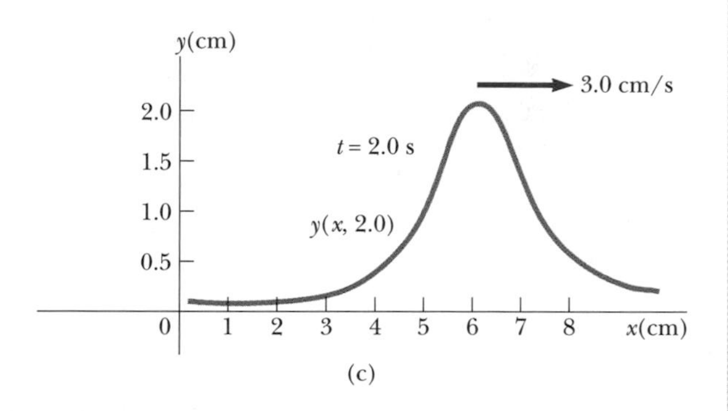

Figure 16.6 (Example 16.1) Graphs of the function $y(x, t) = 2/[(x - 3.0t)^2 + 1]$ at (a) $t = 0$, (b) $t = 1.0$ s, and (c) $t = 2.0$ s.

16.2 Sinusoidal Waves

In this section, we introduce an important wave function whose shape is shown in Figure 16.7. The wave represented by this curve is called a **sinusoidal wave** because the curve is the same as that of the function sin θ plotted against θ. On a rope, a sinusoidal wave could be established by shaking the end of the rope up and down in simple harmonic motion.

The sinusoidal wave is the simplest example of a periodic continuous wave and can be used to build more complex waves (see Section 18.8). The brown curve in Figure 16.7 represents a snapshot of a traveling sinusoidal wave at $t = 0$, and the blue curve represents a snapshot of the wave at some later time t. Notice two types of motion that can be seen in your mind. First, the entire waveform in Figure 16.7 moves to the right, so that the brown curve moves toward the right and eventually reaches the position of the blue curve. This is the motion of the *wave.* If we focus on one element of the medium, such as the element at $x = 0$, we see that each element moves up and down along the y axis in simple harmonic motion. This is the motion of the *elements of the medium.* It is important to differentiate between the motion of the wave and the motion of the elements of the medium.

Figure 16.8a shows a snapshot of a wave moving through a medium. Figure 16.8b shows a graph of the position of one element of the medium as a function of time. The

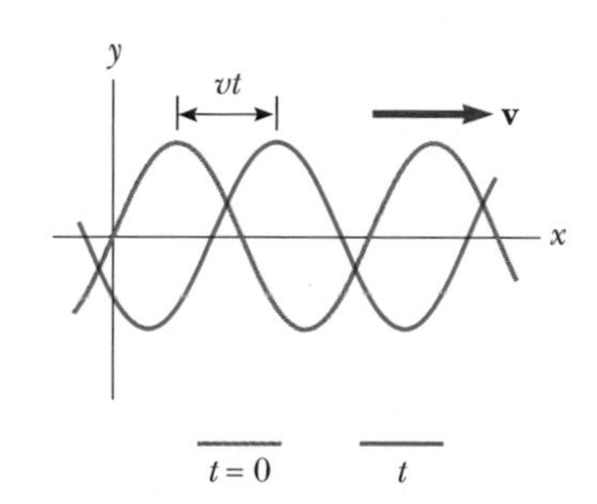

Active Figure 16.7 A one-dimensional sinusoidal wave traveling to the right with a speed v. The brown curve represents a snapshot of the wave at $t = 0$, and the blue curve represents a snapshot at some later time t.

At the Active Figures link at http://www.pse6.com, you can watch the wave move and take snapshots of it at various times.

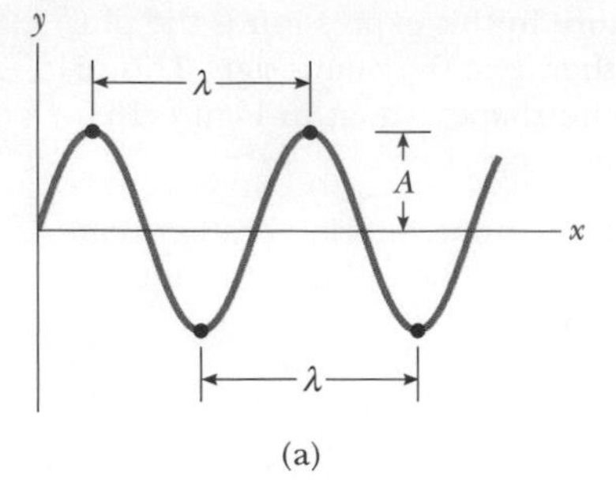

(a)

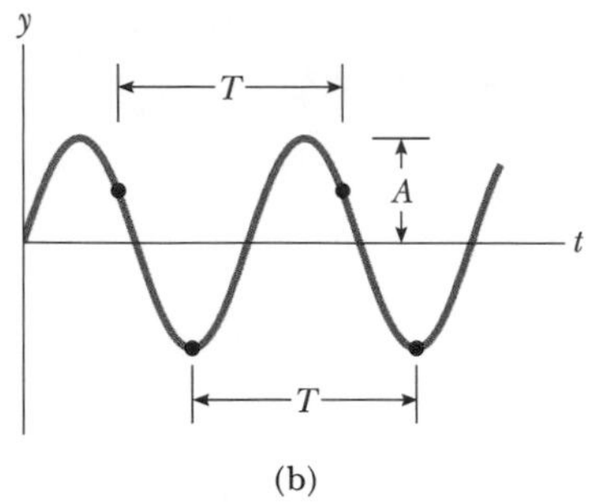

(b)

Active Figure 16.8 (a) The wavelength λ of a wave is the distance between adjacent crests or adjacent troughs. (b) The period T of a wave is the time interval required for the wave to travel one wavelength.

At the Active Figures link at http://www.pse6.com, you can change the parameters to see the effect on the wave function.

PITFALL PREVENTION

16.1 What's the Difference Between Figure 16.8a and 16.8b?

Notice the visual similarity between Figures 16.8a and 16.8b. The shapes are the same, but (a) is a graph of vertical position versus horizontal position while (b) is vertical position versus time. Figure 16.8a is a pictorial representation of the wave *for a series of particles of the medium*—this is what you would see at an instant of time. Figure 16.8b is a graphical representation of the position of *one element of the medium* as a function of time. The fact that both figures have the identical shape represents Equation 16.1—a wave is the *same* function of both x and t.

point at which the displacement of the element from its normal position is highest is called the **crest** of the wave. The distance from one crest to the next is called the **wavelength** λ (Greek lambda). More generally, **the wavelength is the minimum distance between any two identical points (such as the crests) on adjacent waves,** as shown in Figure 16.8a.

If you count the number of seconds between the arrivals of two adjacent crests at a given point in space, you are measuring the **period** T of the waves. In general, **the period is the time interval required for two identical points (such as the crests) of adjacent waves to pass by a point.** The period of the wave is the same as the period of the simple harmonic oscillation of one element of the medium.

The same information is more often given by the inverse of the period, which is called the **frequency** f. In general, **the frequency of a periodic wave is the number of crests (or troughs, or any other point on the wave) that pass a given point in a unit time interval.** The frequency of a sinusoidal wave is related to the period by the expression

$$f = \frac{1}{T} \tag{16.3}$$

The frequency of the wave is the same as the frequency of the simple harmonic oscillation of one element of the medium. The most common unit for frequency, as we learned in Chapter 15, is second^{-1}, or **hertz** (Hz). The corresponding unit for T is seconds.

The maximum displacement from equilibrium of an element of the medium is called the **amplitude** A of the wave.

Waves travel with a specific speed, and this speed depends on the properties of the medium being disturbed. For instance, sound waves travel through room-temperature air with a speed of about 343 m/s (781 mi/h), whereas they travel through most solids with a speed greater than 343 m/s.

Consider the sinusoidal wave in Figure 16.8a, which shows the position of the wave at $t = 0$. Because the wave is sinusoidal, we expect the wave function at this instant to be expressed as $y(x, 0) = A \sin ax$, where A is the amplitude and a is a constant to be determined. At $x = 0$, we see that $y(0, 0) = A \sin a(0) = 0$, consistent with Figure 16.8a. The next value of x for which y is zero is $x = \lambda/2$. Thus,

$$y\left(\frac{\lambda}{2}, 0\right) = A \sin a\left(\frac{\lambda}{2}\right) = 0$$

For this to be true, we must have $a(\lambda/2) = \pi$, or $a = 2\pi/\lambda$. Thus, the function describing the positions of the elements of the medium through which the sinusoidal wave is traveling can be written

$$y(x, 0) = A \sin\left(\frac{2\pi}{\lambda}x\right) \tag{16.4}$$

where the constant A represents the wave amplitude and the constant λ is the wavelength. We see that the vertical position of an element of the medium is the same whenever x is increased by an integral multiple of λ. If the wave moves to the right with a speed v, then the wave function at some later time t is

$$y(x, t) = A \sin\left[\frac{2\pi}{\lambda}(x - vt)\right] \tag{16.5}$$

That is, the traveling sinusoidal wave moves to the right a distance vt in the time t, as shown in Figure 16.7. Note that the wave function has the form $f(x - vt)$ (Eq. 16.1). If the wave were traveling to the left, the quantity $x - vt$ would be replaced by $x + vt$, as we learned when we developed Equations 16.1 and 16.2.

By definition, the wave travels a distance of one wavelength in one period T. Therefore, the wave speed, wavelength, and period are related by the expression

$$v = \frac{\lambda}{T} \tag{16.6}$$

Substituting this expression for v into Equation 16.5, we find that

$$y = A \sin\left[2\pi\left(\frac{x}{\lambda} - \frac{t}{T}\right)\right] \tag{16.7}$$

This form of the wave function shows the *periodic* nature of y. (We will often use y rather than $y(x, t)$ as a shorthand notation.) At any given time t, y has the *same* value at the positions x, $x + \lambda$, $x + 2\lambda$, and so on. Furthermore, at any given position x, the value of y is the same at times t, $t + T$, $t + 2T$, and so on.

We can express the wave function in a convenient form by defining two other quantities, the **angular wave number** k (usually called simply the **wave number**) and the **angular frequency** ω:

$$k \equiv \frac{2\pi}{\lambda} \tag{16.8}$$

Angular wave number

$$\omega \equiv \frac{2\pi}{T} \tag{16.9}$$

Angular frequency

Using these definitions, we see that Equation 16.7 can be written in the more compact form

$$y = A \sin(kx - \omega t) \tag{16.10}$$

Wave function for a sinusoidal wave

Using Equations 16.3, 16.8, and 16.9, we can express the wave speed v originally given in Equation 16.6 in the alternative forms

$$v = \frac{\omega}{k} \tag{16.11}$$

Speed of a sinusoidal wave

$$v = \lambda f \tag{16.12}$$

The wave function given by Equation 16.10 assumes that the vertical position y of an element of the medium is zero at $x = 0$ and $t = 0$. This need not be the case. If it is not, we generally express the wave function in the form

$$y = A \sin(kx - \omega t + \phi) \tag{16.13}$$

General expression for a sinusoidal wave

where ϕ is the **phase constant,** just as we learned in our study of periodic motion in Chapter 15. This constant can be determined from the initial conditions.

Quick Quiz 16.3 A sinusoidal wave of frequency f is traveling along a stretched string. The string is brought to rest, and a second traveling wave of frequency $2f$ is established on the string. The wave speed of the second wave is (a) twice that of the first wave (b) half that of the first wave (c) the same as that of the first wave (d) impossible to determine.

Quick Quiz 16.4 Consider the waves in Quick Quiz 16.3 again. The wavelength of the second wave is (a) twice that of the first wave (b) half that of the first wave (c) the same as that of the first wave (d) impossible to determine.

Quick Quiz 16.5 Consider the waves in Quick Quiz 16.3 again. The amplitude of the second wave is (a) twice that of the first wave (b) half that of the first wave (c) the same as that of the first wave (d) impossible to determine.

Example 16.2 A Traveling Sinusoidal Wave

A sinusoidal wave traveling in the positive x direction has an amplitude of 15.0 cm, a wavelength of 40.0 cm, and a frequency of 8.00 Hz. The vertical position of an element of the medium at $t = 0$ and $x = 0$ is also 15.0 cm, as shown in Figure 16.9.

(A) Find the wave number k, period T, angular frequency ω, and speed v of the wave.

Solution Using Equations 16.8, 16.3, 16.9, and 16.12, we find the following:

$$k = \frac{2\pi}{\lambda} = \frac{2\pi \text{ rad}}{40.0 \text{ cm}} = 0.157 \text{ rad/cm}$$

$$T = \frac{1}{f} = \frac{1}{8.00 \text{ s}^{-1}} = 0.125 \text{ s}$$

$$\omega = 2\pi f = 2\pi(8.00 \text{ s}^{-1}) = 50.3 \text{ rad/s}$$

$$v = \lambda f = (40.0 \text{ cm})(8.00 \text{ s}^{-1}) = 320 \text{ cm/s}$$

(B) Determine the phase constant ϕ, and write a general expression for the wave function.

Solution Because $A = 15.0$ cm and because $y = 15.0$ cm at $x = 0$ and $t = 0$, substitution into Equation 16.13 gives

$$15.0 = (15.0) \sin \phi \quad \text{or} \quad \sin \phi = 1$$

We may take the principal value $\phi = \pi/2$ rad (or 90°). Hence, the wave function is of the form

$$y = A \sin\left(kx - \omega t + \frac{\pi}{2}\right) = A \cos(kx - \omega t)$$

By inspection, we can see that the wave function must have this form, noting that the cosine function has the same shape as the sine function displaced by 90°. Substituting the values for A, k, and ω into this expression, we obtain

$$y = (15.0 \text{ cm}) \cos(0.157x - 50.3t)$$

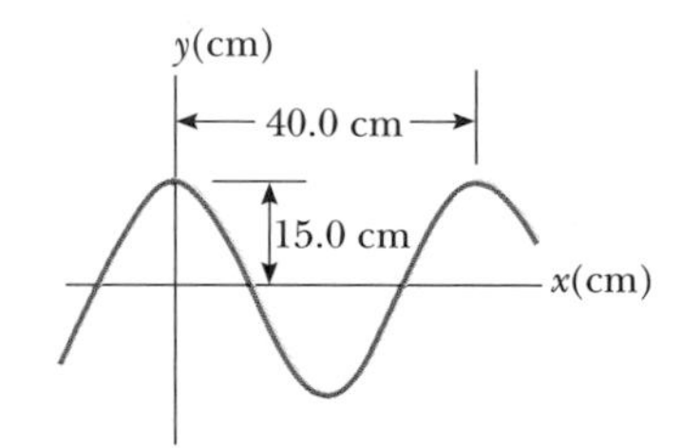

Figure 16.9 (Example 16.2) A sinusoidal wave of wavelength $\lambda = 40.0$ cm and amplitude $A = 15.0$ cm. The wave function can be written in the form $y = A\cos(kx - \omega t)$.

Sinusoidal Waves on Strings

In Figure 16.1, we demonstrated how to create a pulse by jerking a taut string up and down once. To create a series of such pulses—a wave—we can replace the hand with an oscillating blade. If the wave consists of a series of identical waveforms, whatever their shape, the relationships $f = 1/T$ and $v = f\lambda$ among speed, frequency, period, and wavelength hold true. We can make more definite statements about the wave function if the source of the waves vibrates in simple harmonic motion. Figure 16.10 represents snapshots of the wave created in this way at intervals of $T/4$. Because the end of the blade oscillates in simple harmonic motion, **each element of the string, such as that at *P*, also oscillates vertically with simple harmonic motion.** This must be the case because each element follows the simple harmonic motion of the blade. Therefore, every element of the string can be treated as a simple harmonic oscillator vibrating with a frequency equal to the frequency of oscillation of the blade.[2] Note that although each element oscillates in the y direction, the wave travels in the x direction with a speed v. Of course, this is the definition of a transverse wave.

If the wave at $t = 0$ is as described in Figure 16.10b, then the wave function can be written as

$$y = A \sin(kx - \omega t)$$

[2] In this arrangement, we are assuming that a string element always oscillates in a vertical line. The tension in the string would vary if an element were allowed to move sideways. Such motion would make the analysis very complex.

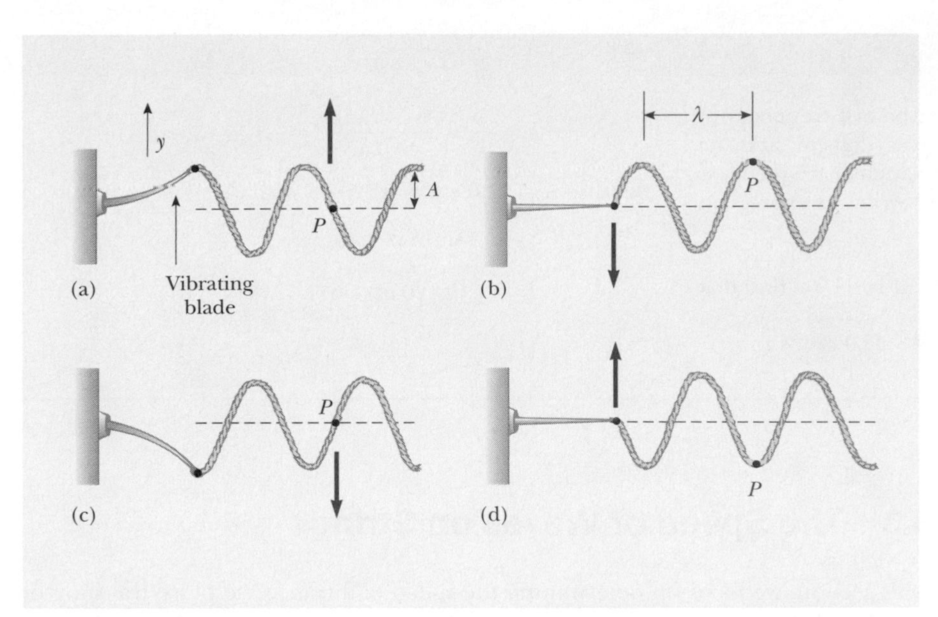

Active Figure 16.10 One method for producing a sinusoidal wave on a string. The left end of the string is connected to a blade that is set into oscillation. Every element of the string, such as that at point *P*, oscillates with simple harmonic motion in the vertical direction.

At the Active Figures link at http://www.pse6.com, you can adjust the frequency of the blade.

We can use this expression to describe the motion of any element of the string. An element at point *P* (or any other element of the string) moves only vertically, and so its x coordinate remains constant. Therefore, the **transverse speed** v_y (not to be confused with the wave speed v) and the **transverse acceleration** a_y of elements of the string are

$$v_y = \left.\frac{dy}{dt}\right]_{x = \text{constant}} = \frac{\partial y}{\partial t} = -\omega A \cos(kx - \omega t) \qquad (16.14)$$

$$a_y = \left.\frac{dv_y}{dt}\right]_{x = \text{constant}} = \frac{\partial v_y}{\partial t} = -\omega^2 A \sin(kx - \omega t) \qquad (16.15)$$

In these expressions, we must use partial derivatives (see Section 8.5) because y depends on both x and t. In the operation $\partial y/\partial t$, for example, we take a derivative with respect to t while holding x constant. The maximum values of the transverse speed and transverse acceleration are simply the absolute values of the coefficients of the cosine and sine functions:

$$v_{y,\,\max} = \omega A \qquad (16.16)$$

$$a_{y,\,\max} = \omega^2 A \qquad (16.17)$$

The transverse speed and transverse acceleration of elements of the string do not reach their maximum values simultaneously. The transverse speed reaches its maximum value (ωA) when $y = 0$, whereas the magnitude of the transverse acceleration reaches its maximum value ($\omega^2 A$) when $y = \pm A$. Finally, Equations 16.16 and 16.17 are identical in mathematical form to the corresponding equations for simple harmonic motion, Equations 15.17 and 15.18.

PITFALL PREVENTION

16.2 Two Kinds of Speed/Velocity

Do not confuse v, the speed of the wave as it propagates along the string, with v_y, the transverse velocity of a point on the string. The speed v is constant while v_y varies sinusoidally.

Quick Quiz 16.6 The amplitude of a wave is doubled, with no other changes made to the wave. As a result of this doubling, which of the following statements is correct? (a) The speed of the wave changes. (b) The frequency of the wave changes. (c) The maximum transverse speed of an element of the medium changes. (d) All of these are true. (e) None of these is true.

Example 16.3 A Sinusoidally Driven String

The string shown in Figure 16.10 is driven at a frequency of 5.00 Hz. The amplitude of the motion is 12.0 cm, and the wave speed is 20.0 m/s. Determine the angular frequency ω and wave number k for this wave, and write an expression for the wave function.

Solution Using Equations 16.3, 16.9, and 16.11, we find that

$$\omega = \frac{2\pi}{T} = 2\pi f = 2\pi(5.00 \text{ Hz}) = \boxed{31.4 \text{ rad/s}}$$

$$k = \frac{\omega}{v} = \frac{31.4 \text{ rad/s}}{20.0 \text{ m/s}} = \boxed{1.57 \text{ rad/m}}$$

Because $A = 12.0$ cm $= 0.120$ m, we have

$$y = A \sin(kx - \omega t)$$

$$= \boxed{(0.120 \text{ m}) \sin(1.57x - 31.4t)}$$

16.3 The Speed of Waves on Strings

In this section, we focus on determining the speed of a transverse pulse traveling on a taut string. Let us first conceptually predict the parameters that determine the speed. If a string under tension is pulled sideways and then released, the tension is responsible for accelerating a particular element of the string back toward its equilibrium position. According to Newton's second law, the acceleration of the element increases with increasing tension. If the element returns to equilibrium more rapidly due to this increased acceleration, we would intuitively argue that the wave speed is greater. Thus, we expect the wave speed to increase with increasing tension.

Likewise, the wave speed should decrease as the mass per unit length of the string increases. This is because it is more difficult to accelerate a massive element of the string than a light element. If the tension in the string is T and its mass per unit length is μ (Greek mu), then as we shall show, the wave speed is

Speed of a wave on a stretched string

$$v = \sqrt{\frac{T}{\mu}} \tag{16.18}$$

First, let us verify that this expression is dimensionally correct. The dimensions of T are ML/T^2, and the dimensions of μ are M/L. Therefore, the dimensions of T/μ are L^2/T^2; hence, the dimensions of $\sqrt{T/\mu}$ are L/T, the dimensions of speed. No other combination of T and μ is dimensionally correct, and if we assume that these are the only variables relevant to the situation, the speed must be proportional to $\sqrt{T/\mu}$.

 **PITFALL PREVENTION**

16.3 Multiple *T*'s

Do not confuse the T in Equation 16.18 for the tension with the symbol T used in this chapter for the period of a wave. The context of the equation should help you to identify which quantity is meant. There simply aren't enough letters in the alphabet to assign a unique letter to each variable!

Now let us use a mechanical analysis to derive Equation 16.18. Consider a pulse moving on a taut string to the right with a uniform speed v measured relative to a stationary frame of reference. Instead of staying in this reference frame, it is more convenient to choose as our reference frame one that moves along with the pulse with the same speed as the pulse, so that the pulse is at rest within the frame. This change of reference frame is permitted because Newton's laws are valid in either a stationary frame or one that moves with constant velocity. In our new reference frame, all elements of the string move to the left—a given element of the string initially to the right of the pulse moves to the left, rises up and follows the shape of the pulse, and then continues to move to the left. Figure 16.11a shows such an element at the instant it is located at the top of the pulse.

The small element of the string of length Δs shown in Figure 16.11a, and magnified in Figure 16.11b, forms an approximate arc of a circle of radius R. In our moving frame of reference (which is moving to the right at a speed v along with the pulse), the shaded element is moving to the left with a speed v. This element has a centripetal acceleration equal to v^2/R, which is supplied by components of the force $\mathbf{T}$ whose magnitude is the tension in the string. The force $\mathbf{T}$ acts on both sides of the element and is tangent to the arc, as shown in Figure 16.11b. The horizontal components of $\mathbf{T}$ cancel, and each vertical component $T \sin \theta$ acts radially toward the center of the arc. Hence, the total

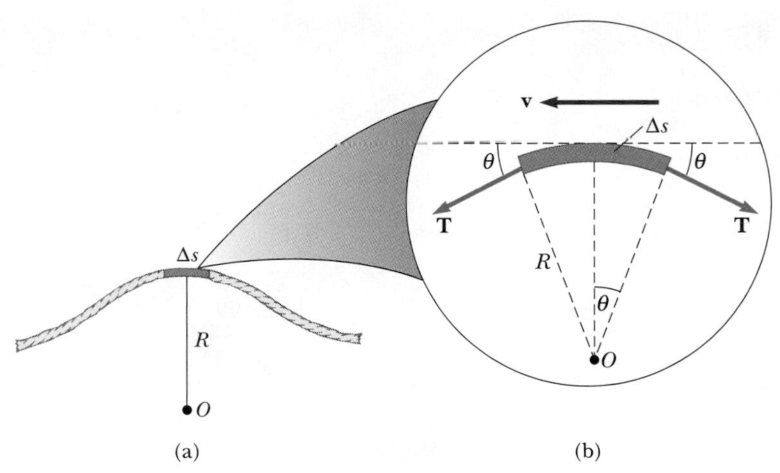

Figure 16.11 (a) To obtain the speed v of a wave on a stretched string, it is convenient to describe the motion of a small element of the string in a moving frame of reference. (b) In the moving frame of reference, the small element of length Δs moves to the left with speed v. The net force on the element is in the radial direction because the horizontal components of the tension force cancel.

radial force on the element is $2T \sin \theta$. Because the element is small, θ is small, and we can use the small-angle approximation $\sin \theta \approx \theta$. Therefore, the total radial force is

$$F_r = 2T \sin \theta \approx 2T\theta$$

The element has a mass $m = \mu \, \Delta s$. Because the element forms part of a circle and subtends an angle 2θ at the center, $\Delta s = R(2\theta)$, we find that

$$m = \mu \, \Delta s = 2\mu R\theta$$

If we apply Newton's second law to this element in the radial direction, we have

$$F_r = ma = \frac{mv^2}{R}$$

$$2T\theta = \frac{2\mu R\theta v^2}{R} \longrightarrow v = \sqrt{\frac{T}{\mu}}$$

This expression for v is Equation 16.18.

Notice that this derivation is based on the assumption that the pulse height is small relative to the length of the string. Using this assumption, we were able to use the approximation $\sin \theta \approx \theta$. Furthermore, the model assumes that the tension T is not affected by the presence of the pulse; thus, T is the same at all points on the string. Finally, this proof does *not* assume any particular shape for the pulse. Therefore, we conclude that a pulse of *any shape* travels along the string with speed $v = \sqrt{T/\mu}$ without any change in pulse shape.

Quick Quiz 16.7 Suppose you create a pulse by moving the free end of a taut string up and down once with your hand beginning at $t = 0$. The string is attached at its other end to a distant wall. The pulse reaches the wall at time t. Which of the following actions, taken by itself, decreases the time interval that it takes for the pulse to reach the wall? More than one choice may be correct. (a) moving your hand more quickly, but still only up and down once by the same amount (b) moving your hand more slowly, but still only up and down once by the same amount (c) moving your hand a greater distance up and down in the same amount of time (d) moving your hand a lesser distance up and down in the same amount of time (e) using a heavier string of the same length and under the same tension (f) using a lighter string of the same length and under the same tension (g) using a string of the same linear mass density but under decreased tension (h) using a string of the same linear mass density but under increased tension

Example 16.4 The Speed of a Pulse on a Cord

A uniform cord has a mass of 0.300 kg and a length of 6.00 m (Fig. 16.12). The cord passes over a pulley and supports a 2.00-kg object. Find the speed of a pulse traveling along this cord.

Solution The tension T in the cord is equal to the weight of the suspended 2.00-kg object:

$$T = mg = (2.00 \text{ kg})(9.80 \text{ m/s}^2) = 19.6 \text{ N}$$

(This calculation of the tension neglects the small mass of the cord. Strictly speaking, the cord can never be exactly horizontal, and therefore the tension is not uniform.) The mass per unit length μ of the cord is

$$\mu = \frac{m}{\ell} = \frac{0.300 \text{ kg}}{6.00 \text{ m}} = 0.050\,0 \text{ kg/m}$$

Therefore, the wave speed is

$$v = \sqrt{\frac{T}{\mu}} = \sqrt{\frac{19.6 \text{ N}}{0.050\,0 \text{ kg/m}}} = 19.8 \text{ m/s}$$

What If? What if the block were swinging back and forth between maximum angles of ± 20° with respect to the vertical? What range of wave speeds would this create on the horizontal cord?

Answer Figure 16.13 shows the swinging block at three positions—its highest position, its lowest position, and an arbitrary position. Summing the forces on the block in the radial direction when the block is at an arbitrary position, Newton's second law gives

$$(1) \qquad \sum F = T - mg\cos\theta = m\frac{v_{\text{block}}^2}{L}$$

where the acceleration of the block is centripetal, L is the length of the vertical piece of string, and v_{block} is the instantaneous speed of the block at the arbitrary position. Now consider conservation of mechanical energy for the block–Earth system. We define the zero of gravitational potential energy for the system when the block is at its lowest point, point Ⓒ in Figure 16.13. Equating the mechanical energy of the system when the block is at Ⓐ to the mechanical energy when the block is at an arbitrary position Ⓑ, we have,

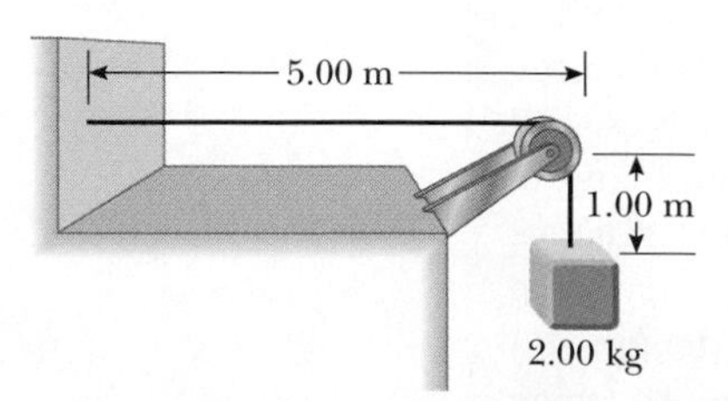

Figure 16.12 (Example 16.4) The tension T in the cord is maintained by the suspended object. The speed of any wave traveling along the cord is given by $v = \sqrt{T/\mu}$.

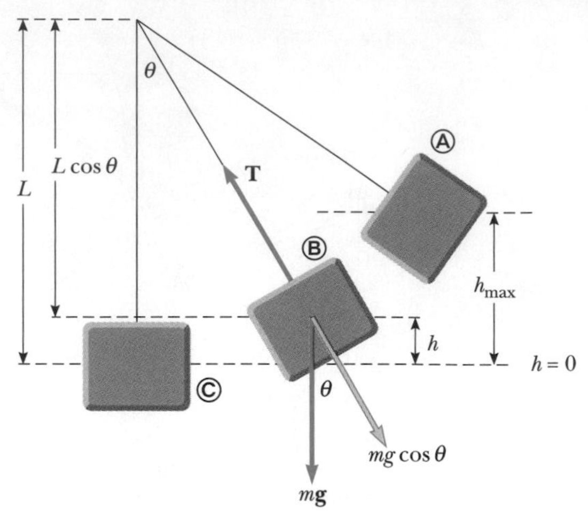

Figure 16.13 (Example 16.4) If the block swings back and forth, the tension in the cord changes, which causes a variation in the wave speed on the horizontal section of cord in Figure 16.12. The forces on the block when it is at arbitrary position Ⓑ are shown. Position Ⓐ is the highest position and Ⓒ is the lowest. (The maximum angle is exaggerated for clarity.)

$$E_{\text{A}} = E_{\text{B}}$$

$$mgh_{\text{max}} = mgh + \tfrac{1}{2}mv_{\text{block}}^2$$

$$mv_{\text{block}}^2 = 2mg(h_{\text{max}} - h)$$

Substituting this into Equation (1), we find an expression for T as a function of angle θ and height h:

$$T - mg\cos\theta = \frac{2mg(h_{\text{max}} - h)}{L}$$

$$T = mg\left[\cos\theta + \frac{2}{L}(h_{\text{max}} - h)\right]$$

The maximum value of T occurs when $\theta = 0$ and $h = 0$:

$$T_{\text{max}} = mg\left[\cos 0 + \frac{2}{L}(h_{\text{max}} - 0)\right] = mg\left(1 + \frac{2h_{\text{max}}}{L}\right)$$

The minimum value of T occurs when $h = h_{\text{max}}$ and $\theta = \theta_{\text{max}}$:

$$T_{\text{min}} = mg\left[\cos\theta_{\text{max}} + \frac{2}{L}(h_{\text{max}} - h_{\text{max}})\right] = mg\cos\theta_{\text{max}}$$

Now we find the maximum and minimum values of the wave speed v, using the fact that, as we see from Figure 16.13, h and θ are related by $h = L - L\cos\theta$:

$$v_{\text{max}} = \sqrt{\frac{T_{\text{max}}}{\mu}} = \sqrt{\frac{mg[1 + (2h_{\text{max}}/L)]}{\mu}}$$

$$= \sqrt{\frac{mg\{1 + [2(L - L\cos\theta_{\text{max}})/L]\}}{\mu}}$$

$$= \sqrt{\frac{mg(3 - 2\cos\theta_{\max})}{\mu}}$$

$$= \sqrt{\frac{(2.00 \text{ kg})(9.80 \text{ m/s}^2)(3 - 2\cos 20°)}{0.050\,0 \text{ kg/m}}} = 21.0 \text{ m/s}$$

$$v_{\min} = \sqrt{\frac{T_{\min}}{\mu}} = \sqrt{\frac{mg\cos\theta_{\max}}{\mu}}$$

$$= \sqrt{\frac{(2.00 \text{ kg})(9.80 \text{ m/s}^2)(\cos 20°)}{0.050\,0 \text{ kg/m}}} = 19.2 \text{ m/s}$$

Example 16.5 Rescuing the Hiker **Interactive**

An 80.0-kg hiker is trapped on a mountain ledge following a storm. A helicopter rescues the hiker by hovering above him and lowering a cable to him. The mass of the cable is 8.00 kg, and its length is 15.0 m. A chair of mass 70.0 kg is attached to the end of the cable. The hiker attaches himself to the chair, and the helicopter then accelerates upward. Terrified by hanging from the cable in midair, the hiker tries to signal the pilot by sending transverse pulses up the cable. A pulse takes 0.250 s to travel the length of the cable. What is the acceleration of the helicopter?

Solution To conceptualize this problem, imagine the effect of the acceleration of the helicopter on the cable. The higher the upward acceleration, the larger is the tension in the cable. In turn, the larger the tension, the higher is the speed of pulses on the cable. Thus, we categorize this problem as a combination of one involving Newton's laws and one involving the speed of pulses on a string. To analyze the problem, we use the time interval for the pulse to travel from the hiker to the helicopter to find the speed of the pulses on the cable:

$$v = \frac{\Delta x}{\Delta t} = \frac{15.0 \text{ m}}{0.250 \text{ s}} = 60.0 \text{ m/s}$$

The speed of pulses on the cable is given by Equation 16.18, which allows us to find the tension in the cable:

$$v = \sqrt{\frac{T}{\mu}} \longrightarrow T = \mu v^2 = \left(\frac{8.00 \text{ kg}}{15.0 \text{ m}}\right)(60.0 \text{ m/s})^2$$

$$T = 1.92 \times 10^3 \text{ N}$$

Newton's second law relates the tension in the cable to the acceleration of the hiker and the chair, which is the same as the acceleration of the helicopter:

$$\sum F = ma \longrightarrow T - mg = ma$$

$$a = \frac{T}{m} - g = \frac{1.92 \times 10^3 \text{ N}}{150.0 \text{ kg}} - 9.80 \text{ m/s}^2$$

$$= 3.00 \text{ m/s}^2$$

To finalize this problem, note that a real cable has stiffness in addition to tension. Stiffness tends to return a wire to its original straight-line shape even when it is not under tension. For example, a piano wire straightens if released from a curved shape; package wrapping string does not.

Stiffness represents a restoring force in addition to tension, and increases the wave speed. Consequently, for a real cable, the speed of 60.0 m/s that we determined is most likely associated with a tension lower than 1.92×10^3 N and a correspondingly smaller acceleration of the helicopter.

Investigate this situation at the Interactive Worked Example link at **http://www.pse6.com.**

16.4 Reflection and Transmission

We have discussed waves traveling through a uniform medium. We now consider how a traveling wave is affected when it encounters a change in the medium. For example, consider a pulse traveling on a string that is rigidly attached to a support at one end as in Figure 16.14. When the pulse reaches the support, a severe change in the medium occurs—the string ends. The result of this change is that the pulse undergoes **reflection**—that is, the pulse moves back along the string in the opposite direction.

Note that the reflected pulse is *inverted*. This inversion can be explained as follows. When the pulse reaches the fixed end of the string, the string produces an upward force on the support. By Newton's third law, the support must exert an equal-magnitude and oppositely directed (downward) reaction force on the string. This downward force causes the pulse to invert upon reflection.

Now consider another case: this time, the pulse arrives at the end of a string that is free to move vertically, as in Figure 16.15. The tension at the free end is maintained because the string is tied to a ring of negligible mass that is free to slide vertically on a smooth post without friction. Again, the pulse is reflected, but this time it is not inverted. When it reaches the post, the pulse exerts a force on the free end of the string, causing the ring to accelerate upward. The ring rises as high as the incoming pulse,

At the Active Figures link at **http://www.pse6.com,** *you can adjust the linear mass density of the string and the transverse direction of the initial pulse.*

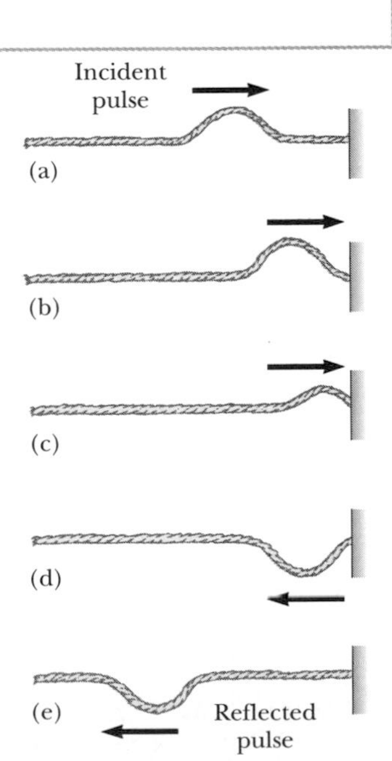

Active Figure 16.14 The reflection of a traveling pulse at the fixed end of a stretched string. The reflected pulse is inverted, but its shape is otherwise unchanged.

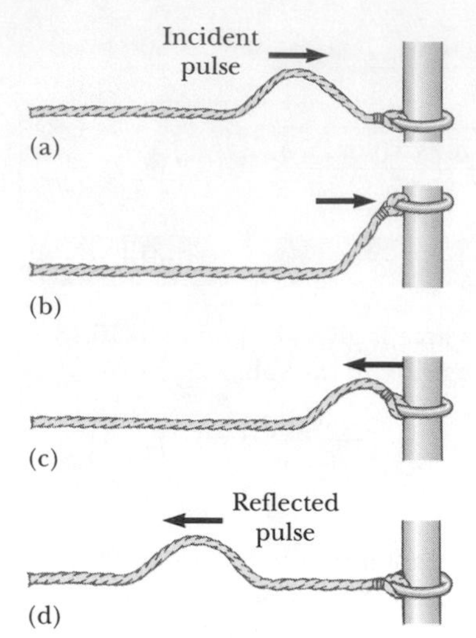

Active Figure 16.15 The reflection of a traveling pulse at the free end of a stretched string. The reflected pulse is not inverted.

At the Active Figures link at http://www.pse6.com, you can adjust the linear mass density of the string and the transverse direction of the initial pulse.

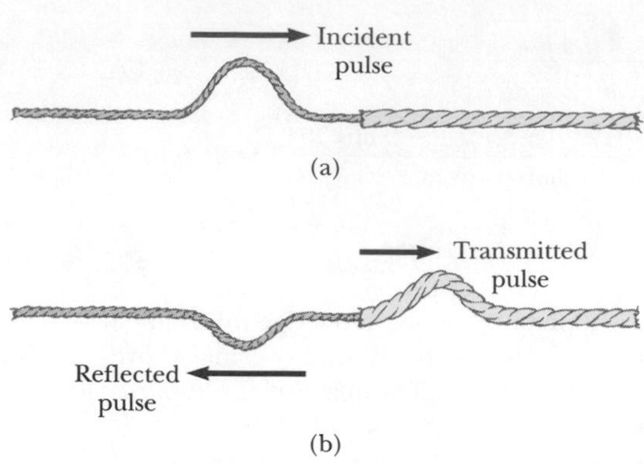

Figure 16.16 (a) A pulse traveling to the right on a light string attached to a heavier string. (b) Part of the incident pulse is reflected (and inverted), and part is transmitted to the heavier string. See Figure 16.17 for an animation available for both figures at the Active Figures link.

and then the downward component of the tension force pulls the ring back down. This movement of the ring produces a reflected pulse that is not inverted and that has the same amplitude as the incoming pulse.

Finally, we may have a situation in which the boundary is intermediate between these two extremes. In this case, part of the energy in the incident pulse is reflected and part undergoes **transmission**—that is, some of the energy passes through the boundary. For instance, suppose a light string is attached to a heavier string, as in Figure 16.16. When a pulse traveling on the light string reaches the boundary between the two, part of the pulse is reflected and inverted and part is transmitted to the heavier string. The reflected pulse is inverted for the same reasons described earlier in the case of the string rigidly attached to a support.

Note that the reflected pulse has a smaller amplitude than the incident pulse. In Section 16.5, we show that the energy carried by a wave is related to its amplitude. According to the principle of the conservation of energy, when the pulse breaks up into a reflected pulse and a transmitted pulse at the boundary, the sum of the energies of these two pulses must equal the energy of the incident pulse. Because the reflected pulse contains only part of the energy of the incident pulse, its amplitude must be smaller.

When a pulse traveling on a heavy string strikes the boundary between the heavy string and a lighter one, as in Figure 16.17, again part is reflected and part is transmitted. In this case, the reflected pulse is not inverted.

In either case, the relative heights of the reflected and transmitted pulses depend on the relative densities of the two strings. If the strings are identical, there is no discontinuity at the boundary and no reflection takes place.

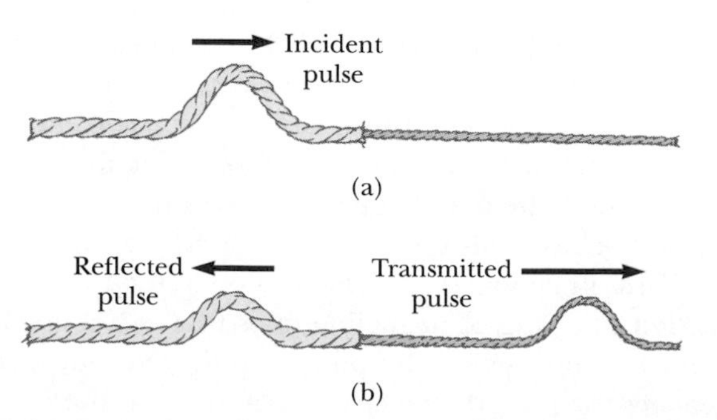

Active Figure 16.17 (a) A pulse traveling to the right on a heavy string attached to a lighter string. (b) The incident pulse is partially reflected and partially transmitted, and the reflected pulse is not inverted.

At the Active Figures link at http://www.pse6.com, you can adjust the linear mass densities of the strings and the transverse direction of the initial pulse.

According to Equation 16.18, the speed of a wave on a string increases as the mass per unit length of the string decreases. In other words, a wave travels more slowly on a heavy string than on a light string if both are under the same tension. The following general rules apply to reflected waves: **when a wave or pulse travels from medium A to medium B and $v_A > v_B$ (that is, when B is denser than A), it is inverted upon reflection. When a wave or pulse travels from medium A to medium B and $v_A < v_B$ (that is, when A is denser than B), it is not inverted upon reflection.**

16.5 Rate of Energy Transfer by Sinusoidal Waves on Strings

Waves transport energy when they propagate through a medium. We can easily demonstrate this by hanging an object on a stretched string and then sending a pulse down the string, as in Figure 16.18a. When the pulse meets the suspended object, the object is momentarily displaced upward, as in Figure 16.18b. In the process, energy is transferred to the object and appears as an increase in the gravitational potential energy of the object–Earth system. This section examines the rate at which energy is transported along a string. We shall assume a one-dimensional sinusoidal wave in the calculation of the energy transferred.

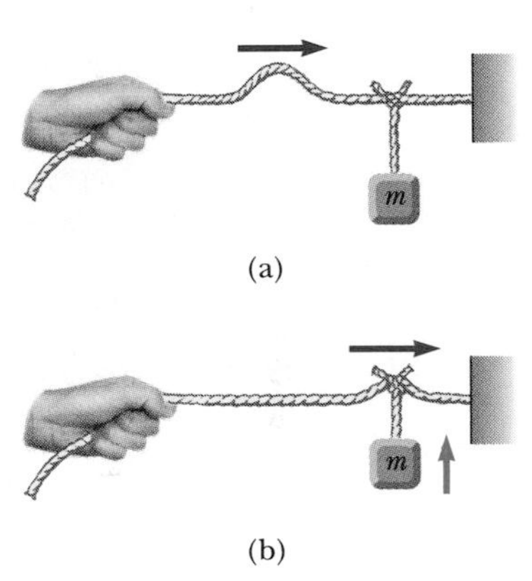

Figure 16.18 (a) A pulse traveling to the right on a stretched string that has an object suspended from it. (b) Energy is transmitted to the suspended object when the pulse arrives.

Consider a sinusoidal wave traveling on a string (Fig. 16.19). The source of the energy is some external agent at the left end of the string, which does work in producing the oscillations. We can consider the string to be a nonisolated system. As the external agent performs work on the end of the string, moving it up and down, energy enters the system of the string and propagates along its length. Let us focus our attention on an element of the string of length Δx and mass Δm. Each such element moves vertically with simple harmonic motion. Thus, we can model each element of the string as a simple harmonic oscillator, with the oscillation in the y direction. All elements have the same angular frequency ω and the same amplitude A. The kinetic energy K associated with a moving particle is $K = \frac{1}{2}mv^2$. If we apply this equation to an element of length Δx and mass Δm, we see that the kinetic energy ΔK of this element is

$$\Delta K = \tfrac{1}{2}(\Delta m)\, v_y^{\,2}$$

where v_y is the transverse speed of the element. If μ is the mass per unit length of the string, then the mass Δm of the element of length Δx is equal to $\mu\, \Delta x$. Hence, we can express the kinetic energy of an element of the string as

$$\Delta K = \tfrac{1}{2}(\mu\, \Delta x)\, v_y^{\,2} \qquad (16.19)$$

As the length of the element of the string shrinks to zero, this becomes a differential relationship:

$$dK = \tfrac{1}{2}(\mu\, dx)\, v_y^{\,2}$$

We substitute for the general transverse speed of a simple harmonic oscillator using Equation 16.14:

$$\begin{aligned} dK &= \tfrac{1}{2}\mu[\omega A \cos(kx - \omega t)]^2\, dx \\ &= \tfrac{1}{2}\mu\omega^2 A^2 \cos^2(kx - \omega t)\, dx \end{aligned}$$

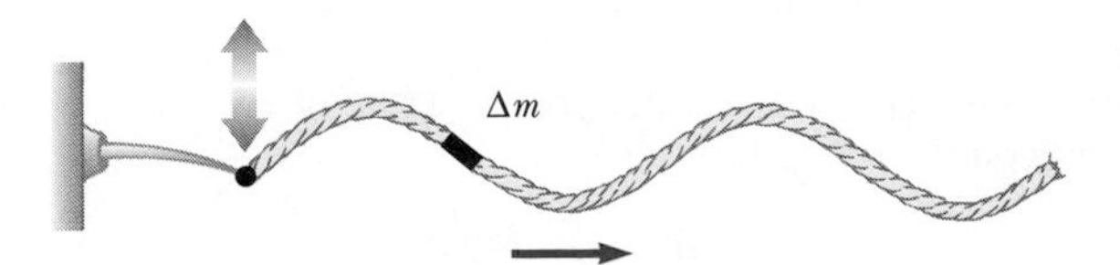

Figure 16.19 A sinusoidal wave traveling along the x axis on a stretched string. Every element moves vertically, and every element has the same total energy.

If we take a snapshot of the wave at time $t = 0$, then the kinetic energy of a given element is

$$dK = \tfrac{1}{2}\mu\omega^2 A^2 \cos^2 kx \, dx$$

Let us integrate this expression over all the string elements in a wavelength of the wave, which will give us the total kinetic energy K_λ in one wavelength:

$$K_\lambda = \int dK = \int_0^\lambda \tfrac{1}{2}\mu\omega^2 A^2 \cos^2 kx \, dx = \tfrac{1}{2}\mu\omega^2 A^2 \int_0^\lambda \cos^2 kx \, dx$$

$$= \tfrac{1}{2}\mu\omega^2 A^2 \left[\tfrac{1}{2}x + \frac{1}{4k}\sin 2kx\right]_0^\lambda = \tfrac{1}{2}\mu\omega^2 A^2 \left[\tfrac{1}{2}\lambda\right] = \tfrac{1}{4}\mu\omega^2 A^2 \lambda$$

In addition to kinetic energy, each element of the string has potential energy associated with it due to its displacement from the equilibrium position and the restoring forces from neighboring elements. A similar analysis to that above for the total potential energy U_λ in one wavelength will give exactly the same result:

$$U_\lambda = \tfrac{1}{4}\mu\omega^2 A^2 \lambda$$

The total energy in one wavelength of the wave is the sum of the potential and kinetic energies:

$$E_\lambda = U_\lambda + K_\lambda = \tfrac{1}{2}\mu\omega^2 A^2 \lambda \qquad (16.20)$$

As the wave moves along the string, this amount of energy passes by a given point on the string during a time interval of one period of the oscillation. Thus, the power, or rate of energy transfer, associated with the wave is

$$\mathcal{P} = \frac{\Delta E}{\Delta t} = \frac{E_\lambda}{T} = \frac{\tfrac{1}{2}\mu\omega^2 A^2 \lambda}{T} = \tfrac{1}{2}\mu\omega^2 A^2 \left(\frac{\lambda}{T}\right)$$

Power of a wave

$$\mathcal{P} = \tfrac{1}{2}\mu\omega^2 A^2 v \qquad (16.21)$$

This expression shows that the rate of energy transfer by a sinusoidal wave on a string is proportional to (a) the square of the frequency, (b) the square of the amplitude, and (c) the wave speed. In fact: **the rate of energy transfer in any sinusoidal wave is proportional to the square of the angular frequency and to the square of the amplitude.**

Quick Quiz 16.8 Which of the following, taken by itself, would be most effective in increasing the rate at which energy is transferred by a wave traveling along a string? (a) reducing the linear mass density of the string by one half (b) doubling the wavelength of the wave (c) doubling the tension in the string (d) doubling the amplitude of the wave

Example 16.6 Power Supplied to a Vibrating String

A taut string for which $\mu = 5.00 \times 10^{-2}$ kg/m is under a tension of 80.0 N. How much power must be supplied to the string to generate sinusoidal waves at a frequency of 60.0 Hz and an amplitude of 6.00 cm?

Solution The wave speed on the string is, from Equation 16.18,

$$v = \sqrt{\frac{T}{\mu}} = \sqrt{\frac{80.0 \text{ N}}{5.00 \times 10^{-2} \text{ kg/m}}} = 40.0 \text{ m/s}$$

Because $f = 60.0$ Hz, the angular frequency ω of the sinusoidal waves on the string has the value

$$\omega = 2\pi f = 2\pi(60.0 \text{ Hz}) = 377 \text{ s}^{-1}$$

Using these values in Equation 16.21 for the power, with $A = 6.00 \times 10^{-2}$ m, we obtain

$$\begin{aligned}\mathcal{P} &= \tfrac{1}{2}\mu\omega^2 A^2 v \\ &= \tfrac{1}{2}(5.00 \times 10^{-2}\ \text{kg/m})(377\ \text{s}^{-1})^2 \\ &\quad \times (6.00 \times 10^{-2}\ \text{m})^2(40.0\ \text{m/s}) \\ &= 512\ \text{W}\end{aligned}$$

What If? What if the string is to transfer energy at a rate of 1 000 W? What must be the required amplitude if all other parameters remain the same?

Answer We set up a ratio of the new and old power, reflecting only a change in the amplitude:

$$\frac{\mathcal{P}_{\text{new}}}{\mathcal{P}_{\text{old}}} = \frac{\frac{1}{2}\mu\omega^2 A_{\text{new}}^2 v}{\frac{1}{2}\mu\omega^2 A_{\text{old}}^2 v} = \frac{A_{\text{new}}^2}{A_{\text{old}}^2}$$

Solving for the new amplitude,

$$\begin{aligned}A_{\text{new}} &= A_{\text{old}}\sqrt{\frac{\mathcal{P}_{\text{new}}}{\mathcal{P}_{\text{old}}}} = (6.00\ \text{cm})\sqrt{\frac{1\,000\ \text{W}}{512\ \text{W}}} \\ &= 8.39\ \text{cm}\end{aligned}$$

16.6 The Linear Wave Equation

In Section 16.1 we introduced the concept of the wave function to represent waves traveling on a string. All wave functions $y(x, t)$ represent solutions of an equation called the *linear wave equation.* This equation gives a complete description of the wave motion, and from it one can derive an expression for the wave speed. Furthermore, the linear wave equation is basic to many forms of wave motion. In this section, we derive this equation as applied to waves on strings.

Suppose a traveling wave is propagating along a string that is under a tension T. Let us consider one small string element of length Δx (Fig. 16.20). The ends of the element make small angles θ_A and θ_B with the x axis. The net force acting on the element in the vertical direction is

$$\sum F_y = T \sin\theta_B - T\sin\theta_A = T(\sin\theta_B - \sin\theta_A)$$

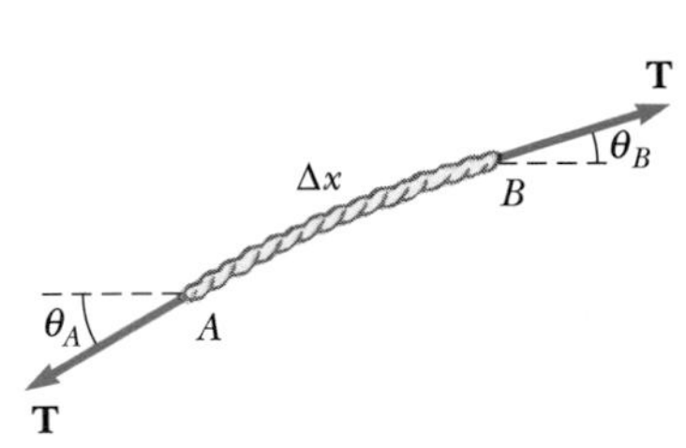

Figure 16.20 An element of a string under tension T.

Because the angles are small, we can use the small-angle approximation $\sin\theta \approx \tan\theta$ to express the net force as

$$\sum F_y \approx T(\tan\theta_B - \tan\theta_A) \qquad (16.22)$$

Imagine undergoing an infinitesimal displacement outward from the end of the rope element in Figure 16.20 along the blue line representing the force **T**. This displacement has infinitesimal x and y components and can be represented by the vector $dx\hat{\mathbf{i}} + dy\hat{\mathbf{j}}$. The tangent of the angle with respect to the x axis for this displacement is dy/dx. Because we are evaluating this tangent at a particular instant of time, we need to express this in partial form as $\partial y/\partial x$. Substituting for the tangents in Equation 16.22 gives

$$\sum F_y \approx T\left[\left(\frac{\partial y}{\partial x}\right)_B - \left(\frac{\partial y}{\partial x}\right)_A\right] \qquad (16.23)$$

We now apply Newton's second law to the element, with the mass of the element given by $m = \mu\,\Delta x$:

$$\sum F_y = ma_y = \mu\,\Delta x\left(\frac{\partial^2 y}{\partial t^2}\right) \qquad (16.24)$$

Combining Equation 16.23 with Equation 16.24, we obtain

$$\mu\,\Delta x\left(\frac{\partial^2 y}{\partial t^2}\right) = T\left[\left(\frac{\partial y}{\partial x}\right)_B - \left(\frac{\partial y}{\partial x}\right)_A\right]$$

$$\frac{\mu}{T}\frac{\partial^2 y}{\partial t^2} = \frac{(\partial y/\partial x)_B - (\partial y/dx)_A}{\Delta x} \qquad (16.25)$$

The right side of this equation can be expressed in a different form if we note that the partial derivative of any function is defined as

$$\frac{\partial f}{\partial x} \equiv \lim_{\Delta x \to 0} \frac{f(x + \Delta x) - f(x)}{\Delta x}$$

If we associate $f(x + \Delta x)$ with $(\partial y/\partial x)_B$ and $f(x)$ with $(\partial y/\partial x)_A$, we see that, in the limit $\Delta x \to 0$, Equation 16.25 becomes

Linear wave equation for a string

$$\frac{\mu}{T}\frac{\partial^2 y}{\partial t^2} = \frac{\partial^2 y}{\partial x^2} \qquad (16.26)$$

This is the linear wave equation as it applies to waves on a string.

We now show that the sinusoidal wave function (Eq. 16.10) represents a solution of the linear wave equation. If we take the sinusoidal wave function to be of the form $y(x, t) = A \sin(kx - \omega t)$, then the appropriate derivatives are

$$\frac{\partial^2 y}{\partial t^2} = -\omega^2 A \sin(kx - \omega t)$$

$$\frac{\partial^2 y}{\partial x^2} = -k^2 A \sin(kx - \omega t)$$

Substituting these expressions into Equation 16.26, we obtain

$$-\frac{\mu\omega^2}{T}\sin(kx - \omega t) = -k^2 \sin(kx - \omega t)$$

This equation must be true for all values of the variables x and t in order for the sinusoidal wave function to be a solution of the wave equation. Both sides of the equation depend on x and t through the same function $\sin(kx - \omega t)$. Because this function divides out, we do indeed have an identity, provided that

$$k^2 = \frac{\mu}{T}\omega^2$$

Using the relationship $v = \omega/k$ (Eq. 16.11) in this expression, we see that

$$v^2 = \frac{\omega^2}{k^2} = \frac{T}{\mu}$$

$$v = \sqrt{\frac{T}{\mu}}$$

which is Equation 16.18. This derivation represents another proof of the expression for the wave speed on a taut string.

The linear wave equation (Eq. 16.26) is often written in the form

Linear wave equation in general

$$\frac{\partial^2 y}{\partial x^2} = \frac{1}{v^2}\frac{\partial^2 y}{\partial t^2} \qquad (16.27)$$

This expression applies in general to various types of traveling waves. For waves on strings, y represents the vertical position of elements of the string. For sound waves, y corresponds to longitudinal position of elements of air from equilibrium or variations in either the pressure or the density of the gas through which the sound waves are propagating. In the case of electromagnetic waves, y corresponds to electric or magnetic field components.

We have shown that the sinusoidal wave function (Eq. 16.10) is one solution of the linear wave equation (Eq. 16.27). Although we do not prove it here, the linear wave equation is satisfied by *any* wave function having the form $y = f(x \pm vt)$. Furthermore, we have seen that the linear wave equation is a direct consequence of Newton's second law applied to any element of a string carrying a traveling wave.

SUMMARY

Take a practice test for this chapter by clicking on the Practice Test link at http://www.pse6.com.

A **transverse wave** is one in which the elements of the medium move in a direction *perpendicular* to the direction of propagation. An example is a wave on a taut string. A **longitudinal wave** is one in which the elements of the medium move in a direction *parallel* to the direction of propagation. Sound waves in fluids are longitudinal.

Any one-dimensional wave traveling with a speed v in the x direction can be represented by a wave function of the form

$$y(x, t) = f(x \pm vt) \qquad (16.1, 16.2)$$

where the positive sign applies to a wave traveling in the negative x direction and the negative sign applies to a wave traveling in the positive x direction. The shape of the wave at any instant in time (a snapshot of the wave) is obtained by holding t constant.

The **wave function** for a one-dimensional sinusoidal wave traveling to the right can be expressed as

$$y = A \sin\left[\frac{2\pi}{\lambda}(x - vt)\right] = A \sin(kx - \omega t) \qquad (16.5, 16.10)$$

where A is the **amplitude,** λ is the **wavelength,** k is the **angular wave number,** and ω is the **angular frequency.** If T is the **period** and f the **frequency,** v, k, and ω can be written

$$v = \frac{\lambda}{T} = \lambda f \qquad (16.6, 16.12)$$

$$k \equiv \frac{2\pi}{\lambda} \qquad (16.8)$$

$$\omega \equiv \frac{2\pi}{T} = 2\pi f \qquad (16.3, 16.9)$$

The speed of a wave traveling on a taut string of mass per unit length μ and tension T is

$$v = \sqrt{\frac{T}{\mu}} \qquad (16.18)$$

A wave is totally or partially reflected when it reaches the end of the medium in which it propagates or when it reaches a boundary where its speed changes discontinuously. If a wave traveling on a string meets a fixed end, the wave is reflected and inverted. If the wave reaches a free end, it is reflected but not inverted.

The **power** transmitted by a sinusoidal wave on a stretched string is

$$\mathcal{P} = \tfrac{1}{2}\mu\omega^2 A^2 v \qquad (16.21)$$

Wave functions are solutions to a differential equation called the **linear wave equation:**

$$\frac{\partial^2 y}{\partial x^2} = \frac{1}{v^2}\frac{\partial^2 y}{\partial t^2} \qquad (16.27)$$

QUESTIONS

1. Why is a pulse on a string considered to be transverse?

2. How would you create a longitudinal wave in a stretched spring? Would it be possible to create a transverse wave in a spring?

3. By what factor would you have to multiply the tension in a stretched string in order to double the wave speed?

4. When traveling on a taut string, does a pulse always invert upon reflection? Explain.

5. Does the vertical speed of a segment of a horizontal taut string, through which a wave is traveling, depend on the wave speed?

6. If you shake one end of a taut rope steadily three times each second, what would be the period of the sinusoidal wave set up in the rope?

7. A vibrating source generates a sinusoidal wave on a string under constant tension. If the power delivered to the string is doubled, by what factor does the amplitude change? Does the wave speed change under these circumstances?

8. Consider a wave traveling on a taut rope. What is the difference, if any, between the speed of the wave and the speed of a small segment of the rope?

9. If a long rope is hung from a ceiling and waves are sent up the rope from its lower end, they do not ascend with constant speed. Explain.

10. How do transverse waves differ from longitudinal waves?

11. When all the strings on a guitar are stretched to the same tension, will the speed of a wave along the most massive bass string be faster, slower, or the same as the speed of a wave on the lighter strings?

12. If one end of a heavy rope is attached to one end of a light rope, the speed of a wave will change as the wave goes from the heavy rope to the light one. Will it increase or decrease? What happens to the frequency? To the wavelength?

13. If you stretch a rubber hose and pluck it, you can observe a pulse traveling up and down the hose. What happens to the speed of the pulse if you stretch the hose more tightly? What happens to the speed if you fill the hose with water?

14. In a longitudinal wave in a spring, the coils move back and forth in the direction of wave motion. Does the speed of the wave depend on the maximum speed of each coil?

15. Both longitudinal and transverse waves can propagate through a solid. A wave on the surface of a liquid can involve both longitudinal and transverse motion of elements of the medium. On the other hand, a wave propagating through the volume of a fluid must be purely longitudinal, not transverse. Why?

16. In an earthquake both S (transverse) and P (longitudinal) waves propagate from the focus of the earthquake. The focus is in the ground below the epicenter on the surface. The S waves travel through the Earth more slowly than the P waves (at about 5 km/s versus 8 km/s). By detecting the time of arrival of the waves, how can one determine the distance to the focus of the quake? How many detection stations are necessary to locate the focus unambiguously?

17. In mechanics, massless strings are often assumed. Why is this not a good assumption when discussing waves on strings?

PROBLEMS

1, 2, 3 = straightforward, intermediate, challenging □ = full solution available in the *Student Solutions Manual and Study Guide*

= coached solution with hints available at http://www.pse6.com = computer useful in solving problem

= paired numerical and symbolic problems

Section 16.1 Propagation of a Disturbance

1. At $t = 0$, a transverse pulse in a wire is described by the function

$$y = \frac{6}{x^2 + 3}$$

where x and y are in meters. Write the function $y(x, t)$ that describes this pulse if it is traveling in the positive x direction with a speed of 4.50 m/s.

2. Ocean waves with a crest-to-crest distance of 10.0 m can be described by the wave function

$$y(x, t) = (0.800 \text{ m}) \sin[0.628(x - vt)]$$

where $v = 1.20$ m/s. (a) Sketch $y(x, t)$ at $t = 0$. (b) Sketch $y(x, t)$ at $t = 2.00$ s. Note that the entire wave form has shifted 2.40 m in the positive x direction in this time interval.

3. A pulse moving along the x axis is described by

$$y(x, t) = 5.00e^{-(x+5.00t)^2}$$

where x is in meters and t is in seconds. Determine (a) the direction of the wave motion, and (b) the speed of the pulse.

4. Two points A and B on the surface of the Earth are at the same longitude and 60.0° apart in latitude. Suppose that an earthquake at point A creates a P wave that reaches point B by traveling straight through the body of the Earth at a constant speed of 7.80 km/s. The earthquake also radiates a *Rayleigh wave,* which travels across the surface of the Earth in an analogous way to a surface wave on water, at 4.50 km/s. (a) Which of these two seismic waves arrives at B first? (b) What is the time difference between the arrivals of the two waves at B? Take the radius of the Earth to be 6 370 km.

5. S and P waves, simultaneously radiated from the hypocenter of an earthquake, are received at a seismographic station 17.3 s apart. Assume the waves have traveled over the same path at speeds of 4.50 km/s and 7.80 km/s. Find the distance from the seismograph to the hypocenter of the quake.

Section 16.2 Sinusoidal Waves

6. For a certain transverse wave, the distance between two successive crests is 1.20 m, and eight crests pass a given point along the direction of travel every 12.0 s. Calculate the wave speed.

7. A sinusoidal wave is traveling along a rope. The oscillator that generates the wave completes 40.0 vibrations in 30.0 s. Also, a given maximum travels 425 cm along the rope in 10.0 s. What is the wavelength?

8. When a particular wire is vibrating with a frequency of 4.00 Hz, a transverse wave of wavelength 60.0 cm is produced. Determine the speed of waves along the wire.

9. A wave is described by $y = (2.00 \text{ cm}) \sin(kx - \omega t)$, where $k = 2.11$ rad/m, $\omega = 3.62$ rad/s, x is in meters, and t is in seconds. Determine the amplitude, wavelength, frequency, and speed of the wave.

10. A sinusoidal wave on a string is described by

$$y = (0.51 \text{ cm}) \sin(kx - \omega t)$$

where $k = 3.10$ rad/cm and $\omega = 9.30$ rad/s. How far does a wave crest move in 10.0 s? Does it move in the positive or negative x direction?

11. Consider further the string shown in Figure 16.10 and treated in Example 16.3. Calculate (a) the maximum transverse speed and (b) the maximum transverse acceleration of a point on the string.

12. Consider the sinusoidal wave of Example 16.2, with the wave function

$$y = (15.0 \text{ cm}) \cos(0.157x - 50.3t).$$

At a certain instant, let point A be at the origin and point B be the first point along the x axis where the wave is 60.0° out of phase with point A. What is the coordinate of point B?

13. A sinusoidal wave is described by

$$y = (0.25 \text{ m}) \sin(0.30x - 40t)$$

where x and y are in meters and t is in seconds. Determine for this wave the (a) amplitude, (b) angular frequency, (c) angular wave number, (d) wavelength, (e) wave speed, and (f) direction of motion.

14. (a) Plot y versus t at $x = 0$ for a sinusoidal wave of the form $y = (15.0 \text{ cm}) \cos(0.157x - 50.3t)$, where x and y are in centimeters and t is in seconds. (b) Determine the period of vibration from this plot and compare your result with the value found in Example 16.2.

15. (a) Write the expression for y as a function of x and t for a sinusoidal wave traveling along a rope in the *negative* x direction with the following characteristics: $A = 8.00$ cm, $\lambda = 80.0$ cm, $f = 3.00$ Hz, and $y(0, t) = 0$ at $t = 0$. (b) **What If?** Write the expression for y as a function of x and t for the wave in part (a) assuming that $y(x, 0) = 0$ at the point $x = 10.0$ cm.

16. A sinusoidal wave traveling in the $-x$ direction (to the left) has an amplitude of 20.0 cm, a wavelength of 35.0 cm, and a frequency of 12.0 Hz. The transverse position of an element of the medium at $t = 0$, $x = 0$ is $y = -3.00$ cm, and the element has a positive velocity here. (a) Sketch the wave at $t = 0$. (b) Find the angular wave number, period, angular frequency, and wave speed of the wave. (c) Write an expression for the wave function $y(x, t)$.

17. A transverse wave on a string is described by the wave function

$$y = (0.120 \text{ m}) \sin[(\pi x/8) + 4\pi t]$$

(a) Determine the transverse speed and acceleration at $t = 0.200$ s for the point on the string located at $x = 1.60$ m. (b) What are the wavelength, period, and speed of propagation of this wave?

18. A transverse sinusoidal wave on a string has a period $T = 25.0$ ms and travels in the negative x direction with a speed of 30.0 m/s. At $t = 0$, a particle on the string at $x = 0$ has a transverse position of 2.00 cm and is traveling downward with a speed of 2.00 m/s. (a) What is the amplitude of the wave? (b) What is the initial phase angle? (c) What is the maximum transverse speed of the string? (d) Write the wave function for the wave.

19. A sinusoidal wave of wavelength 2.00 m and amplitude 0.100 m travels on a string with a speed of 1.00 m/s to the right. Initially, the left end of the string is at the origin. Find (a) the frequency and angular frequency, (b) the angular wave number, and (c) the wave function for this wave. Determine the equation of motion for (d) the left end of the string and (e) the point on the string at $x = 1.50$ m to the right of the left end. (f) What is the maximum speed of any point on the string?

20. A wave on a string is described by the wave function $y = (0.100 \text{ m}) \sin(0.50x - 20t)$. (a) Show that a particle in the string at $x = 2.00$ m executes simple harmonic motion. (b) Determine the frequency of oscillation of this particular point.

Section 16.3 The Speed of Waves on Strings

21. A telephone cord is 4.00 m long. The cord has a mass of 0.200 kg. A transverse pulse is produced by plucking one end of the taut cord. The pulse makes four trips down and back along the cord in 0.800 s. What is the tension in the cord?

22. Transverse waves with a speed of 50.0 m/s are to be produced in a taut string. A 5.00-m length of string with a total mass of 0.060 0 kg is used. What is the required tension?

23. A piano string having a mass per unit length equal to 5.00×10^{-3} kg/m is under a tension of 1 350 N. Find the speed of a wave traveling on this string.

24. A transverse traveling wave on a taut wire has an amplitude of 0.200 mm and a frequency of 500 Hz. It travels with a speed of 196 m/s. (a) Write an equation in SI units of the form $y = A \sin(kx - \omega t)$ for this wave. (b) The mass per unit length of this wire is 4.10 g/m. Find the tension in the wire.

25. An astronaut on the Moon wishes to measure the local value of the free-fall acceleration by timing pulses traveling down a wire that has an object of large mass suspended from it. Assume a wire has a mass of 4.00 g and a length of 1.60 m, and that a 3.00-kg object is suspended from it. A pulse requires 36.1 ms to traverse the length of the wire. Calculate g_{Moon} from these data. (You may ignore the mass of the wire when calculating the tension in it.)

26. Transverse pulses travel with a speed of 200 m/s along a taut copper wire whose diameter is 1.50 mm. What is the tension in the wire? (The density of copper is 8.92 g/cm^3.)

27. Transverse waves travel with a speed of 20.0 m/s in a string under a tension of 6.00 N. What tension is required for a wave speed of 30.0 m/s in the same string?

28. A simple pendulum consists of a ball of mass M hanging from a uniform string of mass m and length L, with $m \ll M$. If the period of oscillations for the pendulum is T, determine the speed of a transverse wave in the string when the pendulum hangs at rest.

29. The elastic limit of the steel forming a piece of wire is equal to 2.70×10^8 Pa. What is the maximum speed at which transverse wave pulses can propagate along this wire without exceeding this stress? (The density of steel is 7.86×10^3 km/m^3.)

30. **Review problem.** A light string with a mass per unit length of 8.00 g/m has its ends tied to two walls separated by a distance equal to three fourths of the length of the string (Fig. P16.30). An object of mass m is suspended from the center of the string, putting a tension in the string. (a) Find an expression for the transverse wave speed in the string as a function of the mass of the hanging object. (b) What should be the mass of the object suspended from the string in order to produce a wave speed of 60.0 m/s?

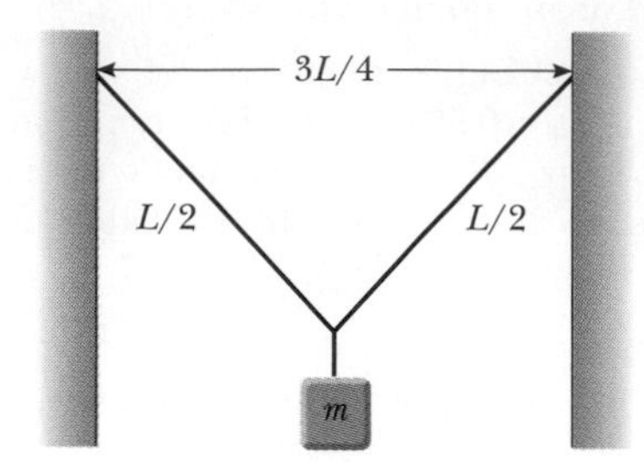

Figure P16.30

31. A 30.0-m steel wire and a 20.0-m copper wire, both with 1.00-mm diameters, are connected end to end and stretched to a tension of 150 N. How long does it take a transverse wave to travel the entire length of the two wires?

32. **Review problem.** A light string of mass m and length L has its ends tied to two walls that are separated by the distance D. Two objects, each of mass M, are suspended from the string as in Figure P16.32. If a wave pulse is sent from point A, how long does it take to travel to point B?

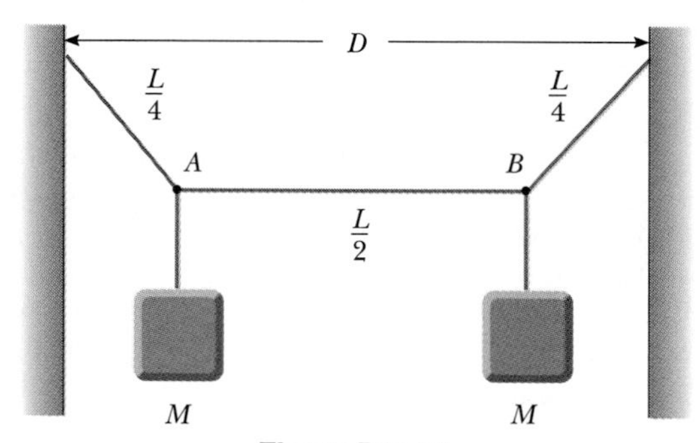

Figure P16.32

33. A student taking a quiz finds on a reference sheet the two equations

$$f = 1/T \quad \text{and} \quad v = \sqrt{T/\mu}$$

She has forgotten what T represents in each equation. (a) Use dimensional analysis to determine the units required for T in each equation. (b) Identify the physical quantity each T represents.

Section 16.5 Rate of Energy Transfer by Sinusoidal Waves on Strings

34. A taut rope has a mass of 0.180 kg and a length of 3.60 m. What power must be supplied to the rope in order to generate sinusoidal waves having an amplitude of 0.100 m and a wavelength of 0.500 m and traveling with a speed of 30.0 m/s?

35. A two-dimensional water wave spreads in circular ripples. Show that the amplitude A at a distance r from the initial disturbance is proportional to $1/\sqrt{r}$. (*Suggestion*: Consider the energy carried by one outward-moving ripple.)

36. Transverse waves are being generated on a rope under constant tension. By what factor is the required power increased or decreased if (a) the length of the rope is doubled and the angular frequency remains constant, (b) the amplitude is doubled and the angular frequency is halved, (c) both the wavelength and the amplitude are doubled, and (d) both the length of the rope and the wavelength are halved?

37. Sinusoidal waves 5.00 cm in amplitude are to be transmitted along a string that has a linear mass density of 4.00×10^{-2} kg/m. If the source can deliver a maximum power of 300 W and the string is under a tension of 100 N, what is the highest frequency at which the source can operate?

38. It is found that a 6.00-m segment of a long string contains four complete waves and has a mass of 180 g. The string is vibrating sinusoidally with a frequency of 50.0 Hz and a peak-to-valley distance of 15.0 cm. (The "peak-to-valley" distance is the vertical distance from the farthest positive position to the farthest negative position.) (a) Write the function that describes this wave traveling in the positive x direction. (b) Determine the power being supplied to the string.

39. A sinusoidal wave on a string is described by the equation

$$y = (0.15 \text{ m}) \sin(0.80x - 50t)$$

where x and y are in meters and t is in seconds. If the mass per unit length of this string is 12.0 g/m, determine (a) the speed of the wave, (b) the wavelength, (c) the frequency, and (d) the power transmitted to the wave.

40. The wave function for a wave on a taut string is

$$y(x, t) = (0.350 \text{ m}) \sin(10\pi t - 3\pi x + \pi/4)$$

where x is in meters and t in seconds. (a) What is the average rate at which energy is transmitted along the string if the linear mass density is 75.0 g/m? (b) What is the energy contained in each cycle of the wave?

41. A horizontal string can transmit a maximum power $\mathcal{P}_0$ (without breaking) if a wave with amplitude A and angular frequency ω is traveling along it. In order to increase this maximum power, a student folds the string and uses this "double string" as a medium. Determine the maximum power that can be transmitted along the "double string," assuming that the tension is constant.

42. In a region far from the epicenter of an earthquake, a seismic wave can be modeled as transporting energy in a single direction without absorption, just as a string wave does. Suppose the seismic wave moves from granite into mudfill with similar density but with a much lower bulk modulus. Assume the speed of the wave gradually drops by a factor of 25.0, with negligible reflection of the wave. Will the amplitude of the ground shaking increase or decrease? By

what factor? This phenomenon led to the collapse of part of the Nimitz Freeway in Oakland, California, during the Loma Prieta earthquake of 1989.

Section 16.6 The Linear Wave Equation

43. (a) Evaluate A in the scalar equality $(7 + 3)4 = A$. (b) Evaluate A, B, and C in the vector equality $7.00\hat{\mathbf{i}} + 3.00\hat{\mathbf{k}} = A\hat{\mathbf{i}} + B\hat{\mathbf{j}} + C\hat{\mathbf{k}}$. Explain how you arrive at the answers to convince a student who thinks that you cannot solve a single equation for three different unknowns. (c) **What If?** The functional equality or identity

$$A + B\cos(Cx + Dt + E) = (7.00 \text{ mm}) \cos(3x + 4t + 2)$$

is true for all values of the variables x and t, which are measured in meters and in seconds, respectively. Evaluate the constants A, B, C, D, and E. Explain how you arrive at the answers.

44. Show that the wave function $y = e^{b(x-vt)}$ is a solution of the linear wave equation (Eq. 16.27), where b is a constant.

45. Show that the wave function $y = \ln[b(x - vt)]$ is a solution to Equation 16.27, where b is a constant.

46. (a) Show that the function $y(x, t) = x^2 + v^2t^2$ is a solution to the wave equation. (b) Show that the function in part (a) can be written as $f(x + vt) + g(x - vt)$, and determine the functional forms for f and g. (c) **What If?** Repeat parts (a) and (b) for the function $y(x, t) = \sin(x)\cos(vt)$.

Additional Problems

47. "The wave" is a particular type of pulse that can propagate through a large crowd gathered at a sports arena to watch a soccer or American football match (Figure P16.47). The elements of the medium are the spectators, with zero position corresponding to their being seated and maximum position corresponding to their standing and raising their arms. When a large fraction of the spectators participate in the wave motion, a somewhat stable pulse shape can develop. The wave speed depends on people's reaction time, which is typically on the order of 0.1 s. Estimate the order of magnitude, in minutes, of the time required for such a pulse to make one circuit around a large sports stadium. State the quantities you measure or estimate and their values.

Gregg Adams/Getty Images

Figure P16.47

48. A traveling wave propagates according to the expression $y = (4.0 \text{ cm}) \sin(2.0x - 3.0t)$, where x is in centimeters and t is in seconds. Determine (a) the amplitude, (b) the wavelength, (c) the frequency, (d) the period, and (e) the direction of travel of the wave.

49. The wave function for a traveling wave on a taut string is (in SI units)

$$y(x, t) = (0.350 \text{ m}) \sin(10\pi t - 3\pi x + \pi/4)$$

(a) What are the speed and direction of travel of the wave? (b) What is the vertical position of an element of the string at $t = 0$, $x = 0.100$ m? (c) What are the wavelength and frequency of the wave? (d) What is the maximum magnitude of the transverse speed of the string?

50. A transverse wave on a string is described by the equation

$$y(x, t) = (0.350 \text{ m}) \sin[(1.25 \text{ rad/m})x + (99.6 \text{ rad/s})t]$$

Consider the element of the string at $x = 0$. (a) What is the time interval between the first two instants when this element has a position of $y = 0.175$ m? (b) What distance does the wave travel during this time interval?

51. Motion picture film is projected at 24.0 frames per second. Each frame is a photograph 19.0 mm high. At what constant speed does the film pass into the projector?

52. Review problem. A block of mass M, supported by a string, rests on an incline making an angle θ with the horizontal (Fig. P16.52). The length of the string is L, and its mass is $m \ll M$. Derive an expression for the time interval required for a transverse wave to travel from one end of the string to the other.

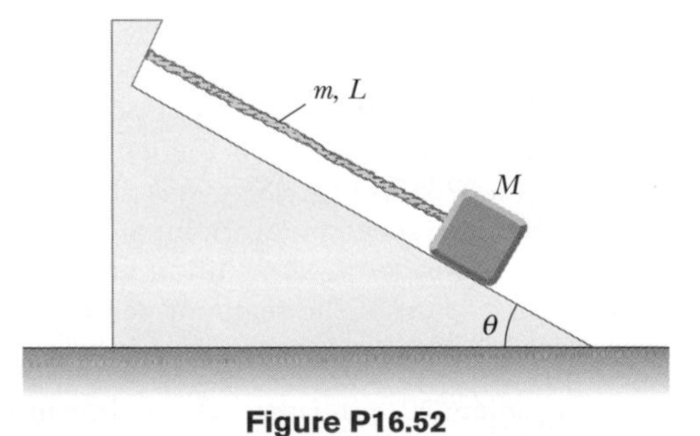

Figure P16.52

53. Review problem. A 2.00-kg block hangs from a rubber cord, being supported so that the cord is not stretched. The unstretched length of the cord is 0.500 m, and its mass is 5.00 g. The "spring constant" for the cord is 100 N/m. The block is released and stops at the lowest point. (a) Determine the tension in the cord when the block is at this lowest point. (b) What is the length of the cord in this "stretched" position? (c) Find the speed of a transverse wave in the cord if the block is held in this lowest position.

54. Review problem. A block of mass M hangs from a rubber cord. The block is supported so that the cord is not stretched. The unstretched length of the cord is L_0 and its mass is m, much less than M. The "spring constant" for the cord is k. The block is released and stops at the lowest point. (a) Determine the tension in the string when the block is at this lowest point. (b) What is the length of the cord in this "stretched" position? (c) Find the speed of a transverse wave in the cord if the block is held in this lowest position.

55. (a) Determine the speed of transverse waves on a string under a tension of 80.0 N if the string has a length of 2.00 m and a mass of 5.00 g. (b) Calculate the power required to generate these waves if they have a wavelength of 16.0 cm and an amplitude of 4.00 cm.

56. A sinusoidal wave in a rope is described by the wave function

$$y = (0.20 \text{ m}) \sin(0.75\pi x + 18\pi t)$$

where x and y are in meters and t is in seconds. The rope has a linear mass density of 0.250 kg/m. If the tension in the rope is provided by an arrangement like the one illustrated in Figure 16.12, what is the value of the suspended mass?

57. A block of mass 0.450 kg is attached to one end of a cord of mass 0.003 20 kg; the other end of the cord is attached to a fixed point. The block rotates with constant angular speed in a circle on a horizontal frictionless table. Through what angle does the block rotate in the time that a transverse wave takes to travel along the string from the center of the circle to the block?

58. A wire of density ρ is tapered so that its cross-sectional area varies with x according to

$$A = (1.0 \times 10^{-3}\, x + 0.010) \text{ cm}^2$$

(a) If the wire is subject to a tension T, derive a relationship for the speed of a wave as a function of position. (b) **What If?** If the wire is aluminum and is subject to a tension of 24.0 N, determine the speed at the origin and at $x = 10.0$ m.

59. A rope of total mass m and length L is suspended vertically. Show that a transverse pulse travels the length of the rope in a time interval $\Delta t = 2\sqrt{L/g}$. (*Suggestion:* First find an expression for the wave speed at any point a distance x from the lower end by considering the tension in the rope as resulting from the weight of the segment below that point.)

60. If an object of mass M is suspended from the bottom of the rope in Problem 59, (a) show that the time interval for a transverse pulse to travel the length of the rope is

$$\Delta t = 2\sqrt{\frac{L}{mg}}\left(\sqrt{M + m} - \sqrt{M}\right)$$

What If? (b) Show that this reduces to the result of Problem 59 when $M = 0$. (c) Show that for $m \ll M$, the expression in part (a) reduces to

$$\Delta t = \sqrt{\frac{mL}{Mg}}$$

61. It is stated in Problem 59 that a pulse travels from the bottom to the top of a hanging rope of length L in a time interval $\Delta t = 2\sqrt{L/g}$. Use this result to answer the following questions. (It is not necessary to set up any new integrations.) (a) How long does it take for a pulse to travel halfway up the rope? Give your answer as a fraction of the quantity $2\sqrt{L/g}$. (b) A pulse starts traveling up the rope. How far has it traveled after a time interval $\sqrt{L/g}$?

62. Determine the speed and direction of propagation of each of the following sinusoidal waves, assuming that x and y are measured in meters and t in seconds.

(a) $y = 0.60 \cos(3.0x - 15t + 2)$

(b) $y = 0.40 \cos(3.0x + 15t - 2)$

(c) $y = 1.2 \sin(15t + 2.0x)$

(d) $y = 0.20 \sin[12t - (x/2) + \pi]$

63. An aluminum wire is clamped at each end under zero tension at room temperature. The tension in the wire is increased by reducing the temperature, which results in a decrease in the wire's equilibrium length. What strain ($\Delta L/L$) results in a transverse wave speed of 100 m/s? Take the cross-sectional area of the wire to be 5.00×10^{-6} m^2, the density to be 2.70×10^3 kg/m^3, and Young's modulus to be 7.00×10^{10} N/m^2.

64. If a loop of chain is spun at high speed, it can roll along the ground like a circular hoop without slipping or collapsing. Consider a chain of uniform linear mass density μ whose center of mass travels to the right at a high speed v_0. (a) Determine the tension in the chain in terms of μ and v_0. (b) If the loop rolls over a bump, the resulting deformation of the chain causes two transverse pulses to propagate along the chain, one moving clockwise and one moving counterclockwise. What is the speed of the pulses traveling along the chain? (c) Through what angle does each pulse travel during the time it takes the loop to make one revolution?

65. (a) Show that the speed of longitudinal waves along a spring of force constant k is $v = \sqrt{kL/\mu}$, where L is the unstretched length of the spring and μ is the mass per unit length. (b) A spring with a mass of 0.400 kg has an unstretched length of 2.00 m and a force constant of 100 N/m. Using the result you obtained in (a), determine the speed of longitudinal waves along this spring.

66. A string of length L consists of two sections. The left half has mass per unit length $\mu = \mu_0/2$, while the right has a mass per unit length $\mu' = 3\mu = 3\mu_0/2$. Tension in the string is T_0. Notice from the data given that this string has the same total mass as a uniform string of length L and mass per unit length μ_0. (a) Find the speeds v and v' at which transverse pulses travel in the two sections. Express the speeds in terms of T_0 and μ_0, and also as multiples of the speed $v_0 = (T_0/\mu_0)^{1/2}$. (b) Find the time interval required for a pulse to travel from one end of the string to the other. Give your result as a multiple of $\Delta t_0 = L/v_0$.

67. A pulse traveling along a string of linear mass density μ is described by the wave function

$$y = [A_0 e^{-bx}] \sin(kx - \omega t)$$

where the factor in brackets before the sine function is said to be the amplitude. (a) What is the power $\mathcal{P}(x)$ carried by this wave at a point x? (b) What is the power carried by this wave at the origin? (c) Compute the ratio $\mathcal{P}(x)/\mathcal{P}(0)$.

68. An earthquake on the ocean floor in the Gulf of Alaska produces a *tsunami* (sometimes incorrectly called a "tidal wave") that reaches Hilo, Hawaii, 4 450 km away, in a time interval of 9 h 30 min. Tsunamis have enormous wavelengths (100 to 200 km), and the propagation speed for these waves is $v \approx \sqrt{gd}$, where d is the average depth of the water. From the information given, find the average wave speed and the average ocean depth between Alaska and Hawaii. (This method was used in 1856 to estimate the average depth of the Pacific Ocean long before soundings were made to give a direct determination.)

69. A string on a musical instrument is held under tension T and extends from the point $x = 0$ to the point $x = L$. The string is overwound with wire in such a way that its mass per unit length $\mu(x)$ increases uniformly from μ_0 at $x = 0$ to μ_L at $x = L$. (a) Find an expression for $\mu(x)$ as a function of x over the range $0 \leq x \leq L$. (b) Show that the time interval required for a transverse pulse to travel the length of the string is given by

$$\Delta t = \frac{2L\left(\mu_L + \mu_0 + \sqrt{\mu_L \mu_0}\right)}{3\sqrt{T}\left(\sqrt{\mu_L} + \sqrt{\mu_0}\right)}$$

Answers to Quick Quizzes

16.1 (b). It is longitudinal because the disturbance (the shift of position of the people) is parallel to the direction in which the wave travels.

16.2 (a). It is transverse because the people stand up and sit down (vertical motion), whereas the wave moves either to the left or to the right.

16.3 (c). The wave speed is determined by the medium, so it is unaffected by changing the frequency.

16.4 (b). Because the wave speed remains the same, the result of doubling the frequency is that the wavelength is half as large.

16.5 (d). The amplitude of a wave is unrelated to the wave speed, so we cannot determine the new amplitude without further information.

16.6 (c). With a larger amplitude, an element of the string has more energy associated with its simple harmonic motion, so the element passes through the equilibrium position with a higher maximum transverse speed.

16.7 Only answers (f) and (h) are correct. (a) and (b) affect the transverse speed of a particle of the string, but not the wave speed along the string. (c) and (d) change the amplitude. (e) and (g) increase the time interval by decreasing the wave speed.

16.8 (d). Doubling the amplitude of the wave causes the power to be larger by a factor of 4. In (a), halving the linear mass density of the string causes the power to change by a factor of 0.71—the rate decreases. In (b), doubling the wavelength of the wave halves the frequency and causes the power to change by a factor of 0.25—the rate decreases. In (c), doubling the tension in the string changes the wave speed and causes the power to change by a factor of 1.4—not as large as in part (d).

Chapter 17

Sound Waves

CHAPTER OUTLINE

17.1 Speed of Sound Waves

17.2 Periodic Sound Waves

17.3 Intensity of Periodic Sound Waves

17.4 The Doppler Effect

17.5 Digital Sound Recording

17.6 Motion Picture Sound

▲ *Human ears have evolved to detect sound waves and interpret them as music or speech. Some animals, such as this young bat-eared fox, have ears adapted for the detection of very weak sounds. (Getty Images)*

Sound waves are the most common example of longitudinal waves. They travel through any material medium with a speed that depends on the properties of the medium. As the waves travel through air, the elements of air vibrate to produce changes in density and pressure along the direction of motion of the wave. If the source of the sound waves vibrates sinusoidally, the pressure variations are also sinusoidal. The mathematical description of sinusoidal sound waves is very similar to that of sinusoidal string waves, which were discussed in the previous chapter.

Sound waves are divided into three categories that cover different frequency ranges. (1) *Audible waves* lie within the range of sensitivity of the human ear. They can be generated in a variety of ways, such as by musical instruments, human voices, or loudspeakers. (2) *Infrasonic waves* have frequencies below the audible range. Elephants can use infrasonic waves to communicate with each other, even when separated by many kilometers. (3) *Ultrasonic waves* have frequencies above the audible range. You may have used a "silent" whistle to retrieve your dog. The ultrasonic sound it emits is easily heard by dogs, although humans cannot detect it at all. Ultrasonic waves are also used in medical imaging.

We begin this chapter by discussing the speed of sound waves and then wave intensity, which is a function of wave amplitude. We then provide an alternative description of the intensity of sound waves that compresses the wide range of intensities to which the ear is sensitive into a smaller range for convenience. We investigate the effects of the motion of sources and/or listeners on the frequency of a sound. Finally, we explore digital reproduction of sound, focusing in particular on sound systems used in modern motion pictures.

17.1 Speed of Sound Waves

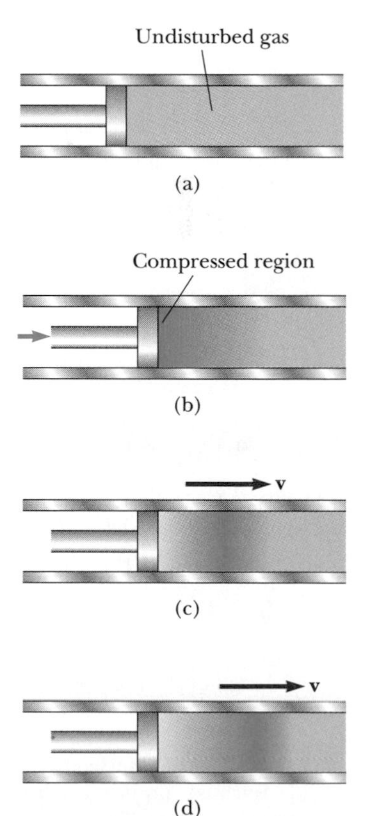

Figure 17.1 Motion of a longitudinal pulse through a compressible gas. The compression (darker region) is produced by the moving piston.

Let us describe pictorially the motion of a one-dimensional longitudinal pulse moving through a long tube containing a compressible gas (Fig. 17.1). A piston at the left end can be moved to the right to compress the gas and create the pulse. Before the piston is moved, the gas is undisturbed and of uniform density, as represented by the uniformly shaded region in Figure 17.1a. When the piston is suddenly pushed to the right (Fig. 17.1b), the gas just in front of it is compressed (as represented by the more heavily shaded region); the pressure and density in this region are now higher than they were before the piston moved. When the piston comes to rest (Fig. 17.1c), the compressed region of the gas continues to move to the right, corresponding to a longitudinal pulse traveling through the tube with speed v. Note that the piston speed does *not* equal v. Furthermore, the compressed region does not "stay with" the piston as the piston moves, because the speed of the wave is usually greater than the speed of the piston.

The speed of sound waves in a medium depends on the compressibility and density of the medium. If the medium is a liquid or a gas and has a bulk modulus B (see

Section 12.4) and density ρ, the speed of sound waves in that medium is

$$v = \sqrt{\frac{B}{\rho}} \tag{17.1}$$

It is interesting to compare this expression with Equation 16.18 for the speed of transverse waves on a string, $v = \sqrt{T/\mu}$. In both cases, the wave speed depends on an elastic property of the medium—bulk modulus B or string tension T—and on an inertial property of the medium—ρ or μ. In fact, the *speed of all mechanical waves* follows an expression of the general form

$$v = \sqrt{\frac{\text{elastic property}}{\text{inertial property}}}$$

For longitudinal sound waves in a solid rod of material, for example, the speed of sound depends on Young's modulus Y and the density ρ. Table 17.1 provides the speed of sound in several different materials.

The speed of sound also depends on the temperature of the medium. For sound traveling through air, the relationship between wave speed and medium temperature is

$$v = (331 \text{ m/s})\sqrt{1 + \frac{T_C}{273°\text{C}}}$$

where 331 m/s is the speed of sound in air at 0°C, and T_C is the air temperature in degrees Celsius. Using this equation, one finds that at 20°C the speed of sound in air is approximately 343 m/s.

This information provides a convenient way to estimate the distance to a thunderstorm. You count the number of seconds between seeing the flash of lightning and hearing the thunder. Dividing this time by 3 gives the approximate distance to the lightning in kilometers, because 343 m/s is approximately $\frac{1}{3}$ km/s. Dividing the time in seconds by 5 gives the approximate distance to the lightning in miles, because the speed of sound in ft/s (1 125 ft/s) is approximately $\frac{1}{5}$ mi/s.

Table 17.1

Speed of Sound in Various Media

Medium	v (m/s)
Gases	
Hydrogen (0°C)	1 286
Helium (0°C)	972
Air (20°C)	343
Air (0°C)	331
Oxygen (0°C)	317
Liquids at 25°C	
Glycerol	1 904
Seawater	1 533
Water	1 493
Mercury	1 450
Kerosene	1 324
Methyl alcohol	1 143
Carbon tetrachloride	926
Solids[a]	
Pyrex glass	5 640
Iron	5 950
Aluminum	6 420
Brass	4 700
Copper	5 010
Gold	3 240
Lucite	2 680
Lead	1 960
Rubber	1 600

[a] Values given are for propagation of longitudinal waves in bulk media. Speeds for longitudinal waves in thin rods are smaller, and speeds of transverse waves in bulk are smaller yet.

Quick Quiz 17.1 The speed of sound in air is a function of (a) wavelength (b) frequency (c) temperature (d) amplitude.

Example 17.1 Speed of Sound in a Liquid **Interactive**

(A) Find the speed of sound in water, which has a bulk modulus of 2.1×10^9 N/m² at a temperature of 0°C and a density of 1.00×10^3 kg/m³.

Solution Using Equation 17.1, we find that

$$v_{\text{water}} = \sqrt{\frac{B}{\rho}} = \sqrt{\frac{2.1 \times 10^9 \text{ N/m}^2}{1.00 \times 10^3 \text{ kg/m}^3}} = 1.4 \text{ km/s}$$

In general, sound waves travel more slowly in liquids than in solids because liquids are more compressible than solids. Note that the speed of sound in water is lower at 0°C than at 25°C (Table 17.1).

(B) Dolphins use sound waves to locate food. Experiments have shown that a dolphin can detect a 7.5-cm target 110 m away, even in murky water. For a bit of "dinner" at that distance, how much time passes between the moment the dolphin emits a sound pulse and the moment the dolphin hears its reflection and thereby detects the distant target?

Solution The total distance covered by the sound wave as it travels from dolphin to target and back is 2×110 m = 220 m. From Equation 2.2, we have, for 25°C water

$$\Delta t = \frac{\Delta x}{v_x} = \frac{220 \text{ m}}{1\,533 \text{ m/s}} = 0.14 \text{ s}$$

At the Interactive Worked Example link at **http://www.pse6.com,** *you can compare the speed of sound through the various media found in Table 17.1.*

17.2 Periodic Sound Waves

This section will help you better comprehend the nature of sound waves. An important fact for understanding how our ears work is that *pressure variations control what we hear.*

One can produce a one-dimensional periodic sound wave in a long, narrow tube containing a gas by means of an oscillating piston at one end, as shown in Figure 17.2. The darker parts of the colored areas in this figure represent regions where the gas is compressed and thus the density and pressure are above their equilibrium values. A compressed region is formed whenever the piston is pushed into the tube. This compressed region, called a **compression,** moves through the tube as a pulse, continuously compressing the region just in front of itself. When the piston is pulled back, the gas in front of it expands, and the pressure and density in this region fall below their equilibrium values (represented by the lighter parts of the colored areas in Fig. 17.2). These low-pressure regions, called **rarefactions,** also propagate along the tube, following the compressions. Both regions move with a speed equal to the speed of sound in the medium.

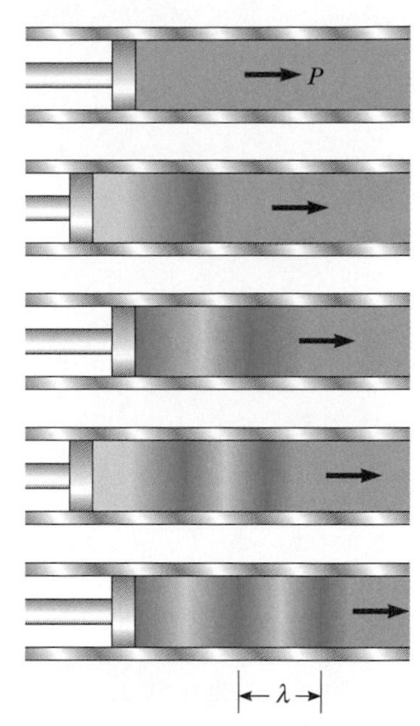

Active Figure 17.2 A longitudinal wave propagating through a gas-filled tube. The source of the wave is an oscillating piston at the left.

At the Active Figures link at http://www.pse6.com, you can adjust the frequency of the piston.

As the piston oscillates sinusoidally, regions of compression and rarefaction are continuously set up. The distance between two successive compressions (or two successive rarefactions) equals the wavelength λ. As these regions travel through the tube, any small element of the medium moves with simple harmonic motion parallel to the direction of the wave. If $s(x, t)$ is the position of a small element relative to its equilibrium position,[1] we can express this harmonic position function as

$$s(x, t) = s_{max} \cos(kx - \omega t) \tag{17.2}$$

where **s_{max} is the maximum position of the element relative to equilibrium.** This is often called the **displacement amplitude** of the wave. The parameter k is the wave number and ω is the angular frequency of the piston. Note that the displacement of the element is along x, in the direction of propagation of the sound wave, which means we are describing a longitudinal wave.

The variation in the gas pressure ΔP measured from the equilibrium value is also periodic. For the position function in Equation 17.2, ΔP is given by

$$\Delta P = \Delta P_{max} \sin(kx - \omega t) \tag{17.3}$$

where **the pressure amplitude ΔP_{max}**—which is the **maximum change in pressure from the equilibrium value**—is given by

$$\Delta P_{max} = \rho v \omega s_{max} \tag{17.4}$$

Thus, we see that a sound wave may be considered as either a displacement wave or a pressure wave. A comparison of Equations 17.2 and 17.3 shows that **the pressure wave is 90° out of phase with the displacement wave.** Graphs of these functions are shown in Figure 17.3. Note that the pressure variation is a maximum when the displacement from equilibrium is zero, and the displacement from equilibrium is a maximum when the pressure variation is zero.

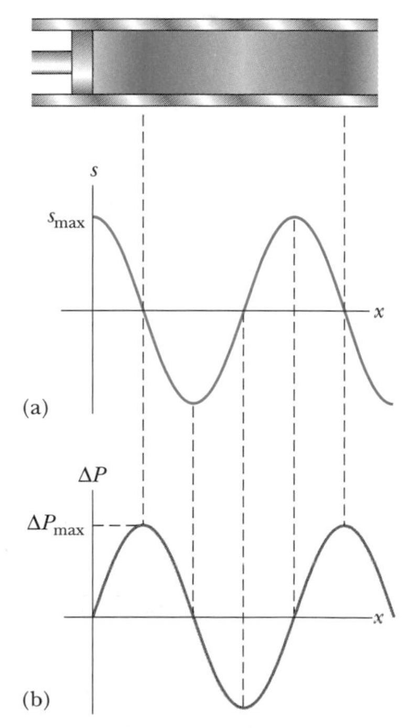

Figure 17.3 (a) Displacement amplitude and (b) pressure amplitude versus position for a sinusoidal longitudinal wave.

Quick Quiz 17.2 If you blow across the top of an empty soft-drink bottle, a pulse of sound travels down through the air in the bottle. At the moment the pulse reaches the bottom of the bottle, the correct descriptions of the displacement of elements of air from their equilibrium positions and the pressure of the air at this point are (a) the displacement and pressure are both at a maximum (b) the displacement and pressure are both at a minimum (c) the displacement is zero and the pressure is a maximum (d) the displacement is zero and the pressure is a minimum.

[1] We use $s(x, t)$ here instead of $y(x, t)$ because the displacement of elements of the medium is not perpendicular to the x direction.

Derivation of Equation 17.3

Consider a thin disk-shaped element of gas whose circular cross section is parallel to the piston in Figure 17.2. This element will undergo changes in position, pressure, and density as a sound wave propagates through the gas. From the definition of bulk modulus (see Eq. 12.8), the pressure variation in the gas is

$$\Delta P = -B\frac{\Delta V}{V_i}$$

The element has a thickness Δx in the horizontal direction and a cross-sectional area A, so its volume is $V_i = A\Delta x$. The change in volume ΔV accompanying the pressure change is equal to $A\Delta s$, where Δs is the difference between the value of s at $x + \Delta x$ and the value of s at x. Hence, we can express ΔP as

$$\Delta P = -B\frac{\Delta V}{V_i} = -B\frac{A\Delta s}{A\Delta x} = -B\frac{\Delta s}{\Delta x}$$

As Δx approaches zero, the ratio $\Delta s/\Delta x$ becomes $\partial s/\partial x$. (The partial derivative indicates that we are interested in the variation of s with position at a *fixed* time.) Therefore,

$$\Delta P = -B\frac{\partial s}{\partial x}$$

If the position function is the simple sinusoidal function given by Equation 17.2, we find that

$$\Delta P = -B\frac{\partial}{\partial x}[s_{\max}\cos(kx - \omega t)] = Bs_{\max}k\sin(kx - \omega t)$$

Because the bulk modulus is given by $B = \rho v^2$ (see Eq. 17.1), the pressure variation reduces to

$$\Delta P = \rho v^2 s_{\max} k\sin(kx - \omega t)$$

From Equation 16.11, we can write $k = \omega/v$; hence, ΔP can be expressed as

$$\Delta P = \rho v\omega s_{\max}\sin(kx - \omega t)$$

Because the sine function has a maximum value of 1, we see that the maximum value of the pressure variation is $\Delta P_{\max} = \rho v\omega s_{\max}$ (see Eq. 17.4), and we arrive at Equation 17.3:

$$\Delta P = \Delta P_{\max}\sin(kx - \omega t)$$

17.3 Intensity of Periodic Sound Waves

In the preceding chapter, we showed that a wave traveling on a taut string transports energy. The same concept applies to sound waves. Consider an element of air of mass Δm and width Δx in front of a piston oscillating with a frequency ω, as shown in Figure 17.4.

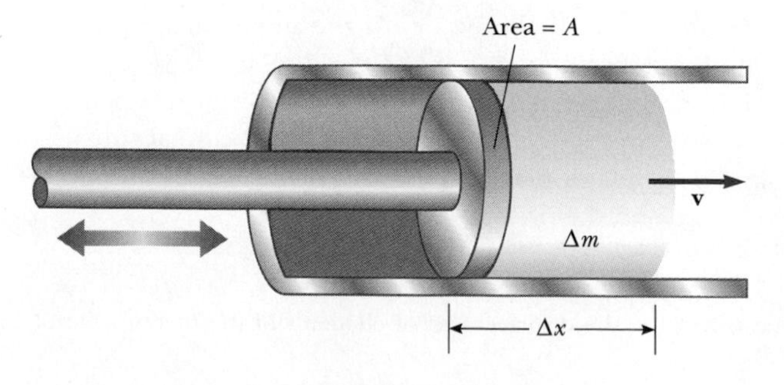

Figure 17.4 An oscillating piston transfers energy to the air in the tube, causing the element of air of width Δx and mass Δm to oscillate with an amplitude $s_{\max}$.

The piston transmits energy to this element of air in the tube, and the energy is propagated away from the piston by the sound wave. To evaluate the rate of energy transfer for the sound wave, we shall evaluate the kinetic energy of this element of air, which is undergoing simple harmonic motion. We shall follow a procedure similar to that in Section 16.5, in which we evaluated the rate of energy transfer for a wave on a string.

As the sound wave propagates away from the piston, the position of any element of air in front of the piston is given by Equation 17.2. To evaluate the kinetic energy of this element of air, we need to know its speed. We find the speed by taking the time derivative of Equation 17.2:

$$v(x, t) = \frac{\partial}{\partial t} s(x, t) = \frac{\partial}{\partial t}[s_{max} \cos(kx - \omega t)] = -\omega s_{max} \sin(kx - \omega t)$$

Imagine that we take a "snapshot" of the wave at $t = 0$. The kinetic energy of a given element of air at this time is

$$\Delta K = \tfrac{1}{2}\Delta m(v)^2 = \tfrac{1}{2}\Delta m(-\omega s_{max} \sin kx)^2 = \tfrac{1}{2}\rho A\, \Delta x(-\omega s_{max} \sin kx)^2$$
$$= \tfrac{1}{2}\rho A\, \Delta x(\omega s_{max})^2 \sin^2 kx$$

where A is the cross-sectional area of the element and $A\Delta x$ is its volume. Now, as in Section 16.5, we integrate this expression over a full wavelength to find the total kinetic energy in one wavelength. Letting the element of air shrink to infinitesimal thickness, so that $\Delta x \rightarrow dx$, we have

$$K_\lambda = \int dK = \int_0^\lambda \tfrac{1}{2}\rho A(\omega s_{max})^2 \sin^2 kx\, dx = \tfrac{1}{2}\rho A(\omega s_{max})^2 \int_0^\lambda \sin^2 kx\, dx$$
$$= \tfrac{1}{2}\rho A(\omega s_{max})^2(\tfrac{1}{2}\lambda) = \tfrac{1}{4}\rho A(\omega s_{max})^2\lambda$$

As in the case of the string wave in Section 16.5, the total potential energy for one wavelength has the same value as the total kinetic energy; thus, the total mechanical energy for one wavelength is

$$E_\lambda = K_\lambda + U_\lambda = \tfrac{1}{2}\rho A(\omega s_{max})^2\lambda$$

As the sound wave moves through the air, this amount of energy passes by a given point during one period of oscillation. Hence, the rate of energy transfer is

$$\mathcal{P} = \frac{\Delta E}{\Delta t} = \frac{E_\lambda}{T} = \frac{\tfrac{1}{2}\rho A(\omega s_{max})^2\lambda}{T} = \tfrac{1}{2}\rho A(\omega s_{max})^2\left(\frac{\lambda}{T}\right) = \tfrac{1}{2}\rho A v(\omega s_{max})^2$$

where v is the speed of sound in air.

We define the **intensity** I of a wave, or the power per unit area, to be the rate at which the energy being transported by the wave transfers through a unit area A perpendicular to the direction of travel of the wave:

$$I \equiv \frac{\mathcal{P}}{A} \qquad (17.5)$$

In the present case, therefore, the intensity is

Intensity of a sound wave

$$I = \frac{\mathcal{P}}{A} = \tfrac{1}{2}\rho v(\omega s_{max})^2$$

Thus, we see that the intensity of a periodic sound wave is proportional to the square of the displacement amplitude and to the square of the angular frequency (as in the case of a periodic string wave). This can also be written in terms of the pressure amplitude $\Delta P_{\max}$; in this case, we use Equation 17.4 to obtain

$$I = \frac{\Delta P_{\max}^2}{2\rho v} \tag{17.6}$$

Now consider a point source emitting sound waves equally in all directions. From everyday experience, we know that the intensity of sound decreases as we move farther from the source. We identify an imaginary sphere of radius r centered on the source. When a source emits sound equally in all directions, we describe the result as a **spherical wave.** The average power $\mathcal{P}_{\text{av}}$ emitted by the source must be distributed uniformly over this spherical surface of area $4\pi r^2$. Hence, the wave intensity at a distance r from the source is

Inverse-square behavior of intensity for a point source

$$I = \frac{\mathcal{P}_{\text{av}}}{A} = \frac{\mathcal{P}_{\text{av}}}{4\pi r^2} \tag{17.7}$$

This inverse-square law, which is reminiscent of the behavior of gravity in Chapter 13, states that the intensity decreases in proportion to the square of the distance from the source.

Quick Quiz 17.3 An *ear trumpet* is a cone-shaped shell, like a megaphone, that was used before hearing aids were developed to help persons who were hard of hearing. The small end of the cone was held in the ear, and the large end was aimed toward the source of sound as in Figure 17.5. The ear trumpet increases the intensity of sound because (a) it increases the speed of sound (b) it reflects sound back toward the source (c) it gathers sound that would normally miss the ear and concentrates it into a smaller area (d) it increases the density of the air.

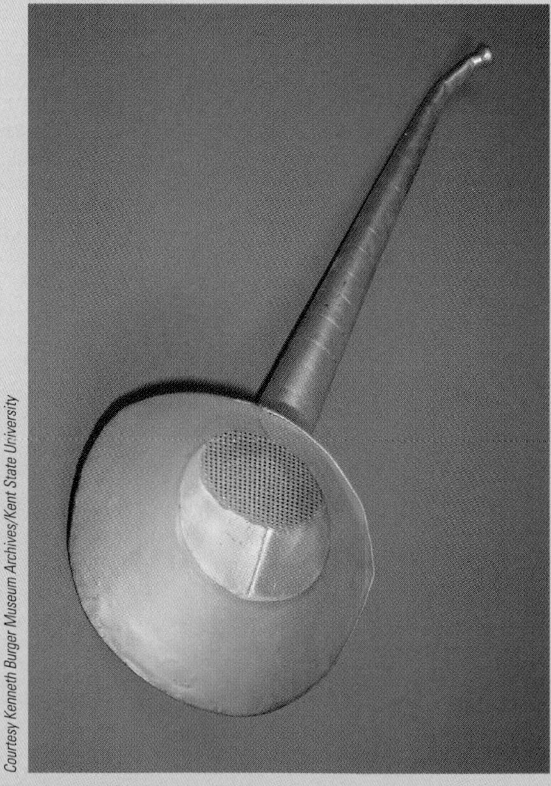

Courtesy Kenneth Burger Museum Archives/Kent State University

Figure 17.5 (Quick Quiz 17.3) An ear trumpet, used before hearing aids to make sounds intense enough for people who were hard of hearing. You can simulate the effect of an ear trumpet by cupping your hands behind your ears.

Quick Quiz 17.4 A vibrating guitar string makes very little sound if it is not mounted on the guitar. But if this vibrating string is attached to the guitar body, so that the body of the guitar vibrates, the sound is higher in intensity. This is because (a) the power of the vibration is spread out over a larger area (b) the energy leaves the guitar at a higher rate (c) the speed of sound is higher in the material of the guitar body (d) none of these.

Example 17.2 Hearing Limits

The faintest sounds the human ear can detect at a frequency of 1 000 Hz correspond to an intensity of about 1.00×10^{-12} W/m²—the so-called *threshold of hearing*. The loudest sounds the ear can tolerate at this frequency correspond to an intensity of about 1.00 W/m²—the *threshold of pain*. Determine the pressure amplitude and displacement amplitude associated with these two limits.

Solution First, consider the faintest sounds. Using Equation 17.6 and taking $v = 343$ m/s as the speed of sound waves in air and $\rho = 1.20$ kg/m³ as the density of air, we obtain

$$\Delta P_{\text{max}} = \sqrt{2\rho v I}$$
$$= \sqrt{2(1.20\ \text{kg/m}^3)(343\ \text{m/s})(1.00 \times 10^{-12}\ \text{W/m}^2)}$$
$$= 2.87 \times 10^{-5}\ \text{N/m}^2$$

Because atmospheric pressure is about 10^5 N/m², this result tells us that the ear is sensitive to pressure fluctuations as small as 3 parts in 10^{10}!

We can calculate the corresponding displacement amplitude by using Equation 17.4, recalling that $\omega = 2\pi f$ (see Eqs. 16.3 and 16.9):

$$s_{\text{max}} = \frac{\Delta P_{\text{max}}}{\rho v \omega} = \frac{2.87 \times 10^{-5}\ \text{N/m}^2}{(1.20\ \text{kg/m}^3)(343\ \text{m/s})(2\pi \times 1\ 000\ \text{Hz})}$$
$$= 1.11 \times 10^{-11}\ \text{m}$$

This is a remarkably small number! If we compare this result for s_{max} with the size of an atom (about 10^{-10} m), we see that the ear is an extremely sensitive detector of sound waves.

In a similar manner, one finds that the loudest sounds the human ear can tolerate correspond to a pressure amplitude of 28.7 N/m² and a displacement amplitude equal to 1.11×10^{-5} m.

Example 17.3 Intensity Variations of a Point Source

A point source emits sound waves with an average power output of 80.0 W.

(A) Find the intensity 3.00 m from the source.

Solution A point source emits energy in the form of spherical waves. Using Equation 17.7, we have

$$I = \frac{\mathcal{P}_{\text{av}}}{4\pi r^2} = \frac{80.0\ \text{W}}{4\pi(3.00\ \text{m})^2} = 0.707\ \text{W/m}^2$$

an intensity that is close to the threshold of pain.

(B) Find the distance at which the intensity of the sound is 1.00×10^{-8} W/m².

Solution Using this value for I in Equation 17.7 and solving for r, we obtain

$$r = \sqrt{\frac{\mathcal{P}_{\text{av}}}{4\pi I}} = \sqrt{\frac{80.0\ \text{W}}{4\pi(1.00 \times 10^{-8}\ \text{W/m}^2)}}$$
$$= 2.52 \times 10^4\ \text{m}$$

which equals about 16 miles!

Sound Level in Decibels

Example 17.2 illustrates the wide range of intensities the human ear can detect. Because this range is so wide, it is convenient to use a logarithmic scale, where the **sound level** β (Greek beta) is defined by the equation

$$\beta \equiv 10 \log\left(\frac{I}{I_0}\right) \qquad (17.8)$$

Sound level in decibels

The constant I_0 is the *reference intensity*, taken to be at the threshold of hearing ($I_0 = 1.00 \times 10^{-12}$ W/m²), and I is the intensity in watts per square meter to which the sound level β corresponds, where β is measured[2] in **decibels** (dB). On this scale,

[2] The unit *bel* is named after the inventor of the telephone, Alexander Graham Bell (1847–1922). The prefix *deci-* is the SI prefix that stands for 10^{-1}.

Table 17.2

Sound Levels

Source of Sound	β (dB)
Nearby jet airplane	150
Jackhammer; machine gun	130
Siren; rock concert	120
Subway; power mower	100
Busy traffic	80
Vacuum cleaner	70
Normal conversation	50
Mosquito buzzing	40
Whisper	30
Rustling leaves	10
Threshold of hearing	0

the threshold of pain ($I = 1.00\ \text{W/m}^2$) corresponds to a sound level of $\beta = 10 \log[(1\ \text{W/m}^2)/(10^{-12}\ \text{W/m}^2)] = 10\ \log(10^{12}) = 120\ \text{dB}$, and the threshold of hearing corresponds to $\beta = 10 \log[(10^{-12}\ \text{W/m}^2)/(10^{-12}\ \text{W/m}^2)] = 0\ \text{dB}$.

Prolonged exposure to high sound levels may seriously damage the ear. Ear plugs are recommended whenever sound levels exceed 90 dB. Recent evidence suggests that "noise pollution" may be a contributing factor to high blood pressure, anxiety, and nervousness. Table 17.2 gives some typical sound-level values.

Quick Quiz 17.5 A violin plays a melody line and is then joined by a second violin, playing at the same intensity as the first violin, in a repeat of the same melody. With both violins playing, what physical parameter has doubled compared to the situation with only one violin playing? (a) wavelength (b) frequency (c) intensity (d) sound level in dB (e) none of these.

Quick Quiz 17.6 Increasing the intensity of a sound by a factor of 100 causes the sound level to increase by (a) 100 dB (b) 20 dB (c) 10 dB (d) 2 dB.

Example 17.4 Sound Levels

Two identical machines are positioned the same distance from a worker. The intensity of sound delivered by each machine at the location of the worker is $2.0 \times 10^{-7}\ \text{W/m}^2$. Find the sound level heard by the worker

(A) when one machine is operating

(B) when both machines are operating.

Solution

(A) The sound level at the location of the worker with one machine operating is calculated from Equation 17.8:

$$\beta_1 = 10 \log\left(\frac{2.0 \times 10^{-7}\ \text{W/m}^2}{1.00 \times 10^{-12}\ \text{W/m}^2}\right) = 10 \log(2.0 \times 10^5)$$

$$= 53\ \text{dB}$$

(B) When both machines are operating, the intensity is doubled to $4.0 \times 10^{-7}\ \text{W/m}^2$; therefore, the sound level now is

$$\beta_2 = 10 \log\left(\frac{4.0 \times 10^{-7}\ \text{W/m}^2}{1.00 \times 10^{-12}\ \text{W/m}^2}\right) = 10 \log(4.0 \times 10^5)$$

$$= 56\ \text{dB}$$

From these results, we see that when the intensity is doubled, the sound level increases by only 3 dB.

What If? ***Loudness*** **is a psychological response to a sound and depends on both the intensity and the frequency of the sound. As a rule of thumb, a doubling in loudness is approximately associated with an increase in sound level of 10 dB. (Note that this rule of thumb is relatively inaccurate at very low or very high frequencies.) If the loudness of the**

machines in this example is to be doubled, how many machines must be running?

Answer Using the rule of thumb, a doubling of loudness corresponds to a sound level increase of 10 dB. Thus,

$$\beta_2 - \beta_1 = 10\text{ dB} = 10\log\left(\frac{I_2}{I_0}\right) - 10\log\left(\frac{I_1}{I_0}\right) = 10\log\left(\frac{I_2}{I_1}\right)$$

$$\log\left(\frac{I_2}{I_1}\right) = 1$$

$$I_2 = 10I_1$$

Thus, ten machines must be operating to double the loudness.

Loudness and Frequency

The discussion of sound level in decibels relates to a *physical* measurement of the strength of a sound. Let us now consider how we describe the *psychological* "measurement" of the strength of a sound.

Of course, we don't have meters in our bodies that can read out numerical values of our reactions to stimuli. We have to "calibrate" our reactions somehow by comparing different sounds to a reference sound. However, this is not easy to accomplish. For example, earlier we mentioned that the threshold intensity is 10^{-12} W/m^2, corresponding to an intensity level of 0 dB. In reality, this value is the threshold only for a sound of frequency 1 000 Hz, which is a standard reference frequency in acoustics. If we perform an experiment to measure the threshold intensity at other frequencies, we find a distinct variation of this threshold as a function of frequency. For example, at 100 Hz, a sound must have an intensity level of about 30 dB in order to be just barely audible! Unfortunately, there is no simple relationship between physical measurements and psychological "measurements." The 100-Hz, 30-dB sound is psychologically "equal" to the 1 000-Hz, 0-dB sound (both are just barely audible) but they are not physically equal (30 dB $\neq$ 0 dB).

By using test subjects, the human response to sound has been studied, and the results are shown in Figure 17.6 (the white area), along with the approximate frequency and sound-level ranges of other sound sources. The lower curve of the white area corresponds to the threshold of hearing. Its variation with frequency is clear from this diagram. Note that humans are sensitive to frequencies ranging from about 20 Hz to about 20 000 Hz. The upper bound of the white area is the threshold of pain. Here the

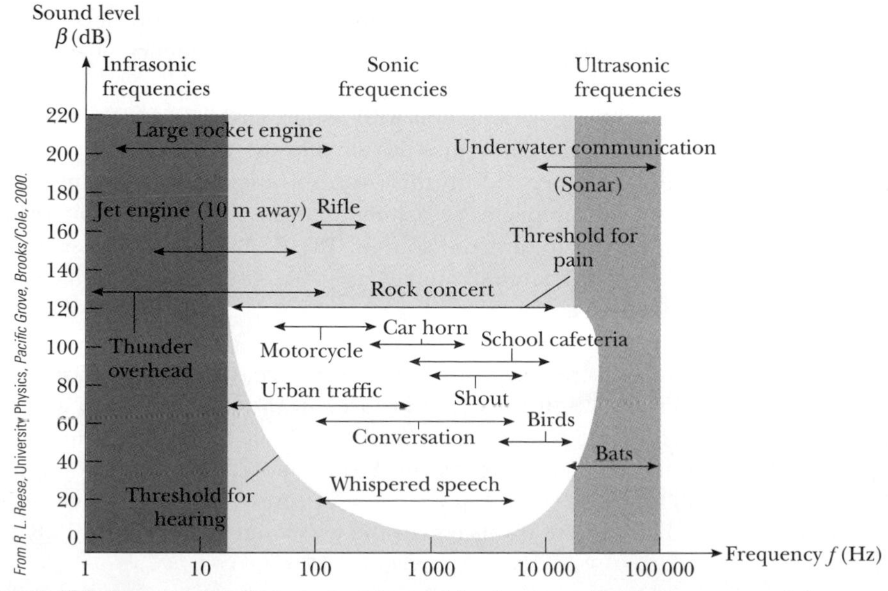

From R. L. Reese, University Physics, Pacific Grove, Brooks/Cole, 2000.

Figure 17.6 Approximate frequency and sound level ranges of various sources and that of normal human hearing, shown by the white area.

boundary of the white area is straight, because the psychological response is relatively independent of frequency at this high sound level.

The most dramatic change with frequency is in the lower left region of the white area, for low frequencies and low intensity levels. Our ears are particularly insensitive in this region. If you are listening to your stereo and the bass (low frequencies) and treble (high frequencies) sound balanced at a high volume, try turning the volume down and listening again. You will probably notice that the bass seems weak, which is due to the insensitivity of the ear to low frequencies at low sound levels, as shown in Figure 17.6.

17.4 The Doppler Effect

Perhaps you have noticed how the sound of a vehicle's horn changes as the vehicle moves past you. The frequency of the sound you hear as the vehicle approaches you is higher than the frequency you hear as it moves away from you. This is one example of the **Doppler effect.**[3]

To see what causes this apparent frequency change, imagine you are in a boat that is lying at anchor on a gentle sea where the waves have a period of $T = 3.0$ s. This means that every 3.0 s a crest hits your boat. Figure 17.7a shows this situation, with the water waves moving toward the left. If you set your watch to $t = 0$ just as one crest hits, the watch reads 3.0 s when the next crest hits, 6.0 s when the third crest hits, and so on. From these observations you conclude that the wave frequency is $f = 1/T = 1/(3.0 \text{ s}) = 0.33$ Hz. Now suppose you start your motor and head directly into the oncoming waves, as in Figure 17.7b. Again you set your watch to $t = 0$ as a crest hits the front of your boat. Now, however, because you are moving toward the next wave crest as it moves toward you, it hits you less than 3.0 s after the first hit. In other words, the period you observe is shorter than the 3.0-s period you observed when you were stationary. Because $f = 1/T$, you observe a higher wave frequency than when you were at rest.

If you turn around and move in the same direction as the waves (see Fig. 17.7c), you observe the opposite effect. You set your watch to $t = 0$ as a crest hits the back of the boat. Because you are now moving away from the next crest, more than 3.0 s has elapsed on your watch by the time that crest catches you. Thus, you observe a lower frequency than when you were at rest.

These effects occur because the *relative* speed between your boat and the waves depends on the direction of travel and on the speed of your boat. When you are moving toward the right in Figure 17.7b, this relative speed is higher than that of the wave speed, which leads to the observation of an increased frequency. When you turn around and move to the left, the relative speed is lower, as is the observed frequency of the water waves.

Let us now examine an analogous situation with sound waves, in which the water waves become sound waves, the water becomes the air, and the person on the boat becomes an observer listening to the sound. In this case, an observer O is moving and a sound source S is stationary. For simplicity, we assume that the air is also stationary and that the observer moves directly toward the source (Fig. 17.8). The observer moves with a speed v_O toward a stationary point source ($v_S = 0$), where *stationary* means at rest with respect to the medium, air.

If a point source emits sound waves and the medium is uniform, the waves move at the same speed in all directions radially away from the source; this is a spherical wave, as was mentioned in Section 17.3. It is useful to represent these waves with a series of circular arcs concentric with the source, as in Figure 17.8. Each arc represents a surface over which the phase of the wave is constant. For example, the surface could pass through the crests of all waves. We call such a surface of constant phase a **wave front.** The distance between adjacent wave fronts equals the wavelength λ. In Figure 17.8, the

[3] Named after the Austrian physicist Christian Johann Doppler (1803–1853), who in 1842 predicted the effect for both sound waves and light waves.

(a)

(b)

(c)

Figure 17.7 (a) Waves moving toward a stationary boat. The waves travel to the left, and their source is far to the right of the boat, out of the frame of the photograph. (b) The boat moving toward the wave source. (c) The boat moving away from the wave source.

circles are the intersections of these three-dimensional wave fronts with the two-dimensional paper.

We take the frequency of the source in Figure 17.8 to be f, the wavelength to be λ, and the speed of sound to be v. If the observer were also stationary, he or she would detect wave fronts at a rate f. (That is, when $v_O = 0$ and $v_S = 0$, the observed frequency equals the source frequency.) When the observer moves toward the source, the speed of the waves relative to the observer is $v' = v + v_O$, as in the case of the boat, but the

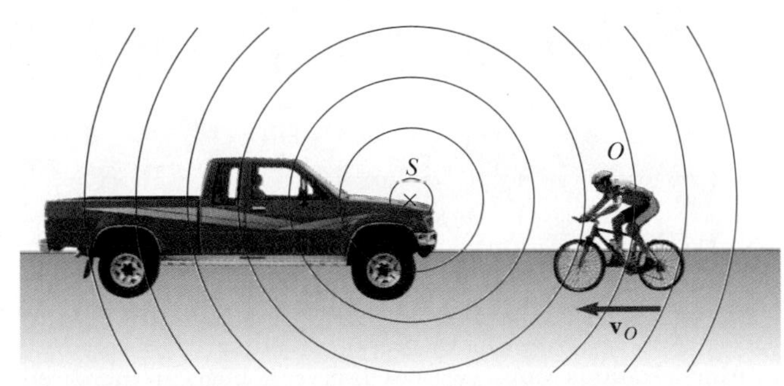

Active Figure 17.8 An observer O (the cyclist) moves with a speed v_O toward a stationary point source S, the horn of a parked truck. The observer hears a frequency f' that is greater than the source frequency.

At the Active Figures link at http://www.pse6.com, you can adjust the speed of the observer.

wavelength λ is unchanged. Hence, using Equation 16.12, $v = \lambda f$, we can say that the frequency f' heard by the observer is *increased* and is given by

$$f' = \frac{v'}{\lambda} = \frac{v + v_O}{\lambda}$$

Because $\lambda = v/f$, we can express f' as

$$f' = \left(\frac{v + v_O}{v}\right)f \qquad \text{(observer moving toward source)} \qquad (17.9)$$

If the observer is moving away from the source, the speed of the wave relative to the observer is $v' = v - v_O$. The frequency heard by the observer in this case is *decreased* and is given by

$$f' = \left(\frac{v - v_O}{v}\right)f \qquad \text{(observer moving away from source)} \qquad (17.10)$$

In general, whenever an observer moves with a speed v_O relative to a stationary source, the frequency heard by the observer is given by Equation 17.9, with a sign convention: a positive value is substituted for v_O when the observer moves toward the source and a negative value is substituted when the observer moves away from the source.

Now consider the situation in which the source is in motion and the observer is at rest. If the source moves directly toward observer A in Figure 17.9a, the wave fronts heard by the observer are closer together than they would be if the source were not moving. As a result, the wavelength λ' measured by observer A is shorter than the wavelength λ of the source. During each vibration, which lasts for a time interval T (the period), the source moves a distance $v_ST = v_S/f$ and the wavelength is *shortened* by this amount. Therefore, the observed wavelength λ' is

$$\lambda' = \lambda - \Delta\lambda = \lambda - \frac{v_S}{f}$$

Because $\lambda = v/f$, the frequency f' heard by observer A is

$$f' = \frac{v}{\lambda'} = \frac{v}{\lambda - (v_S/f)} = \frac{v}{(v/f) - (v_S/f)}$$

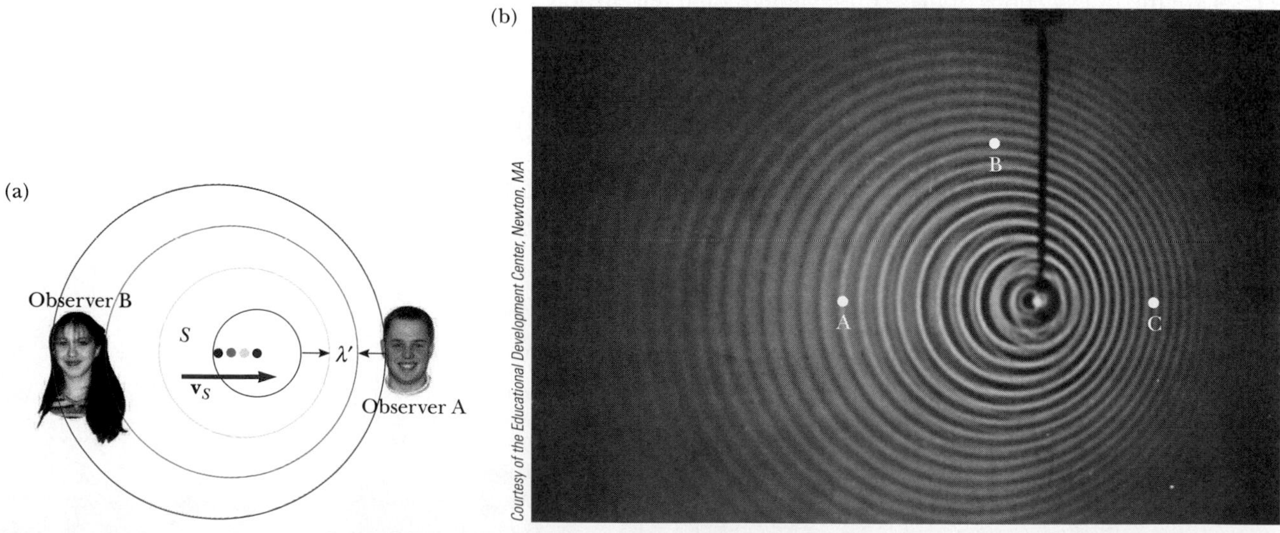

Active Figure 17.9 (a) A source S moving with a speed v_S toward a stationary observer A and away from a stationary observer B. Observer A hears an increased frequency, and observer B hears a decreased frequency. (b) The Doppler effect in water, observed in a ripple tank. A point source is moving to the right with speed v_S. Letters shown in the photo refer to Quick Quiz 17.7.

At the Active Figures link at http://www.pse6.com, you can adjust the speed of the source.

$$f' = \left(\frac{v}{v - v_S}\right)f \qquad \text{(source moving toward observer)} \qquad (17.11)$$

That is, the observed frequency is *increased* whenever the source is moving toward the observer.

When the source moves away from a stationary observer, as is the case for observer B in Figure 17.9a, the observer measures a wavelength λ' that is *greater* than λ and hears a *decreased* frequency:

$$f' = \left(\frac{v}{v + v_S}\right)f \qquad \text{(source moving away from observer)} \qquad (17.12)$$

We can express the general relationship for the observed frequency when a source is moving and an observer is at rest as Equation 17.11, with the same sign convention applied to v_S as was applied to v_O: a positive value is substituted for v_S when the source moves toward the observer and a negative value is substituted when the source moves away from the observer.

Finally, we find the following general relationship for the observed frequency:

General Doppler-shift expression

$$f' = \left(\frac{v + v_O}{v - v_S}\right)f \qquad (17.13)$$

In this expression, the signs for the values substituted for v_O and v_S depend on the direction of the velocity. A positive value is used for motion of the observer or the source *toward* the other, and a negative sign for motion of one *away from* the other.

A convenient rule concerning signs for you to remember when working with all Doppler-effect problems is as follows:

The word *toward* is associated with an *increase* in observed frequency. The words *away from* are associated with a *decrease* in observed frequency.

Although the Doppler effect is most typically experienced with sound waves, it is a phenomenon that is common to all waves. For example, the relative motion of source and observer produces a frequency shift in light waves. The Doppler effect is used in police radar systems to measure the speeds of motor vehicles. Likewise, astronomers use the effect to determine the speeds of stars, galaxies, and other celestial objects relative to the Earth.

PITFALL PREVENTION

17.1 Doppler Effect Does Not Depend on Distance

Many people think that the Doppler effect depends on the distance between the source and the observer. While the intensity of a sound varies as the distance changes, the apparent frequency depends only on the relative speed of source and observer. As you listen to an approaching source, you will detect increasing intensity but constant frequency. As the source passes, you will hear the frequency suddenly drop to a new constant value and the intensity begin to decrease.

Quick Quiz 17.7 Consider detectors of water waves at three locations A, B, and C in Figure 17.9b. Which of the following statements is true? (a) The wave speed is highest at location A. (b) The wave speed is highest at location C. (c) The detected wavelength is largest at location B. (c) The detected wavelength is largest at location C. (e) The detected frequency is highest at location C. (f) The detected frequency is highest at location A.

Quick Quiz 17.8 You stand on a platform at a train station and listen to a train approaching the station at a constant velocity. While the train approaches, but before it arrives, you hear (a) the intensity and the frequency of the sound both increasing (b) the intensity and the frequency of the sound both decreasing (c) the intensity increasing and the frequency decreasing (d) the intensity decreasing and the frequency increasing (e) the intensity increasing and the frequency remaining the same (f) the intensity decreasing and the frequency remaining the same.

Example 17.5 The Broken Clock Radio

Your clock radio awakens you with a steady and irritating sound of frequency 600 Hz. One morning, it malfunctions and cannot be turned off. In frustration, you drop the clock radio out of your fourth-story dorm window, 15.0 m from the ground. Assume the speed of sound is 343 m/s.

(A) As you listen to the falling clock radio, what frequency do you hear just before you hear the radio striking the ground?

(B) At what rate does the frequency that you hear change with time just before you hear the radio striking the ground?

Solution

(A) In conceptualizing the problem, note that the speed of the radio increases as it falls. Thus, it is a source of sound moving away from you with an increasing speed. We categorize this problem as one in which we must combine our understanding of falling objects with that of the frequency shift due to the Doppler effect. To analyze the problem, we identify the clock radio as a moving source of sound for which the Doppler-shifted frequency is given by

$$f' = \left(\frac{v}{v - v_S}\right)f$$

The speed of the source of sound is given by Equation 2.9 for a falling object:

$$v_S = v_{yi} + a_y t = 0 - gt = -gt$$

Thus, the Doppler-shifted frequency of the falling clock radio is

$$(1) \qquad f' = \left(\frac{v}{v - (-gt)}\right)f = \left(\frac{v}{v + gt}\right)f$$

The time at which the radio strikes the ground is found from Equation 2.12:

$$y_f = y_i + v_{yi}t - \tfrac{1}{2}gt^2$$

$$-15.0 \text{ m} = 0 + 0 - \tfrac{1}{2}(9.80 \text{ m/s}^2)t^2$$

$$t = 1.75 \text{ s}$$

Thus, the Doppler-shifted frequency just as the radio strikes the ground is

$$f' = \left(\frac{v}{v + gt}\right)f$$

$$= \left(\frac{343 \text{ m/s}}{343 \text{ m/s} + (9.80 \text{ m/s}^2)(1.75 \text{ s})}\right)(600 \text{ Hz})$$

$$= \boxed{571 \text{ Hz}}$$

(B) The rate at which the frequency changes is found by differentiating Equation (1) with respect to t:

$$\frac{df'}{dt} = \frac{d}{dt}\left(\frac{vf}{v + gt}\right) = \frac{-vg}{(v + gt)^2}f$$

$$= \frac{-(343 \text{ m/s})(9.80 \text{ m/s}^2)}{[343 \text{ m/s} + (9.80 \text{ m/s}^2)(1.75 \text{ s})]^2}(600 \text{ Hz})$$

$$= \boxed{-15.5 \text{ Hz/s}}$$

To finalize this problem, consider the following **What If?**

What If? Suppose you live on the eighth floor instead of the fourth floor. If you repeat the radio-dropping activity, does the frequency shift in part (A) and the rate of change of frequency in part (B) of this example double?

Answer The doubled height does not give a time at which the radio lands that is twice the time found in part (A). From Equation 2.12:

$$y_f = y_i + v_{yi}t - \tfrac{1}{2}gt^2$$

$$-30.0 \text{ m} = 0 + 0 - \tfrac{1}{2}(9.80 \text{ m/s}^2)t^2$$

$$t = 2.47 \text{ s}$$

The new frequency heard just before you hear the radio strike the ground is

$$f' = \left(\frac{v}{v + gt}\right)f$$

$$= \left(\frac{343 \text{ m/s}}{343 \text{ m/s} + (9.80 \text{ m/s}^2)(2.47 \text{ s})}\right)(600 \text{ Hz})$$

$$= \boxed{560 \text{ Hz}}$$

The frequency shift heard on the fourth floor is 600 Hz − 571 Hz = 29 Hz, while the frequency shift heard from the eighth floor is 600 Hz − 560 Hz = 40 Hz, which is not twice as large.

The new rate of change of frequency is

$$\frac{df'}{dt} = \frac{-vg}{(v + gt)^2}f$$

$$= \frac{-(343 \text{ m/s})(9.80 \text{ m/s}^2)}{[343 \text{ m/s} + (9.80 \text{ m/s}^2)(2.47 \text{ s})]^2}(600 \text{ Hz})$$

$$= -15.0 \text{ Hz/s}$$

Note that this value is actually *smaller* in magnitude than the previous value of − 15.5 Hz/s!

Example 17.6 Doppler Submarines

Interactive

A submarine (sub A) travels through water at a speed of 8.00 m/s, emitting a sonar wave at a frequency of 1 400 Hz. The speed of sound in the water is 1 533 m/s. A second submarine (sub B) is located such that both submarines are traveling directly toward one another. The second submarine is moving at 9.00 m/s.

(A) What frequency is detected by an observer riding on sub B as the subs approach each other?

(B) The subs barely miss each other and pass. What frequency is detected by an observer riding on sub B as the subs recede from each other?

Solution

(A) We use Equation 17.13 to find the Doppler-shifted frequency. As the two submarines approach each other, the observer in sub B hears the frequency

$$f' = \left(\frac{v + v_O}{v - v_S}\right)f$$

$$= \left(\frac{1\,533\,\text{m/s} + (+9.00\,\text{m/s})}{1\,533\,\text{m/s} - (+8.00\,\text{m/s})}\right)(1\,400\,\text{Hz}) = 1\,416\,\text{Hz}$$

(B) As the two submarines recede from each other, the observer in sub B hears the frequency

$$f' = \left(\frac{v + v_O}{v - v_S}\right)f$$

$$= \left(\frac{1\,533\,\text{m/s} + (-9.00\,\text{m/s})}{1\,533\,\text{m/s} - (-8.00\,\text{m/s})}\right)(1\,400\,\text{Hz}) = 1\,385\,\text{Hz}$$

What If? While the subs are approaching each other, some of the sound from sub A will reflect from sub B and return to sub A. If this sound were to be detected by an observer on sub A, what is its frequency?

Answer The sound of apparent frequency 1 416 Hz found in part (A) will be reflected from a moving source (sub B) and then detected by a moving observer (sub A). Thus, the frequency detected by sub A is

$$f'' = \left(\frac{v + v_O}{v - v_S}\right)f'$$

$$= \left(\frac{1\,533\text{ m/s} + (+8.00\text{ m/s})}{1\,533\text{ m/s} - (+9.00\text{ m/s})}\right)(1\,416\text{ Hz}) = 1\,432\text{ Hz}$$

This technique is used by police officers to measure the speed of a moving car. Microwaves are emitted from the police car and reflected by the moving car. By detecting the Doppler-shifted frequency of the reflected microwaves, the police officer can determine the speed of the moving car.

At the Interactive Worked Example link at http://www.pse6.com, *you can alter the relative speeds of the submarines and observe the Doppler-shifted frequency.*

Shock Waves

Now consider what happens when the speed v_S of a source *exceeds* the wave speed v. This situation is depicted graphically in Figure 17.10a. The circles represent spherical wave fronts emitted by the source at various times during its motion. At $t = 0$, the source is at S_0, and at a later time t, the source is at S_n. At the time t, the wave front

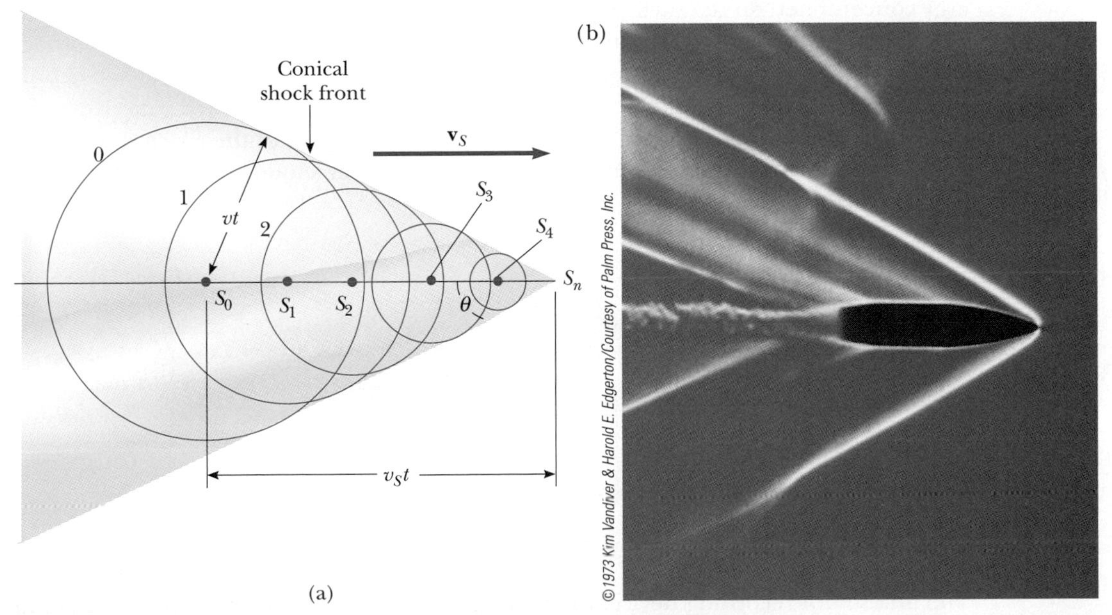

Figure 17.10 (a) A representation of a shock wave produced when a source moves from S_0 to S_n with a speed v_S, which is greater than the wave speed v in the medium. The envelope of the wave fronts forms a cone whose apex half-angle is given by $\sin\theta = v/v_S$. (b) A stroboscopic photograph of a bullet moving at supersonic speed through the hot air above a candle. Note the shock wave in the vicinity of the bullet.

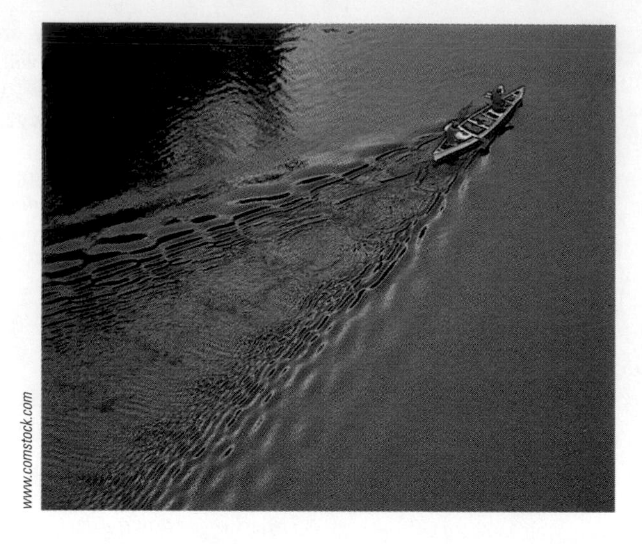

www.comstock.com

Figure 17.11 The **V**-shaped bow wave of a boat is formed because the boat speed is greater than the speed of the water waves it generates. A bow wave is analogous to a shock wave formed by an airplane traveling faster than sound.

centered at S_0 reaches a radius of vt. In this same time interval, the source travels a distance v_St to S_n. At the instant the source is at S_n, waves are just beginning to be generated at this location, and hence the wave front has zero radius at this point. The tangent line drawn from S_n to the wave front centered on S_0 is tangent to all other wave fronts generated at intermediate times. Thus, we see that the envelope of these wave fronts is a cone whose apex half-angle θ (the "Mach angle") is given by

$$\sin\theta = \frac{vt}{v_St} = \frac{v}{v_S}$$

The ratio v_S/v is referred to as the *Mach number*, and the conical wave front produced when $v_S > v$ (supersonic speeds) is known as a *shock wave*. An interesting analogy to shock waves is the **V**-shaped wave fronts produced by a boat (the bow wave) when the boat's speed exceeds the speed of the surface-water waves (Fig. 17.11).

Jet airplanes traveling at supersonic speeds produce shock waves, which are responsible for the loud "sonic boom" one hears. The shock wave carries a great deal of energy concentrated on the surface of the cone, with correspondingly great pressure variations. Such shock waves are unpleasant to hear and can cause damage to buildings when aircraft fly supersonically at low altitudes. In fact, an airplane flying at supersonic speeds produces a double boom because two shock waves are formed, one from the nose of the plane and one from the tail. People near the path of the space shuttle as it glides toward its landing point often report hearing what sounds like two very closely spaced cracks of thunder.

Quick Quiz 17.9 An airplane flying with a constant velocity moves from a cold air mass into a warm air mass. Does the Mach number (a) increase (b) decrease (c) stay the same?

17.5 Digital Sound Recording

The first sound recording device, the phonograph, was invented by Thomas Edison in the nineteenth century. Sound waves were recorded in early phonographs by encoding the sound waveforms as variations in the depth of a continuous groove cut in tin foil wrapped around a cylinder. During playback, as a needle followed along the groove of the rotating cylinder, the needle was pushed back and forth according to the sound

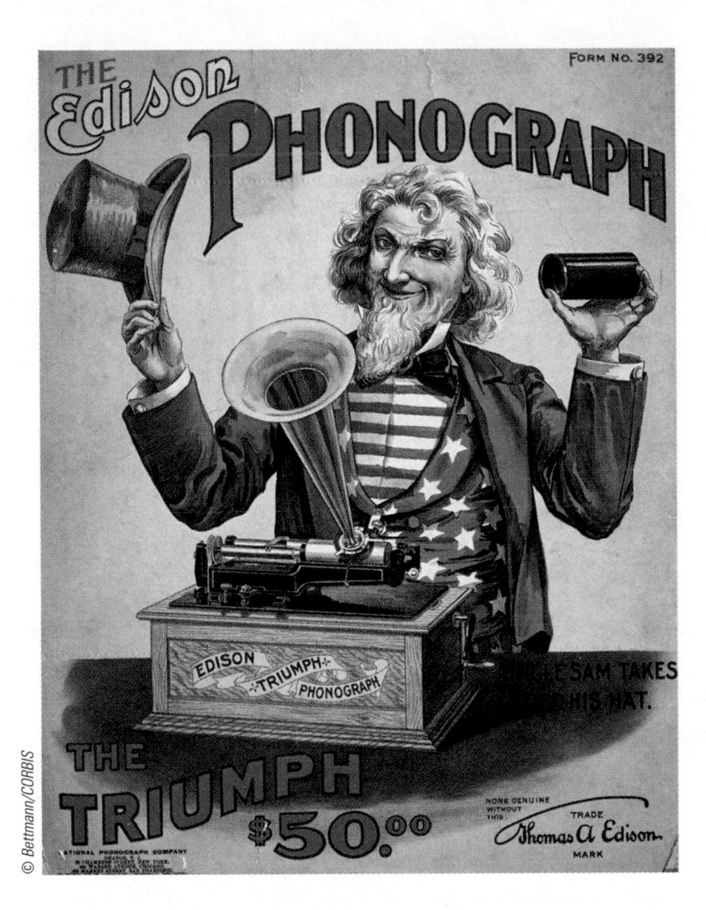

Figure 17.12 An Edison phonograph. Sound information is recorded in a groove on a rotating cylinder of wax. A needle follows the groove and vibrates according to the sound information. A diaphragm and a horn make the sound intense enough to hear.

waves encoded on the record. The needle was attached to a diaphragm and a horn (Fig. 17.12), which made the sound loud enough to be heard.

As the development of the phonograph continued, sound was recorded on cardboard cylinders coated with wax. During the last decade of the nineteenth century and the first half of the twentieth century, sound was recorded on disks made of shellac and clay. In 1948, the plastic phonograph disk was introduced and dominated the recording industry market until the advent of compact discs in the 1980s.

There are a number of problems with phonograph records. As the needle follows along the groove of the rotating phonograph record, the needle is pushed back and forth according to the sound waves encoded on the record. By Newton's third law, the needle also pushes on the plastic. As a result, the recording quality diminishes with each playing as small pieces of plastic break off and the record wears away.

Another problem occurs at high frequencies. The wavelength of the sound on the record is so small that natural bumps and graininess in the plastic create signals as loud as the sound signal, resulting in noise. The noise is especially noticeable during quiet passages in which high frequencies are being played. This is handled electronically by a process known as *pre-emphasis.* In this process, the high frequencies are recorded with more intensity than they actually have, which increases the amplitude of the vibrations and overshadows the sources of noise. Then, an *equalization circuit* in the playback system is used to reduce the intensity of the high-frequency sounds, which also reduces the intensity of the noise.

Example 17.7 Wavelengths on a Phonograph Record

Consider a 10 000-Hz sound recorded on a phonograph record which rotates at $33\frac{1}{3}$ rev/min. How far apart are the crests of the wave for this sound on the record

(A) at the outer edge of the record, 6.0 inches from the center?

(B) at the inner edge, 1.0 inch from the center?

Solution

(A) The linear speed v of a point at the outer edge of the record is $2\pi r/T$ where T is the period of the rotation and r

is the distance from the center. We first find T:

$$T = \frac{1}{f} = \frac{1}{33.33 \text{ rev/min}} = 0.030 \text{ min}\left(\frac{60 \text{ s}}{1 \text{ min}}\right) = 1.8 \text{ s}$$

Now, the linear speed at the outer edge is

$$v = \frac{2\pi r}{T} = \frac{2\pi(6.0 \text{ in.})}{1.8 \text{ s}} = 21 \text{ in./s}\left(\frac{2.54 \text{ cm}}{1 \text{ in.}}\right)$$
$$= 53 \text{ cm/s}$$

Thus, the wave on the record is moving past the needle at this speed. The wavelength is

$$\lambda = \frac{v}{f} = \frac{53 \text{ cm/s}}{10\,000 \text{ Hz}} = 5.3 \times 10^{-5} \text{ m}$$
$$= 53 \ \mu\text{m}$$

(B) The linear speed at the inner edge is

$$v = \frac{2\pi r}{T} = \frac{2\pi(1.0 \text{ in.})}{1.8 \text{ s}} = 3.5 \text{ in./s}\left(\frac{2.54 \text{ cm}}{1 \text{ in.}}\right)$$
$$= 8.9 \text{ cm/s}$$

The wavelength is

$$\lambda = \frac{v}{f} = \frac{8.9 \text{ cm/s}}{10\,000 \text{ Hz}} = 8.9 \times 10^{-6} \text{ m}$$
$$= 8.9 \ \mu\text{m}$$

Thus, the problem with noise interfering with the recorded sound is more severe at the inner edge of the disk than at the outer edge.

Digital Recording

In digital recording, information is converted to binary code (ones and zeroes), similar to the dots and dashes of Morse code. First, the waveform of the sound is *sampled,* typically at the rate of 44 100 times per second. Figure 17.13 illustrates this process. The sampling frequency is much higher than the upper range of hearing, about 20 000 Hz, so all frequencies of sound are sampled at this rate. During each sampling, the pressure of the wave is measured and converted to a voltage. Thus, there are 44 100 numbers associated with each second of the sound being sampled.

These measurements are then converted to *binary numbers,* which are numbers expressed using base 2 rather than base 10. Table 17.3 shows some sample binary numbers. Generally, voltage measurements are recorded in 16-bit "words," where each bit is a one or a zero. Thus, the number of different voltage levels that can be assigned codes is $2^{16} = 65\,536$. The number of bits in one second of sound is $16 \times 44\,100 = 705\,600$. It is these strings of ones and zeroes, in 16-bit words, that are recorded on the surface of a compact disc.

Figure 17.14 shows a magnification of the surface of a compact disc. There are two types of areas that are detected by the laser playback system—*lands* and *pits.* The lands are untouched regions of the disc surface that are highly reflective. The pits, which are areas burned into the surface, scatter light rather than reflecting it back to the detection system. The playback system samples the reflected light 705 600 times per second. When the laser moves from a pit to a flat or from a flat to a pit, the reflected light changes during the sampling and the bit is recorded as a one. If there is no change during the sampling, the bit is recorded as a zero.

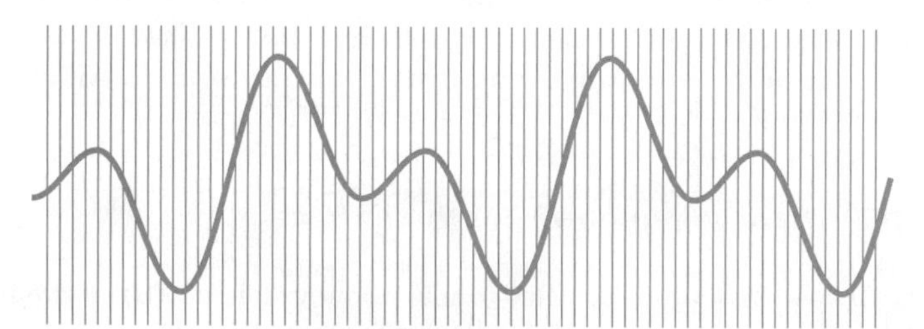

Figure 17.13 Sound is digitized by electronically sampling the sound waveform at periodic intervals. During each time interval between the blue lines, a number is recorded for the average voltage during the interval. The sampling rate shown here is much slower than the actual sampling rate of 44 100 samples per second.

Table 17.3

Sample Binary Numbers

Number in Base 10	Number in Binary	Sum
1	0000000000000001	1
2	0000000000000010	2 + 0
3	0000000000000011	2 + 1
10	0000000000001010	8 + 0 + 2 + 0
37	0000000000100101	32 + 0 + 0 + 4 + 0 + 1
275	0000000100010011	256 + 0 + 0 + 0 + 16 + 0 + 0 + 2 + 1

The binary numbers read from the CD are converted back to voltages, and the waveform is reconstructed, as shown in Figure 17.15. Because the sampling rate is so high—44 100 voltage readings each second—the fact that the waveform is constructed from step-wise discrete voltages is not evident in the sound.

The advantage of digital recording is in the high fidelity of the sound. With analog recording, any small imperfection in the record surface or the recording equipment can cause a distortion of the waveform. If all peaks of a maximum in a waveform are clipped off so as to be only 90% as high, for example, this will have a major effect on the spectrum of the sound in an analog recording. With digital recording, however, it takes a major imperfection to turn a one into a zero. If an imperfection causes the magnitude of a one to be 90% of the original value, it still registers as a one, and there is no distortion. Another advantage of digital recording is that the information is extracted optically, so that there is no mechanical wear on the disc.

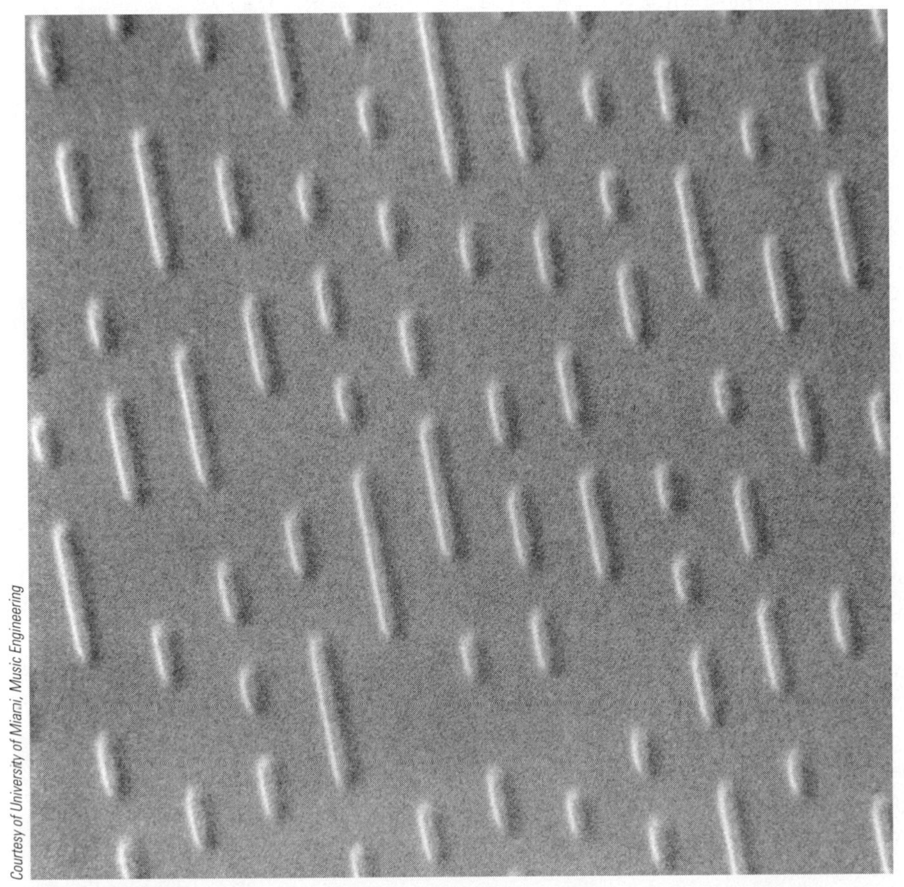

Courtesy of University of Miami, Music Engineering

Figure 17.14 The surface of a compact disc, showing the pits. Transitions between pits and lands correspond to ones. Regions without transitions correspond to zeroes.

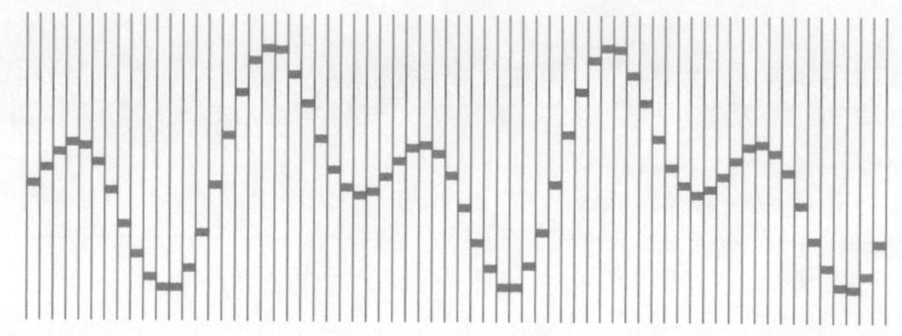

Figure 17.15 The reconstruction of the sound wave sampled in Figure 17.13. Notice that the reconstruction is step-wise, rather than the continuous waveform in Figure 17.13.

Example 17.8 How Big Are the Pits?

In Example 10.2, we mentioned that the speed with which the CD surface passes the laser is 1.3 m/s. What is the average length of the audio track on a CD associated with each bit of the audio information?

Solution In one second, a 1.3-m length of audio track passes by the laser. This length includes 705 600 bits of audio information. Thus, the average length per bit is

$$\frac{1.3 \text{ m}}{705\,600 \text{ bits}} = 1.8 \times 10^{-6} \text{ m/bit}$$

$$= \boxed{1.8\ \mu\text{m/bit}}$$

The average length per bit of *total* information on the CD is smaller than this because there is additional information on the disc besides the audio information. This information includes error correction codes, song numbers, timing codes, etc. As a result, the shortest length per bit is actually about 0.8 μm.

Example 17.9 What's the Number?

Consider the photograph of the compact disc surface in Figure 17.14. Audio data undergoes complicated processing in order to reduce a variety of errors in reading the data. Thus, an audio "word" is not laid out linearly on the disc. Suppose that data has been read from the disc, the error encoding has been removed, and the resulting audio word is

1 0 1 1 1 0 1 1 1 0 1 1 1 0 1 1

What is the decimal number represented by this 16-bit word?

Solution We convert each of these bits to a power of 2 and add the results:

$1 \times 2^{15} = 32\,768$
$0 \times 2^{14} = 0$
$1 \times 2^{13} = 8\,192$
$1 \times 2^{12} = 4\,096$
$1 \times 2^{11} = 2\,048$
$0 \times 2^{10} = 0$
$1 \times 2^{9} = 512$
$1 \times 2^{8} = 256$
$1 \times 2^{7} = 128$
$0 \times 2^{6} = 0$
$1 \times 2^{5} = 32$
$1 \times 2^{4} = 16$
$1 \times 2^{3} = 8$
$0 \times 2^{2} = 0$
$1 \times 2^{1} = 2$
$1 \times 2^{0} = 1$

$\text{sum} = \boxed{48\,059}$

This number is converted by the CD player into a voltage, representing one of the 44 100 values that will be used to build one second of the electronic waveform that represents the recorded sound.

17.6 Motion Picture Sound

Another interesting application of digital sound is the soundtrack in a motion picture. Early twentieth-century movies recorded sound on phonograph records, which were synchronized with the action on the screen. Beginning with early newsreel films, the *variable-area optical soundtrack* process was introduced, in which sound was recorded on an optical track on the film. The width of the transparent portion of the track varied according to the sound wave that was recorded. A photocell detecting light passing through the track converted the varying light intensity to a sound wave. As with phonograph recording, there are a number of difficulties with this recording system. For example, dirt or fingerprints on the film cause fluctuations in intensity and loss of fidelity.

Digital recording on film first appeared with *Dick Tracy* (1990), using the Cinema Digital Sound (CDS) system. This system suffered from lack of an analog backup system in case of equipment failure and is no longer used in the film industry. It did, however, introduce the use of 5.1 channels of sound—Left, Center, Right, Right Surround, Left Surround, and Low Frequency Effects (LFE). The LFE channel, which is the "0.1

channel" of 5.1, carries very low frequencies for dramatic sound from explosions, earthquakes, and the like.

Current motion pictures are produced with three systems of digital sound recording:

Dolby Digital; In this format, 5.1 channels of digital sound are optically stored between the sprocket holes of the film. There is an analog optical backup in case the digital system fails. The first film to use this technique was *Batman Returns* (1992).

DTS (Digital Theater Sound); 5.1 channels of sound are stored on a separate CD-ROM which is synchronized to the film print by time codes on the film. There is an analog optical backup in case the digital system fails. The first film to use this technique was *Jurassic Park* (1993).

SDDS (Sony Dynamic Digital Sound); Eight full channels of digital sound are optically stored outside the sprocket holes on both sides of film. There is an analog optical backup in case the digital system fails. The first film to use this technique was *Last Action Hero* (1993). The existence of information on both sides of the tape is a system of redundancy—in case one side is damaged, the system will still operate. SDDS employs a full-spectrum LFE channel and two additional channels (left center and right center behind the screen). In Figure 17.16, showing a section of SDDS film, both the analog optical soundtrack and the dual digital soundtracks can be seen.

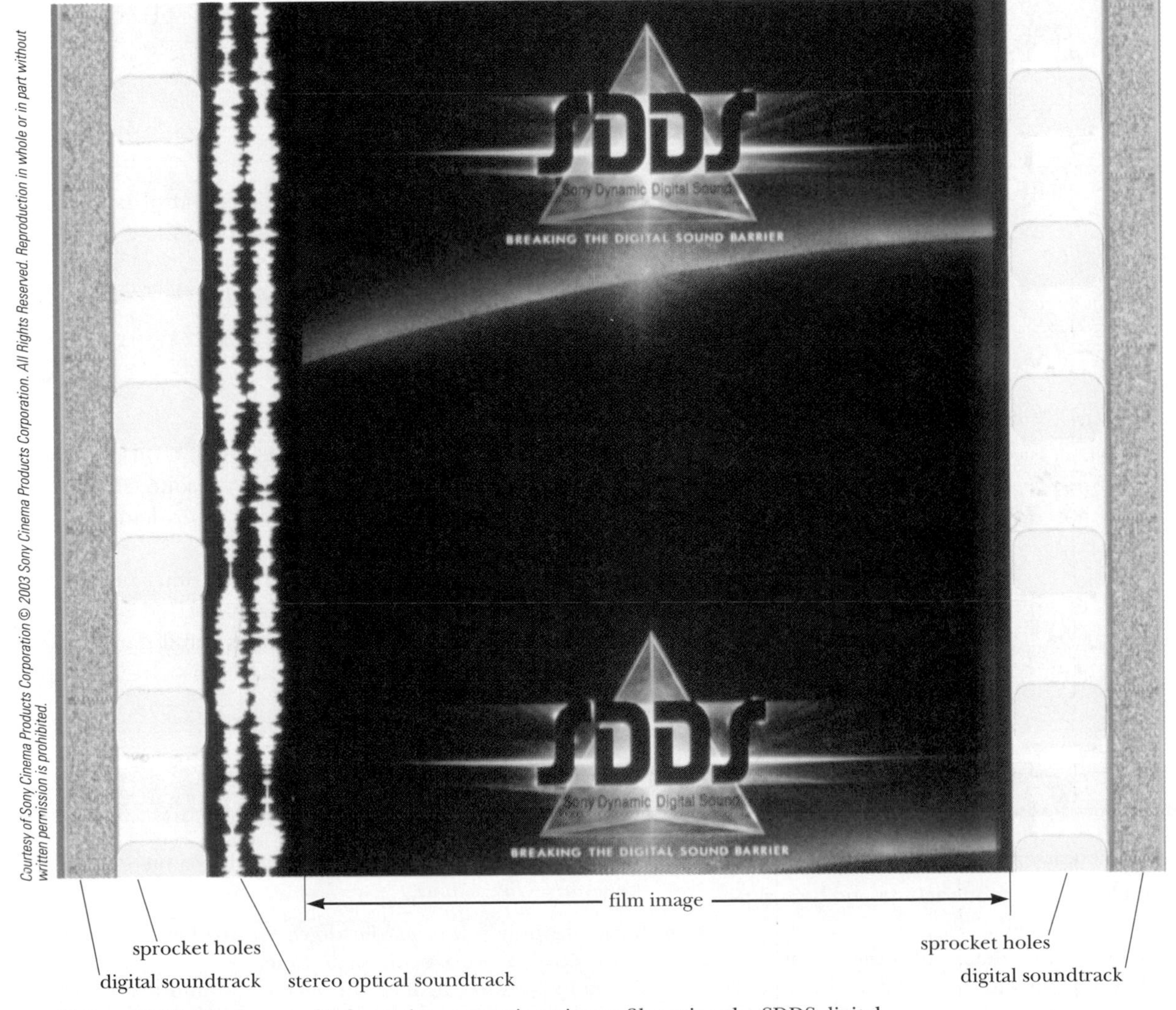

Figure 17.16 The layout of information on motion picture film using the SDDS digital sound system.

SUMMARY

Take a practice test for this chapter by clicking on the Practice Test link at http://www.pse6.com.

Sound waves are longitudinal and travel through a compressible medium with a speed that depends on the elastic and inertial properties of that medium. The speed of sound in a liquid or gas having a bulk modulus B and density ρ is

$$v = \sqrt{\frac{B}{\rho}} \tag{17.1}$$

For sinusoidal sound waves, the variation in the position of an element of the medium is given by

$$s(x, t) = s_{max} \cos(kx - \omega t) \tag{17.2}$$

and the variation in pressure from the equilibrium value is

$$\Delta P = \Delta P_{max} \sin(kx - \omega t) \tag{17.3}$$

where ΔP_{max} is the **pressure amplitude.** The pressure wave is 90° out of phase with the displacement wave. The relationship between s_{max} and ΔP_{max} is given by

$$\Delta P_{max} = \rho v \omega s_{max} \tag{17.4}$$

The intensity of a periodic sound wave, which is the power per unit area, is

$$I \equiv \frac{\mathcal{P}}{A} = \frac{\Delta P_{max}^2}{2\rho v} \tag{17.5, 17.6}$$

The sound level of a sound wave, in decibels, is given by

$$\beta \equiv 10 \log\left(\frac{I}{I_0}\right) \tag{17.8}$$

The constant I_0 is a reference intensity, usually taken to be at the threshold of hearing (1.00×10^{-12} W/m^2), and I is the intensity of the sound wave in watts per square meter.

The change in frequency heard by an observer whenever there is relative motion between a source of sound waves and the observer is called the **Doppler effect.** The observed frequency is

$$f' = \left(\frac{v + v_O}{v - v_S}\right) f \tag{17.13}$$

In this expression, the signs for the values substituted for v_O and v_S depend on the direction of the velocity. A positive value for the velocity of the observer or source is substituted if the velocity of one is toward the other, while a negative value represents a velocity of one away from the other.

In digital recording of sound, the sound waveform is sampled 44 100 times per second. The pressure of the wave for each sampling is measured and converted to a binary number. In playback, these binary numbers are read and used to build the original waveform.

QUESTIONS

1. Why are sound waves characterized as longitudinal?
2. If an alarm clock is placed in a good vacuum and then activated, no sound is heard. Explain.
3. A sonic ranger is a device that determines the distance to an object by sending out an ultrasonic sound pulse and measuring how long it takes for the wave to return after it reflects from the object. Typically these devices cannot reliably detect an object that is less than half a meter from the sensor. Why is that?
4. A friend sitting in her car far down the road waves to you and beeps her horn at the same time. How far away must she be for you to calculate the speed of sound to two significant figures by measuring the time it takes for the sound to reach you?

5. If the wavelength of sound is reduced by a factor of 2, what happens to its frequency? Its speed?

6. By listening to a band or orchestra, how can you determine that the speed of sound is the same for all frequencies?

7. In Example 17.3 we found that a point source with a power output of 80 W produces sound with an intensity of $1.00 \times 10^{-8}\ W/m^2$, which corresponds to 40 dB, at a distance of about 16 miles. Why do you suppose you cannot normally hear a rock concert that is going on 16 miles away? (See Table 17.2.)

8. If the distance from a point source is tripled, by what factor does the intensity decrease?

9. *The Tunguska Event.* On June 30, 1908, a meteor burned up and exploded in the atmosphere above the Tunguska River valley in Siberia. It knocked down trees over thousands of square kilometers and started a forest fire, but apparently caused no human casualties. A witness sitting on his doorstep outside the zone of falling trees recalled events in the following sequence: He saw a moving light in the sky, brighter than the sun and descending at a low angle to the horizon. He felt his face become warm. He felt the ground shake. An invisible agent picked him up and immediately dropped him about a meter farther away from where the light had been. He heard a very loud protracted rumbling. Suggest an explanation for these observations and for the order in which they happened.

10. Explain how the Doppler effect with microwaves is used to determine the speed of an automobile.

11. Explain what happens to the frequency of the echo of your car horn as you move in a vehicle toward the wall of a canyon. What happens to the frequency as you move away from the wall?

12. Of the following sounds, which is most likely to have a sound level of 60 dB: a rock concert, the turning of a page in this textbook, normal conversation, or a cheering crowd at a football game?

13. Estimate the decibel level of each of the sounds in the previous question.

14. A binary star system consists of two stars revolving about their common center of mass. If we observe the light reaching us from one of these stars as it makes one complete revolution, what does the Doppler effect predict will happen to this light?

15. How can an object move with respect to an observer so that the sound from it is not shifted in frequency?

16. Suppose the wind blows. Does this cause a Doppler effect for sound propagating through the air? Is it like a moving source or a moving observer?

17. Why is it not possible to use sonar (sound waves) to determine the speed of an object traveling faster than the speed of sound?

18. Why is it so quiet after a snowfall?

19. Why is the intensity of an echo less than that of the original sound?

20. A loudspeaker built into the exterior wall of an airplane produces a large-amplitude burst of vibration at 200 Hz, then a burst at 300 Hz, and then a burst at 400 Hz (Boop . . . baap . . . beep), all while the plane is flying faster than the speed of sound. Describe qualitatively what an observer hears if she is in front of the plane, close to its flight path. **What If?** What will the observer hear if the pilot uses the loudspeaker to say, “How are you?”

21. In several cases, a nearby star has been found to have a large planet orbiting about it, although the planet could not be seen. Using the ideas of a system rotating about its center of mass and of the Doppler shift for light (which is in several ways similar to the Doppler effect for sound), explain how an astronomer could determine the presence of the invisible planet.

PROBLEMS

1, 2, 3 = straightforward, intermediate, challenging ☐ = full solution available in the *Student Solutions Manual and Study Guide*

www = coached solution with hints available at http://www.pse6.com ▭ = computer useful in solving problem

▭ = paired numerical and symbolic problems

Section 17.1 Speed of Sound Waves

1. Suppose that you hear a clap of thunder 16.2 s after seeing the associated lightning stroke. The speed of sound waves in air is 343 m/s, and the speed of light is 3.00×10^8 m/s. How far are you from the lightning stroke?

2. Find the speed of sound in mercury, which has a bulk modulus of approximately $2.80 \times 10^{10}\ N/m^2$ and a density of 13 600 kg/m^3.

3. A flowerpot is knocked off a balcony 20.0 m above the sidewalk and falls toward an unsuspecting 1.75-m-tall man who is standing below. How close to the sidewalk can the flower pot fall before it is too late for a warning shouted from the balcony to reach the man in time? Assume that the man below requires 0.300 s to respond to the warning.

4. The speed of sound in air (in m/s) depends on temperature according to the approximate expression

$$v = 331.5 + 0.607T_C$$

where T_C is the Celsius temperature. In dry air the temperature decreases about 1°C for every 150 m rise in altitude. (a) Assuming this change is constant up to an altitude of 9 000 m, how long will it take the sound from an airplane flying at 9 000 m to reach the ground on a day when the ground temperature is 30°C? (b) **What If?** Compare this to the time interval required if the air were a constant 30°C. Which time interval is longer?

5. A cowboy stands on horizontal ground between two parallel vertical cliffs. He is not midway between the cliffs. He fires a shot and hears its echoes. The second echo arrives 1.92 s after the first and 1.47 s before the third. Consider only the sound traveling parallel to the ground and reflecting from the cliffs. Take the speed of sound as 340 m/s. (a) What is the distance between the cliffs? (b) **What If?** If he can hear a fourth echo, how long after the third echo does it arrive?

6. A rescue plane flies horizontally at a constant speed searching for a disabled boat. When the plane is directly above the boat, the boat's crew blows a loud horn. By the time the plane's sound detector perceives the horn's sound, the plane has traveled a distance equal to half its altitude above the ocean. If it takes the sound 2.00 s to reach the plane, determine (a) the speed of the plane and (b) its altitude. Take the speed of sound to be 343 m/s.

Section 17.2 Periodic Sound Waves

> *Note*: Use the following values as needed unless otherwise specified: the equilibrium density of air at 20°C is $\rho = 1.20\ \text{kg/m}^3$. The speed of sound in air is $v = 343\ \text{m/s}$. Pressure variations ΔP are measured relative to atmospheric pressure, $1.013 \times 10^5\ \text{N/m}^2$.
> Problem 70 in Chapter 2 can also be assigned with this section.

7. A bat (Fig. P17.7) can detect very small objects, such as an insect whose length is approximately equal to one wavelength of the sound the bat makes. If a bat emits chirps at a frequency of 60.0 kHz, and if the speed of sound in air is 340 m/s, what is the smallest insect the bat can detect?

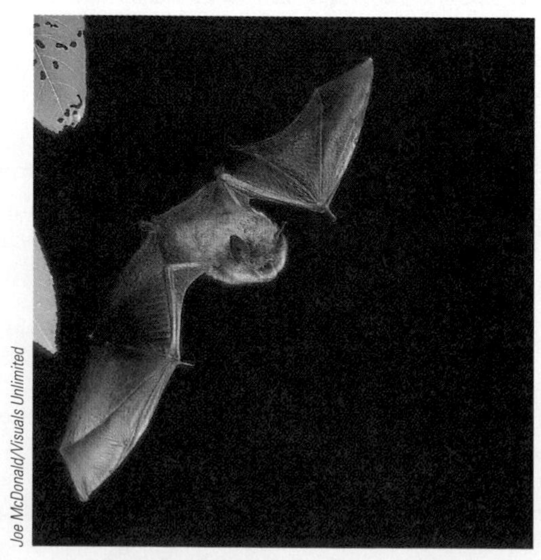

Figure P17.7 Problems 7 and 60.

8. An ultrasonic tape measure uses frequencies above 20 MHz to determine dimensions of structures such as buildings. It does this by emitting a pulse of ultrasound into air and then measuring the time for an echo to return from a reflecting surface whose distance away is to be measured. The distance is displayed as a digital read-out. For a tape measure that emits a pulse of ultrasound with a frequency of 22.0 MHz, (a) What is the distance to an object from which the echo pulse returns after 24.0 ms when the air temperature is 26°C? (b) What should be the duration of the emitted pulse if it is to include 10 cycles of the ultrasonic wave? (c) What is the spatial length of such a pulse?

9. Ultrasound is used in medicine both for diagnostic imaging and for therapy. For diagnosis, short pulses of ultrasound are passed through the patient's body. An echo reflected from a structure of interest is recorded, and from the time delay for the return of the echo the distance to the structure can be determined. A single transducer emits and detects the ultrasound. An image of the structure is obtained by reducing the data with a computer. With sound of low intensity, this technique is noninvasive and harmless. It is used to examine fetuses, tumors, aneurysms, gallstones, and many other structures. A Doppler ultrasound unit is used to study blood flow and functioning of the heart. To reveal detail, the wavelength of the reflected ultrasound must be small compared to the size of the object reflecting the wave. For this reason, frequencies in the range 1.00 to 20.0 MHz are used. What is the range of wavelengths corresponding to this range of frequencies? The speed of ultrasound in human tissue is about 1 500 m/s (nearly the same as the speed of sound in water).

10. A sound wave in air has a pressure amplitude equal to $4.00 \times 10^{-3}\ \text{N/m}^2$. Calculate the displacement amplitude of the wave at a frequency of 10.0 kHz.

11. A sinusoidal sound wave is described by the displacement wave function

$$s(x, t) = (2.00\ \mu\text{m}) \cos[(15.7\ \text{m}^{-1})x - (858\ \text{s}^{-1})t]$$

(a) Find the amplitude, wavelength, and speed of this wave. (b) Determine the instantaneous displacement from equilibrium of the elements of air at the position $x = 0.050\,0$ m at $t = 3.00$ ms. (c) Determine the maximum speed of the element's oscillatory motion.

12. As a certain sound wave travels through the air, it produces pressure variations (above and below atmospheric pressure) given by $\Delta P = 1.27 \sin(\pi x - 340\pi t)$ in SI units. Find (a) the amplitude of the pressure variations, (b) the frequency, (c) the wavelength in air, and (d) the speed of the sound wave.

13. Write an expression that describes the pressure variation as a function of position and time for a sinusoidal sound wave in air, if $\lambda = 0.100$ m and $\Delta P_{max} = 0.200\ \text{N/m}^2$.

14. Write the function that describes the displacement wave corresponding to the pressure wave in Problem 13.

15. An experimenter wishes to generate in air a sound wave that has a displacement amplitude of 5.50×10^{-6} m. The pressure amplitude is to be limited to $0.840\ \text{N/m}^2$. What is the minimum wavelength the sound wave can have?

16. The tensile stress in a thick copper bar is 99.5% of its elastic breaking point of 13.0×10^{10} N/m^2. If a 500-Hz sound wave is transmitted through the material, (a) what displacement amplitude will cause the bar to break? (b) What is the maximum speed of the elements of copper at this moment? (c) What is the sound intensity in the bar?

17. Prove that sound waves propagate with a speed given by Equation 17.1. Proceed as follows. In Figure 17.3, consider a thin cylindrical layer of air in the cylinder, with face area A and thickness Δx. Draw a free-body diagram of this thin layer. Show that $\Sigma F_x = ma_x$ implies that $-[\partial(\Delta P)/\partial x]A\,\Delta x = \rho A\,\Delta x(\partial^2 s/\partial t^2)$. By substituting $\Delta P = -B(\partial s/\partial x)$, obtain the wave equation for sound, $(B/\rho)(\partial^2 s/\partial x^2) = (\partial^2 s/\partial t^2)$. To a mathematical physicist, this equation demonstrates the existence of sound waves and determines their speed. As a physics student, you must take another step or two. Substitute into the wave equation the trial solution $s(x, t) = s_{max}\cos(kx - \omega t)$. Show that this function satisfies the wave equation provided that $\omega/k = \sqrt{B/\rho}$. This result reveals that sound waves exist provided that they move with the speed $v = f\lambda = (2\pi f)(\lambda/2\pi) = \omega/k = \sqrt{B/\rho}$.

Section 17.3 Intensity of Periodic Sound Waves

18. The area of a typical eardrum is about 5.00×10^{-5} m^2. Calculate the sound power incident on an eardrum at (a) the threshold of hearing and (b) the threshold of pain.

19. Calculate the sound level in decibels of a sound wave that has an intensity of 4.00 μW/m^2.

20. A vacuum cleaner produces sound with a measured sound level of 70.0 dB. (a) What is the intensity of this sound in W/m^2? (b) What is the pressure amplitude of the sound?

21. The intensity of a sound wave at a fixed distance from a speaker vibrating at 1.00 kHz is 0.600 W/m^2. (a) Determine the intensity if the frequency is increased to 2.50 kHz while a constant displacement amplitude is maintained. (b) Calculate the intensity if the frequency is reduced to 0.500 kHz and the displacement amplitude is doubled.

22. The intensity of a sound wave at a fixed distance from a speaker vibrating at a frequency f is I. (a) Determine the intensity if the frequency is increased to f' while a constant displacement amplitude is maintained. (b) Calculate the intensity if the frequency is reduced to $f/2$ and the displacement amplitude is doubled.

23. The most soaring vocal melody is in Johann Sebastian Bach's *Mass in B minor.* A portion of the score for the Credo section, number 9, bars 25 to 33, appears in Figure P17.23. The repeating syllable O in the phrase "resurrectionem mortuorum" (the resurrection of the dead) is seamlessly passed from basses to tenors to altos to first sopranos, like a baton in a relay. Each voice carries the melody up in a run of an octave or more. Together they carry it from D below middle C to A above a tenor's high C. In concert pitch, these notes are now assigned frequencies of 146.8 Hz and 880.0 Hz. (a) Find the wavelengths of the initial and final notes. (b) Assume that the choir sings the melody with a uniform sound level of 75.0 dB. Find the pressure amplitudes of the initial and final notes. (c) Find the displacement amplitudes of the initial and final notes. (d) **What If?** In Bach's time, before the invention of the tuning fork, frequencies were assigned to notes as a matter of immediate local convenience. Assume that the rising melody was sung starting from 134.3 Hz and ending at 804.9 Hz. How would the answers to parts (a) through (c) change?

24. The tube depicted in Figure 17.2 is filled with air at 20°C and equilibrium pressure 1 atm. The diameter of the tube is 8.00 cm. The piston is driven at a frequency of 600 Hz with an amplitude of 0.120 cm. What power must be supplied to maintain the oscillation of the piston?

25. A family ice show is held at an enclosed arena. The skaters perform to music with level 80.0 dB. This is too loud for your baby, who yells at 75.0 dB. (a) What total sound intensity engulfs you? (b) What is the combined sound level?

26. Consider sinusoidal sound waves propagating in these three different media: air at 0°C, water, and iron. Use densities and speeds from Tables 14.1 and 17.1. Each wave has the same intensity I_0 and the same angular frequency ω_0. (a) Compare the values of the wavelength in the three media. (b) Compare the values of the displacement amplitude in the three media. (c) Compare the values of the pressure amplitude in the three media. (d) For values of $\omega_0 = 2\,000\,\pi$ rad/s and $I_0 = 1.00 \times 10^{-6}$ W/m^2, evaluate the wavelength, displacement amplitude, and pressure amplitude in each of the three media.

27. The power output of a certain public address speaker is 6.00 W. Suppose it broadcasts equally in all directions. (a) Within what distance from the speaker would the sound be painful to the ear? (b) At what distance from the speaker would the sound be barely audible?

Figure P17.23 Bass (blue), tenor (green), alto (brown), and first soprano (red) parts for a portion of Bach's *Mass in B minor.* For emphasis, the line we choose to call the melody is printed in black. Parts for the second soprano, violins, viola, flutes, oboes, and continuo are omitted. The tenor part is written as it is sung.

28. Show that the difference between decibel levels β_1 and β_2 of a sound is related to the ratio of the distances r_1 and r_2 from the sound source by

$$\beta_2 - \beta_1 = 20\log\left(\frac{r_1}{r_2}\right)$$

29. A firework charge is detonated many meters above the ground. At a distance of 400 m from the explosion, the acoustic pressure reaches a maximum of 10.0 N/m^2. Assume that the speed of sound is constant at 343 m/s throughout the atmosphere over the region considered, that the ground absorbs all the sound falling on it, and that the air absorbs sound energy as described by the rate 7.00 dB/km. What is the sound level (in dB) at 4.00 km from the explosion?

30. A loudspeaker is placed between two observers who are 110 m apart, along the line connecting them. If one observer records a sound level of 60.0 dB and the other records a sound level of 80.0 dB, how far is the speaker from each observer?

31. Two small speakers emit sound waves of different frequencies. Speaker *A* has an output of 1.00 mW, and speaker *B* has an output of 1.50 mW. Determine the sound level (in dB) at point *C* (Fig. P17.31) if (a) only speaker *A* emits sound, (b) only speaker *B* emits sound, and (c) both speakers emit sound.

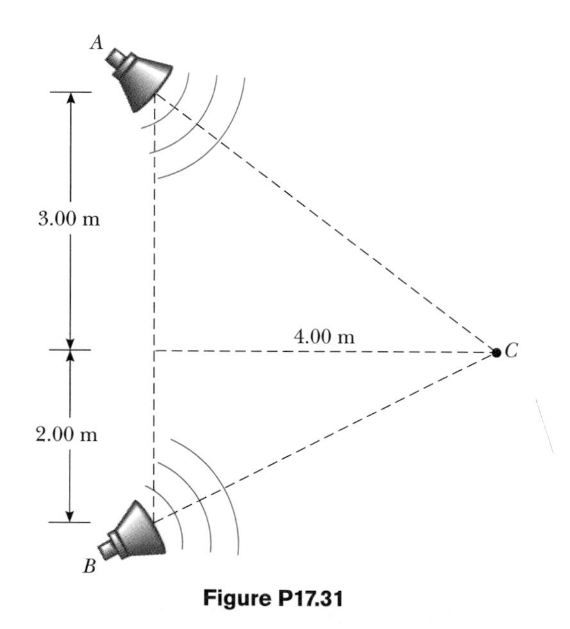

Figure P17.31

32. A jackhammer, operated continuously at a construction site, behaves as a point source of spherical sound waves. A construction supervisor stands 50.0 m due north of this sound source and begins to walk due west. How far does she have to walk in order for the amplitude of the wave function to drop by a factor of 2.00?

33. The sound level at a distance of 3.00 m from a source is 120 dB. At what distance will the sound level be (a) 100 dB and (b) 10.0 dB?

34. A fireworks rocket explodes at a height of 100 m above the ground. An observer on the ground directly under the explosion experiences an average sound intensity of 7.00×10^{-2} W/m^2 for 0.200 s. (a) What is the total sound energy of the explosion? (b) What is the sound level in decibels heard by the observer?

35. As the people sing in church, the sound level everywhere inside is 101 dB. No sound is transmitted through the massive walls, but all the windows and doors are open on a summer morning. Their total area is 22.0 m^2. (a) How much sound energy is radiated in 20.0 min? (b) Suppose the ground is a good reflector and sound radiates uniformly in all horizontal and upward directions. Find the sound level 1 km away.

36. The smallest change in sound level that a person can distinguish is approximately 1 dB. When you are standing next to your power lawnmower as it is running, can you hear the steady roar of your neighbor's lawnmower? Perform an order-of-magnitude calculation to substantiate your answer, stating the data you measure or estimate.

Section 17.4 The Doppler Effect

37. A train is moving parallel to a highway with a constant speed of 20.0 m/s. A car is traveling in the same direction as the train with a speed of 40.0 m/s. The car horn sounds at a frequency of 510 Hz, and the train whistle sounds at a frequency of 320 Hz. (a) When the car is behind the train, what frequency does an occupant of the car observe for the train whistle? (b) After the car passes and is in front of the train, what frequency does a train passenger observe for the car horn?

38. Expectant parents are thrilled to hear their unborn baby's heartbeat, revealed by an ultrasonic motion detector. Suppose the fetus's ventricular wall moves in simple harmonic motion with an amplitude of 1.80 mm and a frequency of 115 per minute. (a) Find the maximum linear speed of the heart wall. Suppose the motion detector in contact with the mother's abdomen produces sound at 2 000 000.0 Hz, which travels through tissue at 1.50 km/s. (b) Find the maximum frequency at which sound arrives at the wall of the baby's heart. (c) Find the maximum frequency at which reflected sound is received by the motion detector. By electronically "listening" for echoes at a frequency different from the broadcast frequency, the motion detector can produce beeps of audible sound in synchronization with the fetal heartbeat.

39. Standing at a crosswalk, you hear a frequency of 560 Hz from the siren of an approaching ambulance. After the ambulance passes, the observed frequency of the siren is 480 Hz. Determine the ambulance's speed from these observations.

40. A block with a speaker bolted to it is connected to a spring having spring constant $k = 20.0$ N/m as in Figure P17.40. The total mass of the block and speaker is 5.00 kg, and the amplitude of this unit's motion is 0.500 m. (a) If the speaker emits sound waves of frequency 440 Hz, determine the highest and lowest frequencies heard by the person to the right of the speaker. (b) If the maximum sound level heard by the person is 60.0 dB when he is closest to the speaker, 1.00 m away, what is the minimum sound level heard by the observer? Assume that the speed of sound is 343 m/s.

Figure P17.40

41. A tuning fork vibrating at 512 Hz falls from rest and accelerates at 9.80 m/s^2. How far below the point of release is the tuning fork when waves of frequency 485 Hz reach the release point? Take the speed of sound in air to be 340 m/s.

42. At the Winter Olympics, an athlete rides her luge down the track while a bell just above the wall of the chute rings continuously. When her sled passes the bell, she hears the frequency of the bell fall by the musical interval called a minor third. That is, the frequency she hears drops to five sixths of its original value. (a) Find the speed of sound in air at the ambient temperature -10.0°C. (b) Find the speed of the athlete.

43. A siren mounted on the roof of a firehouse emits sound at a frequency of 900 Hz. A steady wind is blowing with a speed of 15.0 m/s. Taking the speed of sound in calm air to be 343 m/s, find the wavelength of the sound (a) upwind of the siren and (b) downwind of the siren. Firefighters are approaching the siren from various directions at 15.0 m/s. What frequency does a firefighter hear (c) if he or she is approaching from an upwind position, so that he or she is moving in the direction in which the wind is blowing? (d) if he or she is approaching from a downwind position and moving against the wind?

44. The Concorde can fly at Mach 1.50, which means the speed of the plane is 1.50 times the speed of sound in air. What is the angle between the direction of propagation of the shock wave and the direction of the plane's velocity?

45. When high-energy charged particles move through a transparent medium with a speed greater than the speed of light in that medium, a shock wave, or bow wave, of light is produced. This phenomenon is called the *Cerenkov effect.* When a nuclear reactor is shielded by a large pool of water, Cerenkov radiation can be seen as a blue glow in the vicinity of the reactor core, due to high-speed electrons moving through the water. In a particular case, the Cerenkov radiation produces a wave front with an apex half-angle of 53.0°. Calculate the speed of the electrons in the water. (The speed of light in water is 2.25×10^8 m/s.)

46. The loop of a circus ringmaster's whip travels at Mach 1.38 (that is, $v_S/v = 1.38$). What angle does the shock wave make with the direction of the whip's motion?

47. A supersonic jet traveling at Mach 3.00 at an altitude of 20 000 m is directly over a person at time $t = 0$ as in Figure P17.47. (a) How long will it be before the person encounters the shock wave? (b) Where will the plane be when it is finally heard? (Assume the speed of sound in air is 335 m/s.)

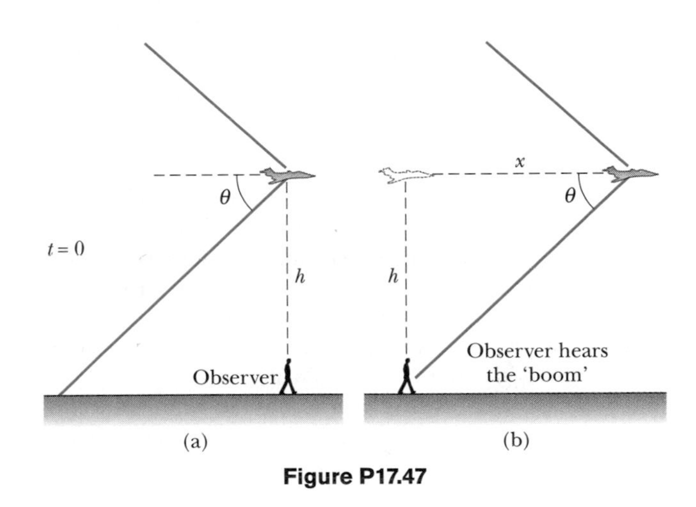

Figure P17.47

Section 17.5 Digital Sound Recording

Section 17.6 Motion Picture Sound

48. This problem represents a possible (but not recommended) way to code instantaneous pressures in a sound wave into 16-bit digital words. Example 17.2 mentions that the pressure amplitude of a 120-dB sound is 28.7 N/m^2. Let this pressure variation be represented by the digital code 65 536. Let zero pressure variation be represented on the recording by the digital word 0. Let other intermediate pressures be represented by digital words of intermediate size, in direct proportion to the pressure. (a) What digital word would represent the maximum pressure in a 40 dB sound? (b) Explain why this scheme works poorly for soft sounds. (c) Explain how this coding scheme would clip off half of the waveform of any sound, ignoring the actual shape of the wave and turning it into a string of zeros. By introducing sharp corners into every recorded waveform, this coding scheme would make everything sound like a buzzer or a kazoo.

49. Only two recording channels are required to give the illusion of sound coming from any point located between two speakers of a stereophonic sound system. If the same signal is recorded in both channels, a listener will hear it coming from a single direction halfway between the two speakers. This "phantom orchestra" illusion can be heard in the two-channel original Broadway cast recording of the song "Do-Re-Mi" from *The Sound of Music* (Columbia Records KOS 2020). Each of the eight singers can be heard at a different location between the loudspeakers. All listeners with normal hearing will agree on their locations. The brain can sense the direction of sound by noting how

much earlier a sound is heard in one ear than in the other. Model your ears as two sensors 19.0 cm apart in a flat screen. If a click from a distant source is heard 210 μs earlier in the left ear than in the right, from what direction does it appear to originate?

50. Assume that a loudspeaker broadcasts sound equally in all directions and produces sound with a level of 103 dB at a distance of 1.60 m from its center. (a) Find its sound power output. (b) If the salesperson claims to be giving you 150 W per channel, he is referring to the electrical power input to the speaker. Find the efficiency of the speaker—that is, the fraction of input power that is converted into useful output power.

Additional Problems

51. A large set of unoccupied football bleachers has solid seats and risers. You stand on the field in front of the bleachers and fire a starter's pistol or sharply clap two wooden boards together once. The sound pulse you produce has no definite frequency and no wavelength. The sound you hear reflected from the bleachers has an identifiable frequency and may remind you of a short toot on a trumpet, or of a buzzer or kazoo. Account for this sound. Compute order-of-magnitude estimates for its frequency, wavelength, and duration, on the basis of data you specify.

52. Many artists sing very high notes in *ad lib* ornaments and cadenzas. The highest note written for a singer in a published score was F-sharp above high C, 1.480 kHz, for Zerbinetta in the original version of Richard Strauss's opera *Ariadne auf Naxos.* (a) Find the wavelength of this sound in air. (b) Suppose people in the fourth row of seats hear this note with level 81.0 dB. Find the displacement amplitude of the sound. (c) **What If?** Because of complaints, Strauss later transposed the note down to F above high C, 1.397 kHz. By what increment did the wavelength change?

53. A sound wave in a cylinder is described by Equations 17.2 through 17.4. Show that $\Delta P = \pm \rho v \omega \sqrt{s_{\max}^2 - s^2}$.

54. On a Saturday morning, pickup trucks and sport utility vehicles carrying garbage to the town dump form a nearly steady procession on a country road, all traveling at 19.7 m/s. From one direction, two trucks arrive at the dump every 3 min. A bicyclist is also traveling toward the dump, at 4.47 m/s. (a) With what frequency do the trucks pass him? (b) **What If?** A hill does not slow down the trucks, but makes the out-of-shape cyclist's speed drop to 1.56 m/s. How often do noisy, smelly, inefficient, garbage-dripping, roadhogging trucks whiz past him now?

55. The ocean floor is underlain by a layer of basalt that constitutes the crust, or uppermost layer, of the Earth in that region. Below this crust is found denser periodotite rock, which forms the Earth's mantle. The boundary between these two layers is called the Mohorovicic discontinuity ("Moho" for short). If an explosive charge is set off at the surface of the basalt, it generates a seismic wave that is reflected back out at the Moho. If the speed of this wave in basalt is 6.50 km/s and the two-way travel time is 1.85 s, what is the thickness of this oceanic crust?

56. For a certain type of steel, stress is always proportional to strain with Young's modulus as shown in Table 12.1. The steel has the density listed for iron in Table 14.1. It will fail by bending permanently if subjected to compressive stress greater than its yield strength $\sigma_y = 400$ MPa. A rod 80.0 cm long, made of this steel, is fired at 12.0 m/s straight at a very hard wall, or at another identical rod moving in the opposite direction. (a) The speed of a one-dimensional compressional wave moving along the rod is given by $\sqrt{Y/\rho}$, where ρ is the density and Y is Young's modulus for the rod. Calculate this speed. (b) After the front end of the rod hits the wall and stops, the back end of the rod keeps moving, as described by Newton's first law, until it is stopped by excess pressure in a sound wave moving back through the rod. How much time elapses before the back end of the rod receives the message that it should stop? (c) How far has the back end of the rod moved in this time? Find (d) the strain in the rod and (e) the stress. (f) If it is not to fail, show that the maximum impact speed a rod can have is given by the expression $\sigma_y/\sqrt{\rho Y}$.

57. To permit measurement of her speed, a skydiver carries a buzzer emitting a steady tone at 1 800 Hz. A friend on the ground at the landing site directly below listens to the amplified sound he receives. Assume that the air is calm and that the sound speed is 343 m/s, independent of altitude. While the skydiver is falling at terminal speed, her friend on the ground receives waves of frequency 2 150 Hz. (a) What is the skydiver's speed of descent? (b) **What If?** Suppose the skydiver can hear the sound of the buzzer reflected from the ground. What frequency does she receive?

58. A train whistle ($f = 400$ Hz) sounds higher or lower in frequency depending on whether it approaches or recedes. (a) Prove that the difference in frequency between the approaching and receding train whistle is

$$\Delta f = \frac{2u/v}{1 - u^2/v^2} f$$

where u is the speed of the train and v is the speed of sound. (b) Calculate this difference for a train moving at a speed of 130 km/h. Take the speed of sound in air to be 340 m/s.

59. Two ships are moving along a line due east. The trailing vessel has a speed relative to a land-based observation point of 64.0 km/h, and the leading ship has a speed of 45.0 km/h relative to that point. The two ships are in a region of the ocean where the current is moving uniformly due west at 10.0 km/h. The trailing ship transmits a sonar signal at a frequency of 1 200.0 Hz. What frequency is monitored by the leading ship? (Use 1 520 m/s as the speed of sound in ocean water.)

60. A bat, moving at 5.00 m/s, is chasing a flying insect (Fig. P17.7). If the bat emits a 40.0 kHz chirp and receives back an echo at 40.4 kHz, at what speed is the insect moving toward or away from the bat? (Take the speed of sound in air to be $v = 340$ m/s.)

61. A supersonic aircraft is flying parallel to the ground. When the aircraft is directly overhead, an observer sees a rocket fired from the aircraft. Ten seconds later the observer hears the sonic boom, followed 2.80 s later by the sound of the rocket engine. What is the Mach number of the aircraft?

62. A police car is traveling east at 40.0 m/s along a straight road, overtaking a car ahead of it moving east at 30.0 m/s. The police car has a malfunctioning siren that is stuck at 1 000 Hz. (a) Sketch the appearance of the wave fronts of the sound produced by the siren. Show the wave fronts both to the east and to the west of the police car. (b) What would be the wavelength in air of the siren sound if the police car were at rest? (c) What is the wavelength in front of the police car? (d) What is it behind the police car? (e) What is the frequency heard by the driver being chased?

63. The speed of a one-dimensional compressional wave traveling along a thin copper rod is 3.56 km/s. A copper bar is given a sharp compressional blow at one end. The sound of the blow, traveling through air at 0°C, reaches the opposite end of the bar 6.40 ms later than the sound transmitted through the metal of the bar. What is the length of the bar?

64. A jet flies toward higher altitude at a constant speed of 1 963 m/s in a direction making an angle θ with the horizontal (Fig. P17.64). An observer on the ground hears the jet for the first time when it is directly overhead. Determine the value of θ if the speed of sound in air is 340 m/s.

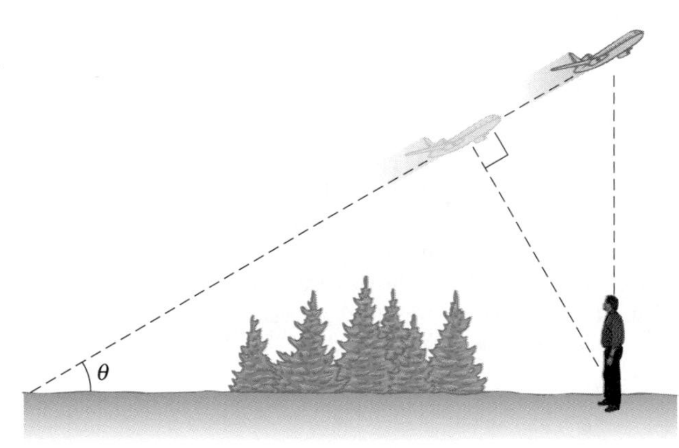

Figure P17.64

65. A meteoroid the size of a truck enters the earth's atmosphere at a speed of 20.0 km/s and is not significantly slowed before entering the ocean. (a) What is the Mach angle of the shock wave from the meteoroid in the atmosphere? (Use 331 m/s as the sound speed.) (b) Assuming that the meteoroid survives the impact with the ocean surface, what is the (initial) Mach angle of the shock wave that the meteoroid produces in the water? (Use the wave speed for seawater given in Table 17.1.)

66. An interstate highway has been built through a poor neighborhood in a city. In the afternoon, the sound level in a rented room is 80.0 dB, as 100 cars pass outside the window every minute. Late at night, when the tenant is working in a factory, the traffic flow is only five cars per minute. What is the average late-night sound level?

67. With particular experimental methods, it is possible to produce and observe in a long thin rod both a longitudinal wave and a transverse wave whose speed depends primarily on tension in the rod. The speed of the longitudinal wave is determined by the Young's modulus and the density of the material as $\sqrt{Y/\rho}$. The transverse wave can be modeled as a wave in a stretched string. A particular metal rod is 150 cm long and has a radius of 0.200 cm and a mass of 50.9 g. Young's modulus for the material is 6.80×10^{10} N/m^2. What must the tension in the rod be if the ratio of the speed of longitudinal waves to the speed of transverse waves is 8.00?

68. A siren creates sound with a level β at a distance d from the speaker. The siren is powered by a battery that delivers a total energy E. Let e represent the efficiency of the siren. (That is, e is equal to the output sound energy divided by the supplied energy). Determine the total time the siren can sound.

69. The Doppler equation presented in the text is valid when the motion between the observer and the source occurs on a straight line, so that the source and observer are moving either directly toward or directly away from each other. If this restriction is relaxed, one must use the more general Doppler equation

$$f' = \left(\frac{v + v_O \cos\theta_O}{v - v_S \cos\theta_S}\right) f$$

where θ_O and θ_S are defined in Figure P17.69a. (a) Show that if the observer and source are moving away from each other, the preceding equation reduces to Equation 17.13 with negative values for both v_O and v_S. (b) Use the preceding equation to solve the following problem. A train moves at a constant speed of 25.0 m/s toward the intersection shown in Figure P17.69b. A car is stopped near the intersection, 30.0 m from the tracks. If the train's horn emits a frequency of 500 Hz, what is the frequency heard by the passengers in the car when the train is 40.0 m from the intersection? Take the speed of sound to be 343 m/s.

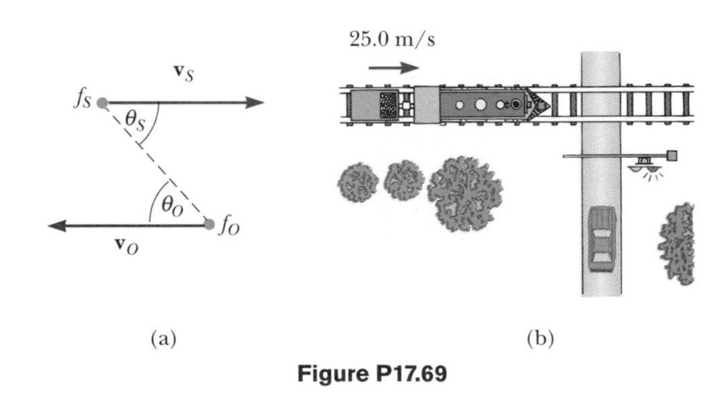

Figure P17.69

70. Equation 17.7 states that, at distance r away from a point source with power $\mathcal{P}_{av}$, the wave intensity is

$$I = \frac{\mathcal{P}_{av}}{4\pi r^2}$$

Study Figure 17.9 and prove that, at distance r straight in front of a point source with power $\mathcal{P}_{av}$ moving with constant speed v_S, the wave intensity is

$$I = \frac{\mathcal{P}_{\text{av}}}{4\pi r^2}\left(\frac{v - v_S}{v}\right)$$

71. Three metal rods are located relative to each other as shown in Figure P17.71, where $L_1 + L_2 = L_3$. The speed of sound in a rod is given by $v = \sqrt{Y/\rho}$, where ρ is the density and Y is Young's modulus for the rod. Values of density and Young's modulus for the three materials are $\rho_1 = 2.70 \times 10^3\ \text{kg/m}^3$, $Y_1 = 7.00 \times 10^{10}\ \text{N/m}^2$, $\rho_2 = 11.3 \times 10^3\ \text{kg/m}^3$, $Y_2 = 1.60 \times 10^{10}\ \text{N/m}^2$, $\rho_3 = 8.80 \times 10^3\ \text{kg/m}^3$, $Y_3 = 11.0 \times 10^{10}\ \text{N/m}^2$. (a) If $L_3 = 1.50$ m, what must the ratio L_1/L_2 be if a sound wave is to travel the length of rods 1 and 2 in the same time as it takes for the wave to travel the length of rod 3? (b) If the frequency of the source is 4.00 kHz, determine the phase difference between the wave traveling along rods 1 and 2 and the one traveling along rod 3.

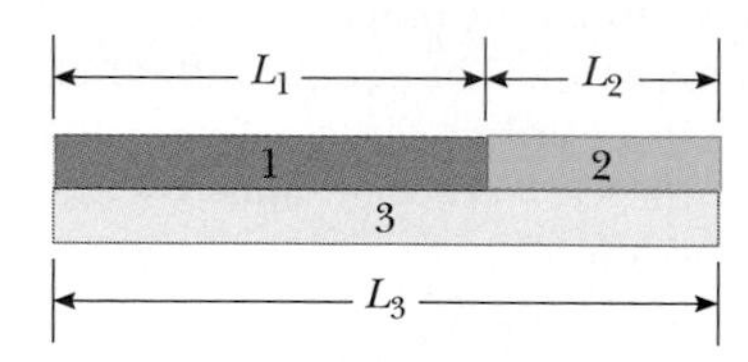

Figure P17.71

72. The smallest wavelength possible for a sound wave in air is on the order of the separation distance between air molecules. Find the order of magnitude of the highest-frequency sound wave possible in air, assuming a wave speed of 343 m/s, density 1.20 kg/m^3, and an average molecular mass of 4.82×10^{-26} kg.

Answers to Quick Quizzes

17.1 (c). Although the speed of a wave is given by the product of its wavelength (a) and frequency (b), it is not affected by changes in either one. The amplitude (d) of a sound wave determines the size of the oscillations of elements of air but does not affect the speed of the wave through the air.

17.2 (c). Because the bottom of the bottle is a rigid barrier, the displacement of elements of air at the bottom is zero. Because the pressure variation is a minimum or a maximum when the displacement is zero, and the pulse is moving downward, the pressure variation at the bottom is a maximum.

17.3 (c). The ear trumpet collects sound waves from the large area of its opening and directs it toward the ear. Most of the sound in this large area would miss the ear in the absence of the trumpet.

17.4 (b). The large area of the guitar body sets many elements of air into oscillation and allows the energy to leave the system by mechanical waves at a much larger rate than from the thin vibrating string.

17.5 (c). The only parameter that adds directly is intensity. Because of the logarithm function in the definition of sound level, sound levels cannot be added directly.

17.6 (b). The factor of 100 is two powers of ten. Thus, the logarithm of 100 is 2, which multiplied by 10 gives 20 dB.

17.7 (e). The wave speed cannot be changed by moving the source, so (a) and (b) are incorrect. The detected wavelength is largest at A, so (c) and (d) are incorrect. Choice (f) is incorrect because the detected frequency is lowest at location A.

17.8 (e). The intensity of the sound increases because the train is moving closer to you. Because the train moves at a constant velocity, the Doppler-shifted frequency remains fixed.

17.9 (b). The Mach number is the ratio of the plane's speed (which does not change) to the speed of sound, which is greater in the warm air than in the cold. The denominator of this ratio increases while the numerator stays constant. Therefore, the ratio as a whole—the Mach number—decreases.

Chapter 18

Superposition and Standing Waves

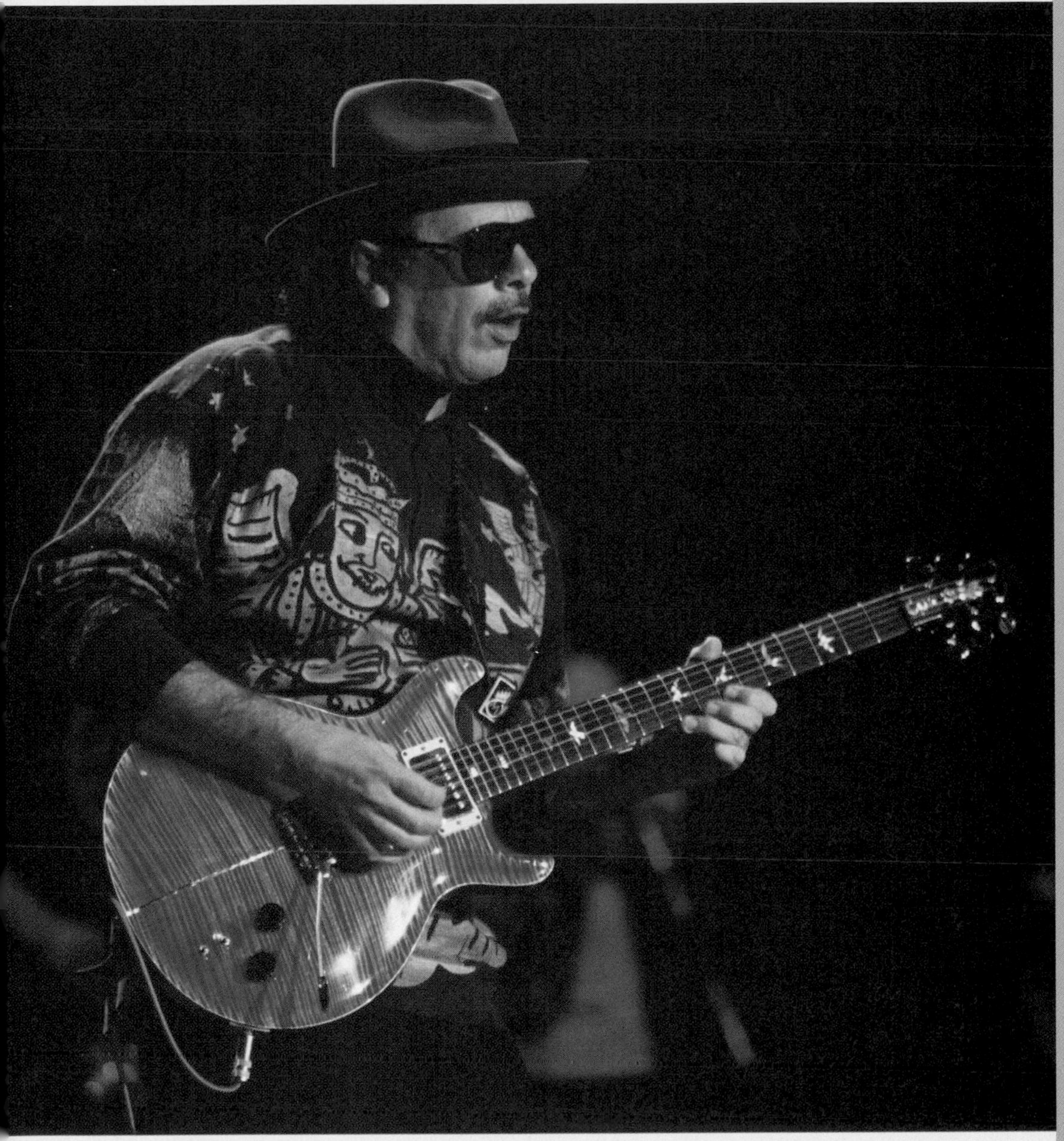

▲ *Guitarist Carlos Santana takes advantage of standing waves on strings. He changes to a higher note on the guitar by pushing the strings against the frets on the fingerboard, shortening the lengths of the portions of the strings that vibrate. (Bettmann/Corbis)*

CHAPTER OUTLINE

18.1 Superposition and Interference

18.2 Standing Waves

18.3 Standing Waves in a String Fixed at Both Ends

18.4 Resonance

18.5 Standing Waves in Air Columns

18.6 Standing Waves in Rods and Membranes

18.7 Beats: Interference in Time

18.8 Nonsinusoidal Wave Patterns

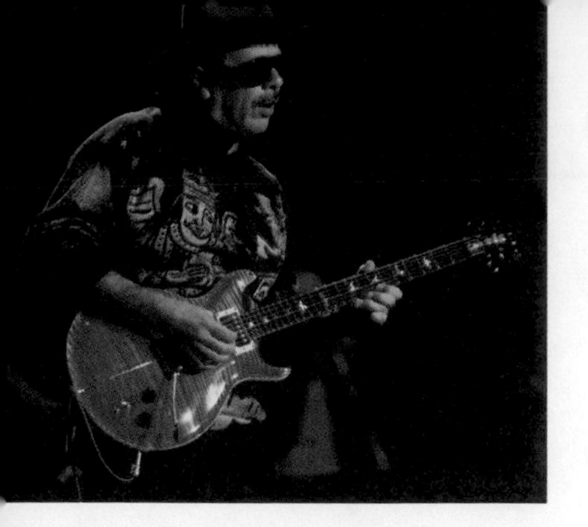

In the previous two chapters, we introduced the wave model. We have seen that waves are very different from particles. A particle is of zero size, while a wave has a characteristic size—the wavelength. Another important difference between waves and particles is that we can explore the possibility of two or more waves combining at one point in the same medium. We can combine particles to form extended objects, but the particles must be at *different* locations. In contrast, two waves can both be present at the same location, and the ramifications of this possibility are explored in this chapter.

When waves are combined, only certain allowed frequencies can exist on systems with boundary conditions—the frequencies are *quantized*. Quantization is a notion that is at the heart of quantum mechanics, a subject that we introduce formally in Chapter 40. There we show that waves under boundary conditions explain many of the quantum phenomena. For our present purposes in this chapter, quantization enables us to understand the behavior of the wide array of musical instruments that are based on strings and air columns.

We also consider the combination of waves having different frequencies and wavelengths. When two sound waves having nearly the same frequency interfere, we hear variations in the loudness called *beats*. The beat frequency corresponds to the rate of alternation between constructive and destructive interference. Finally, we discuss how any nonsinusoidal periodic wave can be described as a sum of sine and cosine functions.

18.1 Superposition and Interference

Many interesting wave phenomena in nature cannot be described by a single traveling wave. Instead, one must analyze complex waves in terms of a combination of traveling waves. To analyze such wave combinations, one can make use of the **superposition principle:**

Superposition principle

> If two or more traveling waves are moving through a medium, the resultant value of the wave function at any point is the algebraic sum of the values of the wave functions of the individual waves.

Waves that obey this principle are called *linear waves*. In the case of mechanical waves, linear waves are generally characterized by having amplitudes much smaller than their wavelengths. Waves that violate the superposition principle are called *nonlinear waves* and are often characterized by large amplitudes. In this book, we deal only with linear waves.

One consequence of the superposition principle is that **two traveling waves can pass through each other without being destroyed or even altered.** For instance,

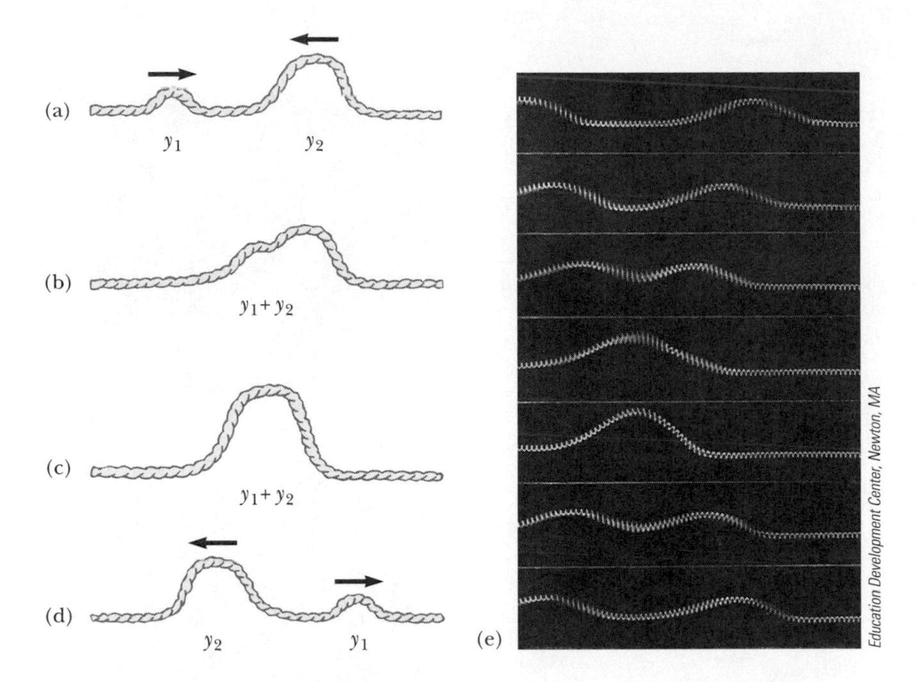

Active Figure 18.1 (a–d) Two pulses traveling on a stretched string in opposite directions pass through each other. When the pulses overlap, as shown in (b) and (c), the net displacement of the string equals the sum of the displacements produced by each pulse. Because each pulse produces positive displacements of the string, we refer to their superposition as *constructive interference.* (e) Photograph of the superposition of two equal, symmetric pulses traveling in opposite directions on a stretched spring.

At the Active Figures link at* http://www.pse6.com, *you can choose the amplitude and orientation of each of the pulses and study the interference between them as they pass each other.

when two pebbles are thrown into a pond and hit the surface at different places, the expanding circular surface waves do not destroy each other but rather pass through each other. The complex pattern that is observed can be viewed as two independent sets of expanding circles. Likewise, when sound waves from two sources move through air, they pass through each other.

Figure 18.1 is a pictorial representation of the superposition of two pulses. The wave function for the pulse moving to the right is y_1, and the wave function for the pulse moving to the left is y_2. The pulses have the same speed but different shapes, and the displacement of the elements of the medium is in the positive y direction for both pulses. When the waves begin to overlap (Fig. 18.1b), the wave function for the resulting complex wave is given by $y_1 + y_2$. When the crests of the pulses coincide (Fig. 18.1c), the resulting wave given by $y_1 + y_2$ has a larger amplitude than that of the individual pulses. The two pulses finally separate and continue moving in their original directions (Fig. 18.1d). Note that the pulse shapes remain unchanged after the interaction, as if the two pulses had never met!

The combination of separate waves in the same region of space to produce a resultant wave is called **interference.** For the two pulses shown in Figure 18.1, the displacement of the elements of the medium is in the positive y direction for both pulses, and the resultant pulse (created when the individual pulses overlap) exhibits an amplitude greater than that of either individual pulse. Because the displacements caused by the two pulses are in the same direction, we refer to their superposition as **constructive interference.**

Constructive interference

Now consider two pulses traveling in opposite directions on a taut string where one pulse is inverted relative to the other, as illustrated in Figure 18.2. In this case, when the pulses begin to overlap, the resultant pulse is given by $y_1 + y_2$, but the values of the function y_2 are negative. Again, the two pulses pass through each other; however, because the displacements caused by the two pulses are in opposite directions, we refer to their superposition as **destructive interference.**

Destructive interference

PITFALL PREVENTION

18.1 Do Waves Really *Interfere*?

In popular usage, the term *interfere* implies that an agent affects a situation in some way so as to preclude something from happening. For example, in American football, *pass interference* means that a defending player has affected the receiver so that he is unable to catch the ball. This is very different from its use in physics, where waves pass through each other and interfere, but do not affect each other in any way. In physics, interference is similar to the notion of *combination* as described in this chapter.

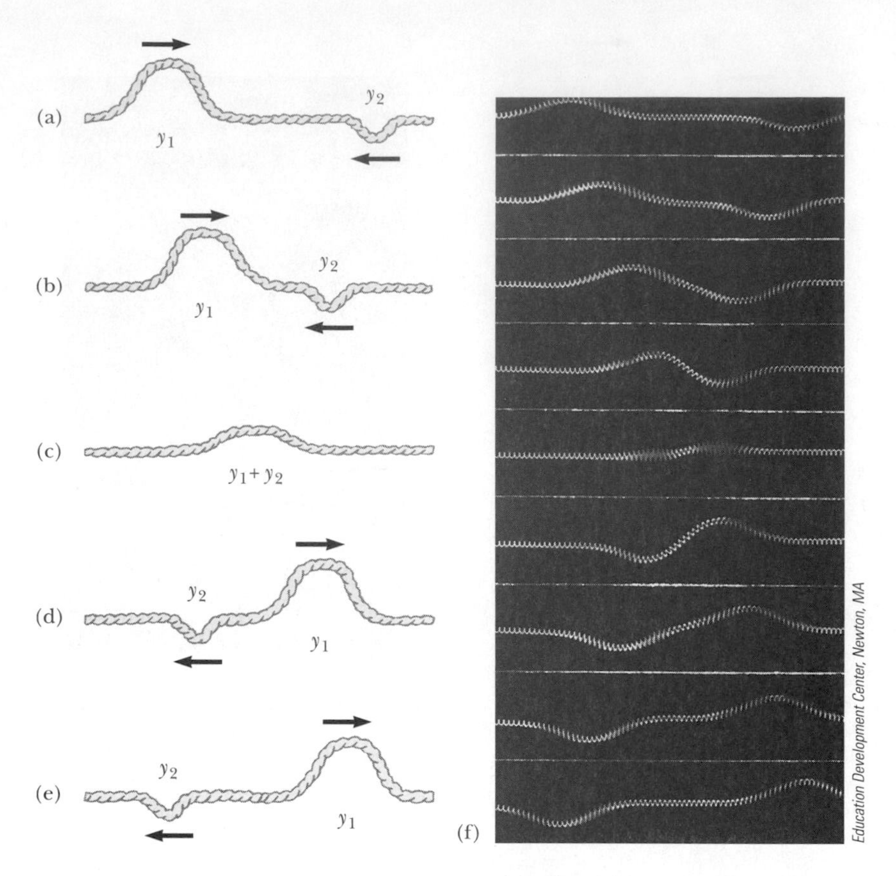

At the Active Figures link at http://www.pse6.com, you can choose the amplitude and orientation of each of the pulses and watch the interference as they pass each other.

Active Figure 18.2 (a–e) Two pulses traveling in opposite directions and having displacements that are inverted relative to each other. When the two overlap in (c), their displacements partially cancel each other. (f) Photograph of the superposition of two symmetric pulses traveling in opposite directions, where one is inverted relative to the other.

Quick Quiz 18.1 Two pulses are traveling toward each other, each at 10 cm/s on a long string, as shown in Figure 18.3. Sketch the shape of the string at $t = 0.6$ s.

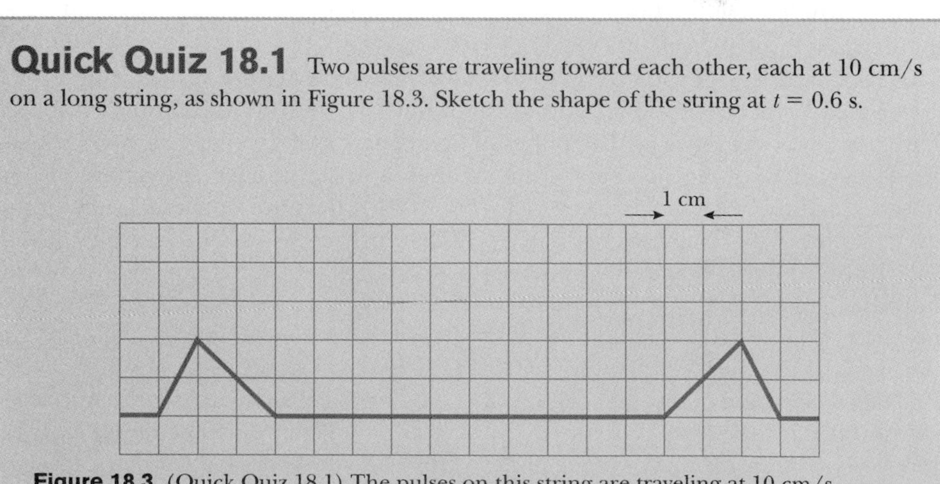

Figure 18.3 (Quick Quiz 18.1) The pulses on this string are traveling at 10 cm/s.

Quick Quiz 18.2 Two pulses move in opposite directions on a string and are identical in shape except that one has positive displacements of the elements of the string and the other has negative displacements. At the moment that the two pulses completely overlap on the string, (a) the energy associated with the pulses has disappeared (b) the string is not moving (c) the string forms a straight line (d) the pulses have vanished and will not reappear.

Superposition of Sinusoidal Waves

Let us now apply the principle of superposition to two sinusoidal waves traveling in the same direction in a linear medium. If the two waves are traveling to the right and have the same frequency, wavelength, and amplitude but differ in phase, we can express their individual wave functions as

$$y_1 = A\sin(kx - \omega t) \qquad y_2 = A\sin(kx - \omega t + \phi)$$

where, as usual, $k = 2\pi/\lambda$, $\omega = 2\pi f$, and ϕ is the phase constant, which we discussed in Section 16.2. Hence, the resultant wave function y is

$$y = y_1 + y_2 = A[\sin(kx - \omega t) + \sin(kx - \omega t + \phi)]$$

To simplify this expression, we use the trigonometric identity

$$\sin a + \sin b = 2\cos\left(\frac{a-b}{2}\right)\sin\left(\frac{a+b}{2}\right)$$

If we let $a = kx - \omega t$ and $b = kx - \omega t + \phi$, we find that the resultant wave function y reduces to

Resultant of two traveling sinusoidal waves

$$y = 2A\cos\left(\frac{\phi}{2}\right)\sin\left(kx - \omega t + \frac{\phi}{2}\right)$$

This result has several important features. The resultant wave function y also is sinusoidal and has the same frequency and wavelength as the individual waves because the sine function incorporates the same values of k and ω that appear in the original wave functions. The amplitude of the resultant wave is $2A\cos(\phi/2)$, and its phase is $\phi/2$. If the phase constant ϕ equals 0, then $\cos(\phi/2) = \cos 0 = 1$, and the amplitude of the resultant wave is $2A$—twice the amplitude of either individual wave. In this case the waves are said to be everywhere *in phase* and thus interfere constructively. That is, the crests and troughs of the individual waves y_1 and y_2 occur at the same positions and combine to form the red curve y of amplitude $2A$ shown in Figure 18.4a. Because the

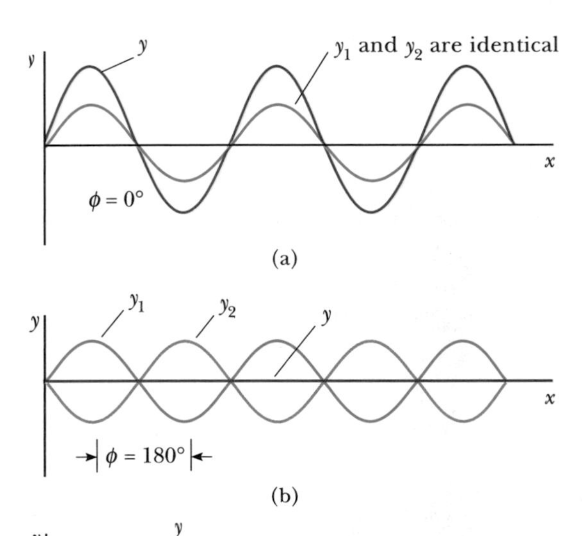

Active Figure 18.4 The superposition of two identical waves y_1 and y_2 (blue and green) to yield a resultant wave (red). (a) When y_1 and y_2 are in phase, the result is constructive interference. (b) When y_1 and y_2 are π rad out of phase, the result is destructive interference. (c) When the phase angle has a value other than 0 or π rad, the resultant wave y falls somewhere between the extremes shown in (a) and (b).

At the Active Figures link at http://www.pse6.com, you can change the phase relationship between the waves and observe the wave representing the superposition.

individual waves are in phase, they are indistinguishable in Figure 18.4a, in which they appear as a single blue curve. In general, constructive interference occurs when $\cos(\phi/2) = \pm 1$. This is true, for example, when $\phi = 0, 2\pi, 4\pi, \ldots$ rad—that is, when ϕ is an *even* multiple of π.

When ϕ is equal to π rad or to any *odd* multiple of π, then $\cos(\phi/2) = \cos(\pi/2) = 0$, and the crests of one wave occur at the same positions as the troughs of the second wave (Fig. 18.4b). Thus, the resultant wave has *zero* amplitude everywhere, as a consequence of destructive interference. Finally, when the phase constant has an arbitrary value other than 0 or an integer multiple of π rad (Fig. 18.4c), the resultant wave has an amplitude whose value is somewhere between 0 and $2A$.

Interference of Sound Waves

One simple device for demonstrating interference of sound waves is illustrated in Figure 18.5. Sound from a loudspeaker S is sent into a tube at point P, where there is a T-shaped junction. Half of the sound energy travels in one direction, and half travels in the opposite direction. Thus, the sound waves that reach the receiver R can travel along either of the two paths. The distance along any path from speaker to receiver is called the **path length** r. The lower path length r_1 is fixed, but the upper path length r_2 can be varied by sliding the U-shaped tube, which is similar to that on a slide trombone. When the difference in the path lengths $\Delta r = |r_2 - r_1|$ is either zero or some integer multiple of the wavelength λ (that is $\Delta r = n\lambda$, where $n = 0, 1, 2, 3, \ldots$), the two waves reaching the receiver at any instant are in phase and interfere constructively, as shown in Figure 18.4a. For this case, a maximum in the sound intensity is detected at the receiver. If the path length r_2 is adjusted such that the path difference $\Delta r = \lambda/2$, $3\lambda/2, \ldots, n\lambda/2$ (for n odd), the two waves are exactly π rad, or 180°, out of phase at the receiver and hence cancel each other. In this case of destructive interference, no sound is detected at the receiver. This simple experiment demonstrates that a phase difference may arise between two waves generated by the same source when they travel along paths of unequal lengths. This important phenomenon will be indispensable in our investigation of the interference of light waves in Chapter 37.

It is often useful to express the path difference in terms of the phase angle ϕ between the two waves. Because a path difference of one wavelength corresponds to a phase angle of 2π rad, we obtain the ratio $\phi/2\pi = \Delta r/\lambda$ or

Relationship between path difference and phase angle

$$\Delta r = \frac{\phi}{2\pi}\lambda \qquad (18.1)$$

Using the notion of path difference, we can express our conditions for constructive and destructive interference in a different way. If the path difference is any even multiple of $\lambda/2$, then the phase angle $\phi = 2n\pi$, where $n = 0, 1, 2, 3, \ldots$, and the interference is constructive. For path differences of odd multiples of $\lambda/2$, $\phi = (2n + 1)\pi$, where $n = 0, 1, 2, 3, \ldots$, and the interference is destructive. Thus, we have the conditions

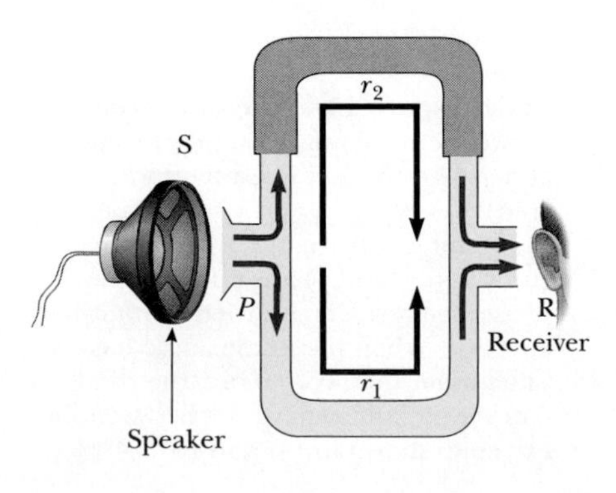

Figure 18.5 An acoustical system for demonstrating interference of sound waves. A sound wave from the speaker (S) propagates into the tube and splits into two parts at point P. The two waves, which combine at the opposite side, are detected at the receiver (R). The upper path length r_2 can be varied by sliding the upper section.

$$\Delta r = (2n)\frac{\lambda}{2} \quad \text{for constructive interference}$$

and (18.2)

$$\Delta r = (2n+1)\frac{\lambda}{2} \quad \text{for destructive interference}$$

This discussion enables us to understand why the speaker wires in a stereo system should be connected properly. When connected the wrong way—that is, when the positive (or red) wire is connected to the negative (or black) terminal on one of the speakers and the other is correctly wired—the speakers are said to be "out of phase"—one speaker cone moves outward while the other moves inward. As a consequence, the sound wave coming from one speaker destructively interferes with the wave coming from the other—along a line midway between the two, a rarefaction region due to one speaker is superposed on a compression region from the other speaker. Although the two sounds probably do not completely cancel each other (because the left and right stereo signals are usually not identical), a substantial loss of sound quality occurs at points along this line.

Example 18.1 Two Speakers Driven by the Same Source

A pair of speakers placed 3.00 m apart are driven by the same oscillator (Fig. 18.6). A listener is originally at point O, which is located 8.00 m from the center of the line connecting the two speakers. The listener then walks to point P, which is a perpendicular distance 0.350 m from O, before reaching the *first minimum* in sound intensity. What is the frequency of the oscillator?

Solution To find the frequency, we must know the wavelength of the sound coming from the speakers. With this information, combined with our knowledge of the speed of sound, we can calculate the frequency. The wavelength can be determined from the interference information given. The first minimum occurs when the two waves reaching the listener at point P are 180° out of phase—in other words, when their path difference Δr equals $\lambda/2$. To calculate the path difference, we must first find the path lengths r_1 and r_2.

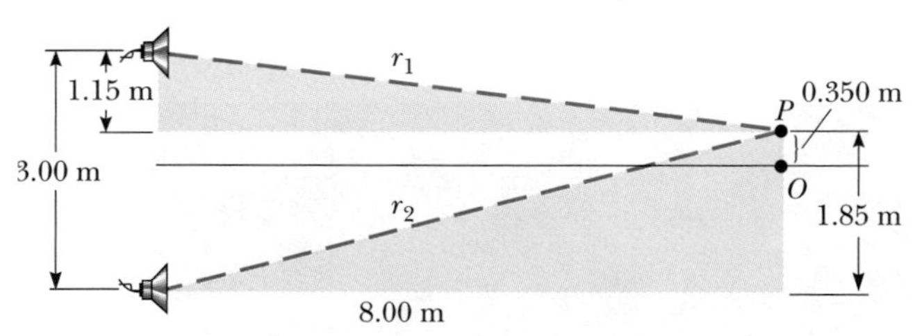

Figure 18.6 (Example 18.1) Two speakers emit sound waves to a listener at P.

Figure 18.6 shows the physical arrangement of the speakers, along with two shaded right triangles that can be drawn on the basis of the lengths described in the problem. From these triangles, we find that the path lengths are

$$r_1 = \sqrt{(8.00 \text{ m})^2 + (1.15 \text{ m})^2} = 8.08 \text{ m}$$

and

$$r_2 = \sqrt{(8.00 \text{ m})^2 + (1.85 \text{ m})^2} = 8.21 \text{ m}$$

Hence, the path difference is $r_2 - r_1 = 0.13$ m. Because we require that this path difference be equal to $\lambda/2$ for the first minimum, we find that $\lambda = 0.26$ m.

To obtain the oscillator frequency, we use Equation 16.12, $v = \lambda f$, where v is the speed of sound in air, 343 m/s:

$$f = \frac{v}{\lambda} = \frac{343 \text{ m/s}}{0.26 \text{ m}} = \boxed{1.3 \text{ kHz}}$$

What If? **What if the speakers were connected out of phase? What happens at point P in Figure 18.6?**

Answer In this situation, the path difference of $\lambda/2$ combines with a phase difference of $\lambda/2$ due to the incorrect wiring to give a full phase difference of λ. As a result, the waves are in phase and there is a *maximum* intensity at point P.

18.2 Standing Waves

The sound waves from the speakers in Example 18.1 leave the speakers in the forward direction, and we considered interference at a point in front of the speakers. Suppose that we turn the speakers so that they face each other and then have them emit sound of the same frequency and amplitude. In this situation, two identical waves travel in

Figure 18.7 Two speakers emit sound waves toward each other. When they overlap, identical waves traveling in opposite directions will combine to form standing waves.

opposite directions in the same medium, as in Figure 18.7. These waves combine in accordance with the superposition principle.

We can analyze such a situation by considering wave functions for two transverse sinusoidal waves having the same amplitude, frequency, and wavelength but traveling in opposite directions in the same medium:

$$y_1 = A\sin(kx - \omega t) \qquad y_2 = A\sin(kx + \omega t)$$

where y_1 represents a wave traveling in the $+x$ direction and y_2 represents one traveling in the $-x$ direction. Adding these two functions gives the resultant wave function y:

$$y = y_1 + y_2 = A\sin(kx - \omega t) + A\sin(kx + \omega t)$$

When we use the trigonometric identity $\sin(a \pm b) = \sin(a)\cos(b) \pm \cos(a)\sin(b)$, this expression reduces to

$$y = (2A\sin kx)\cos\omega t \qquad (18.3)$$

Equation 18.3 represents the wave function of a **standing wave.** A standing wave, such as the one shown in Figure 18.8, is an oscillation pattern *with a stationary outline* that results from the superposition of two identical waves traveling in opposite directions.

Notice that Equation 18.3 does not contain a function of $kx - \omega t$. Thus, it is not an expression for a traveling wave. If we observe a standing wave, we have no sense of motion in the direction of propagation of either of the original waves. If we compare this equation with Equation 15.6, we see that Equation 18.3 describes a special kind of simple harmonic motion. Every element of the medium oscillates in simple harmonic motion with the same frequency ω (according to the $\cos\omega t$ factor in the equation). However, the amplitude of the simple harmonic motion of a given element (given by the factor $2A\sin kx$, the coefficient of the cosine function) depends on the location x of the element in the medium.

The maximum amplitude of an element of the medium has a minimum value of zero when x satisfies the condition $\sin kx = 0$, that is, when

$$kx = \pi, 2\pi, 3\pi, \ldots$$

Because $k = 2\pi/\lambda$, these values for kx give

$$x = \frac{\lambda}{2}, \lambda, \frac{3\lambda}{2}, \ldots = \frac{n\lambda}{2} \qquad n = 0, 1, 2, 3, \ldots \qquad (18.4)$$

These points of zero amplitude are called **nodes.**

PITFALL PREVENTION

18.2 Three Types of Amplitude

We need to distinguish carefully here between the **amplitude of the individual waves,** which is A, and the **amplitude of the simple harmonic motion of the elements of the medium,** which is $2A\sin kx$. A given element in a standing wave vibrates within the constraints of the *envelope* function $2A\sin kx$, where x is that element's position in the medium. This is in contrast to traveling sinusoidal waves, in which all elements oscillate with the same amplitude and the same frequency, and the amplitude A of the wave is the same as the amplitude A of the simple harmonic motion of the elements. Furthermore, we can identify the **amplitude of the standing wave** as $2A$.

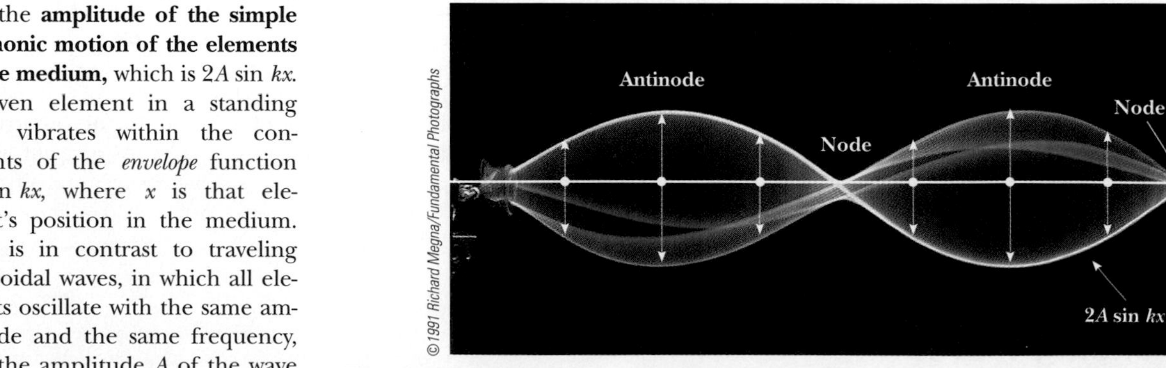

Figure 18.8 Multiflash photograph of a standing wave on a string. The time behavior of the vertical displacement from equilibrium of an individual element of the string is given by $\cos\omega t$. That is, each element vibrates at an angular frequency ω. The amplitude of the vertical oscillation of any elements of the string depends on the horizontal position of the element. Each element vibrates within the confines of the envelope function $2A\sin kx$.

The element with the *greatest* possible displacement from equilibrium has an amplitude of $2A$, and we define this as the amplitude of the standing wave. The positions in the medium at which this maximum displacement occurs are called **antinodes.** The antinodes are located at positions for which the coordinate x satisfies the condition $\sin kx = \pm 1$, that is, when

$$kx = \frac{\pi}{2}, \frac{3\pi}{2}, \frac{5\pi}{2}, \ldots$$

Thus, the positions of the antinodes are given by

$$x = \frac{\lambda}{4}, \frac{3\lambda}{4}, \frac{5\lambda}{4}, \ldots = \frac{n\lambda}{4} \qquad n = 1, 3, 5, \ldots \qquad (18.5)$$

Position of antinodes

In examining Equations 18.4 and 18.5, we note the following important features of the locations of nodes and antinodes:

The distance between adjacent antinodes is equal to $\lambda/2$.
The distance between adjacent nodes is equal to $\lambda/2$.
The distance between a node and an adjacent antinode is $\lambda/4$.

Wave patterns of the elements of the medium produced at various times by two waves traveling in opposite directions are shown in Figure 18.9. The blue and green curves are the wave patterns for the individual traveling waves, and the red curves are the wave patterns for the resultant standing wave. At $t = 0$ (Fig. 18.9a), the two traveling waves are in phase, giving a wave pattern in which each element of the medium is experiencing its maximum displacement from equilibrium. One quarter of a period later, at $t = T/4$ (Fig. 18.9b), the traveling waves have moved one quarter of a wavelength (one to the right and the other to the left). At this time, the traveling waves are out of phase, and each element of the medium is passing through the equilibrium position in its simple harmonic motion. The result is zero displacement for elements at all values of x—that is, the wave pattern is a straight line. At $t = T/2$ (Fig. 18.9c), the traveling waves are again in phase, producing a wave pattern that is inverted relative to the $t = 0$ pattern. In the standing wave, the elements of the medium alternate in time between the extremes shown in Figure 18.9a and c.

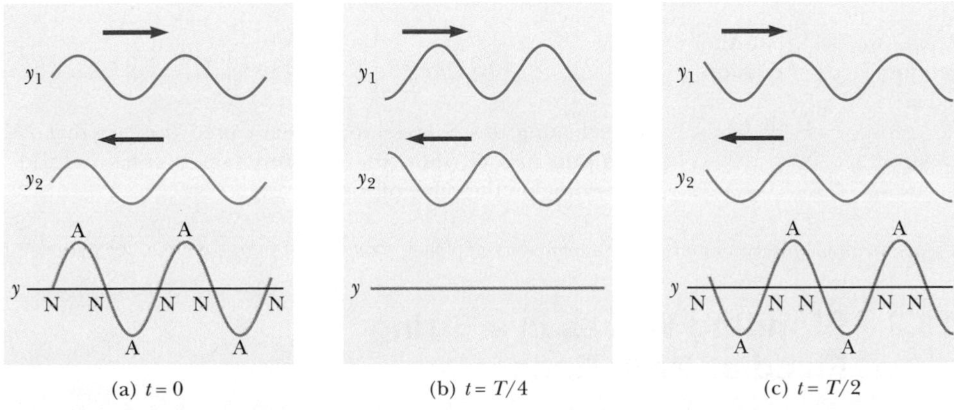

(a) $t = 0$ (b) $t = T/4$ (c) $t = T/2$

Active Figure 18.9 Standing-wave patterns produced at various times by two waves of equal amplitude traveling in opposite directions. For the resultant wave y, the nodes (N) are points of zero displacement, and the antinodes (A) are points of maximum displacement.

At the Active Figures link at http://www.pse6.com, you can choose the wavelength of the waves and see the standing wave that results.

Quick Quiz 18.3 Consider a standing wave on a string as shown in Figure 18.9. Define the velocity of elements of the string as positive if they are moving upward in the figure. At the moment the string has the shape shown by the red curve in Figure 18.9a, the instantaneous velocity of elements along the string (a) is zero for all elements (b) is positive for all elements (c) is negative for all elements (d) varies with the position of the element.

Quick Quiz 18.4 Continuing with the scenario in Quick Quiz 18.3, at the moment the string has the shape shown by the red curve in Figure 18.9b, the instantaneous velocity of elements along the string (a) is zero for all elements (b) is positive for all elements (c) is negative for all elements (d) varies with the position of the element.

Example 18.2 Formation of a Standing Wave

Two waves traveling in opposite directions produce a standing wave. The individual wave functions are

$$y_1 = (4.0 \text{ cm}) \sin(3.0x - 2.0t)$$
$$y_2 = (4.0 \text{ cm}) \sin(3.0x + 2.0t)$$

where x and y are measured in centimeters.

(A) Find the amplitude of the simple harmonic motion of the element of the medium located at $x = 2.3$ cm.

Solution The standing wave is described by Equation 18.3; in this problem, we have $A = 4.0$ cm, $k = 3.0$ rad/cm, and $\omega = 2.0$ rad/s. Thus,

$$y = (2A \sin kx) \cos \omega t = [(8.0 \text{ cm}) \sin 3.0x] \cos 2.0t$$

Thus, we obtain the amplitude of the simple harmonic motion of the element at the position $x = 2.3$ cm by evaluating the coefficient of the cosine function at this position:

$$y_{\text{max}} = (8.0 \text{ cm}) \sin 3.0x|_{x=2.3}$$
$$= (8.0 \text{ cm}) \sin (6.9 \text{ rad}) = 4.6 \text{ cm}$$

(B) Find the positions of the nodes and antinodes if one end of the string is at $x = 0$.

Solution With $k = 2\pi/\lambda = 3.0$ rad/cm, we see that the wavelength is $\lambda = (2\pi/3.0)$ cm. Therefore, from Equation 18.4 we find that the nodes are located at

$$x = n\frac{\lambda}{2} = n\left(\frac{\pi}{3}\right) \text{ cm} \qquad n = 0, 1, 2, 3, \ldots$$

and from Equation 18.5 we find that the antinodes are located at

$$x = n\frac{\lambda}{4} = n\left(\frac{\pi}{6}\right) \text{ cm} \qquad n = 1, 3, 5, \ldots$$

(C) What is the maximum value of the position in the simple harmonic motion of an element located at an antinode?

Solution According to Equation 18.3, the maximum position of an element at an antinode is the amplitude of the standing wave, which is twice the amplitude of the individual traveling waves:

$$y_{\text{max}} = 2A(\sin kx)_{\text{max}} = 2(4.0 \text{ cm})(\pm 1) = \pm 8.0 \text{ cm}$$

where we have used the fact that the maximum value of $\sin kx$ is ± 1. Let us check this result by evaluating the coefficient of our standing-wave function at the positions we found for the antinodes:

$$y_{\text{max}} = (8.0 \text{ cm}) \sin 3.0x|_{x=n(\pi/6)}$$
$$= (8.0 \text{ cm}) \sin\left[3.0n\left(\frac{\pi}{6}\right) \text{rad}\right]$$
$$= (8.0 \text{ cm}) \sin\left[n\left(\frac{\pi}{2}\right) \text{rad}\right] = \pm 8.0 \text{ cm}$$

In evaluating this expression, we have used the fact that n is an odd integer; thus, the sine function is equal to ± 1, depending on the value of n.

18.3 Standing Waves in a String Fixed at Both Ends

Consider a string of length L fixed at both ends, as shown in Figure 18.10. Standing waves are set up in the string by a continuous superposition of waves incident on and reflected from the ends. Note that there is a boundary condition for the waves on the

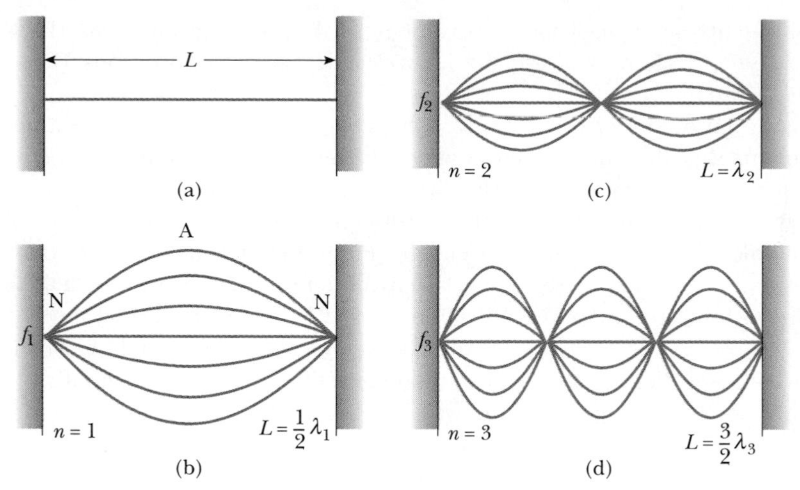

Active Figure 18.10 (a) A string of length L fixed at both ends. The normal modes of vibration form a harmonic series: (b) the fundamental, or first harmonic; (c) the second harmonic; (d) the third harmonic.

At the Active Figures link at http://www.pse6.com, you can choose the mode number and see the corresponding standing wave.

string. The ends of the string, because they are fixed, must necessarily have zero displacement and are, therefore, nodes by definition. The boundary condition results in the string having a number of natural patterns of oscillation, called **normal modes,** each of which has a characteristic frequency that is easily calculated. This situation in which only certain frequencies of oscillation are allowed is called **quantization.** Quantization is a common occurrence when waves are subject to boundary conditions and will be a central feature in our discussions of quantum physics in the extended version of this text.

Figure 18.11 shows one of the normal modes of oscillation of a string fixed at both ends. Except for the nodes, which are always stationary, all elements of the string oscillate vertically with the same frequency but with different amplitudes of simple harmonic motion. Figure 18.11 represents snapshots of the standing wave at various times over one half of a period. The red arrows show the velocities of various elements of the string at various times. As we found in Quick Quizzes 18.3 and 18.4,

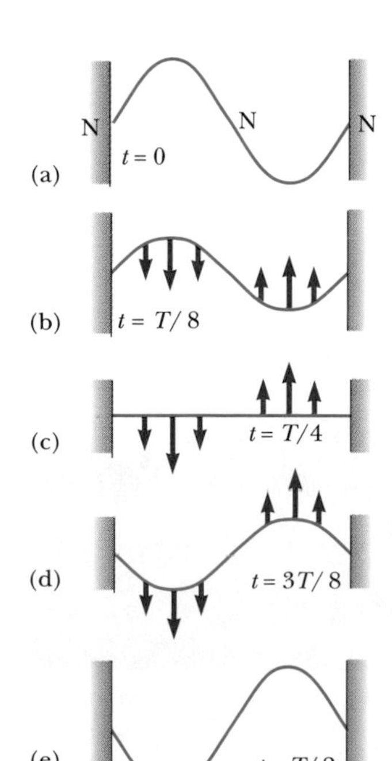

Figure 18.11 A standing-wave pattern in a taut string. The five "snapshots" were taken at intervals of one eighth of the period. (a) At $t = 0$, the string is momentarily at rest. (b) At $t = T/8$, the string is in motion, as indicated by the red arrows, and different parts of the string move in different directions with different speeds. (c) At $t = T/4$, the string is moving but horizontal (undeformed). (d) The motion continues as indicated. (e) At $t = T/2$, the string is again momentarily at rest, but the crests and troughs of (a) are reversed. The cycle continues until ultimately, when a time interval equal to T has passed, the configuration shown in (a) is repeated.

all elements of the string have zero velocity at the extreme positions (Figs. 18.11a and 18.11e) and elements have varying velocities at other positions (Figs. 18.11b through 18.11d).

The normal modes of oscillation for the string can be described by imposing the requirements that the ends be nodes and that the nodes and antinodes be separated by one fourth of a wavelength. The first normal mode that is consistent with the boundary conditions, shown in Figure 18.10b, has nodes at its ends and one antinode in the middle. This is the longest-wavelength mode that is consistent with our requirements. This first normal mode occurs when the length of the string is half the wavelength λ_1, as indicated in Figure 18.10b, or $\lambda_1 = 2L$. The next normal mode (see Fig. 18.10c) of wavelength λ_2 occurs when the wavelength equals the length of the string, that is, when $\lambda_2 = L$. The third normal mode (see Fig. 18.10d) corresponds to the case in which $\lambda_3 = 2L/3$. In general, the wavelengths of the various normal modes for a string of length L fixed at both ends are

Wavelengths of normal modes

$$\lambda_n = \frac{2L}{n} \qquad n = 1, 2, 3, \ldots \tag{18.6}$$

where the index n refers to the nth normal mode of oscillation. These are the *possible* modes of oscillation for the string. The *actual* modes that are excited on a string are discussed shortly.

The natural frequencies associated with these modes are obtained from the relationship $f = v/\lambda$, where the wave speed v is the same for all frequencies. Using Equation 18.6, we find that the natural frequencies f_n of the normal modes are

Frequencies of normal modes as functions of wave speed and length of string

$$f_n = \frac{v}{\lambda_n} = n\frac{v}{2L} \qquad n = 1, 2, 3, \ldots \tag{18.7}$$

These natural frequencies are also called the *quantized frequencies* associated with the vibrating string fixed at both ends.

Because $v = \sqrt{T/\mu}$ (see Eq. 16.18), where T is the tension in the string and μ is its linear mass density, we can also express the natural frequencies of a taut string as

Frequencies of normal modes as functions of string tension and linear mass density

$$f_n = \frac{n}{2L}\sqrt{\frac{T}{\mu}} \qquad n = 1, 2, 3, \ldots \tag{18.8}$$

The lowest frequency f_1, which corresponds to $n = 1$, is called either the **fundamental** or the **fundamental frequency** and is given by

Fundamental frequency of a taut string

$$f_1 = \frac{1}{2L}\sqrt{\frac{T}{\mu}} \tag{18.9}$$

The frequencies of the remaining normal modes are integer multiples of the fundamental frequency. Frequencies of normal modes that exhibit an integer-multiple relationship such as this form a **harmonic series,** and the normal modes are called

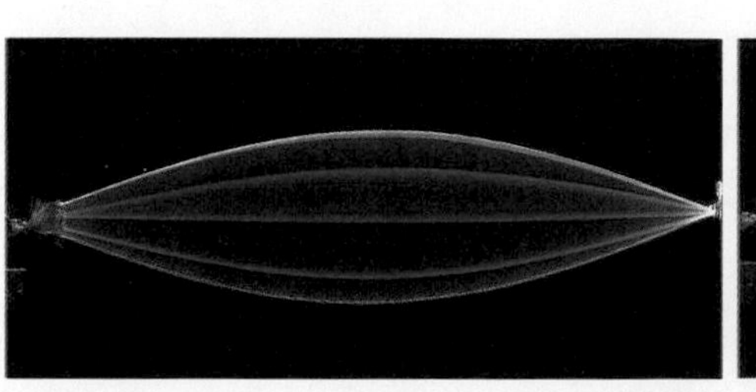

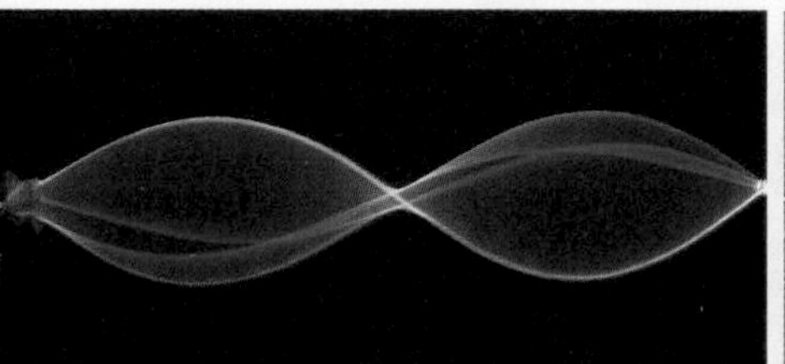

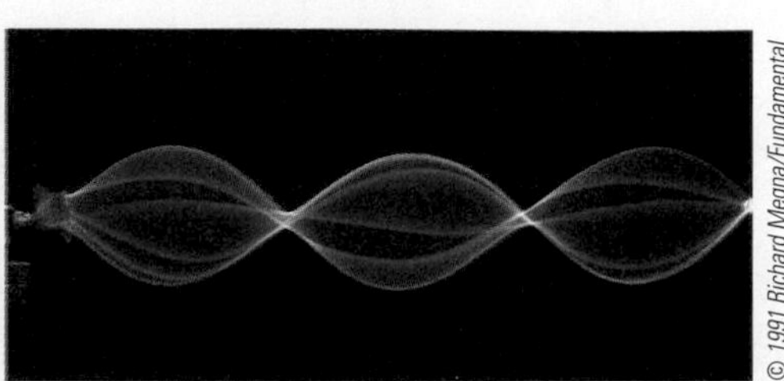

Multiflash photographs of standing-wave patterns in a cord driven by a vibrator at its left end. The single-loop pattern represents the first normal mode (the fundamental), $n = 1$). The double-loop pattern represents the second normal mode ($n = 2$), and the triple-loop pattern represents the third normal mode ($n = 3$).

harmonics. The fundamental frequency f_1 is the frequency of the first harmonic; the frequency $f_2 = 2f_1$ is the frequency of the second harmonic; and the frequency $f_n = nf_1$ is the frequency of the nth harmonic. Other oscillating systems, such as a drumhead, exhibit normal modes, but the frequencies are not related as integer multiples of a fundamental. Thus, we do not use the term *harmonic* in association with these types of systems.

In obtaining Equation 18.6, we used a technique based on the separation distance between nodes and antinodes. We can obtain this equation in an alternative manner. Because we require that the string be fixed at $x = 0$ and $x = L$, the wave function $y(x, t)$ given by Equation 18.3 must be zero at these points for all times. That is, the *boundary conditions* require that $y(0, t) = 0$ and $y(L, t) = 0$ for all values of t. Because the standing wave is described by $y = 2A(\sin kx)\cos \omega t$, the first boundary condition, $y(0, t) = 0$, is automatically satisfied because $\sin kx = 0$ at $x = 0$. To meet the second boundary condition, $y(L, t) = 0$, we require that $\sin kL = 0$. This condition is satisfied when the angle kL equals an integer multiple of π rad. Therefore, the allowed values of k are given by[1]

$$k_nL = n\pi \qquad n = 1, 2, 3, \ldots \tag{18.10}$$

Because $k_n = 2\pi/\lambda_n$, we find that

$$\left(\frac{2\pi}{\lambda_n}\right)L = n\pi \qquad \text{or} \qquad \lambda_n = \frac{2L}{n}$$

which is identical to Equation 18.6.

Let us examine further how these various harmonics are created in a string. If we wish to excite just a single harmonic, we must distort the string in such a way that its distorted shape corresponds to that of the desired harmonic. After being released, the string vibrates at the frequency of that harmonic. This maneuver is difficult to perform, however, and it is not how we excite a string of a musical instrument. If the string is distorted such that its distorted shape is not that of just one harmonic, the resulting vibration includes various harmonics. Such a distortion occurs in musical instruments when the string is plucked (as in a guitar), bowed (as in a cello), or struck (as in a piano). When the string is distorted into a nonsinusoidal shape, only waves that satisfy the boundary conditions can persist on the string. These are the harmonics.

The frequency of a string that defines the musical note that it plays is that of the fundamental. The frequency of the string can be varied by changing either the tension or the string's length. For example, the tension in guitar and violin strings is varied by a screw adjustment mechanism or by tuning pegs located on the neck of the instrument. As the tension is increased, the frequency of the normal modes increases in accordance with Equation 18.8. Once the instrument is "tuned," players vary the frequency by moving their fingers along the neck, thereby changing the length of the oscillating portion of the string. As the length is shortened, the frequency increases because, as Equation 18.8 specifies, the normal-mode frequencies are inversely proportional to string length.

Quick Quiz 18.5 When a standing wave is set up on a string fixed at both ends, (a) the number of nodes is equal to the number of antinodes (b) the wavelength is equal to the length of the string divided by an integer (c) the frequency is equal to the number of nodes times the fundamental frequency (d) the shape of the string at any time is symmetric about the midpoint of the string.

[1] We exclude $n = 0$ because this value corresponds to the trivial case in which no wave exists ($k = 0$).

Example 18.3 Give Me a C Note!

Middle C on a piano has a fundamental frequency of 262 Hz, and the first A above middle C has a fundamental frequency of 440 Hz.

(A) Calculate the frequencies of the next two harmonics of the C string.

Solution Knowing that the frequencies of higher harmonics are integer multiples of the fundamental frequency $f_1 = 262$ Hz, we find that

$$f_2 = 2f_1 = \boxed{524 \text{ Hz}}$$

$$f_3 = 3f_1 = \boxed{786 \text{ Hz}}$$

(B) If the A and C strings have the same linear mass density μ and length L, determine the ratio of tensions in the two strings.

Solution Using Equation 18.9 for the two strings vibrating at their fundamental frequencies gives

$$f_{1A} = \frac{1}{2L}\sqrt{\frac{T_A}{\mu}} \quad \text{and} \quad f_{1C} = \frac{1}{2L}\sqrt{\frac{T_C}{\mu}}$$

Setting up the ratio of these frequencies, we find that

$$\frac{f_{1A}}{f_{1C}} = \sqrt{\frac{T_A}{T_C}}$$

$$\frac{T_A}{T_C} = \left(\frac{f_{1A}}{f_{1C}}\right)^2 = \left(\frac{440}{262}\right)^2 = \boxed{2.82}$$

What If? **What if we look inside a real piano? In this case, the assumption we made in part (B) is only partially true. The string densities are equal, but the length of the A string is only 64 percent of the length of the C string. What is the ratio of their tensions?**

Answer Using Equation 18.8 again, we set up the ratio of frequencies:

$$\frac{f_{1A}}{f_{1C}} = \frac{L_C}{L_A}\sqrt{\frac{T_A}{T_C}} = \left(\frac{100}{64}\right)\sqrt{\frac{T_A}{T_C}}$$

$$\frac{T_A}{T_C} = (0.64)^2\left(\frac{440}{262}\right)^2 = 1.16$$

Example 18.4 Guitar Basics

Interactive

The high E string on a guitar measures 64.0 cm in length and has a fundamental frequency of 330 Hz. By pressing down so that the string is in contact with the first fret (Fig. 18.12), the string is shortened so that it plays an F note that has a frequency of 350 Hz. How far is the fret from the neck end of the string?

Solution Equation 18.7 relates the string's length to the fundamental frequency. With $n = 1$, we can solve for the speed of the wave on the string,

$$v = \frac{2L}{n} f_n = \frac{2(0.640 \text{ m})}{1}(330 \text{ Hz}) = 422 \text{ m/s}$$

Because we have not adjusted the tuning peg, the tension in the string, and hence the wave speed, remain constant. We can again use Equation 18.7, this time solving for L and

Figure 18.12 (Example 18.4) Playing an F note on a guitar.

substituting the new frequency to find the shortened string length:

$$L = n\frac{v}{2f_n} = (1)\frac{422 \text{ m/s}}{2(350 \text{ Hz})} = 0.603 \text{ m} = 60.3 \text{ cm}$$

The difference between this length and the measured length of 64.0 cm is the distance from the fret to the neck end of the string, or $\boxed{3.7 \text{ cm.}}$

What If? **What if we wish to play an F sharp, which we do by pressing down on the second fret from the neck in Figure 18.12? The frequency of F sharp is 370 Hz. Is this fret another 3.7 cm from the neck?**

Answer If you inspect a guitar fingerboard, you will find that the frets are *not* equally spaced. They are far apart near the neck and close together near the opposite end. Consequently, from this observation, we would not expect the F sharp fret to be another 3.7 cm from the end.

Let us repeat the calculation of the string length, this time for the frequency of F sharp:

$$L = n\frac{v}{2f_n} = (1)\frac{422 \text{ m/s}}{2(370 \text{ Hz})} = 0.571 \text{ m}$$

This gives a distance of 0.640 m − 0.571 m = 0.069 m = 6.9 cm from the neck. Subtracting the distance from the neck to the first fret, the separation distance between the first and second frets is 6.9 cm − 3.7 cm = 3.2 cm.

Explore this situation at the Interactive Worked Example link at **http://www.pse6.com.**

Example 18.5 Changing String Vibration with Water

Interactive

One end of a horizontal string is attached to a vibrating blade and the other end passes over a pulley as in Figure 18.13a. A sphere of mass 2.00 kg hangs on the end of the string. The string is vibrating in its second harmonic. A container of water is raised under the sphere so that the sphere is completely submerged. After this is done, the string vibrates in its fifth harmonic, as shown in Figure 18.13b. What is the radius of the sphere?

Solution To conceptualize the problem, imagine what happens when the sphere is immersed in the water. The buoyant force acts upward on the sphere, reducing the tension in the string. The change in tension causes a change in the speed of waves on the string, which in turn causes a change in the wavelength. This altered wavelength results in the string vibrating in its fifth normal mode rather than the second. We categorize the problem as one in which we will need to combine our understanding of Newton's second law, buoyant forces, and standing waves on strings. We begin to analyze the problem by studying Figure 18.13a. Newton's second law applied to the sphere tells us that the tension in the string is equal to the weight of the sphere:

$$\Sigma F = T_1 - mg = 0$$

$$T_1 = mg = (2.00 \text{ kg})(9.80 \text{ m/s}^2) = 19.6 \text{ N}$$

where the subscript 1 is used to indicate initial variables before we immerse the sphere in water. Once the sphere is immersed in water, the tension in the string decreases to T_2. Applying Newton's second law to the sphere again in this situation, we have

$$T_2 + B - mg = 0$$

$$(1) \qquad B = mg - T_2$$

The desired quantity, the radius of the sphere, will appear in the expression for the buoyant force B. Before proceeding in this direction, however, we must evaluate T_2. We do this from the standing wave information. We write the equation for the frequency of a standing wave on a string (Equation 18.8) twice, once before we immerse the sphere and once after, and divide the equations:

$$f = \frac{n_1}{2L}\sqrt{\frac{T_1}{\mu}}, \quad f = \frac{n_2}{2L}\sqrt{\frac{T_2}{\mu}} \quad \longrightarrow \quad 1 = \frac{n_1}{n_2}\sqrt{\frac{T_1}{T_2}}$$

where the frequency f is the same in both cases, because it is determined by the vibrating blade. In addition, the linear mass density μ and the length L of the vibrating portion of the string are the same in both cases. Solving for T_2, we have

$$T_2 = \left(\frac{n_1}{n_2}\right)^2 T_1 = \left(\frac{2}{5}\right)^2 (19.6 \text{ N}) = 3.14 \text{ N}$$

Substituting this into Equation (1), we can evaluate the buoyant force on the sphere:

$$B = mg - T_2 = 19.6 \text{ N} - 3.14 \text{ N} = 16.5 \text{ N}$$

Finally, expressing the buoyant force (Eq. 14.5) in terms of the radius of the sphere, we solve for the radius:

$$B = \rho_{\text{water}} g V_{\text{sphere}} = \rho_{\text{water}} g \left(\tfrac{4}{3}\pi r^3\right)$$

$$r = \sqrt[3]{\frac{3B}{4\pi\rho_{\text{water}} g}} = \sqrt[3]{\frac{3(16.5 \text{ N})}{4\pi(1\,000 \text{ kg/m}^3)(9.80 \text{ m/s}^2)}}$$

$$= 7.38 \times 10^{-2} \text{ m} = \boxed{7.38 \text{ cm}}$$

To finalize this problem, note that only certain radii of the sphere will result in the string vibrating in a normal mode. This is because the speed of waves on the string must be changed to a value such that the length of the string is an integer multiple of half wavelengths. This is a feature of the *quantization* that we introduced earlier in this chapter—the sphere radii that cause the string to vibrate in a normal mode are *quantized.*

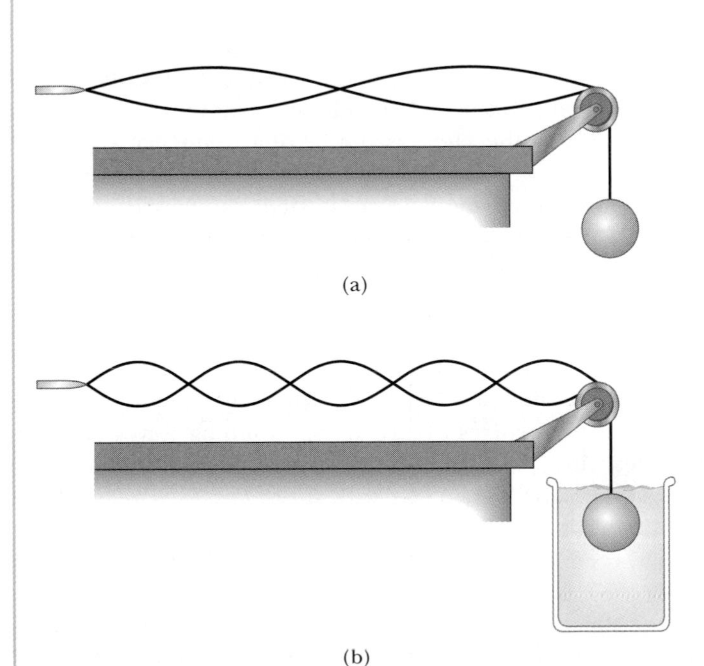

Figure 18.13 (Example 18.5) When the sphere hangs in air, the string vibrates in its second harmonic. When the sphere is immersed in water, the string vibrates in its fifth harmonic.

You can adjust the mass at the Interactive Worked Example link at **http://www.pse6.com.**

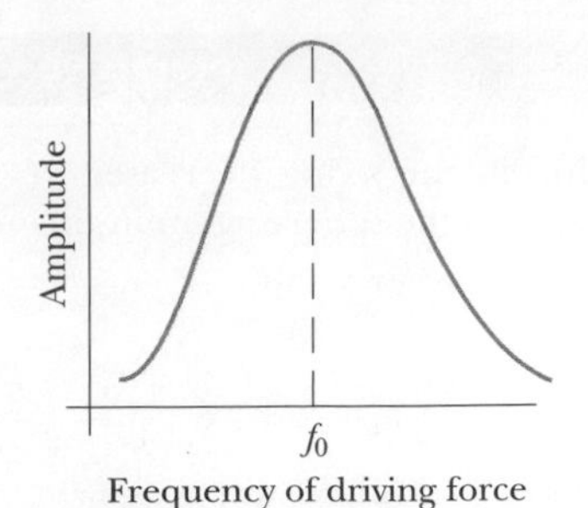

Figure 18.14 Graph of the amplitude (response) versus driving frequency for an oscillating system. The amplitude is a maximum at the resonance frequency f_0.

18.4 Resonance

We have seen that a system such as a taut string is capable of oscillating in one or more normal modes of oscillation. **If a periodic force is applied to such a system, the amplitude of the resulting motion is greatest when the frequency of the applied force is equal to one of the natural frequencies of the system.** We discussed this phenomenon, known as *resonance,* briefly in Section 15.7. Although a block–spring system or a simple pendulum has only one natural frequency, standing-wave systems have a whole set of natural frequencies, such as that given by Equation 18.7 for a string. Because an oscillating system exhibits a large amplitude when driven at any of its natural frequencies, these frequencies are often referred to as **resonance frequencies.**

Figure 18.14 shows the response of an oscillating system to various driving frequencies, where one of the resonance frequencies of the system is denoted by f_0. Note that the amplitude of oscillation of the system is greatest when the frequency of the driving force equals the resonance frequency. The maximum amplitude is limited by friction in the system. If a driving force does work on an oscillating system that is initially at rest, the input energy is used both to increase the amplitude of the oscillation and to overcome the friction force. Once maximum amplitude is reached, the work done by the driving force is used only to compensate for mechanical energy loss due to friction.

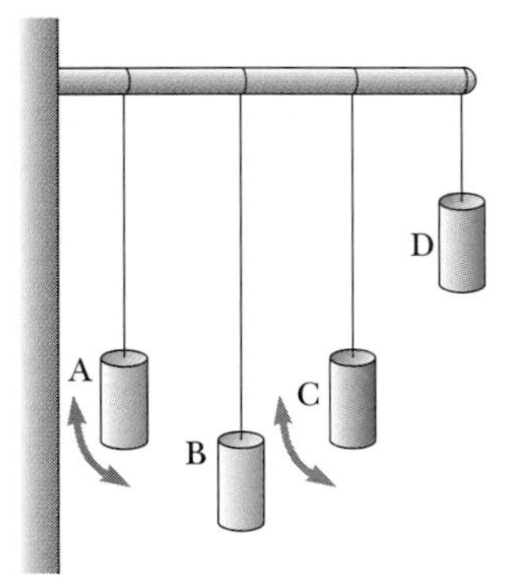

Figure 18.15 An example of resonance. If pendulum A is set into oscillation, only pendulum C, whose length matches that of A, eventually oscillates with large amplitude, or resonates. The arrows indicate motion in a plane perpendicular to the page.

Examples of Resonance

A playground swing is a pendulum having a natural frequency that depends on its length. Whenever we use a series of regular impulses to push a child in a swing, the swing goes higher if the frequency of the periodic force equals the natural frequency of the swing. We can demonstrate a similar effect by suspending pendulums of different lengths from a horizontal support, as shown in Figure 18.15. If pendulum A is set into oscillation, the other pendulums begin to oscillate as a result of waves transmitted along the beam. However, pendulum C, the length of which is close to the length of A, oscillates with a much greater amplitude than pendulums B and D, the lengths of which are much different from that of pendulum A. Pendulum C moves the way it does because its natural frequency is nearly the same as the driving frequency associated with pendulum A.

Next, consider a taut string fixed at one end and connected at the opposite end to an oscillating blade, as illustrated in Figure 18.16. The fixed end is a node, and the end connected to the blade is very nearly a node because the amplitude of the blade's motion is small compared with that of the elements of the string. As the blade oscillates, transverse waves sent down the string are reflected from the fixed end. As we learned in Section 18.3, the string has natural frequencies that are determined by its length, tension, and linear mass density (see Eq. 18.8). When the frequency of the blade equals one of the natural frequencies of the string, standing waves are produced and the string oscillates with a large amplitude. In this resonance case, the wave generated by the oscillating blade is in phase with the reflected wave, and the string absorbs energy from the blade. If the string is driven at a frequency that is not one of its natural frequencies, then the oscillations are of low amplitude and exhibit no stable pattern.

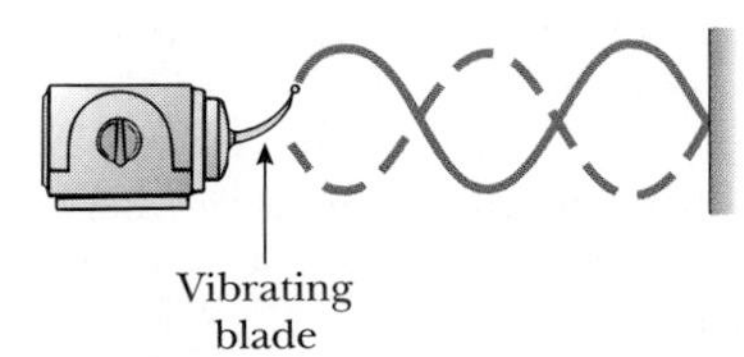

Figure 18.16 Standing waves are set up in a string when one end is connected to a vibrating blade. When the blade vibrates at one of the natural frequencies of the string, large-amplitude standing waves are created.

Once the amplitude of the standing-wave oscillations is a maximum, the mechanical energy delivered by the blade and absorbed by the system is transformed to internal energy because of the damping forces caused by friction in the system. If the applied frequency differs from one of the natural frequencies, energy is transferred to the string at first, but later the phase of the wave becomes such that it forces the blade to receive energy from the string, thereby reducing the energy in the string.

Resonance is very important in the excitation of musical instruments based on air columns. We shall discuss this application of resonance in Section 18.5.

Quick Quiz 18.6 A wine glass can be shattered through resonance by maintaining a certain frequency of a high-intensity sound wave. Figure 18.17a shows a side view of a wine glass vibrating in response to such a sound wave. Sketch the standing-wave pattern in the rim of the glass as seen from above. If an integral number of waves "fit" around the circumference of the vibrating rim, how many wavelengths fit around the rim in Figure 18.17a?

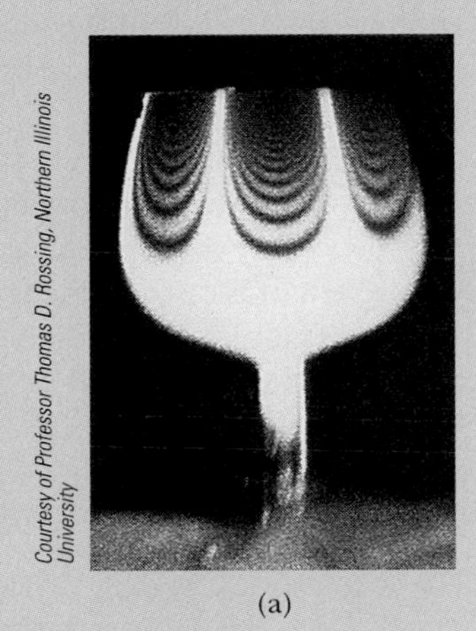

(a)

(b)

Figure 18.17 (Quick Quiz 18.6) (a) Standing-wave pattern in a vibrating wine glass. The glass shatters if the amplitude of vibration becomes too great. (b) A wine glass shattered by the amplified sound of a human voice.

18.5 Standing Waves in Air Columns

Standing waves can be set up in a tube of air, such as that inside an organ pipe, as the result of interference between longitudinal sound waves traveling in opposite directions. The phase relationship between the incident wave and the wave reflected from one end of the pipe depends on whether that end is open or closed. This relationship is analogous to the phase relationships between incident and reflected transverse waves at the end of a string when the end is either fixed or free to move (see Figs. 16.14 and 16.15).

In a pipe closed at one end, **the closed end is a displacement node because the wall at this end does not allow longitudinal motion of the air.** As a result, at a closed end of a pipe, the reflected sound wave is 180° out of phase with the incident wave. Furthermore, because the pressure wave is 90° out of phase with the displacement wave (see Section 17.2), **the closed end of an air column corresponds to a pressure antinode** (that is, a point of maximum pressure variation).

The open end of an air column is approximately a displacement antinode[2] and a pressure node. We can understand why no pressure variation occurs at an open end by noting that the end of the air column is open to the atmosphere; thus, the pressure at this end must remain constant at atmospheric pressure.

You may wonder how a sound wave can reflect from an open end, as there may not appear to be a change in the medium at this point. It is indeed true that the medium

[2] Strictly speaking, the open end of an air column is not exactly a displacement antinode. A compression reaching an open end does not reflect until it passes beyond the end. For a tube of circular cross section, an end correction equal to approximately $0.6R$, where R is the tube's radius, must be added to the length of the air column. Hence, the effective length of the air column is longer than the true length L. We ignore this end correction in this discussion.

through which the sound wave moves is air both inside and outside the pipe. However, sound is a pressure wave, and a compression region of the sound wave is constrained by the sides of the pipe as long as the region is inside the pipe. As the compression region exits at the open end of the pipe, the constraint of the pipe is removed and the compressed air is free to expand into the atmosphere. Thus, there is a change in the *character* of the medium between the inside of the pipe and the outside even though there is no change in the *material* of the medium. This change in character is sufficient to allow some reflection.

With the boundary conditions of nodes or antinodes at the ends of the air column, we have a set of normal modes of oscillation, as we do for the string fixed at both ends. Thus, the air column has quantized frequencies.

The first three normal modes of oscillation of a pipe open at both ends are shown in Figure 18.18a. Note that both ends are displacement antinodes (approximately). In the first normal mode, the standing wave extends between two adjacent antinodes, which is a distance of half a wavelength. Thus, the wavelength is twice the length of the pipe, and the fundamental frequency is $f_1 = v/2L$. As Figure 18.18a shows, the frequencies of the higher harmonics are $2f_1, 3f_1, \ldots$. Thus, we can say that

In a pipe open at both ends, the natural frequencies of oscillation form a harmonic series that includes all integral multiples of the fundamental frequency.

PITFALL PREVENTION

18.3 Sound Waves in Air Are Longitudinal, not Transverse

Note that the standing longitudinal waves are drawn as transverse waves in Figure 18.18. This is because it is difficult to draw longitudinal displacements—they are in the same direction as the propagation. Thus, it is best to interpret the curves in Figure 18.18 as a graphical representation of the waves (our diagrams of string waves are pictorial representations), with the vertical axis representing horizontal displacement of the elements of the medium.

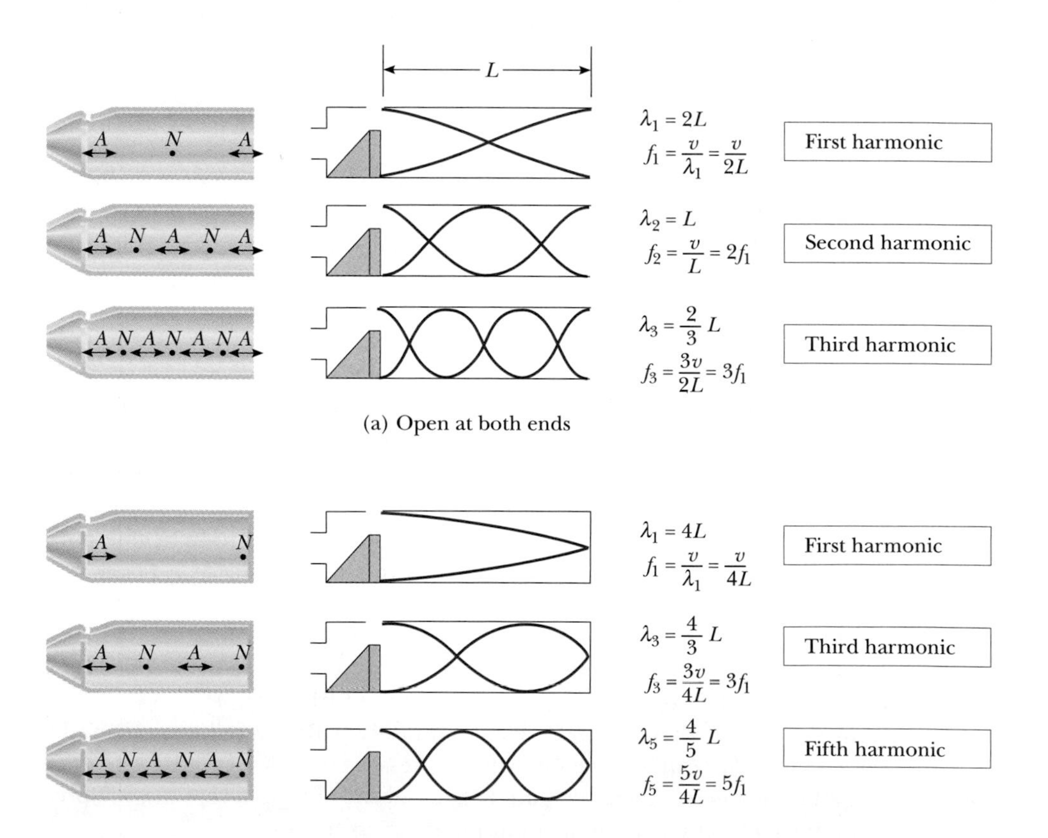

Figure 18.18 Motion of elements of air in standing longitudinal waves in a pipe, along with schematic representations of the waves. In the schematic representations, the structure at the left end has the purpose of exciting the air column into a normal mode. The hole in the upper edge of the column assures that the left end acts as an open end. The graphs represent the displacement amplitudes, not the pressure amplitudes. (a) In a pipe open at both ends, the harmonic series created consists of all integer multiples of the fundamental frequency: $f_1, 2f_1, 3f_1, \ldots$. (b) In a pipe closed at one end and open at the other, the harmonic series created consists of only odd-integer multiples of the fundamental frequency: $f_1, 3f_1, 5f_1, \ldots$.

Because all harmonics are present, and because the fundamental frequency is given by the same expression as that for a string (see Eq. 18.7), we can express the natural frequencies of oscillation as

$$f_n = n\frac{v}{2L} \qquad n = 1, 2, 3, \ldots \tag{18.11}$$

Natural frequencies of a pipe open at both ends

Despite the similarity between Equations 18.7 and 18.11, you must remember that v in Equation 18.7 is the speed of waves on the string, whereas v in Equation 18.11 is the speed of sound in air.

If a pipe is closed at one end and open at the other, the closed end is a displacement node (see Fig. 18.18b). In this case, the standing wave for the fundamental mode extends from an antinode to the adjacent node, which is one fourth of a wavelength. Hence, the wavelength for the first normal mode is $4L$, and the fundamental frequency is $f_1 = v/4L$. As Figure 18.18b shows, the higher-frequency waves that satisfy our conditions are those that have a node at the closed end and an antinode at the open end; this means that the higher harmonics have frequencies $3f_1, 5f_1, \ldots$.

In a pipe closed at one end, the natural frequencies of oscillation form a harmonic series that includes only odd integral multiples of the fundamental frequency.

We express this result mathematically as

$$f_n = n\frac{v}{4L} \qquad n = 1, 3, 5, \ldots \tag{18.12}$$

Natural frequencies of a pipe closed at one end and open at the other

It is interesting to investigate what happens to the frequencies of instruments based on air columns and strings during a concert as the temperature rises. The sound emitted by a flute, for example, becomes sharp (increases in frequency) as it warms up because the speed of sound increases in the increasingly warmer air inside the flute (consider Eq. 18.11). The sound produced by a violin becomes flat (decreases in frequency) as the strings thermally expand because the expansion causes their tension to decrease (see Eq. 18.8).

Musical instruments based on air columns are generally excited by resonance. The air column is presented with a sound wave that is rich in many frequencies. The air column then responds with a large-amplitude oscillation to the frequencies that match the quantized frequencies in its set of harmonics. In many woodwind instruments, the initial rich sound is provided by a vibrating reed. In the brasses, this excitation is provided by the sound coming from the vibration of the player's lips. In a flute, the initial excitation comes from blowing over an edge at the mouthpiece of the instrument. This is similar to blowing across the opening of a bottle with a narrow neck. The sound of the air rushing across the edge has many frequencies, including one that sets the air cavity in the bottle into resonance.

Quick Quiz 18.7 A pipe open at both ends resonates at a fundamental frequency f_{open}. When one end is covered and the pipe is again made to resonate, the fundamental frequency is f_{closed}. Which of the following expressions describes how these two resonant frequencies compare? (a) $f_{\text{closed}} = f_{\text{open}}$ (b) $f_{\text{closed}} = \frac{1}{2} f_{\text{open}}$ (c) $f_{\text{closed}} = 2 f_{\text{open}}$ (d) $f_{\text{closed}} = \frac{3}{2} f_{\text{open}}$

Quick Quiz 18.8 Balboa Park in San Diego has an outdoor organ. When the air temperature increases, the fundamental frequency of one of the organ pipes (a) stays the same (b) goes down (c) goes up (d) is impossible to determine.

Example 18.6 Wind in a Culvert

A section of drainage culvert 1.23 m in length makes a howling noise when the wind blows.

(A) Determine the frequencies of the first three harmonics of the culvert if it is cylindrical in shape and open at both ends. Take $v = 343$ m/s as the speed of sound in air.

Solution The frequency of the first harmonic of a pipe open at both ends is

$$f_1 = \frac{v}{2L} = \frac{343 \text{ m/s}}{2(1.23 \text{ m})} = \boxed{139 \text{ Hz}}$$

Because both ends are open, all harmonics are present; thus,

$$f_2 = 2f_1 = \boxed{278 \text{ Hz}} \quad \text{and} \quad f_3 = 3f_1 = \boxed{417 \text{ Hz}}$$

(B) What are the three lowest natural frequencies of the culvert if it is blocked at one end?

Solution The fundamental frequency of a pipe closed at one end is

$$f_1 = \frac{v}{4L} = \frac{343 \text{ m/s}}{4(1.23 \text{ m})} = \boxed{69.7 \text{ Hz}}$$

In this case, only odd harmonics are present; hence, the next two harmonics have frequencies $f_3 = 3f_1 = \boxed{209 \text{ Hz}}$ and $f_5 = 5f_1 = \boxed{349 \text{ Hz.}}$

(C) For the culvert open at both ends, how many of the harmonics present fall within the normal human hearing range (20 to 20 000 Hz)?

Solution Because all harmonics are present for a pipe open at both ends, we can express the frequency of the highest harmonic heard as $f_n = nf_1$ where n is the number of harmonics that we can hear. For $f_n = 20\,000$ Hz, we find that the number of harmonics present in the audible range is

$$n = \frac{20\,000 \text{ Hz}}{139 \text{ Hz}} = \boxed{143}$$

Only the first few harmonics are of sufficient amplitude to be heard.

Example 18.7 Measuring the Frequency of a Tuning Fork

A simple apparatus for demonstrating resonance in an air column is depicted in Figure 18.19. A vertical pipe open at both ends is partially submerged in water, and a tuning fork vibrating at an unknown frequency is placed near the top of the pipe. The length L of the air column can be adjusted by moving the pipe vertically. The sound waves generated by the fork are reinforced when L corresponds to one of the resonance frequencies of the pipe.

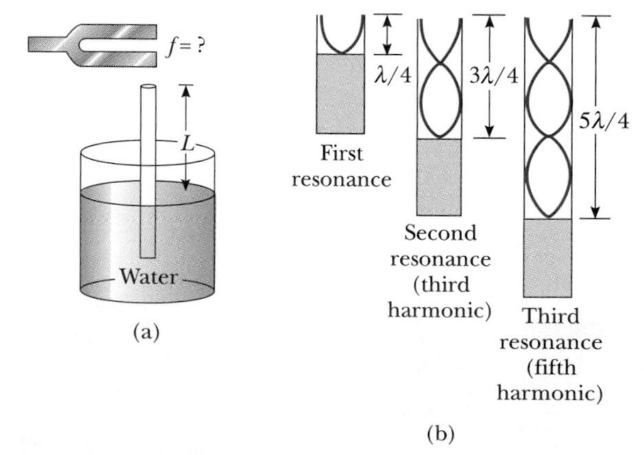

Figure 18.19 (Example 18.7) (a) Apparatus for demonstrating the resonance of sound waves in a pipe closed at one end. The length L of the air column is varied by moving the pipe vertically while it is partially submerged in water. (b) The first three normal modes of the system shown in part (a).

For a certain pipe, the smallest value of L for which a peak occurs in the sound intensity is 9.00 cm. What are

(A) the frequency of the tuning fork

(B) the values of L for the next two resonance frequencies?

Solution

(A) Although the pipe is open at its lower end to allow the water to enter, the water's surface acts like a wall at one end. Therefore, this setup can be modeled as an air column closed at one end, and so the fundamental frequency is given by $f_1 = v/4L$. Taking $v = 343$ m/s for the speed of sound in air and $L = 0.090\,0$ m, we obtain

$$f_1 = \frac{v}{4L} = \frac{343 \text{ m/s}}{4(0.090\,0 \text{ m})} = \boxed{953 \text{ Hz}}$$

Because the tuning fork causes the air column to resonate at this frequency, this must also be the frequency of the tuning fork.

(B) Because the pipe is closed at one end, we know from Figure 18.18b that the wavelength of the fundamental mode is $\lambda = 4L = 4(0.090\ 0 \text{ m}) = 0.360$ m. Because the frequency of the tuning fork is constant, the next two normal modes (see Fig. 18.19b) correspond to lengths of

$$L = 3\lambda/4 = \boxed{0.270 \text{ m}} \quad \text{and} \quad L = 5\lambda/4 = \boxed{0.450 \text{ m.}}$$

18.6 Standing Waves in Rods and Membranes

Standing waves can also be set up in rods and membranes. A rod clamped in the middle and stroked parallel to the rod at one end oscillates, as depicted in Figure 18.20a. The oscillations of the elements of the rod are longitudinal, and so the broken lines in Figure 18.20 represent *longitudinal* displacements of various parts of the rod. For clarity, we have drawn them in the transverse direction, just as we did for air columns. The midpoint is a displacement node because it is fixed by the clamp, whereas the ends are displacement antinodes because they are free to oscillate. The oscillations in this setup are analogous to those in a pipe open at both ends. The broken lines in Figure 18.20a represent the first normal mode, for which the wavelength is $2L$ and the frequency is $f = v/2L$, where v is the speed of longitudinal waves in the rod. Other normal modes may be excited by clamping the rod at different points. For example, the second normal mode (Fig. 18.20b) is excited by clamping the rod a distance $L/4$ away from one end.

Musical instruments that depend on standing waves in rods include triangles, marimbas, xylophones, glockenspiels, chimes, and vibraphones. Other devices that make sounds from bars include music boxes and wind chimes.

Two-dimensional oscillations can be set up in a flexible membrane stretched over a circular hoop, such as that in a drumhead. As the membrane is struck at some point, waves that arrive at the fixed boundary are reflected many times. The resulting sound is not harmonic because the standing waves have frequencies that are *not* related by integer multiples. Without this relationship, the sound may be more correctly described as *noise* than as music. This is in contrast to the situation in wind and stringed instruments, which produce sounds that we describe as musical.

Some possible normal modes of oscillation for a two-dimensional circular membrane are shown in Figure 18.21. While nodes are *points* in one-dimensional standing

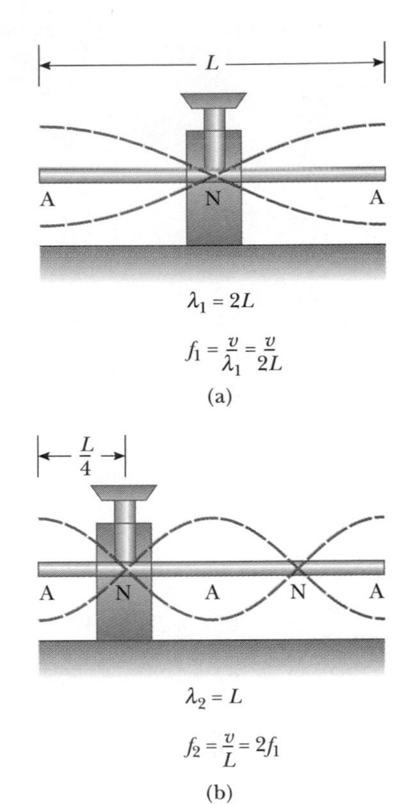

Figure 18.20 Normal-mode longitudinal vibrations of a rod of length L (a) clamped at the middle to produce the first normal mode and (b) clamped at a distance $L/4$ from one end to produce the second normal mode. Note that the broken lines represent oscillations parallel to the rod (longitudinal waves).

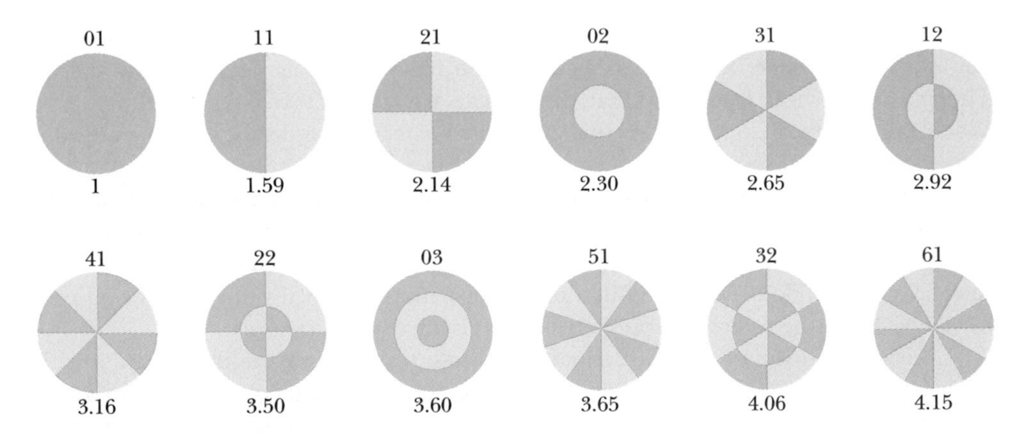

Figure 18.21 Representation of some of the normal modes possible in a circular membrane fixed at its perimeter. The pair of numbers above each pattern corresponds to the number of radial nodes and the number of circular nodes. Below each pattern is a factor by which the frequency of the mode is larger than that of the 01 mode. The frequencies of oscillation do not form a harmonic series because these factors are not integers. In each diagram, elements of the membrane on either side of a nodal line move in opposite directions, as indicated by the colors. (*Adapted from T. D. Rossing,* The Science of Sound, 2nd ed, *Reading, Massachusetts, Addison-Wesley Publishing Co., 1990*)

waves on strings and in air columns, a two-dimensional oscillator has *curves* along which there is no displacement of the elements of the medium. The lowest normal mode, which has a frequency f_1, contains only one nodal curve; this curve runs around the outer edge of the membrane. The other possible normal modes show additional nodal curves that are circles and straight lines across the diameter of the membrane.

18.7 Beats: Interference in Time

The interference phenomena with which we have been dealing so far involve the superposition of two or more waves having the same frequency. Because the amplitude of the oscillation of elements of the medium varies with the position in space of the element, we refer to the phenomenon as *spatial interference.* Standing waves in strings and pipes are common examples of spatial interference.

We now consider another type of interference, one that results from the superposition of two waves having slightly *different* frequencies. In this case, when the two waves are observed at the point of superposition, they are periodically in and out of phase. That is, there is a *temporal* (time) alternation between constructive and destructive interference. As a consequence, we refer to this phenomenon as *interference in time* or *temporal interference.* For example, if two tuning forks of slightly different frequencies are struck, one hears a sound of periodically varying amplitude. This phenomenon is called **beating:**

Definition of beating

Beating is the periodic variation in amplitude at a given point due to the superposition of two waves having slightly different frequencies.

The number of amplitude maxima one hears per second, or the *beat frequency,* equals the difference in frequency between the two sources, as we shall show below. The maximum beat frequency that the human ear can detect is about 20 beats/s. When the beat frequency exceeds this value, the beats blend indistinguishably with the sounds producing them.

A piano tuner can use beats to tune a stringed instrument by "beating" a note against a reference tone of known frequency. The tuner can then adjust the string tension until the frequency of the sound it emits equals the frequency of the reference tone. The tuner does this by tightening or loosening the string until the beats produced by it and the reference source become too infrequent to notice.

Consider two sound waves of equal amplitude traveling through a medium with slightly different frequencies f_1 and f_2. We use equations similar to Equation 16.10 to represent the wave functions for these two waves at a point that we choose as $x = 0$:

$$y_1 = A \cos \omega_1 t = A \cos 2\pi f_1 t$$
$$y_2 = A \cos \omega_2 t = A \cos 2\pi f_2 t$$

Using the superposition principle, we find that the resultant wave function at this point is

$$y = y_1 + y_2 = A(\cos 2\pi f_1 t + \cos 2\pi f_2 t)$$

The trigonometric identity

$$\cos a + \cos b = 2 \cos\left(\frac{a - b}{2}\right) \cos\left(\frac{a + b}{2}\right)$$

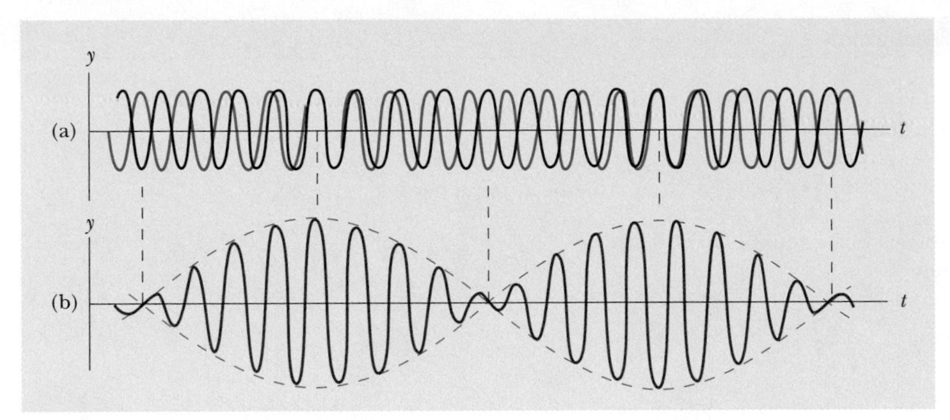

Active Figure 18.22 Beats are formed by the combination of two waves of slightly different frequencies. (a) The individual waves. (b) The combined wave has an amplitude (broken line) that oscillates in time.

At the Active Figures link at http://www.pse6.com, you can choose the two frequencies and see the corresponding beats.

allows us to write the expression for y as

$$y = \left[2A \cos 2\pi \left(\frac{f_1 - f_2}{2}\right) t\right] \cos 2\pi \left(\frac{f_1 + f_2}{2}\right) t \quad (18.13)$$

Resultant of two waves of different frequencies but equal amplitude

Graphs of the individual waves and the resultant wave are shown in Figure 18.22. From the factors in Equation 18.13, we see that the resultant sound for a listener standing at any given point has an effective frequency equal to the average frequency $(f_1 + f_2)/2$ and an amplitude given by the expression in the square brackets:

$$A_{\text{resultant}} = 2A \cos 2\pi \left(\frac{f_1 - f_2}{2}\right) t \quad (18.14)$$

That is, the **amplitude and therefore the intensity of the resultant sound vary in time.** The broken blue line in Figure 18.22b is a graphical representation of Equation 18.14 and is a sine wave varying with frequency $(f_1 - f_2)/2$.

Note that a maximum in the amplitude of the resultant sound wave is detected whenever

$$\cos 2\pi\left(\frac{f_1 - f_2}{2}\right)t = \pm 1$$

This means there are *two* maxima in each period of the resultant wave. Because the amplitude varies with frequency as $(f_1 - f_2)/2$, the number of beats per second, or the beat frequency f_{beat}, is twice this value. That is,

$$f_{\text{beat}} = |f_1 - f_2| \quad (18.15)$$

Beat frequency

For instance, if one tuning fork vibrates at 438 Hz and a second one vibrates at 442 Hz, the resultant sound wave of the combination has a frequency of 440 Hz (the musical note A) and a beat frequency of 4 Hz. A listener would hear a 440-Hz sound wave go through an intensity maximum four times every second.

Quick Quiz 18.9 You are tuning a guitar by comparing the sound of the string with that of a standard tuning fork. You notice a beat frequency of 5 Hz when both sounds are present. You tighten the guitar string and the beat frequency rises to 8 Hz. In order to tune the string exactly to the tuning fork, you should (a) continue to tighten the string (b) loosen the string (c) impossible to determine.

Example 18.8 The Mistuned Piano Strings

Two identical piano strings of length 0.750 m are each tuned exactly to 440 Hz. The tension in one of the strings is then increased by 1.0%. If they are now struck, what is the beat frequency between the fundamentals of the two strings?

Solution We find the ratio of frequencies if the tension in one string is 1.0% larger than the other:

$$\frac{f_2}{f_1} = \frac{(v_2/2L)}{(v_1/2L)} = \frac{v_2}{v_1} = \frac{\sqrt{T_2/\mu}}{\sqrt{T_1/\mu}} = \sqrt{\frac{T_2}{T_1}} = \sqrt{\frac{1.010T_1}{T_1}}$$
$$= 1.005$$

Thus, the frequency of the tightened string is

$$f_2 = 1.005f_1 = 1.005(440\text{ Hz}) = 442\text{ Hz}$$

and the beat frequency is

$$f_{\text{beat}} = 442\text{ Hz} - 440\text{ Hz} = 2\text{ Hz.}$$

18.8 Nonsinusoidal Wave Patterns

The sound wave patterns produced by the majority of musical instruments are nonsinusoidal. Characteristic patterns produced by a tuning fork, a flute, and a clarinet, each playing the same note, are shown in Figure 18.23. Each instrument has its own characteristic pattern. Note, however, that despite the differences in the patterns, each pattern is periodic. This point is important for our analysis of these waves.

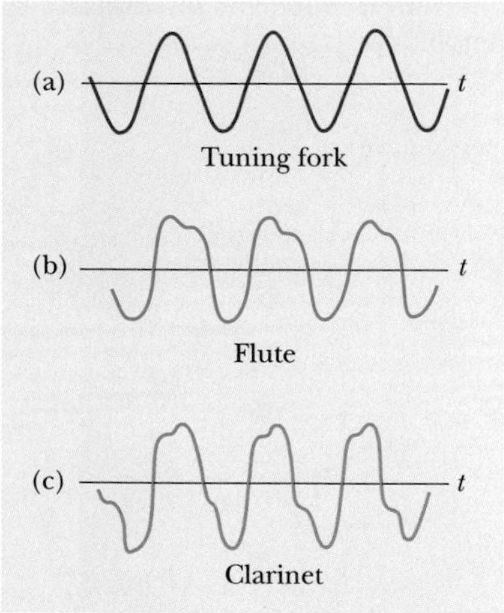

(a) t Tuning fork

(b) t Flute

(c) t Clarinet

Figure 18.23 Sound wave patterns produced by (a) a tuning fork, (b) a flute, and (c) a clarinet, each at approximately the same frequency. (*Adapted from C. A. Culver,* Musical Acoustics, 4th ed., *New York, McGraw-Hill Book Company, 1956, p. 128.*)

It is relatively easy to distinguish the sounds coming from a violin and a saxophone even when they are both playing the same note. On the other hand, an individual untrained in music may have difficulty distinguishing a note played on a clarinet from the same note played on an oboe. We can use the pattern of the sound waves from various sources to explain these effects.

This is in contrast to a musical instrument that makes a noise, such as the drum, in which the combination of frequencies do not form a harmonic series. When frequencies that are integer multiples of a fundamental frequency are combined, the result is a *musical* sound. A listener can assign a pitch to the sound, based on the fundamental frequency. Pitch is a psychological reaction to a sound that allows the listener to place the sound on a scale of low to high (bass to treble). Combinations of frequencies that are not integer multiples of a fundamental result in a *noise,* rather than a musical sound. It is much harder for a listener to assign a pitch to a noise than to a musical sound.

The wave patterns produced by a musical instrument are the result of the superposition of various harmonics. This superposition results in the corresponding richness of musical tones. The human perceptive response associated with various mixtures of harmonics is the *quality* or *timbre* of the sound. For instance, the sound of the trumpet is perceived to have a "brassy" quality (that is, we have learned to associate the adjective *brassy* with that sound); this quality enables us to distinguish the sound of the trumpet from that of the saxophone, whose quality is perceived as "reedy." The clarinet and oboe, however, both contain air columns excited by reeds; because of this similarity, it is more difficult for the ear to distinguish them on the basis of their sound quality.

The problem of analyzing nonsinusoidal wave patterns appears at first sight to be a formidable task. However, if the wave pattern is periodic, it can be represented as closely as desired by the combination of a sufficiently large number of sinusoidal waves that form a harmonic series. In fact, we can represent any periodic function as a series of sine and cosine terms by using a mathematical technique based on **Fourier's theorem.**[3] The corresponding sum of terms that represents the periodic wave pattern

[3] Developed by Jean Baptiste Joseph Fourier (1786–1830).

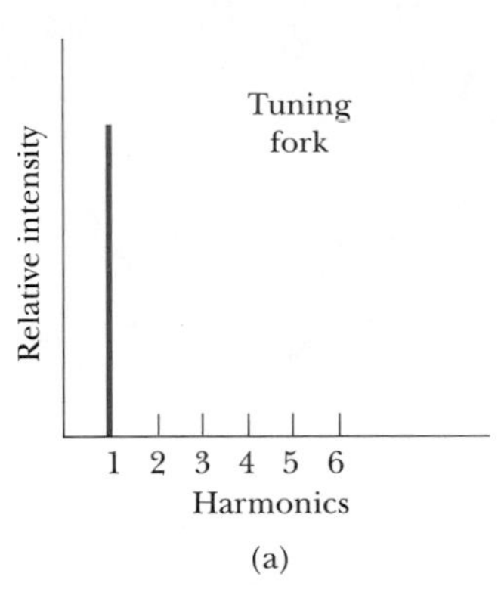

(a)

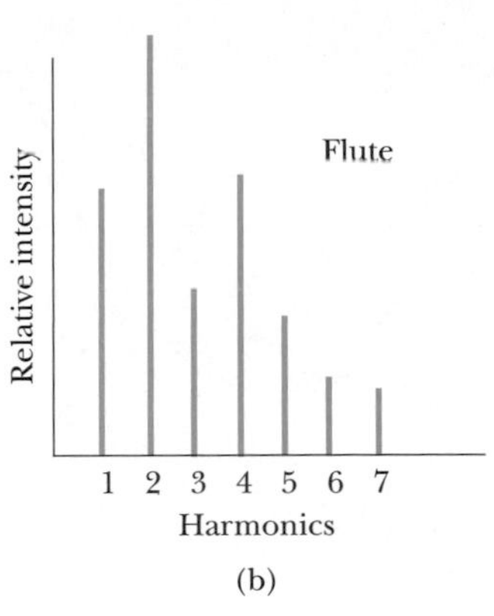

(b)

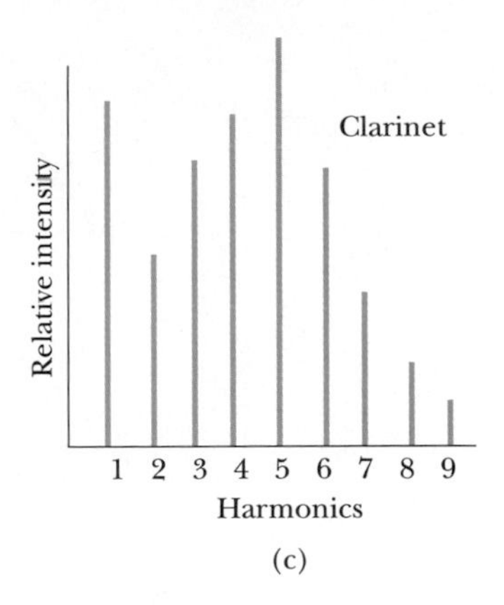

(c)

Figure 18.24 Harmonics of the wave patterns shown in Figure 18.23. Note the variations in intensity of the various harmonics. (*Adapted from C. A. Culver,* Musical Acoustics, 4th ed., *New York, McGraw-Hill Book Company, 1956.*)

is called a **Fourier series.** Let $y(t)$ be any function that is periodic in time with period T, such that $y(t+T) = y(t)$. Fourier's theorem states that this function can be written as

$$y(t) = \sum_n (A_n \sin 2\pi f_n t + B_n \cos 2\pi f_n t) \tag{18.16}$$

Fourier's theorem

where the lowest frequency is $f_1 = 1/T$. The higher frequencies are integer multiples of the fundamental, $f_n = nf_1$, and the coefficients A_n and B_n represent the amplitudes of the various waves. Figure 18.24 represents a harmonic analysis of the wave patterns shown in Figure 18.23. Note that a struck tuning fork produces only one harmonic (the first), whereas the flute and clarinet produce the first harmonic and many higher ones.

Note the variation in relative intensity of the various harmonics for the flute and the clarinet. In general, any musical sound consists of a fundamental frequency f plus other frequencies that are integer multiples of f, all having different intensities.

PITFALL PREVENTION

18.4 Pitch vs. Frequency

Do not confuse the term *pitch* with *frequency*. Frequency is the physical measurement of the number of oscillations per second. Pitch is a psychological reaction to sound that enables a person to place the sound on a scale from high to low, or from treble to bass. Thus, frequency is the stimulus and pitch is the response. Although pitch is related mostly (but not completely) to frequency, they are not the same. A phrase such as "the pitch of the sound" is incorrect because pitch is not a physical property of the sound.

(a)

(b)

(c)

Each musical instrument has its own characteristic sound and mixture of harmonics. Instruments shown are (a) the violin, (b) the saxophone, and (c) the trumpet.

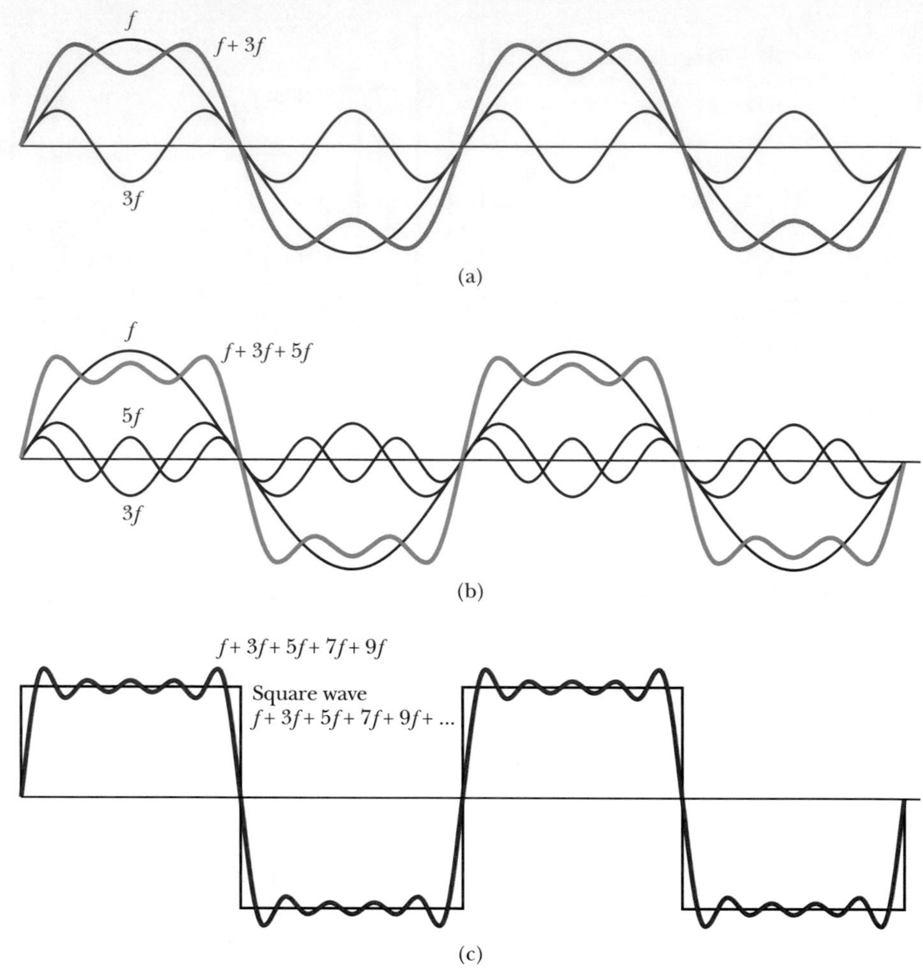

Active Figure 18.25 Fourier synthesis of a square wave, which is represented by the sum of odd multiples of the first harmonic, which has frequency f. (a) Waves of frequency f and $3f$ are added. (b) One more odd harmonic of frequency $5f$ is added. (c) The synthesis curve approaches closer to the square wave when odd frequencies up to $9f$ are added.

At the Active Figures link at* http://www.pse6.com, *you can add in harmonics with frequencies higher than 9f to try to synthesize a square wave.

We have discussed the *analysis* of a wave pattern using Fourier's theorem. The analysis involves determining the coefficients of the harmonics in Equation 18.16 from a knowledge of the wave pattern. The reverse process, called *Fourier synthesis,* can also be performed. In this process, the various harmonics are added together to form a resultant wave pattern. As an example of Fourier synthesis, consider the building of a square wave, as shown in Figure 18.25. The symmetry of the square wave results in only odd multiples of the fundamental frequency combining in its synthesis. In Figure 18.25a, the orange curve shows the combination of f and $3f$. In Figure 18.25b, we have added $5f$ to the combination and obtained the green curve. Notice how the general shape of the square wave is approximated, even though the upper and lower portions are not flat as they should be.

Figure 18.25c shows the result of adding odd frequencies up to $9f$. This approximation (purple curve) to the square wave is better than the approximations in parts a and b. To approximate the square wave as closely as possible, we would need to add all odd multiples of the fundamental frequency, up to infinite frequency.

Using modern technology, we can generate musical sounds electronically by mixing different amplitudes of any number of harmonics. These widely used electronic music synthesizers are capable of producing an infinite variety of musical tones.

SUMMARY

Take a practice test for this chapter by clicking on the Practice Test link at http://www.pse6.com.

The **superposition principle** specifies that when two or more waves move through a medium, the value of the resultant wave function equals the algebraic sum of the values of the individual wave functions.

When two traveling waves having equal amplitudes and frequencies superimpose, the resultant wave has an amplitude that depends on the phase angle ϕ between the two waves. **Constructive interference** occurs when the two waves are in phase, corresponding to $\phi = 0, 2\pi, 4\pi, \ldots$ rad. **Destructive interference** occurs when the two waves are 180° out of phase, corresponding to $\phi = \pi, 3\pi, 5\pi, \ldots$ rad.

Standing waves are formed from the superposition of two sinusoidal waves having the same frequency, amplitude, and wavelength but traveling in opposite directions. The resultant standing wave is described by the wave function

$$y = (2A \sin kx) \cos \omega t \tag{18.3}$$

Hence, the amplitude of the standing wave is $2A$, and the amplitude of the simple harmonic motion of any particle of the medium varies according to its position as $2A \sin kx$. The points of zero amplitude (called **nodes**) occur at $x = n\lambda/2$ ($n = 0, 1, 2, 3, \ldots$). The maximum amplitude points (called **antinodes**) occur at $x = n\lambda/4$ ($n = 1, 3, 5, \ldots$). Adjacent antinodes are separated by a distance $\lambda/2$. Adjacent nodes also are separated by a distance $\lambda/2$.

The natural frequencies of vibration of a taut string of length L and fixed at both ends are quantized and are given by

$$f_n = \frac{n}{2L}\sqrt{\frac{T}{\mu}} \qquad n = 1, 2, 3, \ldots \tag{18.8}$$

where T is the tension in the string and μ is its linear mass density. The natural frequencies of vibration $f_1, 2f_1, 3f_1, \ldots$ form a **harmonic series.**

An oscillating system is in **resonance** with some driving force whenever the frequency of the driving force matches one of the natural frequencies of the system. When the system is resonating, it responds by oscillating with a relatively large amplitude.

Standing waves can be produced in a column of air inside a pipe. If the pipe is open at both ends, all harmonics are present and the natural frequencies of oscillation are

$$f_n = n\frac{v}{2L} \qquad n = 1, 2, 3, \ldots \tag{18.11}$$

If the pipe is open at one end and closed at the other, only the odd harmonics are present, and the natural frequencies of oscillation are

$$f_n = n\frac{v}{4L} \qquad n = 1, 3, 5, \ldots \tag{18.12}$$

The phenomenon of **beating** is the periodic variation in intensity at a given point due to the superposition of two waves having slightly different frequencies.

QUESTIONS

1. Does the phenomenon of wave interference apply only to sinusoidal waves?

2. As oppositely moving pulses of the same shape (one upward, one downward) on a string pass through each other, there is one instant at which the string shows no displacement from the equilibrium position at any point. Has the energy carried by the pulses disappeared at this instant of time? If not, where is it?

3. Can two pulses traveling in opposite directions on the same string reflect from each other? Explain.

4. When two waves interfere, can the amplitude of the resultant wave be greater than either of the two original waves? Under what conditions?

5. For certain positions of the movable section shown in Figure 18.5, no sound is detected at the receiver—a situation corresponding to destructive interference. This suggests that energy is somehow lost. What happens to the energy transmitted by the speaker?

6. When two waves interfere constructively or destructively, is there any gain or loss in energy? Explain.

7. A standing wave is set up on a string, as shown in Figure 18.10. Explain why no energy is transmitted along the string.

8. What limits the amplitude of motion of a real vibrating system that is driven at one of its resonant frequencies?

9. Explain why your voice seems to sound better than usual when you sing in the shower.

10. What is the purpose of the slide on a trombone or of the valves on a trumpet?

11. Explain why all harmonics are present in an organ pipe open at both ends, but only odd harmonics are present in a pipe closed at one end.

12. Explain how a musical instrument such as a piano may be tuned by using the phenomenon of beats.

13. To keep animals away from their cars, some people mount short, thin pipes on the fenders. The pipes give out a high-pitched wail when the cars are moving. How do they create the sound?

14. When a bell is rung, standing waves are set up around the bell's circumference. What boundary conditions must be satisfied by the resonant wavelengths? How does a crack in the bell, such as in the Liberty Bell, affect the satisfying of the boundary conditions and the sound emanating from the bell?

15. An archer shoots an arrow from a bow. Does the string of the bow exhibit standing waves after the arrow leaves? If so, and if the bow is perfectly symmetric so that the arrow leaves from the center of the string, what harmonics are excited?

16. Despite a reasonably steady hand, a person often spills his coffee when carrying it to his seat. Discuss resonance as a possible cause of this difficulty, and devise a means for solving the problem.

17. An airplane mechanic notices that the sound from a twin-engine aircraft rapidly varies in loudness when both engines are running. What could be causing this variation from loud to soft?

18. When the base of a vibrating tuning fork is placed against a chalkboard, the sound that it emits becomes louder. This is because the vibrations of the tuning fork are transmitted to the chalkboard. Because it has a larger area than the tuning fork, the vibrating chalkboard sets more air into vibration. Thus, the chalkboard is a better radiator of sound than the tuning fork. How does this affect the length of time during which the fork vibrates? Does this agree with the principle of conservation of energy?

19. If you wet your finger and lightly run it around the rim of a fine wineglass, a high-frequency sound is heard. Why? How could you produce various musical notes with a set of wineglasses, each of which contains a different amount of water?

20. If you inhale helium from a balloon and do your best to speak normally, your voice will have a comical quacky quality. Explain why this "Donald Duck effect" happens. *Caution*: Helium is an asphyxiating gas and asphyxiation can cause panic. Helium can contain poisonous contaminants.

21. You have a standard tuning fork whose frequency is 262 Hz and a second tuning fork with an unknown frequency. When you tap both of them on the heel of one of your sneakers, you hear beats with a frequency of 4 per second. Thoughtfully chewing your gum, you wonder whether the unknown frequency is 258 Hz or 266 Hz. How can you decide?

PROBLEMS

1, 2, 3 = straightforward, intermediate, challenging ☐ = full solution available in the *Student Solutions Manual and Study Guide*

= coached solution with hints available at http://www.pse6.com = computer useful in solving problem

= paired numerical and symbolic problems

Section 18.1 Superposition and Interference

1. Two waves in one string are described by the wave functions

$$y_1 = 3.0 \cos(4.0x - 1.6t)$$

and

$$y_2 = 4.0 \sin(5.0x - 2.0t)$$

where y and x are in centimeters and t is in seconds. Find the superposition of the waves $y_1 + y_2$ at the points (a) $x = 1.00$, $t = 1.00$, (b) $x = 1.00$, $t = 0.500$, and (c) $x = 0.500$, $t = 0$. (Remember that the arguments of the trigonometric functions are in radians.)

2. Two pulses A and B are moving in opposite directions along a taut string with a speed of 2.00 cm/s. The amplitude of A is twice the amplitude of B. The pulses are shown in Figure P18.2 at $t = 0$. Sketch the shape of the string at $t = 1$, 1.5, 2, 2.5, and 3 s.

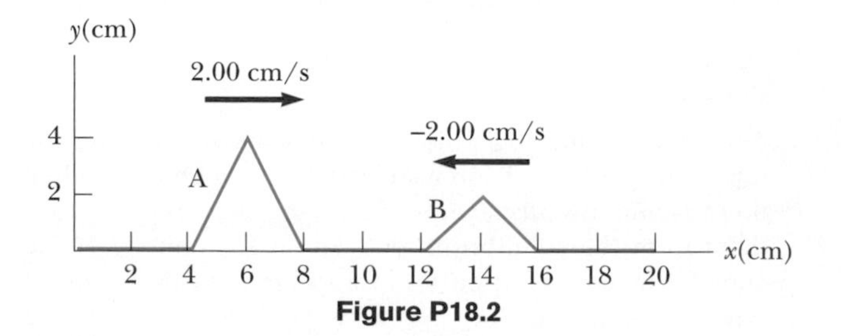

Figure P18.2

3. Two pulses traveling on the same string are described by

$$y_1 = \frac{5}{(3x - 4t)^2 + 2} \quad \text{and} \quad y_2 = \frac{-5}{(3x + 4t - 6)^2 + 2}$$

(a) In which direction does each pulse travel? (b) At what time do the two cancel everywhere? (c) At what point do the two pulses always cancel?

4. Two waves are traveling in the same direction along a stretched string. The waves are 90.0° out of phase. Each wave has an amplitude of 4.00 cm. Find the amplitude of the resultant wave.

5. Two traveling sinusoidal waves are described by the wave functions

$$y_1 = (5.00 \text{ m}) \sin[\pi(4.00x - 1\,200t)]$$

and

$$y_2 = (5.00 \text{ m}) \sin[\pi(4.00x - 1\,200t - 0.250)]$$

where x, y_1, and y_2 are in meters and t is in seconds. (a) What is the amplitude of the resultant wave? (b) What is the frequency of the resultant wave?

6. Two identical sinusoidal waves with wavelengths of 3.00 m travel in the same direction at a speed of 2.00 m/s. The second wave originates from the same point as the first, but at a later time. Determine the minimum possible time interval between the starting moments of the two waves if the amplitude of the resultant wave is the same as that of each of the two initial waves.

7. **Review problem.** A series of pulses, each of amplitude 0.150 m, is sent down a string that is attached to a post at one end. The pulses are reflected at the post and travel back along the string without loss of amplitude. What is the net displacement at a point on the string where two pulses are crossing, (a) if the string is rigidly attached to the post? (b) if the end at which reflection occurs is free to slide up and down?

8. Two loudspeakers are placed on a wall 2.00 m apart. A listener stands 3.00 m from the wall directly in front of one of the speakers. A single oscillator is driving the speakers at a frequency of 300 Hz. (a) What is the phase difference between the two waves when they reach the observer? (b) **What If?** What is the frequency closest to 300 Hz to which the oscillator may be adjusted such that the observer hears minimal sound?

9. Two speakers are driven by the same oscillator whose frequency is 200 Hz. They are located on a vertical pole a distance of 4.00 m from each other. A man walks straight toward the lower speaker in a direction perpendicular to the pole as shown in Figure P18.9. (a) How many times will he hear a minimum in sound intensity, and (b) how far is he from the pole at these moments? Take the speed of sound to be 330 m/s and ignore any sound reflections coming off the ground.

10. Two speakers are driven by the same oscillator whose frequency is f. They are located a distance d from each other on a vertical pole. A man walks straight toward the lower speaker in a direction perpendicular to the pole, as shown in Figure P18.9. (a) How many times will he hear a minimum in sound intensity? (b) How far is he from the pole at these moments? Let v represent the speed of sound, and assume that the ground does not reflect sound.

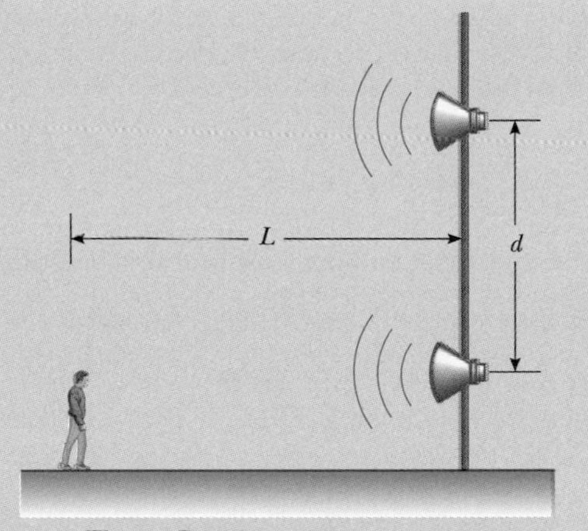

Figure P18.9 Problems 9 and 10.

11. Two sinusoidal waves in a string are defined by the functions

$$y_1 = (2.00 \text{ cm}) \sin(20.0x - 32.0t)$$

and

$$y_2 = (2.00 \text{ cm}) \sin(25.0x - 40.0t)$$

where y_1, y_2, and x are in centimeters and t is in seconds. (a) What is the phase difference between these two waves at the point $x = 5.00$ cm at $t = 2.00$ s? (b) What is the positive x value closest to the origin for which the two phases differ by $\pm\pi$ at $t = 2.00$ s? (This is where the two waves add to zero.)

12. Two identical speakers 10.0 m apart are driven by the same oscillator with a frequency of $f = 21.5$ Hz (Fig. P18.12). (a) Explain why a receiver at point A records a minimum in sound intensity from the two speakers. (b) If the receiver is moved in the plane of the speakers, what path should it take so that the intensity remains at a minimum? That is, deter-

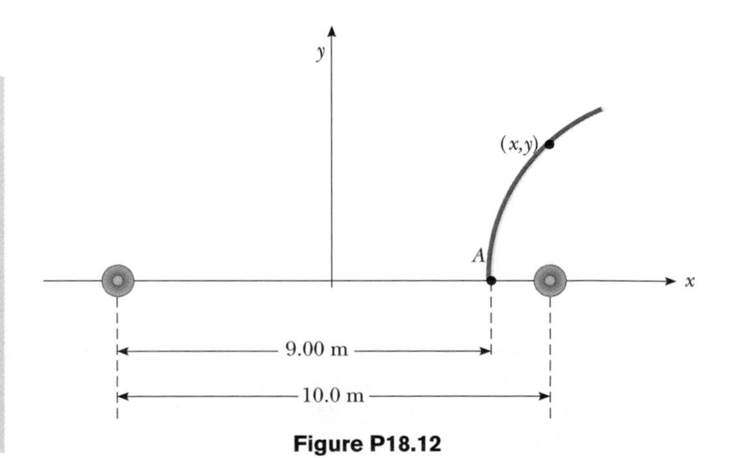

Figure P18.12

mine the relationship between x and y (the coordinates of the receiver) that causes the receiver to record a minimum in sound intensity. Take the speed of sound to be 344 m/s.

Section 18.2 Standing Waves

13. Two sinusoidal waves traveling in opposite directions interfere to produce a standing wave with the wave function

$$y = (1.50 \text{ m}) \sin(0.400x) \cos(200t)$$

where x is in meters and t is in seconds. Determine the wavelength, frequency, and speed of the interfering waves.

14. Two waves in a long string have wave functions given by

$$y_1 = (0.015\,0 \text{ m}) \cos\left(\frac{x}{2} - 40t\right)$$

and

$$y_2 = (0.015\,0 \text{ m}) \cos\left(\frac{x}{2} + 40t\right)$$

where y_1, y_2, and x are in meters and t is in seconds. (a) Determine the positions of the nodes of the resulting standing wave. (b) What is the maximum transverse position of an element of the string at the position $x = 0.400$ m?

15. Two speakers are driven in phase by a common oscillator at 800 Hz and face each other at a distance of 1.25 m. Locate the points along a line joining the two speakers where relative minima of sound pressure amplitude would be expected. (Use $v = 343$ m/s.)

16. Verify by direct substitution that the wave function for a standing wave given in Equation 18.3,

$$y = 2A \sin kx \cos \omega t$$

is a solution of the general linear wave equation, Equation 16.27:

$$\frac{\partial^2 y}{\partial x^2} = \frac{1}{v^2}\frac{\partial^2 y}{\partial t^2}$$

17. Two sinusoidal waves combining in a medium are described by the wave functions

$$y_1 = (3.0 \text{ cm}) \sin \pi(x + 0.60t)$$

and

$$y_2 = (3.0 \text{ cm}) \sin \pi(x - 0.60t)$$

where x is in centimeters and t is in seconds. Determine the *maximum* transverse position of an element of the medium at (a) $x = 0.250$ cm, (b) $x = 0.500$ cm, and (c) $x = 1.50$ cm. (d) Find the three smallest values of x corresponding to antinodes.

18. Two waves that set up a standing wave in a long string are given by the wave functions

$$y_1 = A \sin(kx - \omega t + \phi) \quad \text{and} \quad y_2 = A \sin(kx + \omega t)$$

Show (a) that the addition of the arbitrary phase constant ϕ changes only the position of the nodes and, in particular, (b) that the distance between nodes is still one half the wavelength.

Section 18.3 Standing Waves in a String Fixed at Both Ends

19. Find the fundamental frequency and the next three frequencies that could cause standing-wave patterns on a string that is 30.0 m long, has a mass per length of 9.00×10^{-3} kg/m, and is stretched to a tension of 20.0 N.

20. A string with a mass of 8.00 g and a length of 5.00 m has one end attached to a wall; the other end is draped over a pulley and attached to a hanging object with a mass of 4.00 kg. If the string is plucked, what is the fundamental frequency of vibration?

21. In the arrangement shown in Figure P18.21, an object can be hung from a string (with linear mass density $\mu = 0.002\,00$ kg/m) that passes over a light pulley. The string is connected to a vibrator (of constant frequency f), and the length of the string between point P and the pulley is $L = 2.00$ m. When the mass m of the object is either 16.0 kg or 25.0 kg, standing waves are observed; however, no standing waves are observed with any mass between these values. (a) What is the frequency of the vibrator? (*Note:* The greater the tension in the string, the smaller the number of nodes in the standing wave.) (b) What is the largest object mass for which standing waves could be observed?

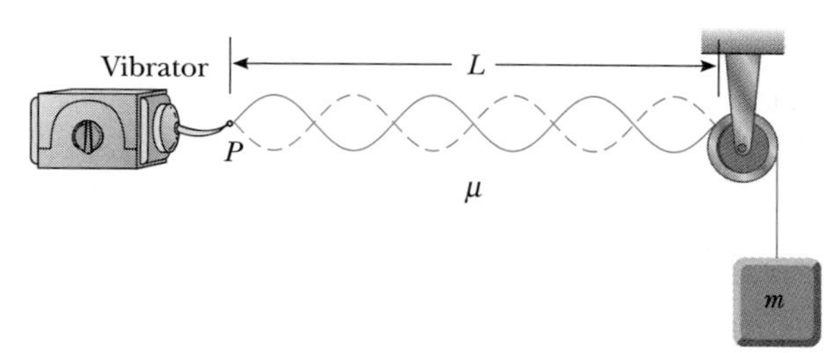

Figure P18.21 Problems 21 and 22.

22. A vibrator, pulley, and hanging object are arranged as in Figure P18.21, with a compound string, consisting of two strings of different masses and lengths fastened together end-to-end. The first string, which has a mass of 1.56 g and a length of 65.8 cm, runs from the vibrator to the junction of the two strings. The second string runs from the junction over the pulley to the suspended 6.93-kg object. The mass and length of the string from the junction to the pulley are, respectively, 6.75 g and 95.0 cm. (a) Find the lowest frequency for which standing waves are observed in both strings, with a node at the junction. The standing wave patterns in the two strings may have different numbers of nodes. (b) What is the total number of nodes observed along the compound string at this frequency, excluding the nodes at the vibrator and the pulley?

23. Example 18.4 tells you that the adjacent notes E, F, and F-sharp can be assigned frequencies of 330 Hz, 350 Hz, and 370 Hz. You might not guess how the pattern continues. The next notes, G, G-sharp, and A, have frequencies of 392 Hz, 416 Hz, and 440 Hz. On the equally tempered or chromatic scale used in Western music, the frequency of each higher note is obtained by multiplying the previous frequency by $\sqrt[12]{2}$. A standard guitar has strings 64.0 cm long and nineteen frets. In Example 18.4, we found the spacings of the first two frets. Calculate the distance between the last two frets.

24. The top string of a guitar has a fundamental frequency of 330 Hz when it is allowed to vibrate as a whole, along all of its 64.0-cm length from the neck to the bridge. A fret is provided for limiting vibration to just the lower two-thirds of the string. (a) If the string is pressed down at this fret and plucked, what is the new fundamental frequency? (b) **What If?** The guitarist can play a "natural harmonic" by gently touching the string at the location of this fret and plucking the string at about one sixth of the way along its length from the bridge. What frequency will be heard then?

25. A string of length L, mass per unit length μ, and tension T is vibrating at its fundamental frequency. What effect will the following have on the fundamental frequency? (a) The length of the string is doubled, with all other factors held constant. (b) The mass per unit length is doubled, with all other factors held constant. (c) The tension is doubled, with all other factors held constant.

26. A 60.000-cm guitar string under a tension of 50.000 N has a mass per unit length of 0.100 00 g/cm. What is the highest resonant frequency that can be heard by a person capable of hearing frequencies up to 20 000 Hz?

27. A cello A-string vibrates in its first normal mode with a frequency of 220 Hz. The vibrating segment is 70.0 cm long and has a mass of 1.20 g. (a) Find the tension in the string. (b) Determine the frequency of vibration when the string vibrates in three segments.

28. A violin string has a length of 0.350 m and is tuned to concert G, with $f_G = 392$ Hz. Where must the violinist place her finger to play concert A, with $f_A = 440$ Hz? If this position is to remain correct to half the width of a finger (that is, to within 0.600 cm), what is the maximum allowable percentage change in the string tension?

29. Review problem. A sphere of mass M is supported by a string that passes over a light horizontal rod of length L (Fig. P18.29). Given that the angle is θ and that f represents the fundamental frequency of standing waves in the portion of the string above the rod, determine the mass of this portion of the string.

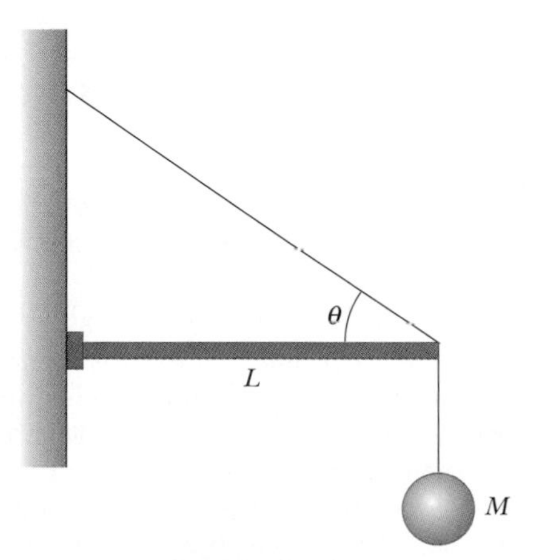

Figure P18.29

30. Review problem. A copper cylinder hangs at the bottom of a steel wire of negligible mass. The top end of the wire is fixed. When the wire is struck, it emits sound with a fundamental frequency of 300 Hz. If the copper cylinder is then submerged in water so that half its volume is below the water line, determine the new fundamental frequency.

31. A standing-wave pattern is observed in a thin wire with a length of 3.00 m. The equation of the wave is

$$y = (0.002 \text{ m}) \sin(\pi x) \cos(100\pi t)$$

where x is in meters and t is in seconds. (a) How many loops does this pattern exhibit? (b) What is the fundamental frequency of vibration of the wire? (c) **What If?** If the original frequency is held constant and the tension in the wire is increased by a factor of 9, how many loops are present in the new pattern?

Section 18.4 Resonance

32. The chains suspending a child's swing are 2.00 m long. At what frequency should a big brother push to make the child swing with largest amplitude?

33. An earthquake can produce a *seiche* in a lake, in which the water sloshes back and forth from end to end with remarkably large amplitude and long period. Consider a seiche produced in a rectangular farm pond, as in the cross-sectional view of Figure P18.33. (The figure is not drawn to scale.) Suppose that the pond is 9.15 m long and of uniform width and depth. You measure that a pulse produced at one end reaches the other end in 2.50 s. (a) What is the wave speed? (b) To produce the seiche, several people stand on the bank at one end and paddle together with snow shovels, moving them in simple harmonic motion. What should be the frequency of this motion?

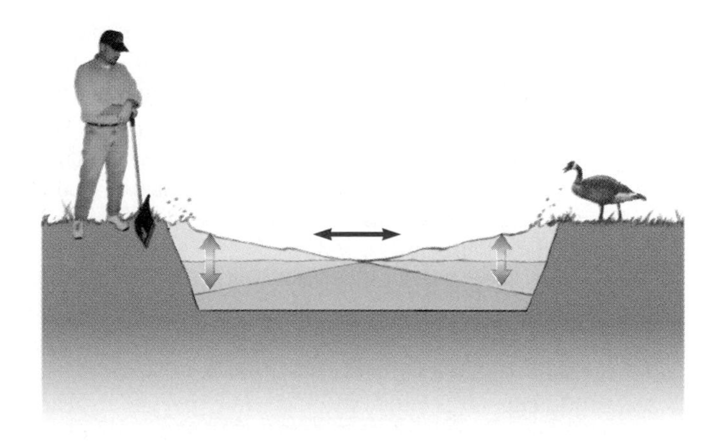

Figure P18.33

34. The Bay of Fundy, Nova Scotia, has the highest tides in the world, as suggested in the photographs on page 452. Assume that in mid-ocean and at the mouth of the bay, the Moon's gravity gradient and the Earth's rotation make the water surface oscillate with an amplitude of a few centimeters and a period of 12 h 24 min. At the head of the bay, the amplitude is several meters. Argue for or against the proposition that the tide is amplified by standing-wave resonance. Assume the bay has a length of 210 km and a uniform depth of 36.1 m. The speed of long-wavelength water waves is given by $\sqrt{gd}$, where d is the water's depth.

35. Standing-wave vibrations are set up in a crystal goblet with four nodes and four antinodes equally spaced around the 20.0-cm circumference of its rim. If transverse waves move around the glass at 900 m/s, an opera singer would have to produce a high harmonic with what frequency to shatter the glass with a resonant vibration?

Section 18.5 Standing Waves in Air Columns

Note: Unless otherwise specified, assume that the speed of sound in air is 343 m/s at 20°C, and is described by

$$v = (331\text{ m/s})\sqrt{1 + \frac{T_C}{273°}}$$

at any Celsius temperature T_C.

36. The overall length of a piccolo is 32.0 cm. The resonating air column vibrates as in a pipe open at both ends. (a) Find the frequency of the lowest note that a piccolo can play, assuming that the speed of sound in air is 340 m/s. (b) Opening holes in the side effectively shortens the length of the resonant column. If the highest note a piccolo can sound is 4 000 Hz, find the distance between adjacent antinodes for this mode of vibration.

37. Calculate the length of a pipe that has a fundamental frequency of 240 Hz if the pipe is (a) closed at one end and (b) open at both ends.

38. The fundamental frequency of an open organ pipe corresponds to middle C (261.6 Hz on the chromatic musical scale). The third resonance of a closed organ pipe has the same frequency. What are the lengths of the two pipes?

39. The windpipe of one typical whooping crane is 5.00 ft long. What is the fundamental resonant frequency of the bird's trachea, modeled as a narrow pipe closed at one end? Assume a temperature of 37°C.

40. Do not stick anything into your ear! Estimate the length of your ear canal, from its opening at the external ear to the eardrum. If you regard the canal as a narrow tube that is open at one end and closed at the other, at approximately what fundamental frequency would you expect your hearing to be most sensitive? Explain why you can hear especially soft sounds just around this frequency.

41. A shower stall measures 86.0 cm × 86.0 cm × 210 cm. If you were singing in this shower, which frequencies would sound the richest (because of resonance)? Assume that the stall acts as a pipe closed at both ends, with nodes at opposite sides. Assume that the voices of various singers range from 130 Hz to 2 000 Hz. Let the speed of sound in the hot shower stall be 355 m/s.

42. As shown in Figure P18.42, water is pumped into a tall vertical cylinder at a volume flow rate R. The radius of the cylinder is r, and at the open top of the cylinder a tuning fork is vibrating with a frequency f. As the water rises, how much time elapses between successive resonances?

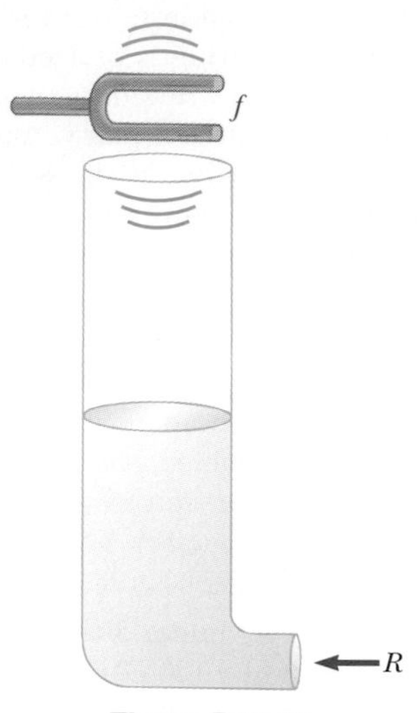

Figure P18.42

43. If two adjacent natural frequencies of an organ pipe are determined to be 550 Hz and 650 Hz, calculate the fundamental frequency and length of this pipe. (Use v = 340 m/s.)

44. A glass tube (open at both ends) of length L is positioned near an audio speaker of frequency f = 680 Hz. For what values of L will the tube resonate with the speaker?

45. An air column in a glass tube is open at one end and closed at the other by a movable piston. The air in the tube is warmed above room temperature, and a 384-Hz tuning fork is held at the open end. Resonance is heard when the piston is 22.8 cm from the open end and again when it is 68.3 cm from the open end. (a) What speed of sound is implied by these data? (b) How far from the open end will the piston be when the next resonance is heard?

46. A tuning fork with a frequency of 512 Hz is placed near the top of the pipe shown in Figure 18.19a. The water level is lowered so that the length L slowly increases from an initial value of 20.0 cm. Determine the next two values of L that correspond to resonant modes.

47. When an open metal pipe is cut into two pieces, the lowest resonance frequency for the air column in one piece is 256 Hz and that for the other is 440 Hz. (a) What resonant frequency would have been produced by the original length of pipe? (b) How long was the original pipe?

48. With a particular fingering, a flute plays a note with frequency 880 Hz at 20.0°C. The flute is open at both ends. (a) Find the air column length. (b) Find the frequency it produces at the beginning of the half-time performance at a late-season American football game, when the ambient temperature is −5.00°C and the musician has not had a chance to warm up the flute.

Section 18.6 Standing Waves in Rods and Membranes

49. An aluminum rod 1.60 m long is held at its center. It is stroked with a rosin-coated cloth to set up a longitudinal vibration. The speed of sound in a thin rod of aluminum is 5 100 m/s. (a) What is the fundamental frequency of the waves established in the rod? (b) What harmonics are set up in the rod held in this manner? (c) **What If?** What would be the fundamental frequency if the rod were made of copper, in which the speed of sound is 3 560 m/s?

50. An aluminum rod is clamped one quarter of the way along its length and set into longitudinal vibration by a variable-frequency driving source. The lowest frequency that produces resonance is 4 400 Hz. The speed of sound in an aluminum rod is 5 100 m/s. Find the length of the rod.

Section 18.7 Beats: Interference in Time

51. In certain ranges of a piano keyboard, more than one string is tuned to the same note to provide extra loudness. For example, the note at 110 Hz has two strings at this frequency. If one string slips from its normal tension of 600 N to 540 N, what beat frequency is heard when the hammer strikes the two strings simultaneously?

52. While attempting to tune the note C at 523 Hz, a piano tuner hears 2 beats/s between a reference oscillator and the string. (a) What are the possible frequencies of the string? (b) When she tightens the string slightly, she hears 3 beats/s. What is the frequency of the string now? (c) By what percentage should the piano tuner now change the tension in the string to bring it into tune?

53. A student holds a tuning fork oscillating at 256 Hz. He walks toward a wall at a constant speed of 1.33 m/s. (a) What beat frequency does he observe between the tuning fork and its echo? (b) How fast must he walk away from the wall to observe a beat frequency of 5.00 Hz?

54. When beats occur at a rate higher than about 20 per second, they are not heard individually but rather as a steady hum, called a *combination tone.* The player of a typical pipe organ can press a single key and make the organ produce sound with different fundamental frequencies. She can select and pull out different stops to make the same key for the note C produce sound at the following frequencies: 65.4 Hz from a so-called eight-foot pipe; $2 \times 65.4 =$ 131 Hz from a four-foot pipe; $3 \times 65.4 = 196$ Hz from a two-and-two-thirds-foot pipe; $4 \times 65.4 = 262$ Hz from a two-foot pipe; or any combination of these. With notes at low frequencies, she obtains sound with the richest quality by pulling out all the stops. When an air leak develops in one of the pipes, that pipe cannot be used. If a leak occurs in an eight-foot pipe, playing a combination of other pipes can create the sensation of sound at the frequency that the eight-foot pipe would produce. Which sets of stops, among those listed, could be pulled out to do this?

Section 18.8 Nonsinusoidal Wave Patterns

55. An A-major chord consists of the notes called A, $C^{\#}$, and E. It can be played on a piano by simultaneously striking strings with fundamental frequencies of 440.00 Hz, 554.37 Hz, and 659.26 Hz. The rich consonance of the chord is associated with near equality of the frequencies of some of the higher harmonics of the three tones. Consider the first five harmonics of each string and determine which harmonics show near equality.

56. Suppose that a flutist plays a 523-Hz C note with first harmonic displacement amplitude $A_1 = 100$ nm. From Figure 18.24b read, by proportion, the displacement amplitudes of harmonics 2 through 7. Take these as the values A_2 through A_7 in the Fourier analysis of the sound, and assume that $B_1 = B_2 = \cdots = B_7 = 0$. Construct a graph of the waveform of the sound. Your waveform will not look exactly like the flute waveform in Figure 18.23b because you simplify by ignoring cosine terms; nevertheless, it produces the same sensation to human hearing.

Additional Problems

57. On a marimba (Fig. P18.57), the wooden bar that sounds a tone when struck vibrates in a transverse standing wave having three antinodes and two nodes. The lowest frequency note is 87.0 Hz, produced by a bar 40.0 cm long. (a) Find the speed of transverse waves on the bar. (b) A resonant pipe suspended vertically below the center of the bar enhances the loudness of the emitted sound. If the pipe is open at the top end only and the speed of sound in air is 340 m/s, what is the length of the pipe required to resonate with the bar in part (a)?

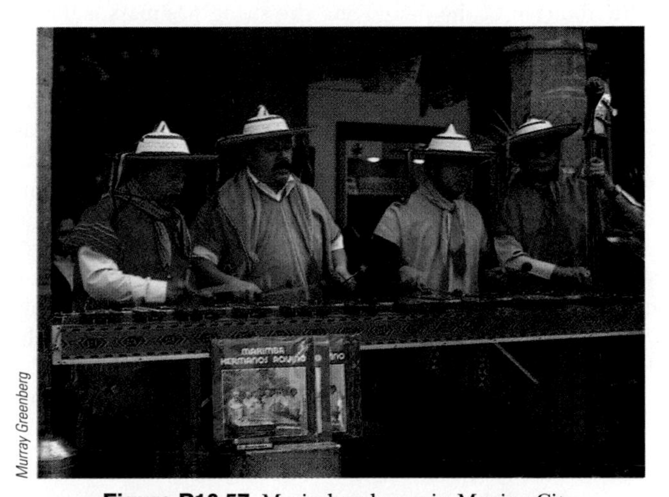

Murray Greenberg

Figure P18.57 Marimba players in Mexico City.

58. A loudspeaker at the front of a room and an identical loudspeaker at the rear of the room are being driven by the same oscillator at 456 Hz. A student walks at a uniform rate of 1.50 m/s along the length of the room. She hears a single tone, repeatedly becoming louder and softer. (a) Model these variations as beats between the Doppler-shifted sounds the student receives. Calculate the number of beats the student hears each second. (b) **What If?** Model the two speakers as producing a standing wave in the room and the student as walking between antinodes. Calculate the number of intensity maxima the student hears each second.

59. Two train whistles have identical frequencies of 180 Hz. When one train is at rest in the station and the other is moving nearby, a commuter standing on the station platform hears beats with a frequency of 2.00 beats/s when the whistles sound at the same time. What are the two possible speeds and directions that the moving train can have?

60. A string fixed at both ends and having a mass of 4.80 g, a length of 2.00 m, and a tension of 48.0 N vibrates in its second ($n = 2$) normal mode. What is the wavelength in air of the sound emitted by this vibrating string?

61. A student uses an audio oscillator of adjustable frequency to measure the depth of a water well. The student hears two successive resonances at 51.5 Hz and 60.0 Hz. How deep is the well?

62. A string has a mass per unit length of 9.00×10^{-3} kg/m and a length of 0.400 m. What must be the tension in the string if its second harmonic has the same frequency as the second resonance mode of a 1.75-m-long pipe open at one end?

63. Two wires are welded together end to end. The wires are made of the same material, but the diameter of one is twice that of the other. They are subjected to a tension of 4.60 N. The thin wire has a length of 40.0 cm and a linear mass density of 2.00 g/m. The combination is fixed at both ends and vibrated in such a way that two antinodes are present, with the node between them being right at the weld. (a) What is the frequency of vibration? (b) How long is the thick wire?

64. Review problem. For the arrangement shown in Figure P18.64, $\theta = 30.0°$, the inclined plane and the small pulley are frictionless, the string supports the object of mass M at the bottom of the plane, and the string has mass m that is small compared to M. The system is in equilibrium and the vertical part of the string has a length h. Standing waves are set up in the vertical section of the string. (a) Find the tension in the string. (b) Model the shape of the string as one leg and the hypotenuse of a right triangle. Find the whole length of the string. (c) Find the mass per unit length of the string. (d) Find the speed of waves on the string. (e) Find the lowest frequency for a standing wave. (f) Find the period of the standing wave having three nodes. (g) Find the wavelength of the standing wave having three nodes. (h) Find the frequency of the beats resulting from the interference of the sound wave of lowest frequency generated by the string with another sound wave having a frequency that is 2.00% greater.

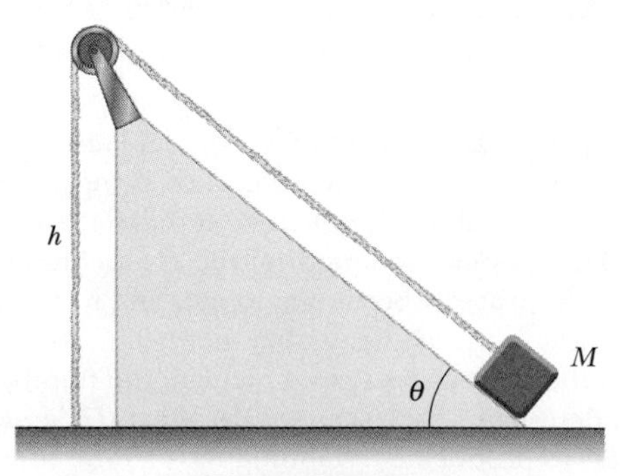

Figure P18.64

65. A standing wave is set up in a string of variable length and tension by a vibrator of variable frequency. Both ends of the string are fixed. When the vibrator has a frequency f, in a string of length L and under tension T, n antinodes are set up in the string. (a) If the length of the string is doubled, by what factor should the frequency be changed so that the same number of antinodes is produced? (b) If the frequency and length are held constant, what tension will produce $n + 1$ antinodes? (c) If the frequency is tripled and the length of the string is halved, by what factor should the tension be changed so that twice as many antinodes are produced?

66. A 0.010 0-kg wire, 2.00 m long, is fixed at both ends and vibrates in its simplest mode under a tension of 200 N. When a vibrating tuning fork is placed near the wire, a beat frequency of 5.00 Hz is heard. (a) What could be the frequency of the tuning fork? (b) What should the tension in the wire be if the beats are to disappear?

67. Two waves are described by the wave functions

$$y_1(x, t) = 5.0 \sin(2.0x - 10t)$$

and

$$y_2(x, t) = 10 \cos(2.0x - 10t)$$

where y_1, y_2, and x are in meters and t is in seconds. Show that the wave resulting from their superposition is also sinusoidal. Determine the amplitude and phase of this sinusoidal wave.

68. The wave function for a standing wave is given in Equation 18.3 as $y = 2A \sin kx \cos \omega t$. (a) Rewrite this wave function in terms of the wavelength λ and the wave speed v of the wave. (b) Write the wave function of the simplest standing-wave vibration of a stretched string of length L. (c) Write the wave function for the second harmonic. (d) Generalize these results and write the wave function for the nth resonance vibration.

69. Review problem. A 12.0-kg object hangs in equilibrium from a string with a total length of $L = 5.00$ m and a linear mass density of $\mu = 0.001\,00$ kg/m. The string is wrapped around two light, frictionless pulleys that are separated by a distance of $d = 2.00$ m (Fig. P18.69a). (a) Determine the tension in the string. (b) At what frequency must the string between the pulleys vibrate in order to form the standing wave pattern shown in Figure P18.69b?

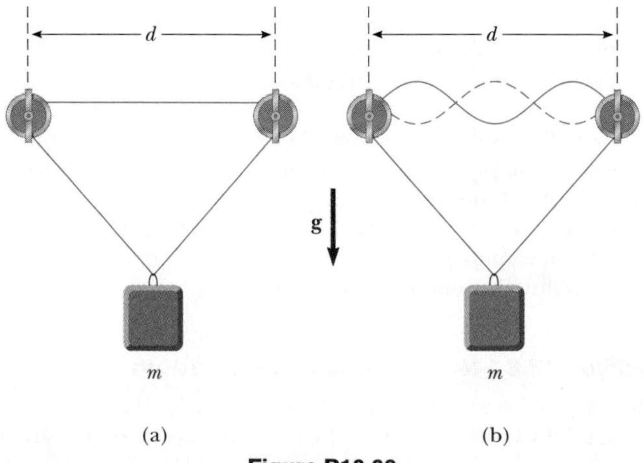

Figure P18.69

70. A quartz watch contains a crystal oscillator in the form of a block of quartz that vibrates by contracting and expanding. Two opposite faces of the block, 7.05 mm apart, are antinodes, moving alternately toward each other and away from each other. The plane halfway between these two faces is a node of the vibration. The speed of sound in quartz is 3.70 km/s. Find the frequency of the vibration. An oscillating electric voltage accompanies the mechanical oscillation—the quartz is described as *piezoelectric.* An electric circuit feeds in energy to maintain the oscillation and also counts the voltage pulses to keep time.

Answers to Quick Quizzes

18.1 The shape of the string at $t = 0.6$ s is shown below.

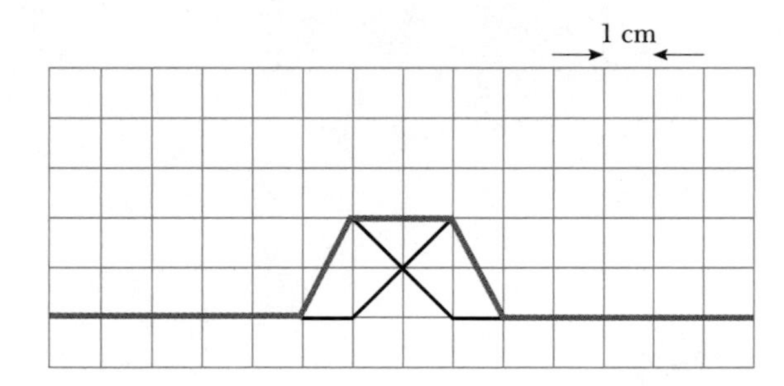

18.2 (c). The pulses completely cancel each other in terms of displacement of elements of the string from equilibrium, but the string is still moving. A short time later, the string will be displaced again and the pulses will have passed each other.

18.3 (a). The pattern shown at the bottom of Figure 18.9a corresponds to the extreme position of the string. All elements of the string have momentarily come to rest.

18.4 (d). Near a nodal point, elements on one side of the point are moving upward at this instant and elements on the other side are moving downward.

18.5 (d). Choice (a) is incorrect because the number of nodes is one greater than the number of antinodes. Choice (b) is only true for half of the modes; it is not true for any odd-numbered mode. Choice (c) would be correct if we replace the word *nodes* with *antinodes.*

18.6 For each natural frequency of the glass, the standing wave must "fit" exactly around the rim. In Figure 18.17a we see three antinodes on the near side of the glass, and thus there must be another three on the far side. This corresponds to three complete waves. In a top view, the wave pattern looks like this (although we have greatly exaggerated the amplitude):

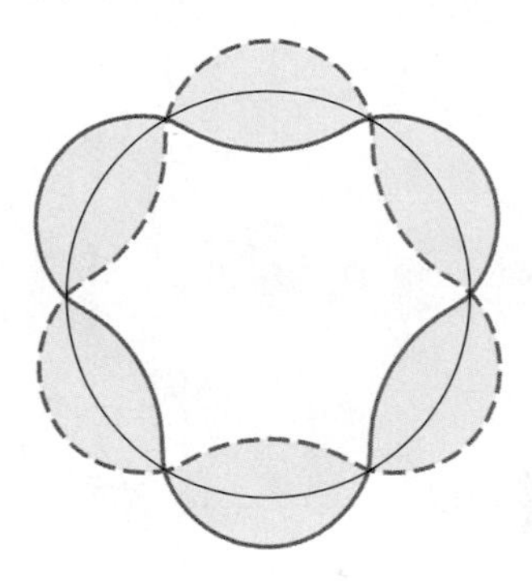

18.7 (b). With both ends open, the pipe has a fundamental frequency given by Equation 18.11: $f_{\text{open}} = v/2L$. With one end closed, the pipe has a fundamental frequency given by Equation 18.12:

$$f_{\text{closed}} = \frac{v}{4L} = \frac{1}{2}\frac{v}{2L} = \frac{1}{2}f_{\text{open}}$$

18.8 (c). The increase in temperature causes the speed of sound to go up. According to Equation 18.11, this will result in an increase in the fundamental frequency of a given organ pipe.

18.9 (b). Tightening the string has caused the frequencies to be farther apart, based on the increase in the beat frequency.

Thermodynamics

We now direct our attention to the study of thermodynamics, which involves situations in which the temperature or state (solid, liquid, gas) of a system changes due to energy transfers. As we shall see, thermodynamics is very successful in explaining the bulk properties of matter and the correlation between these properties and the mechanics of atoms and molecules.

Historically, the development of thermodynamics paralleled the development of the atomic theory of matter. By the 1820s, chemical experiments had provided solid evidence for the existence of atoms. At that time, scientists recognized that a connection between thermodynamics and the structure of matter must exist. In 1827, the botanist Robert Brown reported that grains of pollen suspended in a liquid move erratically from one place to another, as if under constant agitation. In 1905, Albert Einstein used kinetic theory to explain the cause of this erratic motion, which today is known as *Brownian motion*. Einstein explained this phenomenon by assuming that the grains are under constant bombardment by "invisible" molecules in the liquid, which themselves move erratically. This explanation gave scientists insight into the concept of molecular motion and gave credence to the idea that matter is made up of atoms. A connection was thus forged between the everyday world and the tiny, invisible building blocks that make up this world.

Thermodynamics also addresses more practical questions. Have you ever wondered how a refrigerator is able to cool its contents, what types of transformations occur in a power plant or in the engine of your automobile, or what happens to the kinetic energy of a moving object when the object comes to rest? The laws of thermodynamics can be used to provide explanations for these and other phenomena. ■

◀ *The Alyeska oil pipeline near the Tazlina River in Alaska. The oil in the pipeline is warm, and energy transferring from the pipeline could melt environmentally sensitive permafrost in the ground. The finned structures on top of the support posts are thermal radiators that allow the energy to be transferred into the air in order to protect the permafrost. (Topham Picturepoint/The Image Works)*

Chapter 19

Temperature

CHAPTER OUTLINE

19.1 Temperature and the Zeroth Law of Thermodynamics

19.2 Thermometers and the Celsius Temperature Scale

19.3 The Constant-Volume Gas Thermometer and the Absolute Temperature Scale

19.4 Thermal Expansion of Solids and Liquids

19.5 Macroscopic Description of an Ideal Gas

▲ *Why would someone designing a pipeline include these strange loops? Pipelines carrying liquids often contain loops such as these to allow for expansion and contraction as the temperature changes. We will study thermal expansion in this chapter. (Lowell Georgia/CORBIS)*

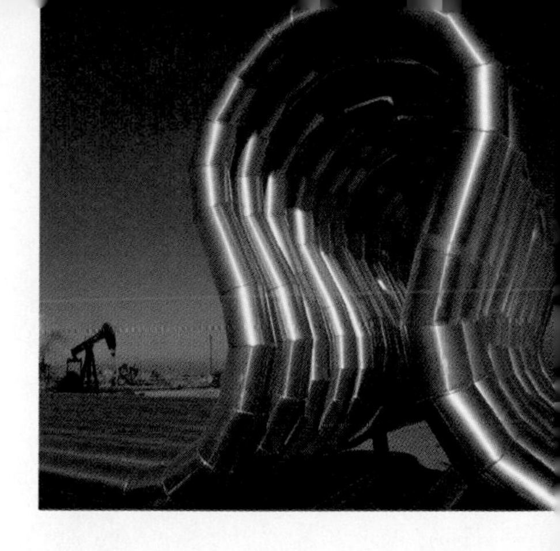

In our study of mechanics, we carefully defined such concepts as *mass*, *force*, and *kinetic energy* to facilitate our quantitative approach. Likewise, a quantitative description of thermal phenomena requires careful definitions of such important terms as *temperature*, *heat*, and *internal energy*. This chapter begins with a discussion of temperature and with a description of one of the laws of thermodynamics (the so-called "zeroth law").

Next, we consider why an important factor when we are dealing with thermal phenomena is the particular substance we are investigating. For example, gases expand appreciably when heated, whereas liquids and solids expand only slightly.

This chapter concludes with a study of ideal gases on the macroscopic scale. Here, we are concerned with the relationships among such quantities as pressure, volume, and temperature. In Chapter 21, we shall examine gases on a microscopic scale, using a model that represents the components of a gas as small particles.

19.1 Temperature and the Zeroth Law of Thermodynamics

We often associate the concept of temperature with how hot or cold an object feels when we touch it. Thus, our senses provide us with a qualitative indication of temperature. However, our senses are unreliable and often mislead us. For example, if we remove a metal ice tray and a cardboard box of frozen vegetables from the freezer, the ice tray feels colder than the box *even though both are at the same temperature.* The two objects feel different because metal transfers energy by heat at a higher rate than cardboard does. What we need is a reliable and reproducible method for measuring the relative hotness or coldness of objects rather than the rate of energy transfer. Scientists have developed a variety of thermometers for making such quantitative measurements.

We are all familiar with the fact that two objects at different initial temperatures eventually reach some intermediate temperature when placed in contact with each other. For example, when hot water and cold water are mixed in a bathtub, the final temperature of the mixture is somewhere between the initial hot and cold temperatures. Likewise, when an ice cube is dropped into a cup of hot coffee, it melts and the coffee's temperature decreases.

To understand the concept of temperature, it is useful to define two often-used phrases: *thermal contact* and *thermal equilibrium.* To grasp the meaning of thermal contact, imagine that two objects are placed in an insulated container such that they interact with each other but not with the environment. If the objects are at different temperatures, energy is exchanged between them, even if they are initially not in physical contact with each other. The energy transfer mechanisms from Chapter 7 that we will focus on are heat and electromagnetic radiation. For purposes of the current discussion, we assume that two objects are in **thermal contact** with each other if energy can be exchanged between them by these processes due to a temperature difference.

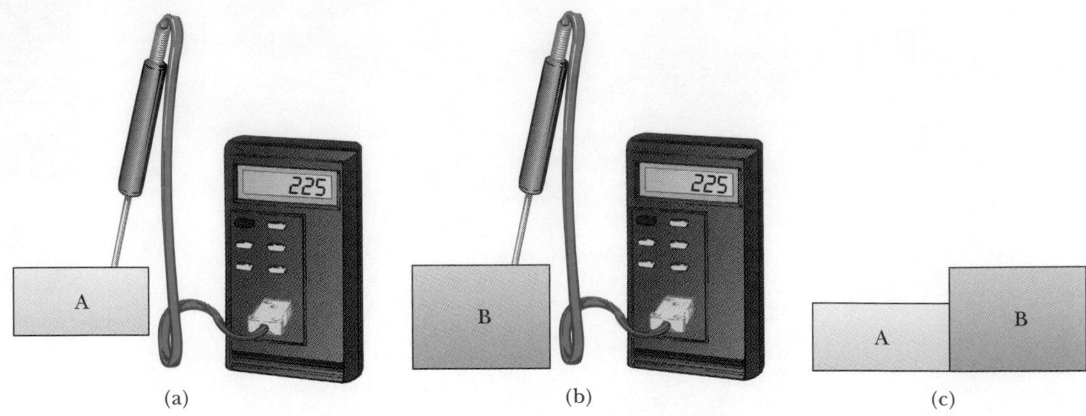

Figure 19.1 The zeroth law of thermodynamics. (a) and (b) If the temperatures of A and B are measured to be the same by placing them in thermal contact with a thermometer (object C), no energy will be exchanged between them when they are placed in thermal contact with each other (c).

Thermal equilibrium is a situation in which two objects would not exchange energy by heat or electromagnetic radiation if they were placed in thermal contact.

Let us consider two objects A and B, which are not in thermal contact, and a third object C, which is our thermometer. We wish to determine whether A and B are in thermal equilibrium with each other. The thermometer (object C) is first placed in thermal contact with object A until thermal equilibrium is reached,[1] as shown in Figure 19.1a. From that moment on, the thermometer's reading remains constant, and we record this reading. The thermometer is then removed from object A and placed in thermal contact with object B, as shown in Figure 19.1b. The reading is again recorded after thermal equilibrium is reached. If the two readings are the same, then object A and object B are in thermal equilibrium with each other. If they are placed in contact with each other as in Figure 19.1c, there is no exchange of energy between them.

We can summarize these results in a statement known as the **zeroth law of thermodynamics** (the law of equilibrium):

Zeroth law of thermodynamics

If objects A and B are separately in thermal equilibrium with a third object C, then A and B are in thermal equilibrium with each other.

This statement can easily be proved experimentally and is very important because it enables us to define temperature. We can think of **temperature** as the property that determines whether an object is in thermal equilibrium with other objects. **Two objects in thermal equilibrium with each other are at the same temperature.** Conversely, if two objects have different temperatures, then they are not in thermal equilibrium with each other.

Quick Quiz 19.1 Two objects, with different sizes, masses, and temperatures, are placed in thermal contact. Energy travels (a) from the larger object to the smaller object (b) from the object with more mass to the one with less (c) from the object at higher temperature to the object at lower temperature.

[1] We assume that negligible energy transfers between the thermometer and object A during the equilibrium process. Without this assumption, which is also made for the thermometer and object B, the measurement of the temperature of an object disturbs the system so that the measured temperature is different from the initial temperature of the object. In practice, whenever you measure a temperature with a thermometer, you measure the disturbed system, not the original system.

19.2 Thermometers and the Celsius Temperature Scale

Thermometers are devices that are used to measure the temperature of a system. All thermometers are based on the principle that some physical property of a system changes as the system's temperature changes. Some physical properties that change with temperature are (1) the volume of a liquid, (2) the dimensions of a solid, (3) the pressure of a gas at constant volume, (4) the volume of a gas at constant pressure, (5) the electric resistance of a conductor, and (6) the color of an object. A temperature scale can be established on the basis of any one of these physical properties.

A common thermometer in everyday use consists of a mass of liquid—usually mercury or alcohol—that expands into a glass capillary tube when heated (Fig. 19.2). In this case the physical property that changes is the volume of a liquid. Any temperature change in the range of the thermometer can be defined as being proportional to the change in length of the liquid column. The thermometer can be calibrated by placing it in thermal contact with some natural systems that remain at constant temperature. One such system is a mixture of water and ice in thermal equilibrium at atmospheric pressure. On the **Celsius temperature scale**, this mixture is defined to have a temperature of zero degrees Celsius, which is written as 0°C; this temperature is called the *ice point* of water. Another commonly used system is a mixture of water and steam in thermal equilibrium at atmospheric pressure; its temperature is 100°C, which is the *steam point* of water. Once the liquid levels in the thermometer have been established at these two points, the length of the liquid column between the two points is divided into 100 equal segments to create the Celsius scale. Thus, each segment denotes a change in temperature of one Celsius degree.

Thermometers calibrated in this way present problems when extremely accurate readings are needed. For instance, the readings given by an alcohol thermometer calibrated at the ice and steam points of water might agree with those given by a mercury thermometer only at the calibration points. Because mercury and alcohol have different thermal expansion properties, when one thermometer reads a temperature of, for example, 50°C, the other may indicate a slightly different value. The discrepancies

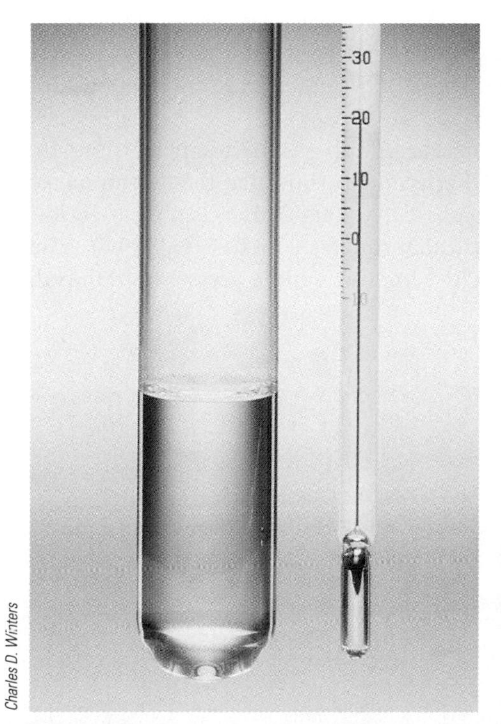

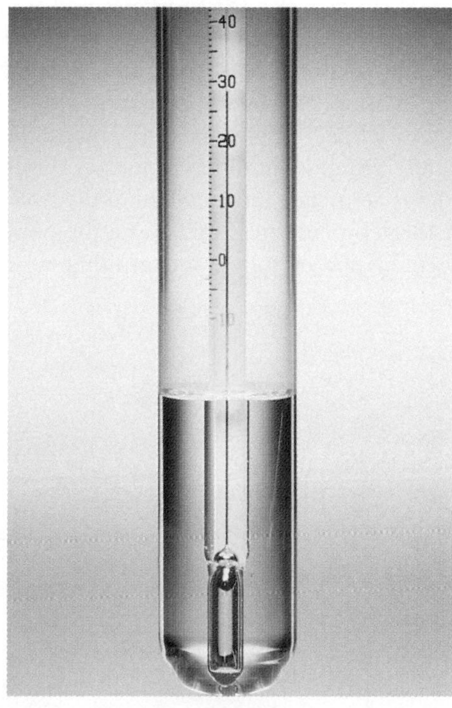

Charles D. Winters

Figure 19.2 As a result of thermal expansion, the level of the mercury in the thermometer rises as the mercury is heated by water in the test tube.

between thermometers are especially large when the temperatures to be measured are far from the calibration points.[2]

An additional practical problem of any thermometer is the limited range of temperatures over which it can be used. A mercury thermometer, for example, cannot be used below the freezing point of mercury, which is $-39°C$, and an alcohol thermometer is not useful for measuring temperatures above 85°C, the boiling point of alcohol. To surmount this problem, we need a universal thermometer whose readings are independent of the substance used in it. The gas thermometer, discussed in the next section, approaches this requirement.

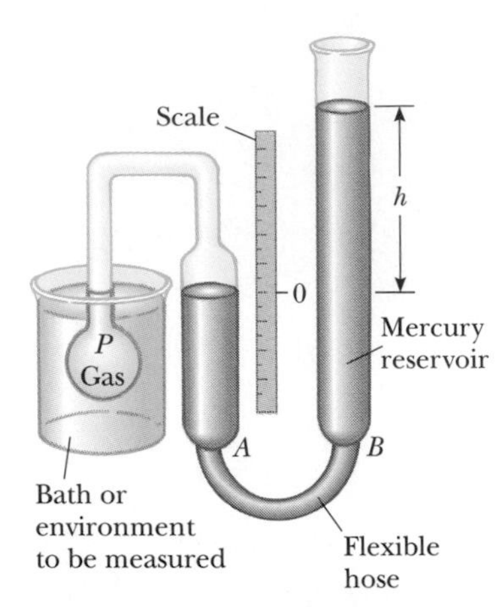

Figure 19.3 A constant-volume gas thermometer measures the pressure of the gas contained in the flask immersed in the bath. The volume of gas in the flask is kept constant by raising or lowering reservoir *B* to keep the mercury level in column *A* constant.

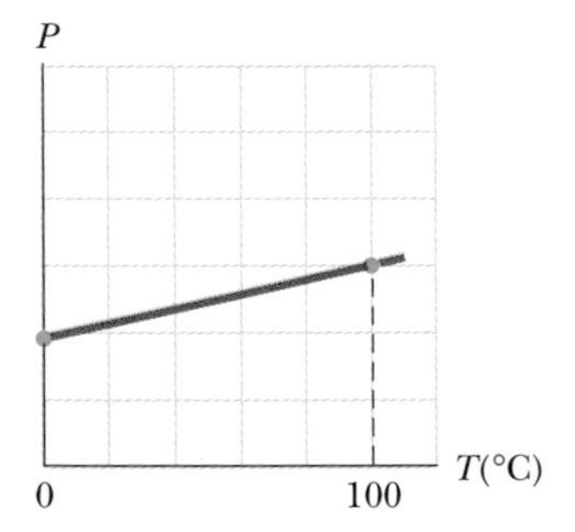

Figure 19.4 A typical graph of pressure versus temperature taken with a constant-volume gas thermometer. The two dots represent known reference temperatures (the ice and steam points of water).

19.3 The Constant-Volume Gas Thermometer and the Absolute Temperature Scale

One version of a gas thermometer is the constant-volume apparatus shown in Figure 19.3. The physical change exploited in this device is the variation of pressure of a fixed volume of gas with temperature. When the constant-volume gas thermometer was developed, it was calibrated by using the ice and steam points of water as follows. (A different calibration procedure, which we shall discuss shortly, is now used.) The flask was immersed in an ice-water bath, and mercury reservoir *B* was raised or lowered until the top of the mercury in column *A* was at the zero point on the scale. The height *h*, the difference between the mercury levels in reservoir *B* and column *A*, indicated the pressure in the flask at 0°C.

The flask was then immersed in water at the steam point, and reservoir *B* was readjusted until the top of the mercury in column *A* was again at zero on the scale; this ensured that the gas's volume was the same as it was when the flask was in the ice bath (hence, the designation "constant volume"). This adjustment of reservoir *B* gave a value for the gas pressure at 100°C. These two pressure and temperature values were then plotted, as shown in Figure 19.4. The line connecting the two points serves as a calibration curve for unknown temperatures. (Other experiments show that a linear relationship between pressure and temperature is a very good assumption.) If we wanted to measure the temperature of a substance, we would place the gas flask in thermal contact with the substance and adjust the height of reservoir *B* until the top of the mercury column in *A* is at zero on the scale. The height of the mercury column indicates the pressure of the gas; knowing the pressure, we could find the temperature of the substance using the graph in Figure 19.4.

Now let us suppose that temperatures are measured with gas thermometers containing different gases at different initial pressures. Experiments show that the thermometer readings are nearly independent of the type of gas used, as long as the gas pressure is low and the temperature is well above the point at which the gas liquefies (Fig. 19.5). The agreement among thermometers using various gases improves as the pressure is reduced.

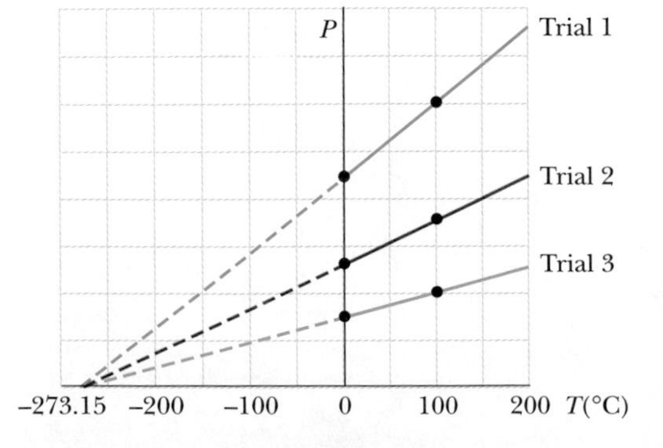

Figure 19.5 Pressure versus temperature for experimental trials in which gases have different pressures in a constant-volume gas thermometer. Note that, for all three trials, the pressure extrapolates to zero at the temperature $-273.15°C$.

[2] Two thermometers that use the same liquid may also give different readings. This is due in part to difficulties in constructing uniform-bore glass capillary tubes.

If we extend the straight lines in Figure 19.5 toward negative temperatures, we find a remarkable result—**in every case, the pressure is zero when the temperature is −273.15°C!** This suggests some special role that this particular temperature must play. It is used as the basis for the **absolute temperature scale**, which sets −273.15°C as its zero point. This temperature is often referred to as **absolute zero**. The size of a degree on the absolute temperature scale is chosen to be identical to the size of a degree on the Celsius scale. Thus, the conversion between these temperatures is

$$T_C = T - 273.15 \tag{19.1}$$

where T_C is the Celsius temperature and T is the absolute temperature.

Because the ice and steam points are experimentally difficult to duplicate, an absolute temperature scale based on two new fixed points was adopted in 1954 by the International Committee on Weights and Measures. The first point is absolute zero. The second reference temperature for this new scale was chosen as the **triple point of water**, which is the single combination of temperature and pressure at which liquid water, gaseous water, and ice (solid water) coexist in equilibrium. This triple point occurs at a temperature of 0.01°C and a pressure of 4.58 mm of mercury. On the new scale, which uses the unit *kelvin*, the temperature of water at the triple point was set at 273.16 kelvins, abbreviated 273.16 K. This choice was made so that the old absolute temperature scale based on the ice and steam points would agree closely with the new scale based on the triple point. This new absolute temperature scale (also called the **Kelvin scale**) employs the SI unit of absolute temperature, the **kelvin**, which is defined to be **1/273.16 of the difference between absolute zero and the temperature of the triple point of water.**

19.1 A Matter of Degree

Note that notations for temperatures in the Kelvin scale do not use the degree sign. The unit for a Kelvin temperature is simply "kelvins" and not "degrees Kelvin."

Figure 19.6 shows the absolute temperature for various physical processes and structures. The temperature of absolute zero (0 K) cannot be achieved, although laboratory experiments incorporating the laser cooling of atoms have come very close.

What would happen to a gas if its temperature could reach 0 K (and it did not liquefy or solidify)? As Figure 19.5 indicates, the pressure it exerts on the walls of its container would be zero. In Chapter 21 we shall show that the pressure of a gas is proportional to the average kinetic energy of its molecules. Thus, according to classical physics, the kinetic energy of the gas molecules would become zero at absolute zero, and molecular motion would cease; hence, the molecules would settle out on the bottom of the container. Quantum theory modifies this prediction and shows that some residual energy, called the *zero-point energy*, would remain at this low temperature.

The Celsius, Fahrenheit, and Kelvin Temperature Scales[3]

Equation 19.1 shows that the Celsius temperature T_C is shifted from the absolute (Kelvin) temperature T by 273.15°. Because the size of a degree is the same on the two scales, a temperature difference of 5°C is equal to a temperature difference of 5 K. The two scales differ only in the choice of the zero point. Thus, the ice-point temperature on the Kelvin scale, 273.15 K, corresponds to 0.00°C, and the Kelvin-scale steam point, 373.15 K, is equivalent to 100.00°C.

A common temperature scale in everyday use in the United States is the **Fahrenheit scale**. This scale sets the temperature of the ice point at 32°F and the temperature of the steam point at 212°F. The relationship between the Celsius and Fahrenheit temperature scales is

$$T_F = \tfrac{9}{5}T_C + 32°\text{F} \tag{19.2}$$

We can use Equations 19.1 and 19.2 to find a relationship between changes in temperature on the Celsius, Kelvin, and Fahrenheit scales:

$$\Delta T_C = \Delta T = \tfrac{5}{9}\,\Delta T_F \tag{19.3}$$

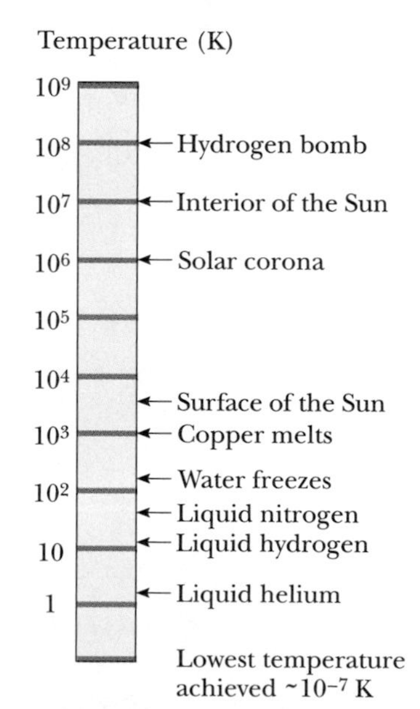

Figure 19.6 Absolute temperatures at which various physical processes occur. Note that the scale is logarithmic.

[3] Named after Anders Celsius (1701–1744), Daniel Gabriel Fahrenheit (1686–1736), and William Thomson, Lord Kelvin (1824–1907), respectively.

Of the three temperature scales that we have discussed, only the Kelvin scale is based on a true zero value of temperature. The Celsius and Fahrenheit scales are based on an arbitrary zero associated with one particular substance—water—on one particular planet—Earth. Thus, if you encounter an equation that calls for a temperature T or involves a ratio of temperatures, you *must* convert all temperatures to kelvins. If the equation contains a change in temperature ΔT, using Celsius temperatures will give you the correct answer, in light of Equation 19.3, but it is always *safest* to convert temperatures to the Kelvin scale.

Quick Quiz 19.2 Consider the following pairs of materials. Which pair represents two materials, one of which is twice as hot as the other? (a) boiling water at 100°C, a glass of water at 50°C (b) boiling water at 100°C, frozen methane at −50°C (c) an ice cube at −20°C, flames from a circus fire-eater at 233°C (d) No pair represents materials one of which is twice as hot as the other

Example 19.1 Converting Temperatures

On a day when the temperature reaches 50°F, what is the temperature in degrees Celsius and in kelvins?

Solution Substituting into Equation 19.2, we obtain

$$T_C = \tfrac{5}{9}(T_F - 32) = \tfrac{5}{9}(50 - 32) = 10°C$$

From Equation 19.1, we find that

$$T = T_C + 273.15 = 10°C + 273.15 = 283\ K$$

A convenient set of weather-related temperature equivalents to keep in mind is that 0°C is (literally) freezing at 32°F, 10°C is cool at 50°F, 20°C is room temperature, 30°C is warm at 86°F, and 40°C is a hot day at 104°F.

Example 19.2 Heating a Pan of Water

A pan of water is heated from 25°C to 80°C. What is the change in its temperature on the Kelvin scale and on the Fahrenheit scale?

Solution From Equation 19.3, we see that the change in temperature on the Celsius scale equals the change on the Kelvin scale. Therefore,

$$\Delta T = \Delta T_C = 80°C - 25°C = 55°C = 55\ K$$

From Equation 19.3, we also find that

$$\Delta T_F = \tfrac{9}{5}\Delta T_C = \tfrac{9}{5}(55°C) = 99°F$$

19.4 Thermal Expansion of Solids and Liquids

Our discussion of the liquid thermometer makes use of one of the best-known changes in a substance: as its temperature increases, its volume increases. This phenomenon, known as **thermal expansion**, has an important role in numerous engineering applications. For example, thermal-expansion joints, such as those shown in Figure 19.7, must be included in buildings, concrete highways, railroad tracks, brick walls, and bridges to compensate for dimensional changes that occur as the temperature changes.

Thermal expansion is a consequence of the change in the *average* separation between the atoms in an object. To understand this, model the atoms as being connected by stiff springs, as discussed in Section 15.3 and shown in Figure 15.12b. At ordinary temperatures, the atoms in a solid oscillate about their equilibrium positions with an amplitude of approximately 10^{-11} m and a frequency of approximately 10^{13} Hz. The average spacing between the atoms is about 10^{-10} m. As the temperature of the solid

(a)

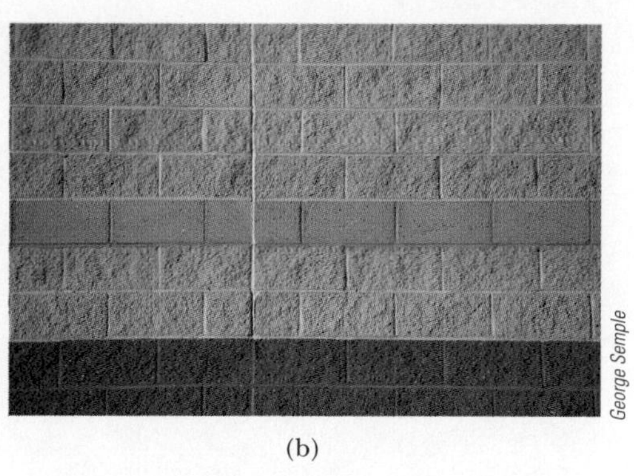

(b)

Figure 19.7 (a) Thermal-expansion joints are used to separate sections of roadways on bridges. Without these joints, the surfaces would buckle due to thermal expansion on very hot days or crack due to contraction on very cold days. (b) The long, vertical joint is filled with a soft material that allows the wall to expand and contract as the temperature of the bricks changes.

increases, the atoms oscillate with greater amplitudes; as a result, the average separation between them increases.[4] Consequently, the object expands.

If thermal expansion is sufficiently small relative to an object's initial dimensions, the change in any dimension is, to a good approximation, proportional to the first power of the temperature change. Suppose that an object has an initial length L_i along some direction at some temperature and that the length increases by an amount ΔL for a change in temperature ΔT. Because it is convenient to consider the fractional change in length per degree of temperature change, we define the **average coefficient of linear expansion** as

$$\alpha \equiv \frac{\Delta L/L_i}{\Delta T}$$

Experiments show that α is constant for small changes in temperature. For purposes of calculation, this equation is usually rewritten as

$$\Delta L = \alpha L_i \Delta T \qquad (19.4)$$

or as

$$L_f - L_i = \alpha L_i (T_f - T_i) \qquad (19.5)$$

where L_f is the final length, T_i and T_f are the initial and final temperatures, and the proportionality constant α is the average coefficient of linear expansion for a given material and has units of $(°\text{C})^{-1}$.

It may be helpful to think of thermal expansion as an effective magnification or as a photographic enlargement of an object. For example, as a metal washer is heated (Fig. 19.8), all dimensions, including the radius of the hole, increase according to Equation 19.4. Notice that this is equivalent to saying that **a cavity in a piece of material expands in the same way as if the cavity were filled with the material.**

Table 19.1 lists the average coefficient of linear expansion for various materials. Note that for these materials α is positive, indicating an increase in length with increasing temperature. This is not always the case. Some substances—calcite ($CaCO_3$) is one example—expand along one dimension (positive α) and contract along another (negative α) as their temperatures are increased.

[4] More precisely, thermal expansion arises from the *asymmetrical* nature of the potential-energy curve for the atoms in a solid, as shown in Figure 15.12a. If the oscillators were truly harmonic, the average atomic separations would not change regardless of the amplitude of vibration.

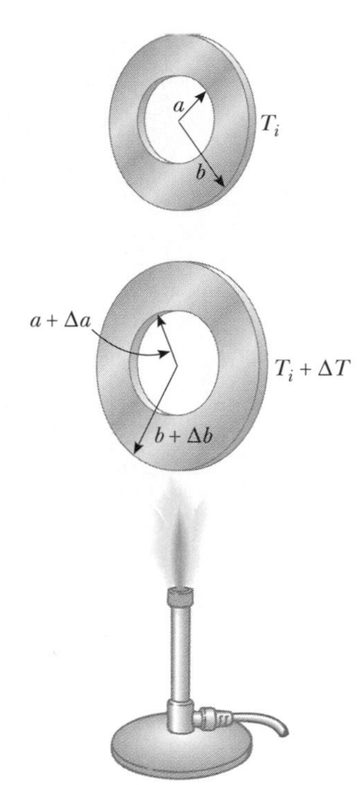

Active Figure 19.8 Thermal expansion of a homogeneous metal washer. As the washer is heated, all dimensions increase. (The expansion is exaggerated in this figure.)

At the Active Figures link at http://www.pse6.com, you can compare expansions for various temperatures of the burner and materials from which the washer is made.

PITFALL PREVENTION

19.2 Do Holes Become Larger or Smaller?

When an object's temperature is raised, every linear dimension increases in size. This includes any holes in the material, which expand in the same way as if the hole were filled with the material, as shown in Figure 19.8. Keep in mind the notion of thermal expansion as being similar to a photographic enlargement.

Table 19.1

Average Expansion Coefficients for Some Materials Near Room Temperature

Material	Average Linear Expansion Coefficient $(\alpha)(°C)^{-1}$	Material	Average Volume Expansion Coefficient $(\beta)(°C)^{-1}$
Aluminum	24×10^{-6}	Alcohol, ethyl	1.12×10^{-4}
Brass and bronze	19×10^{-6}	Benzene	1.24×10^{-4}
Copper	17×10^{-6}	Acetone	1.5×10^{-4}
Glass (ordinary)	9×10^{-6}	Glycerin	4.85×10^{-4}
Glass (Pyrex)	3.2×10^{-6}	Mercury	1.82×10^{-4}
Lead	29×10^{-6}	Turpentine	9.0×10^{-4}
Steel	11×10^{-6}	Gasoline	9.6×10^{-4}
Invar (Ni–Fe alloy)	0.9×10^{-6}	Air[a] at 0°C	3.67×10^{-3}
Concrete	12×10^{-6}	Helium[a]	3.665×10^{-3}

[a] Gases do not have a specific value for the volume expansion coefficient because the amount of expansion depends on the type of process through which the gas is taken. The values given here assume that the gas undergoes an expansion at constant pressure.

Because the linear dimensions of an object change with temperature, it follows that surface area and volume change as well. The change in volume is proportional to the initial volume V_i and to the change in temperature according to the relationship

$$\Delta V = \beta V_i \Delta T \tag{19.6}$$

where β is the **average coefficient of volume expansion**. For a solid, the average coefficient of volume expansion is three times the average linear expansion coefficient: $\beta = 3\alpha$. (This assumes that the average coefficient of linear expansion of the solid is the same in all directions—that is, the material is *isotropic.*)

To see that $\beta = 3\alpha$ for a solid, consider a solid box of dimensions ℓ, w, and h. Its volume at some temperature T_i is $V_i = \ell wh$. If the temperature changes to $T_i + \Delta T$, its volume changes to $V_i + \Delta V$, where each dimension changes according to Equation 19.4. Therefore,

$$\begin{aligned} V_i + \Delta V &= (\ell + \Delta\ell)(w + \Delta w)(h + \Delta h) \\ &= (\ell + \alpha\ell\Delta T)(w + \alpha w\Delta T)(h + \alpha h\Delta T) \\ &= \ell wh(1 + \alpha\Delta T)^3 \\ &= V_i[1 + 3\alpha\Delta T + 3(\alpha\Delta T)^2 + (\alpha\Delta T)^3] \end{aligned}$$

If we now divide both sides by V_i and isolate the term $\Delta V/V_i$, we obtain the fractional change in volume:

$$\frac{\Delta V}{V_i} = 3\alpha\Delta T + 3(\alpha\Delta T)^2 + (\alpha\Delta T)^3$$

Because $\alpha\Delta T \ll 1$ for typical values of ΔT ($<\sim 100°C$), we can neglect the terms $3(\alpha\Delta T)^2$ and $(\alpha\Delta T)^3$. Upon making this approximation, we see that

$$\frac{\Delta V}{V_i} = 3\alpha\Delta T$$

$$3\alpha = \frac{1}{V_i}\frac{\Delta V}{\Delta T}$$

Equation 19.6 shows that the right side of this expression is equal to β, and so we have $3\alpha = \beta$, the relationship we set out to prove. In a similar way, you can show that the change in area of a rectangular plate is given by $\Delta A = 2\alpha A_i\,\Delta T$ (see Problem 55).

As Table 19.1 indicates, each substance has its own characteristic average coefficient of expansion. For example, when the temperatures of a brass rod and a steel rod of

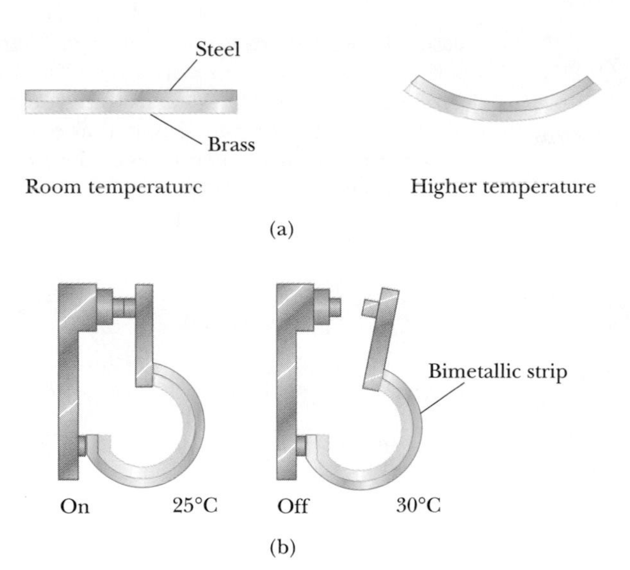

Figure 19.9 (a) A bimetallic strip bends as the temperature changes because the two metals have different expansion coefficients. (b) A bimetallic strip used in a thermostat to break or make electrical contact.

equal length are raised by the same amount from some common initial value, the brass rod expands more than the steel rod does because brass has a greater average coefficient of expansion than steel does. A simple mechanism called a *bimetallic strip* utilizes this principle and is found in practical devices such as thermostats. It consists of two thin strips of dissimilar metals bonded together. As the temperature of the strip increases, the two metals expand by different amounts and the strip bends, as shown in Figure 19.9.

Quick Quiz 19.3 If you are asked to make a very sensitive glass thermometer, which of the following working liquids would you choose? (a) mercury (b) alcohol (c) gasoline (d) glycerin

Quick Quiz 19.4 Two spheres are made of the same metal and have the same radius, but one is hollow and the other is solid. The spheres are taken through the same temperature increase. Which sphere expands more? (a) solid sphere (b) hollow sphere (c) They expand by the same amount. (d) not enough information to say

Example 19.3 Expansion of a Railroad Track

A segment of steel railroad track has a length of 30.000 m when the temperature is 0.0°C.

(A) What is its length when the temperature is 40.0°C?

Solution Making use of Table 19.1 and noting that the change in temperature is 40.0°C, we find that the increase in length is

$$\Delta L = \alpha L_i \, \Delta T = [11 \times 10^{-6} (°\text{C})^{-1}](30.000 \text{ m})(40.0°\text{C})$$
$$= 0.013 \text{ m}$$

If the track is 30.000 m long at 0.0°C, its length at 40.0°C is

30.013 m.

(B) Suppose that the ends of the rail are rigidly clamped at 0.0°C so that expansion is prevented. What is the thermal stress set up in the rail if its temperature is raised to 40.0°C?

Solution The thermal stress will be the same as that in the situation in which we allow the rail to expand freely and then compress it with a mechanical force F back to its original length. From the definition of Young's modulus for a solid (see Eq. 12.6), we have

$$\text{Tensile stress} = \frac{F}{A} = Y \frac{\Delta L}{L_i}$$

Because Y for steel is 20×10^{10} N/m^2 (see Table 12.1), we have

$$\frac{F}{A} = (20 \times 10^{10}\ \text{N/m}^2)\left(\frac{0.013\ \text{m}}{30.000\ \text{m}}\right) = 8.7 \times 10^7\ \text{N/m}^2$$

What If? What if the temperature drops to −40.0°C? What is the length of the unclamped segment?

The expression for the change in length in Equation 19.4 is the same whether the temperature increases or decreases. Thus, if there is an increase in length of 0.013 m when the temperature increases by 40°C, then there is a decrease in length of 0.013 m when the temperature decreases by 40°C. (We assume that α is constant over the entire range of temperatures.) The new length at the colder temperature is 30.000 m − 0.013 m = 29.987 m.

Example 19.4 The Thermal Electrical Short

An electronic device has been poorly designed so that two bolts attached to different parts of the device almost touch each other in its interior, as in Figure 19.10. The steel and brass bolts are at different electric potentials and if they touch, a short circuit will develop, damaging the device. (We will study electric potential in Chapter 25.) If the initial gap between the ends of the bolts is 5.0 μm at 27°C, at what temperature will the bolts touch?

Solution We can conceptualize the situation by imagining that the ends of both bolts expand into the gap between them as the temperature rises. We categorize this as a thermal expansion problem, in which the *sum* of the changes in length of the two bolts must equal the length of the initial gap between the ends. To analyze the problem, we write this condition mathematically:

$$\Delta L_{\text{br}} + \Delta L_{\text{st}} = \alpha_{\text{br}} L_{i,\text{br}} \Delta T + \alpha_{\text{st}} L_{i,\text{st}} \Delta T = 5.0 \times 10^{-6}\ \text{m}$$

Solving for ΔT, we find

$$\Delta T = \frac{5.0 \times 10^{-6}\ \text{m}}{\alpha_{\text{br}} L_{i,\text{br}} + \alpha_{\text{st}} L_{i,\text{st}}}$$

$$= \frac{5.0 \times 10^{-6}\ \text{m}}{(19 \times 10^{-6}\,°\text{C}^{-1})(0.030\ \text{m}) + (11 \times 10^{-6}\,°\text{C}^{-1})(0.010\ \text{m})}$$

$$= 7.4°\text{C}$$

Thus, the temperature at which the bolts touch is 27°C + 7.4°C = 34°C To finalize this problem, note that this temperature is possible if the air conditioning in the building housing the device fails for a long period on a very hot summer day.

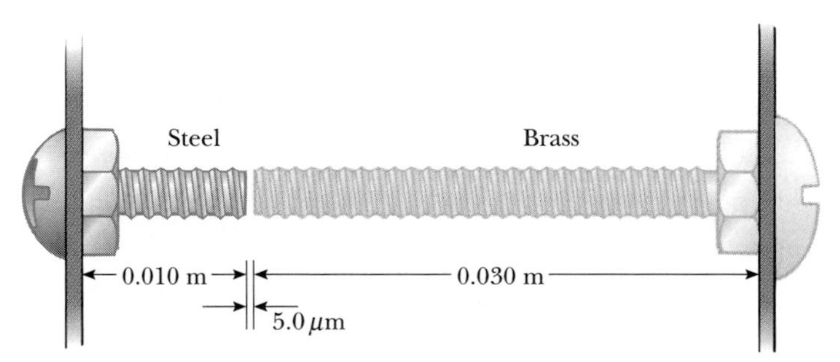

Figure 19.10 (Example 19.4) Two bolts attached to different parts of an electrical device are almost touching when the temperature is 27°C. As the temperature increases, the ends of the bolts move toward each other.

The Unusual Behavior of Water

Liquids generally increase in volume with increasing temperature and have average coefficients of volume expansion about ten times greater than those of solids. Cold water is an exception to this rule, as we can see from its density-versus-temperature curve, shown in Figure 19.11. As the temperature increases from 0°C to 4°C, water contracts and thus its density increases. Above 4°C, water expands with increasing temperature, and so its density decreases. Thus, the density of water reaches a maximum value of 1.000 g/cm^3 at 4°C.

We can use this unusual thermal-expansion behavior of water to explain why a pond begins freezing at the surface rather than at the bottom. When the atmospheric temperature drops from, for example, 7°C to 6°C, the surface water also cools and consequently decreases in volume. This means that the surface water is denser than the water below it, which has not cooled and decreased in volume. As a result, the surface water sinks, and warmer water from below is forced to the surface to be cooled. When the atmospheric temperature is between 4°C and 0°C, however, the surface water expands as it cools, becoming less dense than the water below it. The mixing process

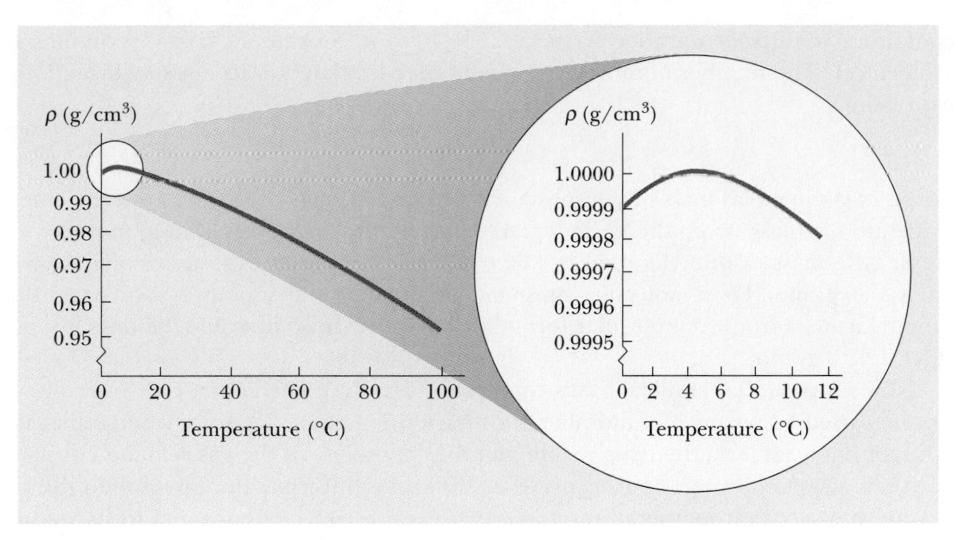

Figure 19.11 The variation in the density of water at atmospheric pressure with temperature. The inset at the right shows that the maximum density of water occurs at 4°C.

stops, and eventually the surface water freezes. As the water freezes, the ice remains on the surface because ice is less dense than water. The ice continues to build up at the surface, while water near the bottom remains at 4°C. If this were not the case, then fish and other forms of marine life would not survive.

19.5 Macroscopic Description of an Ideal Gas

The volume expansion equation $\Delta V = \beta V_i \Delta T$ is based on the assumption that the material has an initial volume V_i before the temperature change occurs. This is the case for solids and liquids because they have a fixed volume at a given temperature.

The case for gases is completely different. The interatomic forces within gases are very weak, and, in many cases, we can imagine these forces to be nonexistent and still make very good approximations. Note that *there is no equilibrium separation* for the atoms and, thus, no "standard" volume at a given temperature. As a result, we cannot express changes in volume ΔV in a process on a gas with Equation 19.6 because we have no defined volume V_i at the beginning of the process. For a gas, the volume is entirely determined by the container holding the gas. Thus, equations involving gases will contain the volume V as a *variable*, rather than focusing on a *change* in the volume from an initial value.

For a gas, it is useful to know how the quantities volume V, pressure P, and temperature T are related for a sample of gas of mass m. In general, the equation that interrelates these quantities, called the *equation of state*, is very complicated. However, if the gas is maintained at a very low pressure (or low density), the equation of state is quite simple and can be found experimentally. Such a low-density gas is commonly referred to as an *ideal gas*.[5]

It is convenient to express the amount of gas in a given volume in terms of the number of moles n. One **mole** of any substance is that amount of the substance that

[5] To be more specific, the assumption here is that the temperature of the gas must not be too low (the gas must not condense into a liquid) or too high, and that the pressure must be low. The concept of an ideal gas implies that the gas molecules do not interact except upon collision, and that the molecular volume is negligible compared with the volume of the container. In reality, an ideal gas does not exist. However, the concept of an ideal gas is very useful because real gases at low pressures behave as ideal gases do.

contains **Avogadro's number** $N_A = 6.022 \times 10^{23}$ of constituent particles (atoms or molecules). The number of moles n of a substance is related to its mass m through the expression

$$n = \frac{m}{M} \tag{19.7}$$

where M is the molar mass of the substance. The molar mass of each chemical element is the atomic mass (from the periodic table, Appendix C) expressed in g/mol. For example, the mass of one He atom is 4.00 u (atomic mass units), so the molar mass of He is 4.00 g/mol. For a molecular substance or a chemical compound, you can add up the molar mass from its molecular formula. The molar mass of stable diatomic oxygen (O_2) is 32.0 g/mol.

Now suppose that an ideal gas is confined to a cylindrical container whose volume can be varied by means of a movable piston, as in Figure 19.12. If we assume that the cylinder does not leak, the mass (or the number of moles) of the gas remains constant. For such a system, experiments provide the following information. First, when the gas is kept at a constant temperature, its pressure is inversely proportional to its volume (Boyle's law). Second, when the pressure of the gas is kept constant, its volume is directly proportional to its temperature (the law of Charles and Gay-Lussac). These observations are summarized by the **equation of state for an ideal gas:**

Equation of state for an ideal gas

$$PV = nRT \tag{19.8}$$

In this expression, known as the **ideal gas law**, R is a constant and n is the number of moles of gas in the sample. Experiments on numerous gases show that as the pressure approaches zero, the quantity PV/nT approaches the same value R for all gases. For this reason, R is called the **universal gas constant**. In SI units, in which pressure is expressed in pascals (1 Pa = 1 N/m^2) and volume in cubic meters, the product PV has units of newton·meters, or joules, and R has the value

$$R = 8.314 \text{ J/mol}\cdot\text{K} \tag{19.9}$$

If the pressure is expressed in atmospheres and the volume in liters (1 L = 10^3 cm^3 = 10^{-3} m^3), then R has the value

$$R = 0.082\,14 \text{ L}\cdot\text{atm/mol}\cdot\text{K}$$

Using this value of R and Equation 19.8, we find that the volume occupied by 1 mol of any gas at atmospheric pressure and at 0°C (273 K) is 22.4 L.

The ideal gas law states that if the volume and temperature of a fixed amount of gas do not change, then the pressure also remains constant. Consider a bottle of champagne that is shaken and then spews liquid when opened, as shown in Figure 19.13.

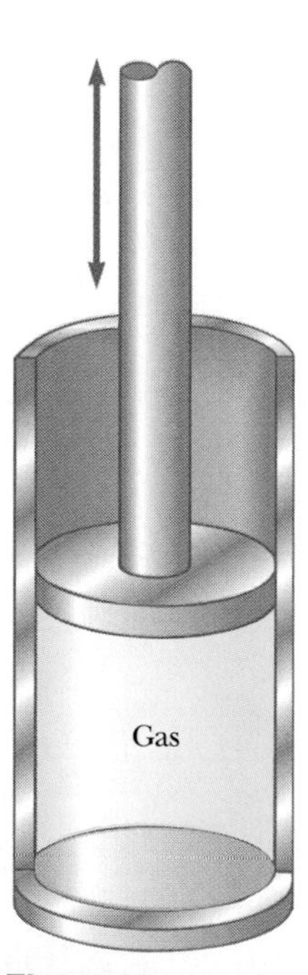

Active Figure 19.12 An ideal gas confined to a cylinder whose volume can be varied by means of a movable piston.

At the Active Figures link at http://www.pse6.com, you can choose to keep either the temperature or the pressure constant and verify Boyle's law and the law of Charles and Gay-Lussac.

Steve Niedorf/Getty Images

Figure 19.13 A bottle of champagne is shaken and opened. Liquid spews out of the opening. A common misconception is that the pressure inside the bottle is increased due to the shaking.

A common misconception is that the pressure inside the bottle is increased when the bottle is shaken. On the contrary, because the temperature of the bottle and its contents remains constant as long as the bottle is sealed, so does the pressure, as can be shown by replacing the cork with a pressure gauge. The correct explanation is as follows. Carbon dioxide gas resides in the volume between the liquid surface and the cork. Shaking the bottle displaces some of this carbon dioxide gas into the liquid, where it forms bubbles, and these bubbles become attached to the inside of the bottle. (No new gas is generated by shaking.) When the bottle is opened, the pressure is reduced; this causes the volume of the bubbles to increase suddenly. If the bubbles are attached to the bottle (beneath the liquid surface), their rapid expansion expels liquid from the bottle. If the sides and bottom of the bottle are first tapped until no bubbles remain beneath the surface, then when the champagne is opened, the drop in pressure will not force liquid from the bottle.

The ideal gas law is often expressed in terms of the total number of molecules N. Because the total number of molecules equals the product of the number of moles n and Avogadro's number N_A, we can write Equation 19.8 as

$$PV = nRT = \frac{N}{N_A}RT$$

$$PV = Nk_BT \qquad (19.10)$$

where k_B is **Boltzmann's constant**, which has the value

$$k_B = \frac{R}{N_A} = 1.38 \times 10^{-23}\ \text{J/K} \qquad (19.11)$$

Boltzmann's constant

It is common to call quantities such as P, V, and T the **thermodynamic variables** of an ideal gas. If the equation of state is known, then one of the variables can always be expressed as some function of the other two.

PITFALL PREVENTION

19.3 So Many k's

There are a variety of physical quantities for which the letter k is used—we have seen two previously, the force constant for a spring (Chapter 15) and the wave number for a mechanical wave (Chapter 16). Boltzmann's constant is another k, and we will see k used for thermal conductivity in Chapter 20 and for an electrical constant in Chapter 23. In order to make some sense of this confusing state of affairs, we will use a subscript for Boltzmann's constant to help us recognize it. In this book, we will see Boltzmann's constant as k_B, but keep in mind that you may see Boltzmann's constant in other resources as simply k.

Quick Quiz 19.5 A common material for cushioning objects in packages is made by trapping bubbles of air between sheets of plastic. This material is more effective at keeping the contents of the package from moving around inside the package on (a) a hot day (b) a cold day (c) either hot or cold days.

Quick Quiz 19.6 A helium-filled rubber balloon is left in a car on a cold winter night. Compared to its size when it was in the warm car the afternoon before, the size the next morning is (a) larger (b) smaller (c) unchanged.

Quick Quiz 19.7 On a winter day, you turn on your furnace and the temperature of the air inside your home increases. Assuming that your home has the normal amount of leakage between inside air and outside air, the number of moles of air in your room at the higher temperature is (a) larger than before (b) smaller than before (c) the same as before.

Example 19.5 How Many Moles of Gas in a Container?

An ideal gas occupies a volume of 100 cm^3 at 20°C and 100 Pa. Find the number of moles of gas in the container.

Solution The quantities given are volume, pressure, and temperature: $V = 100\ \text{cm}^3 = 1.00 \times 10^{-4}\ \text{m}^3$, $P = 100$ Pa, and $T = 20°\text{C} = 293$ K. Using Equation 19.8, we find that

$$n = \frac{PV}{RT} = \frac{(100\ \text{Pa})(1.00 \times 10^{-4}\ \text{m}^3)}{(8.314\ \text{J/mol}\cdot\text{K})(293\ \text{K})}$$

$$= 4.11 \times 10^{-6}\ \text{mol}$$

Example 19.6 Filling a Scuba Tank

A certain scuba tank is designed to hold 66.0 ft^3 of air when it is at atmospheric pressure at 22°C. When this volume of air is compressed to an absolute pressure of 3 000 lb/in.^{2} and stored in a 10.0-L (0.350-ft^3) tank, the air becomes so hot that the tank must be allowed to cool before it can be used. Before the air cools, what is its temperature? (Assume that the air behaves like an ideal gas.)

Solution If no air escapes during the compression, then the number of moles n of air remains constant; therefore, using $PV = nRT$, with n and R constant, we obtain a relationship between the initial and final values:

$$\frac{P_iV_i}{T_i} = \frac{P_fV_f}{T_f}$$

The initial pressure of the air is 14.7 lb/in.^{2}, its final pressure is 3 000 lb/in.^{2}, and the air is compressed from an initial volume of 66.0 ft^3 to a final volume of 0.350 ft^3. The initial temperature, converted to SI units, is 295 K. Solving for T_f, we obtain

$$T_f = \left(\frac{P_fV_f}{P_iV_i}\right)T_i = \frac{(3\,000 \text{ lb/in.}^2)(0.350 \text{ ft}^3)}{(14.7 \text{ lb/in.}^2)(66.0 \text{ ft}^3)}(295 \text{ K})$$

$$= \boxed{319 \text{ K}}$$

Example 19.7 Heating a Spray Can — Interactive

A spray can containing a propellant gas at twice atmospheric pressure (202 kPa) and having a volume of 125.00 cm^3 is at 22°C. It is then tossed into an open fire. When the temperature of the gas in the can reaches 195°C, what is the pressure inside the can? Assume any change in the volume of the can is negligible.

Solution We employ the same approach we used in Example 19.6, starting with the expression

$$(1) \qquad \frac{P_iV_i}{T_i} = \frac{P_fV_f}{T_f}$$

Because the initial and final volumes of the gas are assumed to be equal, this expression reduces to

$$\frac{P_i}{T_i} = \frac{P_f}{T_f}$$

Solving for P_f gives

$$(2) \qquad P_f = \left(\frac{T_f}{T_i}\right)P_i = \left(\frac{468 \text{ K}}{295 \text{ K}}\right)(202 \text{ kPa}) = \boxed{320 \text{ kPa}}$$

Obviously, the higher the temperature, the higher the pressure exerted by the trapped gas. Of course, if the pressure increases sufficiently, the can will explode. Because of this possibility, you should never dispose of spray cans in a fire.

What If? Suppose we include a volume change due to thermal expansion of the steel can as the temperature increases. Does this alter our answer for the final pressure significantly?

Because the thermal expansion coefficient of steel is very small, we do not expect much of an effect on our final answer. The change in the volume of the can is found using Equation 19.6 and the value for α for steel from Table 19.1:

$$\Delta V = \beta V_i\,\Delta T = 3\alpha V_i\,\Delta T$$
$$= 3(11 \times 10^{-6}\,°\text{C}^{-1})(125.00 \text{ cm}^3)(173°\text{C})$$
$$= 0.71 \text{ cm}^3$$

So the final volume of the can is 125.71 cm^3. Starting from Equation (1) again, the equation for the final pressure becomes

$$P_f = \left(\frac{T_f}{T_i}\right)\left(\frac{V_i}{V_f}\right)P_i$$

This differs from Equation (2) only in the factor V_i/V_f. Let us evaluate this factor:

$$\frac{V_i}{V_f} = \frac{125.00 \text{ cm}^3}{125.71 \text{ cm}^3} = 0.994 = 99.4\%$$

Thus, the final pressure will differ by only 0.6% from the value we calculated without considering the thermal expansion of the can. Taking 99.4% of the previous final pressure, the final pressure including thermal expansion is 318 kPa.

Explore this situation at the Interactive Worked Example link at **http://www.pse6.com.**

SUMMARY

Take a practice test for this chapter by clicking on the Practice Test link at **http://www.pse6.com.**

Two objects are in **thermal equilibrium** with each other if they do not exchange energy when in thermal contact.

The **zeroth law of thermodynamics** states that if objects A and B are separately in thermal equilibrium with a third object C, then objects A and B are in thermal equilibrium with each other.

Temperature is the property that determines whether an object is in thermal equilibrium with other objects. **Two objects in thermal equilibrium with each other are at the same temperature.**

The SI unit of absolute temperature is the **kelvin**, which is defined to be the fraction 1/273.16 of the temperature of the triple point of water.

When the temperature of an object is changed by an amount ΔT, its length changes by an amount ΔL that is proportional to ΔT and to its initial length L_i:

$$\Delta L = \alpha L_i \Delta T \tag{19.4}$$

where the constant α is the **average coefficient of linear expansion**. The **average coefficient of volume expansion** β for a solid is approximately equal to 3α.

An **ideal gas** is one for which PV/nT is constant. An ideal gas is described by the **equation of state,**

$$PV = nRT \tag{19.8}$$

where n equals the number of moles of the gas, V is its volume, R is the universal gas constant (8.314 J/mol·K), and T is the absolute temperature. A real gas behaves approximately as an ideal gas if it has a low density.

QUESTIONS

1. Is it possible for two objects to be in thermal equilibrium if they are not in contact with each other? Explain.
2. A piece of copper is dropped into a beaker of water. If the water's temperature rises, what happens to the temperature of the copper? Under what conditions are the water and copper in thermal equilibrium?
3. In describing his upcoming trip to the Moon, and as portrayed in the movie *Apollo 13* (Universal, 1995), astronaut Jim Lovell said, "I'll be walking in a place where there's a 400-degree difference between sunlight and shadow." What is it that is hot in sunlight and cold in shadow? Suppose an astronaut standing on the Moon holds a thermometer in his gloved hand. Is it reading the temperature of the vacuum at the Moon's surface? Does it read any temperature? If so, what object or substance has that temperature?
4. Rubber has a negative average coefficient of linear expansion. What happens to the size of a piece of rubber as it is warmed?
5. Explain why a column of mercury in a thermometer first descends slightly and then rises when the thermometer is placed into hot water.
6. Why should the amalgam used in dental fillings have the same average coefficient of expansion as a tooth? What would occur if they were mismatched?
7. Markings to indicate length are placed on a steel tape in a room that has a temperature of 22°C. Are measurements made with the tape on a day when the temperature is 27°C too long, too short, or accurate? Defend your answer.
8. Determine the number of grams in a mole of the following gases: (a) hydrogen (b) helium (c) carbon monoxide.
9. What does the ideal gas law predict about the volume of a sample of gas at absolute zero? Why is this prediction incorrect?
10. An inflated rubber balloon filled with air is immersed in a flask of liquid nitrogen that is at 77 K. Describe what happens to the balloon, assuming that it remains flexible while being cooled.
11. Two identical cylinders at the same temperature each contain the same kind of gas and the same number of moles of gas. If the volume of cylinder A is three times greater than the volume of cylinder B, what can you say about the relative pressures in the cylinders?
12. After food is cooked in a pressure cooker, why is it very important to cool off the container with cold water before attempting to remove the lid?
13. The shore of the ocean is very rocky at a particular place. The rocks form a cave sloping upward from an underwater opening, as shown in Figure Q19.13a. (a) Inside the cave is

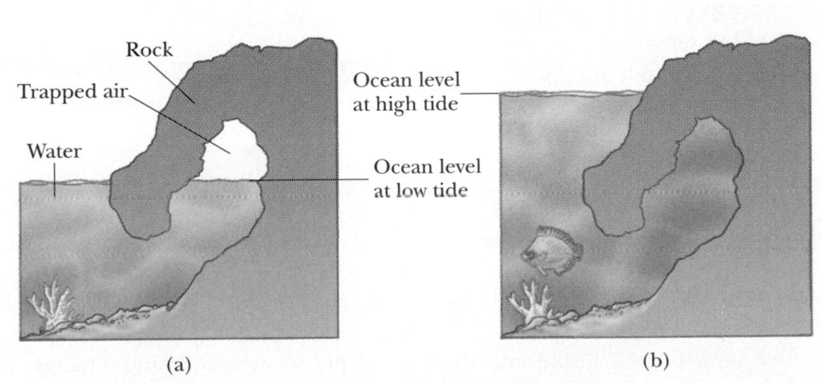

Figure Q19.13

a pocket of trapped air. As the level of the ocean rises and falls with the tides, will the level of water in the cave rise and fall? If so, will it have the same amplitude as that of the ocean? (b) **What If?** Now suppose that the cave is deeper in the water, so that it is completely submerged and filled with water at high tide, as shown in Figure Q19.13b. At low tide, will the level of the water in the cave be the same as that of the ocean?

14. In *Colonization: Second Contact* (Harry Turtledove, Ballantine Publishing Group, 1999), the Earth has been partially settled by aliens from another planet, whom humans call Lizards. Laboratory study by humans of Lizard science requires "shifting back and forth between the metric system and the one the Lizards used, which was also based on powers of ten but used different basic quantities for everything but temperature." Why might temperature be an exception?

15. The pendulum of a certain pendulum clock is made of brass. When the temperature increases, does the period of the clock increase, decrease, or remain the same? Explain.

16. An automobile radiator is filled to the brim with water while the engine is cool. What happens to the water when the engine is running and the water is heated? What do modern automobiles have in their cooling systems to prevent the loss of coolants?

17. Metal lids on glass jars can often be loosened by running hot water over them. How is this possible?

18. When the metal ring and metal sphere in Figure Q19.18 are both at room temperature, the sphere can just be passed through the ring. After the sphere is heated, it cannot be passed through the ring. Explain. **What If?** What if the ring is heated and the sphere is left at room temperature? Does the sphere pass through the ring?

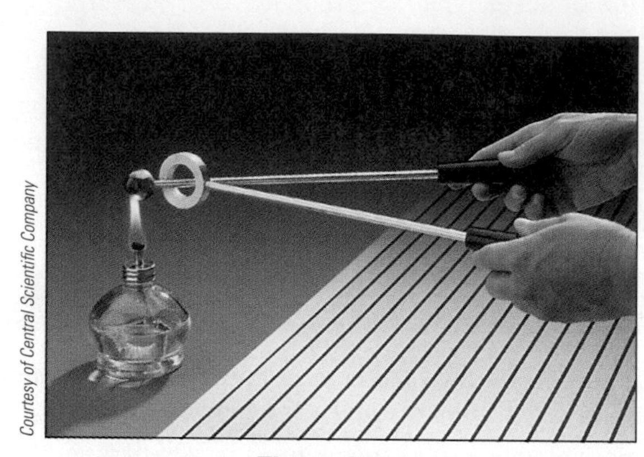

Courtesy of Central Scientific Company

Figure Q19.18

PROBLEMS

1, 2, 3 = straightforward, intermediate, challenging □ = full solution available in the *Student Solutions Manual and Study Guide*

= coached solution with hints available at http://www.pse6.com = computer useful in solving problem

= paired numerical and symbolic problems

Section 19.2 Thermometers and the Celsius Temperature Scale

Section 19.3 The Constant-Volume Gas Thermometer and the Absolute Temperature Scale

1. A constant-volume gas thermometer is calibrated in dry ice (that is, carbon dioxide in the solid state, which has a temperature of $-80.0°C$) and in boiling ethyl alcohol ($78.0°C$). The two pressures are 0.900 atm and 1.635 atm. (a) What Celsius value of absolute zero does the calibration yield? What is the pressure at (b) the freezing point of water and (c) the boiling point of water?

2. In a constant-volume gas thermometer, the pressure at $20.0°C$ is 0.980 atm. (a) What is the pressure at $45.0°C$? (b) What is the temperature if the pressure is 0.500 atm?

3. Liquid nitrogen has a boiling point of $-195.81°C$ at atmospheric pressure. Express this temperature (a) in degrees Fahrenheit and (b) in kelvins.

4. Convert the following to equivalent temperatures on the Celsius and Kelvin scales: (a) the normal human body temperature, $98.6°F$; (b) the air temperature on a cold day, $-5.00°F$.

5. The temperature difference between the inside and the outside of an automobile engine is $450°C$. Express this temperature difference on (a) the Fahrenheit scale and (b) the Kelvin scale.

6. On a Strange temperature scale, the freezing point of water is $-15.0°S$ and the boiling point is $+60.0°S$. Develop a *linear* conversion equation between this temperature scale and the Celsius scale.

7. The melting point of gold is $1\,064°C$, and the boiling point is $2\,660°C$. (a) Express these temperatures in kelvins. (b) Compute the difference between these temperatures in Celsius degrees and kelvins.

Section 19.4 Thermal Expansion of Solids and Liquids

Note: Table 19.1 is available for use in solving problems in this section.

8. The New River Gorge bridge in West Virginia is a steel arch bridge 518 m in length. How much does the total length of the roadway decking change between temperature extremes of $-20.0°C$ and $35.0°C$? The result indicates the size of the expansion joints that must be built into the structure.

9. A copper telephone wire has essentially no sag between poles 35.0 m apart on a winter day when the temperature is $-20.0°C$. How much longer is the wire on a summer day when $T_C = 35.0°C$?

10. The concrete sections of a certain superhighway are designed to have a length of 25.0 m. The sections are poured and cured at 10.0°C. What minimum spacing should the engineer leave between the sections to eliminate buckling if the concrete is to reach a temperature of 50.0°C?

11. A pair of eyeglass frames is made of epoxy plastic. At room temperature (20.0°C), the frames have circular lens holes 2.20 cm in radius. To what temperature must the frames be heated if lenses 2.21 cm in radius are to be inserted in them? The average coefficient of linear expansion for epoxy is $1.30 \times 10^{-4}\ (°C)^{-1}$.

12. Each year thousands of children are badly burned by hot tap water. Figure P19.12 shows a cross-sectional view of an antiscalding faucet attachment designed to prevent such accidents. Within the device, a spring made of material with a high coefficient of thermal expansion controls a movable plunger. When the water temperature rises above a preset safe value, the expansion of the spring causes the plunger to shut off the water flow. If the initial length L of the unstressed spring is 2.40 cm and its coefficient of linear expansion is $22.0 \times 10^{-6}\ (°C)^{-1}$, determine the increase in length of the spring when the water temperature rises by 30.0°C. (You will find the increase in length to be small. For this reason actual devices have a more complicated mechanical design, to provide a greater variation in valve opening for the temperature change anticipated.)

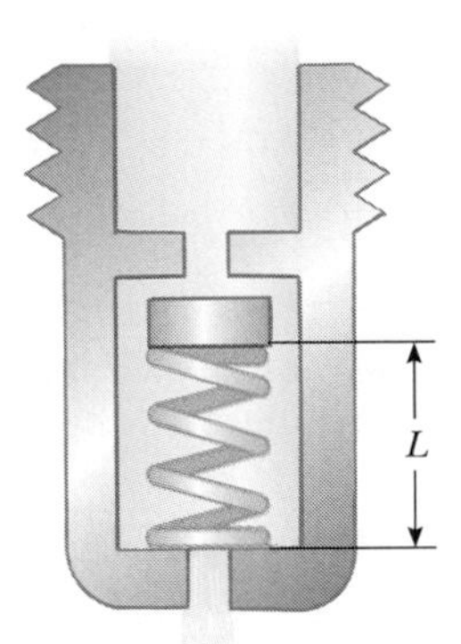

Figure P19.12

13. The active element of a certain laser is made of a glass rod 30.0 cm long by 1.50 cm in diameter. If the temperature of the rod increases by 65.0°C, what is the increase in (a) its length, (b) its diameter, and (c) its volume? Assume that the average coefficient of linear expansion of the glass is $9.00 \times 10^{-6}\ (°C)^{-1}$.

14. **Review problem.** Inside the wall of a house, an L-shaped section of hot-water pipe consists of a straight horizontal piece 28.0 cm long, an elbow, and a straight vertical piece 134 cm long (Figure P19.14). A stud and a second-story floorboard hold stationary the ends of this section of copper pipe. Find the magnitude and direction of the displacement of the pipe elbow when the water flow is turned on, raising the temperature of the pipe from 18.0°C to 46.5°C.

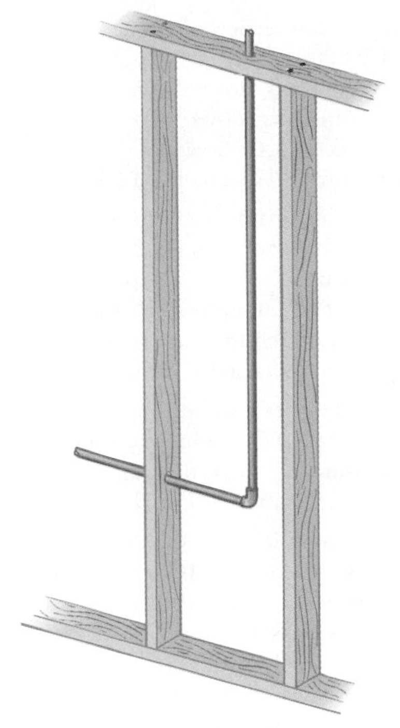

Figure P19.14

15. A brass ring of diameter 10.00 cm at 20.0°C is heated and slipped over an aluminum rod of diameter 10.01 cm at 20.0°C. Assuming the average coefficients of linear expansion are constant, (a) to what temperature must this combination be cooled to separate them? Is this attainable? (b) **What If?** What if the aluminum rod were 10.02 cm in diameter?

16. A square hole 8.00 cm along each side is cut in a sheet of copper. (a) Calculate the change in the area of this hole if the temperature of the sheet is increased by 50.0 K. (b) Does this change represent an increase or a decrease in the area enclosed by the hole?

17. The average coefficient of volume expansion for carbon tetrachloride is $5.81 \times 10^{-4}\ (°C)^{-1}$. If a 50.0-gal steel container is filled completely with carbon tetrachloride when the temperature is 10.0°C, how much will spill over when the temperature rises to 30.0°C?

18. At 20.0°C, an aluminum ring has an inner diameter of 5.000 0 cm and a brass rod has a diameter of 5.050 0 cm. (a) If only the ring is heated, what temperature must it reach so that it will just slip over the rod? (b) **What If?** If both are heated together, what temperature must they both reach so that the ring just slips over the rod? Would this latter process work?

19. A volumetric flask made of Pyrex is calibrated at 20.0°C. It is filled to the 100-mL mark with 35.0°C acetone. (a) What is the volume of the acetone when it cools to 20.0°C? (b) How significant is the change in volume of the flask?

20. A concrete walk is poured on a day when the temperature is 20.0°C in such a way that the ends are unable to move. (a) What is the stress in the cement on a hot day of 50.0°C? (b) Does the concrete fracture? Take Young's modulus for concrete to be 7.00×10^9 N/m^2 and the compressive strength to be 2.00×10^9 N/m^2.

21. A hollow aluminum cylinder 20.0 cm deep has an internal capacity of 2.000 L at 20.0°C. It is completely filled with turpentine and then slowly warmed to 80.0°C. (a) How much turpentine overflows? (b) If the cylinder is then cooled back to 20.0°C, how far below the cylinder's rim does the turpentine's surface recede?

22. A beaker made of ordinary glass contains a lead sphere of diameter 4.00 cm firmly attached to its bottom. At a uniform temperature of −10.0°C, the beaker is filled to the brim with 118 cm^3 of mercury, which completely covers the sphere. How much mercury overflows from the beaker if the temperature is raised to 30.0°C?

23. A steel rod undergoes a stretching force of 500 N. Its cross-sectional area is 2.00 cm^2. Find the change in temperature that would elongate the rod by the same amount as the 500-N force does. Tables 12.1 and 19.1 are available to you.

24. The Golden Gate Bridge in San Francisco has a main span of length 1.28 km—one of the longest in the world. Imagine that a taut steel wire with this length and a cross-sectional area of 4.00×10^{-6} m^2 is laid on the bridge deck with its ends attached to the towers of the bridge, on a summer day when the temperature of the wire is 35.0°C. (a) When winter arrives, the towers stay the same distance apart and the bridge deck keeps the same shape as its expansion joints open. When the temperature drops to −10.0°C, what is the tension in the wire? Take Young's modulus for steel to be 20.0×10^{10} N/m^2. (b) Permanent deformation occurs if the stress in the steel exceeds its elastic limit of 3.00×10^8 N/m^2. At what temperature would this happen? (c) **What If?** How would your answers to (a) and (b) change if the Golden Gate Bridge were twice as long?

25. A certain telescope forms an image of part of a cluster of stars on a square silicon charge-coupled detector (CCD) chip 2.00 cm on each side. A star field is focused on the CCD chip when it is first turned on and its temperature is 20.0°C. The star field contains 5 342 stars scattered uniformly. To make the detector more sensitive, it is cooled to −100°C. How many star images then fit onto the chip? The average coefficient of linear expansion of silicon is 4.68×10^{-6} (°C)$^{-1}$.

Section 19.5 Macroscopic Description of an Ideal Gas

Note: Problem 8 in Chapter 1 can be assigned with this section.

26. Gas is contained in an 8.00-L vessel at a temperature of 20.0°C and a pressure of 9.00 atm. (a) Determine the number of moles of gas in the vessel. (b) How many molecules are there in the vessel?

27. An automobile tire is inflated with air originally at 10.0°C and normal atmospheric pressure. During the process, the air is compressed to 28.0% of its original volume and the temperature is increased to 40.0°C. (a) What is the tire pressure? (b) After the car is driven at high speed, the tire air temperature rises to 85.0°C and the interior volume of the tire increases by 2.00%. What is the new tire pressure (absolute) in pascals?

28. A tank having a volume of 0.100 m^3 contains helium gas at 150 atm. How many balloons can the tank blow up if each filled balloon is a sphere 0.300 m in diameter at an absolute pressure of 1.20 atm?

29. An auditorium has dimensions 10.0 m × 20.0 m × 30.0 m. How many molecules of air fill the auditorium at 20.0°C and a pressure of 101 kPa?

30. Imagine a baby alien playing with a spherical balloon the size of the Earth in the outer solar system. Helium gas inside the balloon has a uniform temperature of 50.0 K due to radiation from the Sun. The uniform pressure of the helium is equal to normal atmospheric pressure on Earth. (a) Find the mass of the gas in the balloon. (b) The baby blows an additional mass of 8.00×10^{20} kg of helium into the balloon. At the same time, she wanders closer to the Sun and the pressure in the balloon doubles. Find the new temperature inside the balloon, whose volume remains constant.

31. Just 9.00 g of water is placed in a 2.00-L pressure cooker and heated to 500°C. What is the pressure inside the container?

32. One mole of oxygen gas is at a pressure of 6.00 atm and a temperature of 27.0°C. (a) If the gas is heated at constant volume until the pressure triples, what is the final temperature? (b) If the gas is heated until both the pressure and volume are doubled, what is the final temperature?

33. The mass of a hot-air balloon and its cargo (not including the air inside) is 200 kg. The air outside is at 10.0°C and 101 kPa. The volume of the balloon is 400 m^3. To what temperature must the air in the balloon be heated before the balloon will lift off? (Air density at 10.0°C is 1.25 kg/m^3.)

34. Your father and your little brother are confronted with the same puzzle. Your father's garden sprayer and your brother's water cannon both have tanks with a capacity of 5.00 L (Figure P19.34). Your father inserts a negligible amount of concentrated insecticide into his tank. They both pour in 4.00 L of water and seal up their tanks, so that they also contain air at atmospheric pressure. Next, each uses a hand-operated piston pump to inject more air, until the absolute pressure in the tank reaches 2.40 atm and it becomes too difficult to move the pump handle. Now each uses his device to spray out water—not air—until the stream becomes feeble, as it does when the pressure in the tank reaches 1.20 atm. Then he must pump it up again, spray again, and so on. In order to spray out all the water, each finds that he must pump up the tank three times. This is the puzzle: most of the water sprays out as a result of the second pumping. The first and the third pumping-up processes seem just as difficult, but result in a disappointingly small amount of water coming out. Account for this phenomenon.

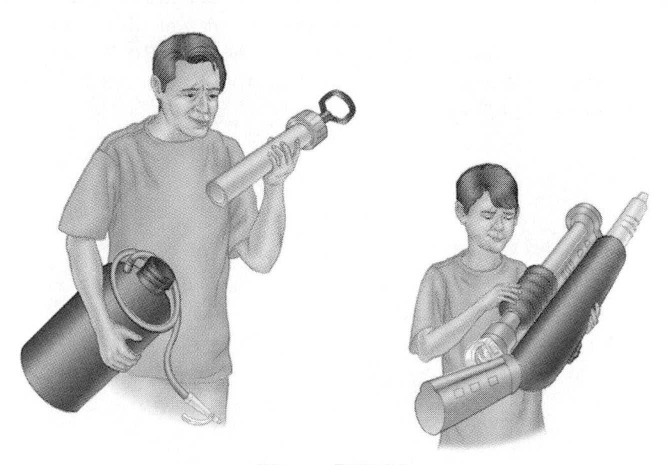

Figure P19.34

35. (a) Find the number of moles in one cubic meter of an ideal gas at 20.0°C and atmospheric pressure. (b) For air, Avogadro's number of molecules has mass 28.9 g. Calculate the mass of one cubic meter of air. Compare the result with the tabulated density of air.

36. The *void fraction* of a porous medium is the ratio of the void volume to the total volume of the material. The void is the hollow space within the material; it may be filled with a fluid. A cylindrical canister of diameter 2.54 cm and height 20.0 cm is filled with activated carbon having a void fraction of 0.765. Then it is flushed with an ideal gas at 25.0°C and pressure 12.5 atm. How many moles of gas are contained in the cylinder at the end of this process?

37. A cube 10.0 cm on each edge contains air (with equivalent molar mass 28.9 g/mol) at atmospheric pressure and temperature 300 K. Find (a) the mass of the gas, (b) its weight, and (c) the force it exerts on each face of the cube. (d) Comment on the physical reason why such a small sample can exert such a great force.

38. At 25.0 m below the surface of the sea (ρ = 1 025 kg/m^3), where the temperature is 5.00°C, a diver exhales an air bubble having a volume of 1.00 cm^3. If the surface temperature of the sea is 20.0°C, what is the volume of the bubble just before it breaks the surface?

39. The pressure gauge on a tank registers the gauge pressure, which is the difference between the interior and exterior pressure. When the tank is full of oxygen (O_2), it contains 12.0 kg of the gas at a gauge pressure of 40.0 atm. Determine the mass of oxygen that has been withdrawn from the tank when the pressure reading is 25.0 atm. Assume that the temperature of the tank remains constant.

40. Estimate the mass of the air in your bedroom. State the quantities you take as data and the value you measure or estimate for each.

41. A popular brand of cola contains 6.50 g of carbon dioxide dissolved in 1.00 L of soft drink. If the evaporating carbon dioxide is trapped in a cylinder at 1.00 atm and 20.0°C, what volume does the gas occupy?

42. In state-of-the-art vacuum systems, pressures as low as 10^{-9} Pa are being attained. Calculate the number of molecules in a 1.00-m^3 vessel at this pressure if the temperature is 27.0°C.

43. A room of volume V contains air having equivalent molar mass M (in g/mol). If the temperature of the room is raised from T_1 to T_2, what mass of air will leave the room? Assume that the air pressure in the room is maintained at P_0.

44. A diving bell in the shape of a cylinder with a height of 2.50 m is closed at the upper end and open at the lower end. The bell is lowered from air into sea water (ρ = 1.025 g/cm^3). The air in the bell is initially at 20.0°C. The bell is lowered to a depth (measured to the bottom of the bell) of 45.0 fathoms or 82.3 m. At this depth the water temperature is 4.0°C, and the bell is in thermal equilibrium with the water. (a) How high does sea water rise in the bell? (b) To what minimum pressure must the air in the bell be raised to expel the water that entered?

Additional Problems

45. A student measures the length of a brass rod with a steel tape at 20.0°C. The reading is 95.00 cm. What will the tape indicate for the length of the rod when the rod and the tape are at (a) − 15.0°C and (b) 55.0°C?

46. The density of gasoline is 730 kg/m^3 at 0°C. Its average coefficient of volume expansion is 9.60×10^{-4}/°C. If 1.00 gal of gasoline occupies 0.003 80 m^3, how many extra kilograms of gasoline would you get if you bought 10.0 gal of gasoline at 0°C rather than at 20.0°C from a pump that is not temperature compensated?

47. A mercury thermometer is constructed as shown in Figure P19.47. The capillary tube has a diameter of 0.004 00 cm,

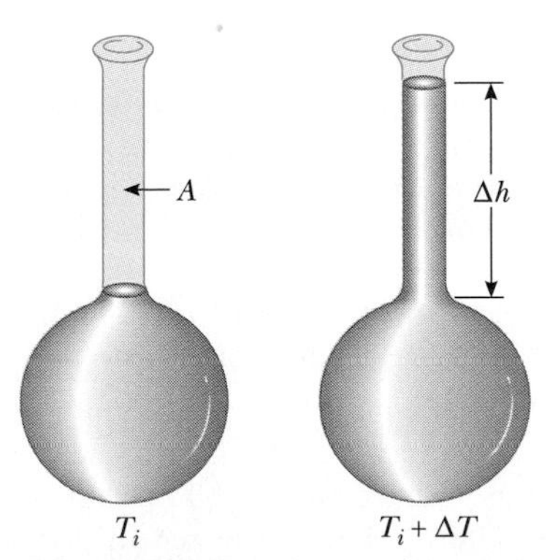

Figure P19.47 Problems 47 and 48.

and the bulb has a diameter of 0.250 cm. Neglecting the expansion of the glass, find the change in height of the mercury column that occurs with a temperature change of 30.0°C.

48. A liquid with a coefficient of volume expansion β just fills a spherical shell of volume V_i at a temperature of T_i (see Fig. P19.47). The shell is made of a material that has an average coefficient of linear expansion α. The liquid is free to expand into an open capillary of area A projecting from the top of the sphere. (a) If the temperature increases by ΔT, show that the liquid rises in the capillary by the amount Δh given by $\Delta h = (V_i/A)(\beta - 3\alpha)\Delta T$. (b) For a typical system, such as a mercury thermometer, why is it a good approximation to neglect the expansion of the shell?

49. **Review problem.** An aluminum pipe, 0.655 m long at 20.0°C and open at both ends, is used as a flute. The pipe is cooled to a low temperature but then is filled with air at 20.0°C as soon as you start to play it. After that, by how much does its fundamental frequency change as the metal rises in temperature from 5.00°C to 20.0°C?

50. A cylinder is closed by a piston connected to a spring of constant 2.00×10^3 N/m (see Fig. P19.50). With the spring relaxed, the cylinder is filled with 5.00 L of gas at a pressure of 1.00 atm and a temperature of 20.0°C. (a) If the piston has a cross-sectional area of 0.010 0 m^2 and negligible mass, how high will it rise when the temperature is raised to 250°C? (b) What is the pressure of the gas at 250°C?

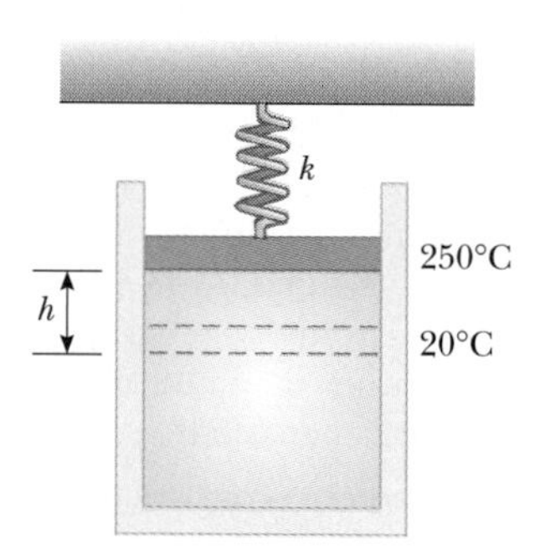

Figure P19.50

51. A liquid has a density ρ. (a) Show that the fractional change in density for a change in temperature ΔT is $\Delta\rho/\rho = -\beta\,\Delta T$. What does the negative sign signify? (b) Fresh water has a maximum density of 1.000 0 g/cm^3 at 4.0°C. At 10.0°C, its density is 0.999 7 g/cm^3. What is β for water over this temperature interval?

52. Long-term space missions require reclamation of the oxygen in the carbon dioxide exhaled by the crew. In one method of reclamation, 1.00 mol of carbon dioxide produces 1.00 mol of oxygen and 1.00 mol of methane as a byproduct. The methane is stored in a tank under pressure and is available to control the attitude of the spacecraft by controlled venting. A single astronaut exhales 1.09 kg of carbon dioxide each day. If the methane generated in the respiration recycling of three astronauts during one week of flight is stored in an originally empty 150-L tank at −45.0°C, what is the final pressure in the tank?

53. A vertical cylinder of cross-sectional area A is fitted with a tight-fitting, frictionless piston of mass m (Fig. P19.53). (a) If n moles of an ideal gas are in the cylinder at a temperature of T, what is the height h at which the piston is in equilibrium under its own weight? (b) What is the value for h if $n = 0.200$ mol, $T = 400$ K, $A = 0.008\,00$ m^2, and $m = 20.0$ kg?

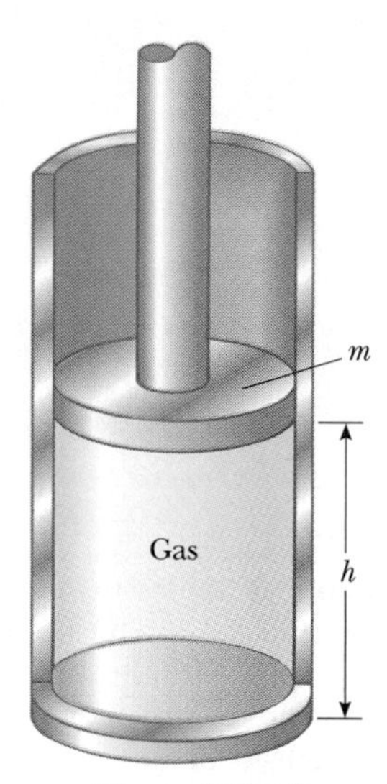

Figure P19.53

54. A bimetallic strip is made of two ribbons of dissimilar metals bonded together. (a) First assume the strip is originally straight. As they are heated, the metal with the greater average coefficient of expansion expands more than the other, forcing the strip into an arc, with the outer radius having a greater circumference (Fig. P19.54a). Derive an expression for the angle of bending θ as a function of the initial length of the strips, their average coefficients of linear expansion, the change in temperature, and the separation of the centers of the strips ($\Delta r = r_2 - r_1$). (b) Show that the angle of bending decreases to zero when ΔT decreases to zero and also when the two average coefficients of expansion become equal. (c) **What If?** What happens if the strip is cooled? (d) Figure P19.54b shows a compact spiral bimetallic strip in a home thermostat. The equation from part (a) applies to it as well, if θ is interpreted as the angle of additional bending caused by a change in temperature. The inner end of the spiral strip is fixed, and the outer end is free to move. Assume the metals are bronze and invar, the thickness of the strip is $2\,\Delta r = 0.500$ mm, and the overall length of the spiral strip is 20.0 cm. Find the angle through which the free end of the strip turns when the temperature changes by one Celsius degree. The free end of the strip supports a capsule partly filled with mercury, visible above the strip in Figure P19.54b. When the capsule tilts, the mercury shifts from one end to the other, to make or break an electrical contact switching the furnace on or off.

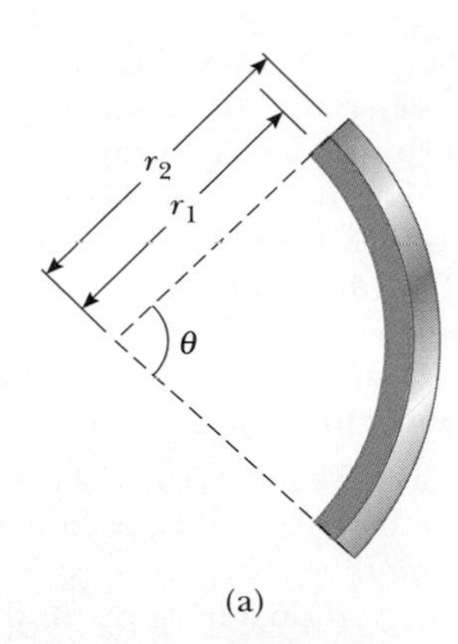

(a)

(b)

Charles D. Winters

Figure P19.54

55. The rectangular plate shown in Figure P19.55 has an area A_i equal to ℓw. If the temperature increases by ΔT, each dimension increases according to the equation $\Delta L = \alpha L_i \Delta T$, where α is the average coefficient of linear expansion. Show that the increase in area is $\Delta A = 2\alpha A_i \Delta T$. What approximation does this expression assume?

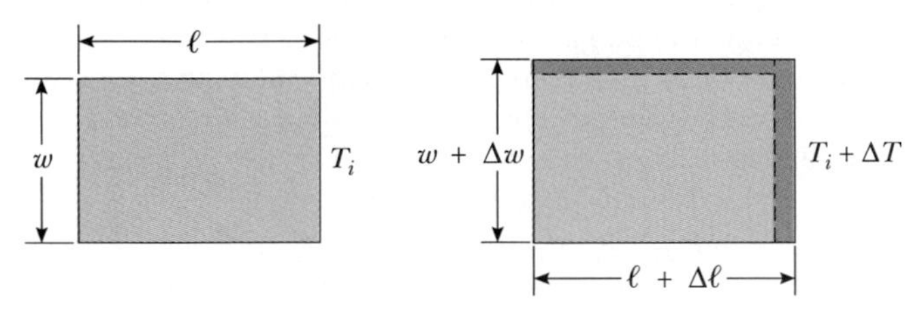

Figure P19.55

56. **Review problem.** A clock with a brass pendulum has a period of 1.000 s at 20.0°C. If the temperature increases to 30.0°C, (a) by how much does the period change, and (b) how much time does the clock gain or lose in one week?

57. **Review problem.** Consider an object with any one of the shapes displayed in Table 10.2. What is the percentage increase in the moment of inertia of the object when it is heated from 0°C to 100°C if it is composed of (a) copper or (b) aluminum? Assume that the average linear expansion coefficients shown in Table 19.1 do not vary between 0°C and 100°C.

58. (a) Derive an expression for the buoyant force on a spherical balloon, submerged in water, as a function of the depth below the surface, the volume of the balloon at the surface, the pressure at the surface, and the density of the water. (Assume water temperature does not change with depth.) (b) Does the buoyant force increase or decrease as the balloon is submerged? (c) At what depth is the buoyant force half the surface value?

59. A copper wire and a lead wire are joined together, end to end. The compound wire has an effective coefficient of linear expansion of $20.0 \times 10^{-6}\ (°C)^{-1}$. What fraction of the length of the compound wire is copper?

60. **Review problem.** Following a collision in outer space, a copper disk at 850°C is rotating about its axis with an angular speed of 25.0 rad/s. As the disk radiates infrared light, its temperature falls to 20.0°C. No external torque acts on the disk. (a) Does the angular speed change as the disk cools off? Explain why. (b) What is its angular speed at the lower temperature?

61. Two concrete spans of a 250-m-long bridge are placed end to end so that no room is allowed for expansion (Fig. P19.61a). If a temperature increase of 20.0°C occurs, what is the height y to which the spans rise when they buckle (Fig. P19.61b)?

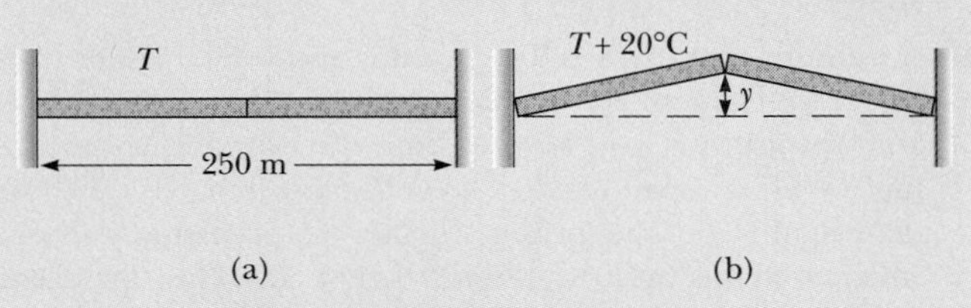

Figure P19.61 Problems 61 and 62.

62. Two concrete spans of a bridge of length L are placed end to end so that no room is allowed for expansion (Fig. P19.61a). If a temperature increase of ΔT occurs, what is the height y to which the spans rise when they buckle (Fig. P19.61b)?

63. (a) Show that the density of an ideal gas occupying a volume V is given by $\rho = PM/RT$, where M is the molar mass. (b) Determine the density of oxygen gas at atmospheric pressure and 20.0°C.

64. (a) Use the equation of state for an ideal gas and the definition of the coefficient of volume expansion, in the form $\beta = (1/V)\ dV/dT$, to show that the coefficient of volume expansion for an ideal gas at constant pressure is given by $\beta = 1/T$, where T is the absolute temperature. (b) What value does this expression predict for β at 0°C? Compare this result with the experimental values for helium and air in Table 19.1. Note that these are much larger than the coefficients of volume expansion for most liquids and solids.

65. Starting with Equation 19.10, show that the total pressure P in a container filled with a mixture of several ideal gases is $P = P_1 + P_2 + P_3 + \cdots$, where $P_1, P_2, \ldots$, are the pressures that each gas would exert if it alone filled the container (these individual pressures are called the *partial pres-*

sures of the respective gases). This result is known as *Dalton's law of partial pressures.*

66. A sample of dry air that has a mass of 100.00 g, collected at sea level, is analyzed and found to consist of the following gases:

nitrogen (N_2) = 75.52 g
oxygen (O_2) = 23.15 g
argon (Ar) = 1.28 g
carbon dioxide (CO_2) = 0.05 g

plus trace amounts of neon, helium, methane, and other gases. (a) Calculate the partial pressure (see Problem 65) of each gas when the pressure is 1.013×10^5 Pa. (b) Determine the volume occupied by the 100-g sample at a temperature of 15.00°C and a pressure of 1.00 atm. What is the density of the air for these conditions? (c) What is the effective molar mass of the air sample?

67. Helium gas is sold in steel tanks. If the helium is used to inflate a balloon, could the balloon lift the spherical tank the helium came in? Justify your answer. Steel will rupture if subjected to tensile stress greater than its yield strength of 5×10^8 N/m^2. *Suggestion:* You may consider a steel shell of radius r and thickness t containing helium at high pressure and on the verge of breaking apart into two hemispheres.

68. A cylinder that has a 40.0-cm radius and is 50.0 cm deep is filled with air at 20.0°C and 1.00 atm (Fig. P19.68a). A 20.0-kg piston is now lowered into the cylinder, compressing the air trapped inside (Fig. P19.68b). Finally, a 75.0-kg man stands on the piston, further compressing the air, which remains at 20°C (Fig. P19.68c). (a) How far down (Δh) does the piston move when the man steps onto it? (b) To what temperature should the gas be heated to raise the piston and man back to h_i?

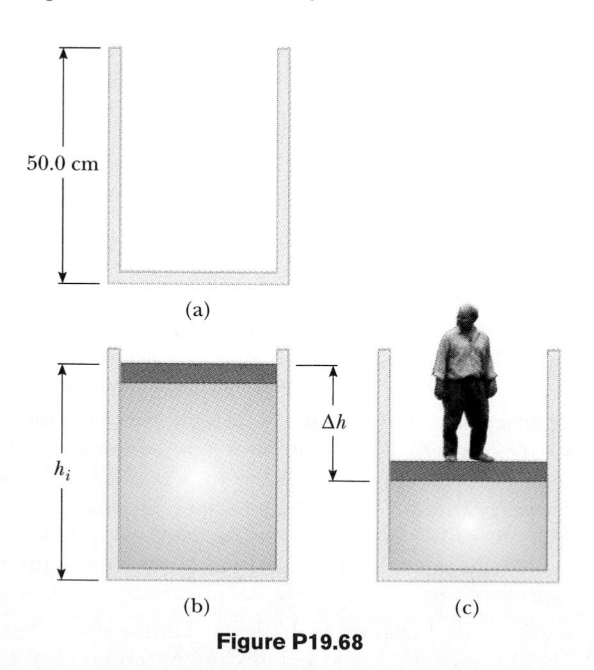

Figure P19.68

69. The relationship $L_f = L_i(1 + \alpha\Delta T)$ is an approximation that works when the average coefficient of expansion is small. If α is large, one must integrate the relationship $dL/dT = \alpha L$ to determine the final length. (a) Assuming that the coefficient of linear expansion is constant as L varies, determine a general expression for the final length. (b) Given a rod of length 1.00 m and a temperature change of 100.0°C, determine the error caused by the approximation when $\alpha = 2.00 \times 10^{-5}$ (°C)$^{-1}$ (a typical value for a metal) and when $\alpha = 0.020\,0$ (°C)$^{-1}$ (an unrealistically large value for comparison).

70. A steel wire and a copper wire, each of diameter 2.000 mm, are joined end to end. At 40.0°C, each has an unstretched length of 2.000 m; they are connected between two fixed supports 4.000 m apart on a tabletop, so that the steel wire extends from $x = -2.000$ m to $x = 0$, the copper wire extends from $x = 0$ to $x = 2.000$ m, and the tension is negligible. The temperature is then lowered to 20.0°C. At this lower temperature, find the tension in the wire and the x coordinate of the junction between the wires. (Refer to Tables 12.1 and 19.1.)

71. **Review problem.** A steel guitar string with a diameter of 1.00 mm is stretched between supports 80.0 cm apart. The temperature is 0.0°C. (a) Find the mass per unit length of this string. (Use the value 7.86×10^3 kg/m^3 for the density.) (b) The fundamental frequency of transverse oscillations of the string is 200 Hz. What is the tension in the string? (c) If the temperature is raised to 30.0°C, find the resulting values of the tension and the fundamental frequency. Assume that both the Young's modulus (Table 12.1) and the average coefficient of expansion (Table 19.1) have constant values between 0.0°C and 30.0°C.

72. In a chemical processing plant, a reaction chamber of fixed volume V_0 is connected to a reservoir chamber of fixed volume $4V_0$ by a passage containing a thermally insulating porous plug. The plug permits the chambers to be at different temperatures. The plug allows gas to pass from either chamber to the other, ensuring that the pressure is the same in both. At one point in the processing, both chambers contain gas at a pressure of 1.00 atm and a temperature of 27.0°C. Intake and exhaust valves to the pair of chambers are closed. The reservoir is maintained at 27.0°C while the reaction chamber is heated to 400°C. What is the pressure in both chambers after this is done?

73. A 1.00-km steel railroad rail is fastened securely at both ends when the temperature is 20.0°C. As the temperature increases, the rail begins to buckle. If its shape is an arc of a vertical circle, find the height h of the center of the rail when the temperature is 25.0°C. You will need to solve a transcendental equation.

74. **Review problem.** A perfectly plane house roof makes an angle θ with the horizontal. When its temperature changes, between T_c before dawn each day to T_h in the middle of each afternoon, the roof expands and contracts uniformly with a coefficient of thermal expansion α_1. Resting on the roof is a flat rectangular metal plate with expansion coefficient α_2, greater than α_1. The length of the plate is L, measured up the slope of the roof. The component of the plate's weight perpendicular to the roof is supported by a normal force uniformly distributed over the area of the plate. The coefficient of kinetic friction between the plate and the roof is μ_k. The plate is always at the same temperature as the roof, so we assume its temperature is continuously changing. Because of the difference in expansion coefficients, each bit of the plate is moving relative to the roof below it, except for points along a certain horizontal line running across the plate. We call this the stationary line. If the temperature is rising, parts of the plate below the stationary line are moving down relative to the roof and feel a force of kinetic friction acting up the roof. Elements of area above the stationary line are sliding up the roof and on them kinetic friction acts downward parallel to the roof. The stationary line occupies no area, so we assume no force of static friction acts on the plate while the temperature is changing. The plate as a whole is very nearly in equilibrium, so the net friction force on it must be equal to the component of its weight acting down the incline. (a) Prove that the stationary line is at a distance of

$$\frac{L}{2}\left(1 - \frac{\tan\theta}{\mu_k}\right)$$

below the top edge of the plate. (b) Analyze the forces that act on the plate when the temperature is falling, and prove that the stationary line is at that same distance above the bottom edge of the plate. (c) Show that the plate steps down the roof like an inchworm, moving each day by the distance

$$\frac{L(\alpha_2 - \alpha_1)(T_h - T_c)\tan\theta}{\mu_k}$$

(d) Evaluate the distance an aluminum plate moves each day if its length is 1.20 m, if the temperature cycles between 4.00°C and 36.0°C, and if the roof has slope 18.5°, coefficient of linear expansion $1.50 \times 10^{-5}\ (°\mathrm{C})^{-1}$, and coefficient of friction 0.420 with the plate. (e) **What If?** What if the expansion coefficient of the plate is less than that of the roof? Will the plate creep up the roof?

Answers to Quick Quizzes

19.1 (c). The direction of the transfer of energy depends only on temperature and not on the size of the object or on which object has more mass.

19.2 (c). The phrase "twice as hot" refers to a ratio of temperatures. When the given temperatures are converted to kelvins, only those in part (c) are in the correct ratio.

19.3 (c). Gasoline has the largest average coefficient of volume expansion.

19.4 (c). A cavity in a material expands in the same way as if it were filled with material.

19.5 (a). On a cold day, the trapped air in the bubbles is reduced in pressure, according to the ideal gas law. Thus, the volume of the bubbles may be smaller than on a hot day, and the package contents can shift more.

19.6 (b). Because of the decreased temperature of the helium, the pressure in the balloon is reduced. The atmospheric pressure around the balloon then compresses it to a smaller size until the pressure in the balloon reaches the atmospheric pressure.

19.7 (b). Because of the increased temperature, the air expands. Consequently, some of the air leaks to the outside, leaving less air in the house.

Chapter 20

Heat and the First Law of Thermodynamics

CHAPTER OUTLINE

20.1 Heat and Internal Energy

20.2 Specific Heat and Calorimetry

20.3 Latent Heat

20.4 Work and Heat in Thermodynamic Processes

20.5 The First Law of Thermodynamics

20.6 Some Applications of the First Law of Thermodynamics

20.7 Energy Transfer Mechanisms

▲ *In this photograph of Bow Lake in Banff National Park, Alberta, we see evidence of water in all three phases. In the lake is liquid water, and solid water in the form of snow appears on the ground. The clouds in the sky consist of liquid water droplets that have condensed from the gaseous water vapor in the air. Changes of a substance from one phase to another are a result of energy transfer. (Jacob Taposchaner/Getty Images)*

Until about 1850, the fields of thermodynamics and mechanics were considered to be two distinct branches of science, and the law of conservation of energy seemed to describe only certain kinds of mechanical systems. However, mid-nineteenth-century experiments performed by the Englishman James Joule and others showed that there was a strong connection between the transfer of energy by heat in thermal processes and the transfer of energy by work in mechanical processes. Today we know that internal energy, which we formally define in this chapter, can be transformed to mechanical energy. Once the concept of energy was generalized from mechanics to include internal energy, the law of conservation of energy emerged as a universal law of nature.

This chapter focuses on the concept of internal energy, the processes by which energy is transferred, the first law of thermodynamics, and some of the important applications of the first law. The first law of thermodynamics is a statement of conservation of energy. It describes systems in which the only energy change is that of internal energy and the transfers of energy are by heat and work. Furthermore, the first law makes no distinction between the results of heat and the results of work. According to the first law, a system's internal energy can be changed by an energy transfer to or from the system either by heat or by work. A major difference in our discussion of work in this chapter from that in the chapters on mechanics is that we will consider work done on *deformable* systems.

20.1 Heat and Internal Energy

At the outset, it is important that we make a major distinction between internal energy and heat. **Internal energy is all the energy of a system that is associated with its microscopic components—atoms and molecules—when viewed from a reference frame at rest with respect to the center of mass of the system.** The last part of this sentence ensures that any bulk kinetic energy of the system due to its motion through space is not included in internal energy. Internal energy includes kinetic energy of random translational, rotational, and vibrational motion of molecules, potential energy within molecules, and potential energy between molecules. It is useful to relate internal energy to the temperature of an object, but this relationship is limited—we show in Section 20.3 that internal energy changes can also occur in the absence of temperature changes.

Heat is defined as the transfer of energy across the boundary of a system due to a temperature difference between the system and its surroundings. When you *heat* a substance, you are transferring energy into it by placing it in contact with surroundings that have a higher temperature. This is the case, for example, when you place a pan of cold water on a stove burner—the burner is at a higher temperature than the water, and so the water gains energy. We shall also use the term *heat* to represent the amount of energy transferred by this method.

Scientists used to think of heat as a fluid called *caloric,* which they believed was transferred between objects; thus, they defined heat in terms of the temperature

PITFALL PREVENTION

20.1 Internal Energy, Thermal Energy, and Bond Energy

In reading other physics books, you may see terms such as *thermal energy* and *bond energy.* Thermal energy can be interpreted as that part of the internal energy associated with random motion of molecules and, therefore, related to temperature. Bond energy is the intermolecular potential energy. Thus,

internal energy = thermal energy + bond energy

While this breakdown is presented here for clarification with regard to other texts, we will not use these terms, because there is no need for them.

PITFALL PREVENTION

20.2 Heat, Temperature, and Internal Energy Are Different

As you read the newspaper or listen to the radio, be alert for incorrectly used phrases including the word *heat,* and think about the proper word to be used in place of *heat.* Incorrect examples include "As the truck braked to a stop, a large amount of heat was generated by friction" and "The heat of a hot summer day . . ."

changes produced in an object during heating. Today we recognize the distinct difference between internal energy and heat. Nevertheless, we refer to quantities using names that do not quite correctly define the quantities but which have become entrenched in physics tradition based on these early ideas. Examples of such quantities are *heat capacity* and *latent heat* (Sections 20.2 and 20.3).

As an analogy to the distinction between heat and internal energy, consider the distinction between work and mechanical energy discussed in Chapter 7. The work done on a system is a measure of the amount of energy transferred to the system from its surroundings, whereas the mechanical energy of the system (kinetic plus potential) is a consequence of the motion and configuration of the system. Thus, when a person does work on a system, energy is transferred from the person to the system. It makes no sense to talk about the work *of* a system—one can refer only to the work done *on* or *by* a system when some process has occurred in which energy has been transferred to or from the system. Likewise, it makes no sense to talk about the heat *of* a system—one can refer to *heat* only when energy has been transferred as a result of a temperature difference. Both heat and work are ways of changing the energy of a system.

It is also important to recognize that the internal energy of a system can be changed even when no energy is transferred by heat. For example, when a gas in an insulated container is compressed by a piston, the temperature of the gas and its internal energy increase, but no transfer of energy by heat from the surroundings to the gas has occurred. If the gas then expands rapidly, it cools and its internal energy decreases, but no transfer of energy by heat from it to the surroundings has taken place. The temperature changes in the gas are due not to a difference in temperature between the gas and its surroundings but rather to the compression and the expansion. In each case, energy is transferred to or from the gas by *work.* The changes in internal energy in these examples are evidenced by corresponding changes in the temperature of the gas.

James Prescott Joule
British physicist (1818–1889)

Joule received some formal education in mathematics, philosophy, and chemistry from John Dalton but was in large part self-educated. Joule's research led to the establishment of the principle of conservation of energy. His study of the quantitative relationship among electrical, mechanical, and chemical effects of heat culminated in his announcement in 1843 of the amount of work required to produce a unit of energy, called the mechanical equivalent of heat. *(By kind permission of the President and Council of the Royal Society)*

Units of Heat

As we have mentioned, early studies of heat focused on the resultant increase in temperature of a substance, which was often water. The early notions of heat based on caloric suggested that the flow of this fluid from one substance to another caused changes in temperature. From the name of this mythical fluid, we have an energy unit related to thermal processes, the **calorie (cal),** which is defined as **the amount of energy transfer necessary to raise the temperature of 1 g of water from 14.5°C to 15.5°C.**[1] (Note that the "Calorie," written with a capital "C" and used in describing the energy content of foods, is actually a kilocalorie.) The unit of energy in the U.S. customary system is the **British thermal unit (Btu),** which is defined as **the amount of energy transfer required to raise the temperature of 1 lb of water from 63°F to 64°F.**

Scientists are increasingly using the SI unit of energy, the *joule,* when describing thermal processes. In this textbook, heat, work, and internal energy are usually measured in joules. (Note that both heat and work are measured in energy units. Do not confuse these two means of energy *transfer* with energy itself, which is also measured in joules.)

The Mechanical Equivalent of Heat

In Chapters 7 and 8, we found that whenever friction is present in a mechanical system, some mechanical energy is lost—in other words, mechanical energy is not conserved in the presence of nonconservative forces. Various experiments show that this lost mechanical energy does not simply disappear but is transformed into internal

[1] Originally, the calorie was defined as the "heat" necessary to raise the temperature of 1 g of water by 1°C. However, careful measurements showed that the amount of energy required to produce a 1°C change depends somewhat on the initial temperature; hence, a more precise definition evolved.

energy. We can perform such an experiment at home by simply hammering a nail into a scrap piece of wood. What happens to all the kinetic energy of the hammer once we have finished? Some of it is now in the nail as internal energy, as demonstrated by the fact that the nail is measurably warmer. Although this connection between mechanical and internal energy was first suggested by Benjamin Thompson, it was Joule who established the equivalence of these two forms of energy.

A schematic diagram of Joule's most famous experiment is shown in Figure 20.1. The system of interest is the water in a thermally insulated container. Work is done on the water by a rotating paddle wheel, which is driven by heavy blocks falling at a constant speed. The temperature of the stirred water increases due to the friction between it and the paddles. If the energy lost in the bearings and through the walls is neglected, then the loss in potential energy associated with the blocks equals the work done by the paddle wheel on the water. If the two blocks fall through a distance h, the loss in potential energy is $2mgh$, where m is the mass of one block; this energy causes the temperature of the water to increase. By varying the conditions of the experiment, Joule found that the loss in mechanical energy $2mgh$ is proportional to the increase in water temperature ΔT. The proportionality constant was found to be approximately $4.18 \text{ J/g} \cdot {}^\circ\text{C}$. Hence, 4.18 J of mechanical energy raises the temperature of 1 g of water by 1°C. More precise measurements taken later demonstrated the proportionality to be $4.186 \text{ J/g} \cdot {}^\circ\text{C}$ when the temperature of the water was raised from 14.5°C to 15.5°C. We adopt this "15-degree calorie" value:

$$1 \text{ cal} \equiv 4.186 \text{ J} \tag{20.1}$$

This equality is known, for purely historical reasons, as the **mechanical equivalent of heat.**

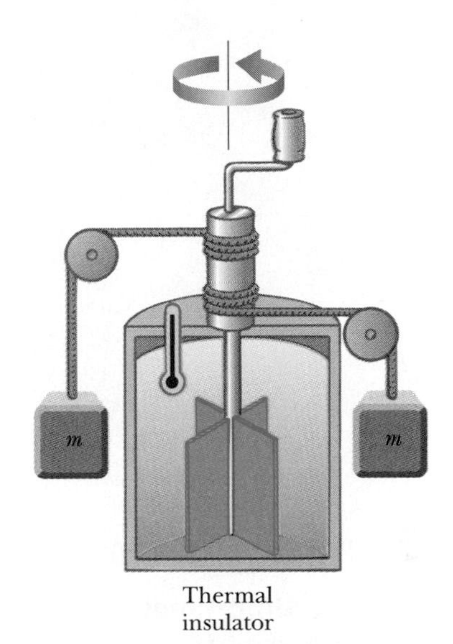

Figure 20.1 Joule's experiment for determining the mechanical equivalent of heat. The falling blocks rotate the paddles, causing the temperature of the water to increase.

Example 20.1 Losing Weight the Hard Way

A student eats a dinner rated at 2 000 Calories. He wishes to do an equivalent amount of work in the gymnasium by lifting a 50.0-kg barbell. How many times must he raise the barbell to expend this much energy? Assume that he raises the barbell 2.00 m each time he lifts it and that he regains no energy when he lowers the barbell.

Solution Because 1 Calorie $= 1.00 \times 10^3$ cal, the total amount of work required to be done on the barbell–Earth system is 2.00×10^6 cal. Converting this value to joules, we have

$$W = (2.00 \times 10^6 \text{ cal})(4.186 \text{ J/cal}) = 8.37 \times 10^6 \text{ J}$$

The work done in lifting the barbell a distance h is equal to mgh, and the work done in lifting it n times is $nmgh$. We equate this to the total work required:

$$W = nmgh = 8.37 \times 10^6 \text{ J}$$

$$n = \frac{W}{mgh} = \frac{8.37 \times 10^6 \text{ J}}{(50.0 \text{ kg})(9.80 \text{ m/s}^2)(2.00 \text{ m})}$$

$$= 8.54 \times 10^3 \text{ times}$$

If the student is in good shape and lifts the barbell once every 5 s, it will take him about 12 h to perform this feat. Clearly, it is much easier for this student to lose weight by dieting.

In reality, the human body is not 100% efficient. Thus, not all of the energy transformed within the body from the dinner transfers out of the body by work done on the barbell. Some of this energy is used to pump blood and perform other functions within the body. Thus, the 2 000 Calories can be worked off in less time than 12 h when these other energy requirements are included.

20.2 Specific Heat and Calorimetry

When energy is added to a system and there is no change in the kinetic or potential energy of the system, the temperature of the system usually rises. (An exception to this statement is the case in which a system undergoes a change of state—also called a *phase transition*—as discussed in the next section.) If the system consists of a sample of a substance, we find that the quantity of energy required to raise the temperature of a given mass of the substance by some amount varies from one substance to another. For example, the quantity of energy required to raise the temperature of 1 kg of water by 1°C is 4 186 J, but the quantity of energy required to raise the temperature of 1 kg of copper by 1°C is only 387 J. In the discussion that follows, we shall use heat as our example of energy transfer, but keep in mind that we could change the temperature of our system by means of any method of energy transfer.

The **heat capacity** C of a particular sample of a substance is defined as the amount of energy needed to raise the temperature of that sample by 1°C. From this definition, we see that if energy Q produces a change ΔT in the temperature of a sample, then

$$Q = C\Delta T \quad (20.2)$$

The **specific heat** c of a substance is the heat capacity per unit mass. Thus, if energy Q transfers to a sample of a substance with mass m and the temperature of the sample changes by ΔT, then the specific heat of the substance is

Specific heat

$$c \equiv \frac{Q}{m\,\Delta T} \quad (20.3)$$

Specific heat is essentially a measure of how thermally insensitive a substance is to the addition of energy. The greater a material's specific heat, the more energy must be added to a given mass of the material to cause a particular temperature change. Table 20.1 lists representative specific heats.

From this definition, we can relate the energy Q transferred between a sample of mass m of a material and its surroundings to a temperature change ΔT as

$$Q = mc\,\Delta T \quad (20.4)$$

PITFALL PREVENTION

20.3 An Unfortunate Choice of Terminology

The name *specific heat* is an unfortunate holdover from the days when thermodynamics and mechanics developed separately. A better name would be *specific energy transfer,* but the existing term is too entrenched to be replaced.

Table 20.1

Specific Heats of Some Substances at 25°C and Atmospheric Pressure

Substance	Specific heat c J/kg·°C	Specific heat c cal/g·°C
Elemental solids		
Aluminum	900	0.215
Beryllium	1 830	0.436
Cadmium	230	0.055
Copper	387	0.092 4
Germanium	322	0.077
Gold	129	0.030 8
Iron	448	0.107
Lead	128	0.030 5
Silicon	703	0.168
Silver	234	0.056
Other solids		
Brass	380	0.092
Glass	837	0.200
Ice (−5°C)	2 090	0.50
Marble	860	0.21
Wood	1 700	0.41
Liquids		
Alcohol (ethyl)	2 400	0.58
Mercury	140	0.033
Water (15°C)	4 186	1.00
Gas		
Steam (100°C)	2 010	0.48

For example, the energy required to raise the temperature of 0.500 kg of water by 3.00°C is (0.500 kg)(4 186 J/kg · °C)(3.00°C) = 6.28×10^3 J. Note that when the temperature increases, Q and ΔT are taken to be positive, and energy transfers into the system. When the temperature decreases, Q and ΔT are negative, and energy transfers out of the system.

Specific heat varies with temperature. However, if temperature intervals are not too great, the temperature variation can be ignored and c can be treated as a constant.[2] For example, the specific heat of water varies by only about 1% from 0°C to 100°C at atmospheric pressure. Unless stated otherwise, we shall neglect such variations.

Measured values of specific heats are found to depend on the conditions of the experiment. In general, measurements made in a constant-pressure process are different from those made in a constant-volume process. For solids and liquids, the difference between the two values is usually no greater than a few percent and is often neglected. Most of the values given in Table 20.1 were measured at atmospheric pressure and room temperature. The specific heats for gases measured at constant pressure are quite different from values measured at constant volume (see Chapter 21).

▲ PITFALL PREVENTION

20.4 Energy Can Be Transferred by Any Method

We will use Q to represent the amount of energy transferred, but keep in mind that the energy transfer in Equation 20.4 could be by *any* of the methods introduced in Chapter 7; it does not have to be heat. For example, repeatedly bending a coat hanger wire raises the temperature at the bending point by *work.*

Quick Quiz 20.1 Imagine you have 1 kg each of iron, glass, and water, and that all three samples are at 10°C. Rank the samples from lowest to highest temperature after 100 J of energy is added to each sample.

Quick Quiz 20.2 Considering the same samples as in Quick Quiz 20.1, rank them from least to greatest amount of energy transferred by heat if each sample increases in temperature by 20°C.

It is interesting to note from Table 20.1 that water has the highest specific heat of common materials. This high specific heat is responsible, in part, for the moderate temperatures found near large bodies of water. As the temperature of a body of water decreases during the winter, energy is transferred from the cooling water to the air by heat, increasing the internal energy of the air. Because of the high specific heat of water, a relatively large amount of energy is transferred to the air for even modest temperature changes of the water. The air carries this internal energy landward when prevailing winds are favorable. For example, the prevailing winds on the West Coast of the United States are toward the land (eastward). Hence, the energy liberated by the Pacific Ocean as it cools keeps coastal areas much warmer than they would otherwise be. This explains why the western coastal states generally have more favorable winter weather than the eastern coastal states, where the prevailing winds do not tend to carry the energy toward land.

Conservation of Energy: Calorimetry

One technique for measuring specific heat involves heating a sample to some known temperature T_x, placing it in a vessel containing water of known mass and temperature $T_w < T_x$, and measuring the temperature of the water after equilibrium has been reached. This technique is called **calorimetry**, and devices in which this energy transfer occurs are called **calorimeters**. If the system of the sample and the water is isolated, the law of the conservation of energy requires that the amount of energy that leaves the sample (of unknown specific heat) equal the amount of energy that enters the water.[3]

[2] The definition given by Equation 20.3 assumes that the specific heat does not vary with temperature over the interval $\Delta T = T_f - T_i$. In general, if c varies with temperature over the interval, then the correct expression for Q is $Q = m\int_{T_i}^{T_f} c\,dT$.

[3] For precise measurements, the water container should be included in our calculations because it also exchanges energy with the sample. Doing so would require a knowledge of its mass and composition, however. If the mass of the water is much greater than that of the container, we can neglect the effects of the container.

PITFALL PREVENTION

20.5 Remember the Negative Sign

It is *critical* to include the negative sign in Equation 20.5. The negative sign in the equation is necessary for consistency with our sign convention for energy transfer. The energy transfer Q_{hot} has a negative value because energy is leaving the hot substance. The negative sign in the equation assures that the right-hand side is a positive number, consistent with the left-hand side, which is positive because energy is entering the cold water.

Conservation of energy allows us to write the mathematical representation of this energy statement as

$$Q_{cold} = -Q_{hot} \quad (20.5)$$

The negative sign in the equation is necessary to maintain consistency with our sign convention for heat.

Suppose m_x is the mass of a sample of some substance whose specific heat we wish to determine. Let us call its specific heat c_x and its initial temperature T_x. Likewise, let m_w, c_w, and T_w represent corresponding values for the water. If T_f is the final equilibrium temperature after everything is mixed, then from Equation 20.4, we find that the energy transfer for the water is $m_wc_w(T_f - T_w)$, which is positive because $T_f > T_w$, and that the energy transfer for the sample of unknown specific heat is $m_xc_x(T_f - T_x)$, which is negative. Substituting these expressions into Equation 20.5 gives

$$m_wc_w(T_f - T_w) = -m_xc_x(T_f - T_x)$$

Solving for c_x gives

$$c_x = \frac{m_wc_w(T_f - T_w)}{m_x(T_x - T_f)}$$

Example 20.2 Cooling a Hot Ingot

A 0.050 0-kg ingot of metal is heated to 200.0°C and then dropped into a beaker containing 0.400 kg of water initially at 20.0°C. If the final equilibrium temperature of the mixed system is 22.4°C, find the specific heat of the metal.

Solution According to Equation 20.5, we can write

$$m_wc_w(T_f - T_w) = -m_xc_x(T_f - T_x)$$

$$(0.400 \text{ kg})(4\,186 \text{ J/kg}\cdot{}^\circ\text{C})(22.4^\circ\text{C} - 20.0^\circ\text{C}) = -(0.050\,0 \text{ kg})(c_x)(22.4^\circ\text{C} - 200.0^\circ\text{C})$$

From this we find that

$$c_x = 453 \text{ J/kg}\cdot{}^\circ\text{C}$$

The ingot is most likely iron, as we can see by comparing this result with the data given in Table 20.1. Note that the temperature of the ingot is initially above the steam point. Thus, some of the water may vaporize when we drop the ingot into the water. We assume that we have a sealed system and that this steam cannot escape. Because the final equilibrium temperature is lower than the steam point, any steam that does result recondenses back into water.

What If? Suppose you are performing an experiment in the laboratory that uses this technique to determine the specific heat of a sample and you wish to decrease the overall uncertainty in your final result for c_x. Of the data given in the text of this example, changing which value would be most effective in decreasing the uncertainty?

Answer The largest experimental uncertainty is associated with the small temperature difference of 2.4°C for $T_f - T_w$. For example, an uncertainty of 0.1°C in each of these two temperature readings leads to an 8% uncertainty in their difference. In order for this temperature difference to be larger experimentally, the most effective change is to *decrease the amount of water.*

Example 20.3 Fun Time for a Cowboy

A cowboy fires a silver bullet with a muzzle speed of 200 m/s into the pine wall of a saloon. Assume that all the internal energy generated by the impact remains with the bullet. What is the temperature change of the bullet?

Solution The kinetic energy of the bullet is

$$K = \tfrac{1}{2}mv^2$$

Because nothing in the environment is hotter than the bullet, the bullet gains no energy by heat. Its temperature increases because the kinetic energy is transformed to extra internal energy when the bullet is stopped by the wall. The temperature change is the same as that which would take place if energy $Q = K$ were transferred by heat from a stove to the bullet. If we imagine this latter process taking place, we can calculate ΔT from Equation 20.4. Using 234 J/kg · °C as the specific heat of silver (see Table 20.1), we obtain

$$(1) \quad \Delta T = \frac{Q}{mc} = \frac{K}{mc} = \frac{\frac{1}{2}\cancel{m}(200 \text{ m/s})^2}{\cancel{m}(234 \text{ J/kg}\cdot{}^\circ\text{C})} = 85.5^\circ\text{C}$$

Note that the result does not depend on the mass of the bullet.

What If? Suppose that the cowboy runs out of silver bullets and fires a lead bullet at the same speed into the wall. Will the temperature change of the bullet be larger or smaller?

Answer Consulting Table 20.1, we find that the specific heat of lead is 128 J/kg · °C, which is smaller than that for silver. Thus, a given amount of energy input will raise lead to a higher temperature than silver and the final temperature of the lead bullet will be larger. In Equation (1), we substitute the new value for the specific heat:

$$\Delta T = \frac{Q}{mc} = \frac{K}{mc} = \frac{\frac{1}{2}m(200\text{ m/s})^2}{m(128\text{ J/kg}\cdot{}^\circ\text{C})} = 156^\circ\text{C}$$

Note that there is no requirement that the silver and lead bullets have the same mass to determine this temperature. The only requirement is that they have the same speed.

20.3 Latent Heat

A substance often undergoes a change in temperature when energy is transferred between it and its surroundings. There are situations, however, in which the transfer of energy does not result in a change in temperature. This is the case whenever the physical characteristics of the substance change from one form to another; such a change is commonly referred to as a **phase change**. Two common phase changes are from solid to liquid (melting) and from liquid to gas (boiling); another is a change in the crystalline structure of a solid. All such phase changes involve a change in internal energy but no change in temperature. The increase in internal energy in boiling, for example, is represented by the breaking of bonds between molecules in the liquid state; this bond breaking allows the molecules to move farther apart in the gaseous state, with a corresponding increase in intermolecular potential energy.

As you might expect, different substances respond differently to the addition or removal of energy as they change phase because their internal molecular arrangements vary. Also, the amount of energy transferred during a phase change depends on the amount of substance involved. (It takes less energy to melt an ice cube than it does to thaw a frozen lake.) If a quantity Q of energy transfer is required to change the phase of a mass m of a substance, the ratio $L \equiv Q/m$ characterizes an important thermal property of that substance. Because this added or removed energy does not result in a temperature change, the quantity L is called the **latent heat** (literally, the "hidden" heat) of the substance. The value of L for a substance depends on the nature of the phase change, as well as on the properties of the substance.

PITFALL PREVENTION

20.6 Signs Are Critical

Sign errors occur very often when students apply calorimetry equations, so we will make this point once again. For phase changes, use the correct explicit sign in Equation 20.6, depending on whether you are adding or removing energy from the substance. In Equation 20.4, there is no explicit sign to consider, but be sure that your ΔT is *always* the final temperature minus the initial temperature. In addition, make sure that you *always* include the negative sign on the right-hand side of Equation 20.5.

From the definition of latent heat, and again choosing heat as our energy transfer mechanism, we find that the energy required to change the phase of a given mass m of a pure substance is

Latent heat

$$Q = \pm mL \qquad (20.6)$$

Latent heat of fusion L_f is the term used when the phase change is from solid to liquid (*to fuse* means "to combine by melting"), and **latent heat of vaporization** L_v is the term used when the phase change is from liquid to gas (the liquid "vaporizes").[4] The latent heats of various substances vary considerably, as data in Table 20.2 show. The positive sign in Equation 20.6 is used when energy enters a system, causing melting or vaporization. The negative sign corresponds to energy leaving a system, such that the system freezes or condenses.

To understand the role of latent heat in phase changes, consider the energy required to convert a 1.00-g cube of ice at $-30.0°C$ to steam at 120.0°C. Figure 20.2 indicates the experimental results obtained when energy is gradually added to the ice. Let us examine each portion of the red curve.

[4] When a gas cools, it eventually *condenses*—that is, it returns to the liquid phase. The energy given up per unit mass is called the *latent heat of condensation* and is numerically equal to the latent heat of vaporization. Likewise, when a liquid cools, it eventually solidifies, and the *latent heat of solidification* is numerically equal to the latent heat of fusion.

Table 20.2

Latent Heats of Fusion and Vaporization

Substance	Melting Point (°C)	Latent Heat of Fusion (J/kg)	Boiling Point (°C)	Latent Heat of Vaporization (J/kg)
Helium	−269.65	5.23×10^3	−268.93	2.09×10^4
Nitrogen	−209.97	2.55×10^4	−195.81	2.01×10^5
Oxygen	−218.79	1.38×10^4	−182.97	2.13×10^5
Ethyl alcohol	−114	1.04×10^5	78	8.54×10^5
Water	0.00	3.33×10^5	100.00	2.26×10^6
Sulfur	119	3.81×10^4	444.60	3.26×10^5
Lead	327.3	2.45×10^4	1 750	8.70×10^5
Aluminum	660	3.97×10^5	2 450	1.14×10^7
Silver	960.80	8.82×10^4	2 193	2.33×10^6
Gold	1 063.00	6.44×10^4	2 660	1.58×10^6
Copper	1 083	1.34×10^5	1 187	5.06×10^6

Part A. On this portion of the curve, the temperature of the ice changes from −30.0°C to 0.0°C. Because the specific heat of ice is 2 090 J/kg · °C, we can calculate the amount of energy added by using Equation 20.4:

$$Q = m_i c_i \, \Delta T = (1.00 \times 10^{-3} \text{ kg})(2\,090 \text{ J/kg} \cdot {}^\circ\text{C})(30.0^\circ\text{C}) = 62.7 \text{ J}$$

Part B. When the temperature of the ice reaches 0.0°C, the ice–water mixture remains at this temperature—even though energy is being added—until all the ice melts. The energy required to melt 1.00 g of ice at 0.0°C is, from Equation 20.6,

$$Q = m_i L_f = (1.00 \times 10^{-3} \text{ kg})(3.33 \times 10^5 \text{ J/kg}) = 333 \text{ J}$$

Thus, we have moved to the 396 J (= 62.7 J + 333 J) mark on the energy axis in Figure 20.2.

Part C. Between 0.0°C and 100.0°C, nothing surprising happens. No phase change occurs, and so all energy added to the water is used to increase its temperature. The amount of energy necessary to increase the temperature from 0.0°C to 100.0°C is

$$Q = m_w c_w \, \Delta T = (1.00 \times 10^{-3} \text{ kg})(4.19 \times 10^3 \text{ J/kg} \cdot {}^\circ\text{C})(100.0^\circ\text{C}) = 419 \text{ J}$$

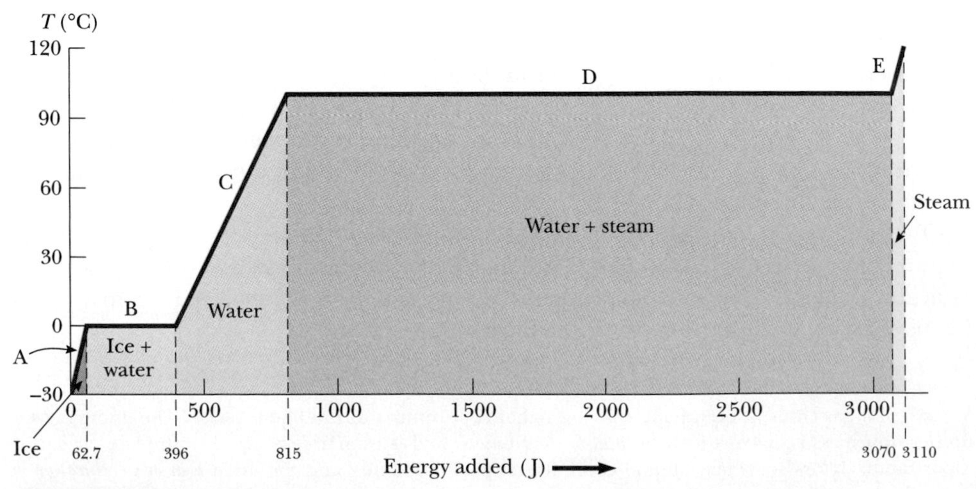

Figure 20.2 A plot of temperature versus energy added when 1.00 g of ice initially at −30.0°C is converted to steam at 120.0°C.

Part D. At 100.0°C, another phase change occurs as the water changes from water at 100.0°C to steam at 100.0°C. Similar to the ice–water mixture in part B, the water–steam mixture remains at 100.0°C—even though energy is being added—until all of the liquid has been converted to steam. The energy required to convert 1.00 g of water to steam at 100.0°C is

$$Q = m_w L_v = (1.00 \times 10^{-3}\ \text{kg})(2.26 \times 10^6\ \text{J/kg}) = 2.26 \times 10^3\ \text{J}$$

Part E. On this portion of the curve, as in parts A and C, no phase change occurs; thus, all energy added is used to increase the temperature of the steam. The energy that must be added to raise the temperature of the steam from 100.0°C to 120.0°C is

$$Q = m_s c_s\ \Delta T = (1.00 \times 10^{-3}\ \text{kg})(2.01 \times 10^3\ \text{J/kg}\cdot{}^\circ\text{C})(20.0^\circ\text{C}) = 40.2\ \text{J}$$

The total amount of energy that must be added to change 1 g of ice at -30.0°C to steam at 120.0°C is the sum of the results from all five parts of the curve, which is 3.11×10^3 J. Conversely, to cool 1 g of steam at 120.0°C to ice at -30.0°C, we must remove 3.11×10^3 J of energy.

Note in Figure 20.2 the relatively large amount of energy that is transferred into the water to vaporize it to steam. Imagine reversing this process—there is a large amount of energy transferred out of steam to condense it into water. This is why a burn to your skin from steam at 100°C is much more damaging than exposure of your skin to water at 100°C. A very large amount of energy enters your skin from the steam and the steam remains at 100°C for a long time while it condenses. Conversely, when your skin makes contact with water at 100°C, the water immediately begins to drop in temperature as energy transfers from the water to your skin.

We can describe phase changes in terms of a rearrangement of molecules when energy is added to or removed from a substance. (For elemental substances in which the atoms do not combine to form molecules, the following discussion should be interpreted in terms of atoms. We use the general term *molecules* to refer to both chemical compounds and elemental substances.) Consider first the liquid-to-gas phase change. The molecules in a liquid are close together, and the forces between them are stronger than those between the more widely separated molecules of a gas. Therefore, work must be done on the liquid against these attractive molecular forces if the molecules are to separate. The latent heat of vaporization is the amount of energy per unit mass that must be added to the liquid to accomplish this separation.

Similarly, for a solid, we imagine that the addition of energy causes the amplitude of vibration of the molecules about their equilibrium positions to become greater as the temperature increases. At the melting point of the solid, the amplitude is great enough to break the bonds between molecules and to allow molecules to move to new positions. The molecules in the liquid also are bound to each other, but less strongly than those in the solid phase. The latent heat of fusion is equal to the energy required per unit mass to transform the bonds among all molecules from the solid-type bond to the liquid-type bond.

As you can see from Table 20.2, the latent heat of vaporization for a given substance is usually somewhat higher than the latent heat of fusion. This is not surprising if we consider that the average distance between molecules in the gas phase is much greater than that in either the liquid or the solid phase. In the solid-to-liquid phase change, we transform solid-type bonds between molecules into liquid-type bonds between molecules, which are only slightly less strong. In the liquid-to-gas phase change, however, we break liquid-type bonds and create a situation in which the molecules of the gas essentially are not bonded to each other. Therefore, it is not surprising that more energy is required to vaporize a given mass of substance than is required to melt it.

Quick Quiz 20.3 Suppose the same process of adding energy to the ice cube is performed as discussed above, but we graph the internal energy of the system as a function of energy input. What would this graph look like?

PITFALL PREVENTION

20.7 Celsius vs. Kelvin

In equations in which T appears—for example, the ideal gas law—the Kelvin temperature *must* be used. In equations involving ΔT, such as calorimetry equations, it is *possible* to use Celsius temperatures, because a change in temperature is the same on both scales. It is *safest,* however, to *consistently* use Kelvin temperatures in all equations involving T or ΔT.

Quick Quiz 20.4 Calculate the slopes for the A, C, and E portions of Figure 20.2. Rank the slopes from least to greatest and explain what this ordering means.

PROBLEM-SOLVING HINTS

Calorimetry Problems

If you have difficulty in solving calorimetry problems, be sure to consider the following points:

- Units of measure must be consistent. For instance, if you are using specific heats measured in J/kg · °C, be sure that masses are in kilograms and temperatures are in Celsius degrees.
- Transfers of energy are given by the equation $Q = mc\ \Delta T$ only for those processes in which no phase changes occur. Use the equations $Q = \pm\, mL_f$ and $Q = \pm\, mL_v$ only when phase changes *are* taking place; be sure to select the proper sign for these equations depending on the direction of energy transfer.
- Often, errors in sign are made when the equation $Q_{\text{cold}} = -Q_{\text{hot}}$ is used. Make sure that you use the negative sign in the equation, and remember that ΔT is always the final temperature minus the initial temperature.

Example 20.4 Cooling the Steam

What mass of steam initially at 130°C is needed to warm 200 g of water in a 100-g glass container from 20.0°C to 50.0°C?

Solution The steam loses energy in three stages. In the first stage, the steam is cooled to 100°C. The energy transfer in the process is

$$\begin{aligned} Q_1 &= m_s c_s\ \Delta T = m_s(2.01 \times 10^3\ \text{J/kg}\cdot{}^\circ\text{C})(-30.0^\circ\text{C}) \\ &= -\,m_s(6.03 \times 10^4\ \text{J/kg}) \end{aligned}$$

where m_s is the unknown mass of the steam.

In the second stage, the steam is converted to water. To find the energy transfer during this phase change, we use $Q = -\,mL_v$, where the negative sign indicates that energy is leaving the steam:

$$Q_2 = -m_s(2.26 \times 10^6\ \text{J/kg})$$

In the third stage, the temperature of the water created from the steam is reduced to 50.0°C. This change requires an energy transfer of

$$\begin{aligned} Q_3 &= m_s c_w \Delta T = m_s(4.19 \times 10^3\ \text{J/kg}\cdot{}^\circ\text{C})(-50.0^\circ\text{C}) \\ &= -\,m_s(2.09 \times 10^5\ \text{J/kg}) \end{aligned}$$

Adding the energy transfers in these three stages, we obtain

$$\begin{aligned} Q_{\text{hot}} &= Q_1 + Q_2 + Q_3 \\ &= -\,m_s[6.03 \times 10^4\ \text{J/kg} + 2.26 \times 10^6\ \text{J/kg} \\ &\quad + 2.09 \times 10^5\ \text{J/kg}] \\ &= -\,m_s(2.53 \times 10^6\ \text{J/kg}) \end{aligned}$$

Now, we turn our attention to the temperature increase of the water and the glass. Using Equation 20.4, we find that

$$\begin{aligned} Q_{\text{cold}} &= (0.200\ \text{kg})(4.19 \times 10^3\ \text{J/kg}\cdot{}^\circ\text{C})(30.0^\circ\text{C}) \\ &\quad + (0.100\ \text{kg})(837\ \text{J/kg}\cdot{}^\circ\text{C})(30.0^\circ\text{C}) \\ &= 2.77 \times 10^4\ \text{J} \end{aligned}$$

Using Equation 20.5, we can solve for the unknown mass:

$$\begin{aligned} Q_{\text{cold}} &= -\,Q_{\text{hot}} \\ 2.77 \times 10^4\ \text{J} &= -[-m_s(2.53 \times 10^6\ \text{J/kg})] \\ m_s &= 1.09 \times 10^{-2}\ \text{kg} = 10.9\ \text{g} \end{aligned}$$

What If? **What if the final state of the system is water at 100°C? Would we need more or less steam? How would the analysis above change?**

Answer More steam would be needed to raise the temperature of the water and glass to 100°C instead of 50.0°C. There would be two major changes in the analysis. First, we would not have a term Q_3 for the steam because the water that condenses from the steam does not cool below 100°C. Second, in Q_{cold}, the temperature change would be 80.0°C instead of 30.0°C. Thus, Q_{hot} becomes

$$\begin{aligned} Q_{\text{hot}} &= Q_1 + Q_2 \\ &= -m_s(6.03 \times 10^4\ \text{J/kg} + 2.26 \times 10^6\ \text{J/kg}) \\ &= -m_s(2.32 \times 10^6\ \text{J/kg}) \end{aligned}$$

and Q_{cold} becomes

$$\begin{aligned} Q_{\text{cold}} &= (0.200\ \text{kg})(4.19 \times 10^3\ \text{J/kg}\cdot{}^\circ\text{C})(80.0^\circ\text{C}) \\ &\quad + (0.100\ \text{kg})(837\ \text{J/kg}\cdot{}^\circ\text{C})(80.0^\circ\text{C}) \\ &= 7.37 \times 10^4\ \text{J} \end{aligned}$$

leading to $m_s = 3.18 \times 10^{-2}\ \text{kg} = 31.8\ \text{g}$.

Example 20.5 Boiling Liquid Helium

Liquid helium has a very low boiling point, 4.2 K, and a very low latent heat of vaporization, 2.09×10^4 J/kg. If energy is transferred to a container of boiling liquid helium from an immersed electric heater at a rate of 10.0 W, how long does it take to boil away 1.00 kg of the liquid?

Solution Because $L_v = 2.09 \times 10^4$ J/kg, we must supply 2.09×10^4 J of energy to boil away 1.00 kg. Because 10.0 W = 10.0 J/s, 10.0 J of energy is transferred to the helium each second. From $\mathcal{P} = \Delta E/\Delta t$, the time interval required to transfer 2.09×10^4 J of energy is

$$\Delta t = \frac{\Delta E}{\mathcal{P}} = \frac{2.09 \times 10^4 \text{ J}}{10.0 \text{ J/s}} = 2.09 \times 10^3 \text{ s} \approx 35 \text{ min}$$

20.4 Work and Heat in Thermodynamic Processes

In the macroscopic approach to thermodynamics, we describe the *state* of a system using such variables as pressure, volume, temperature, and internal energy. As a result, these quantities belong to a category called **state variables**. For any given configuration of the system, we can identify values of the state variables. It is important to note that a *macroscopic state* of an isolated system can be specified only if the system is in thermal equilibrium internally. In the case of a gas in a container, internal thermal equilibrium requires that every part of the gas be at the same pressure and temperature.

A second category of variables in situations involving energy is **transfer variables**. These variables are zero *unless* a process occurs in which energy is transferred across the boundary of the system. Because a transfer of energy across the boundary represents a change in the system, transfer variables are not associated with a given state of the system, but with a *change* in the state of the system. In the previous sections, we discussed heat as a transfer variable. For a given set of conditions of a system, there is no defined value for the heat. We can only assign a value of the heat if energy crosses the boundary by heat, resulting in a change in the system. State variables are characteristic of a system in thermal equilibrium. Transfer variables are characteristic of a process in which energy is transferred between a system and its environment.

In this section, we study another important transfer variable for thermodynamic systems—work. Work performed on particles was studied extensively in Chapter 7, and here we investigate the work done on a deformable system—a gas. Consider a gas contained in a cylinder fitted with a movable piston (Fig. 20.3). At equilibrium, the gas oc-

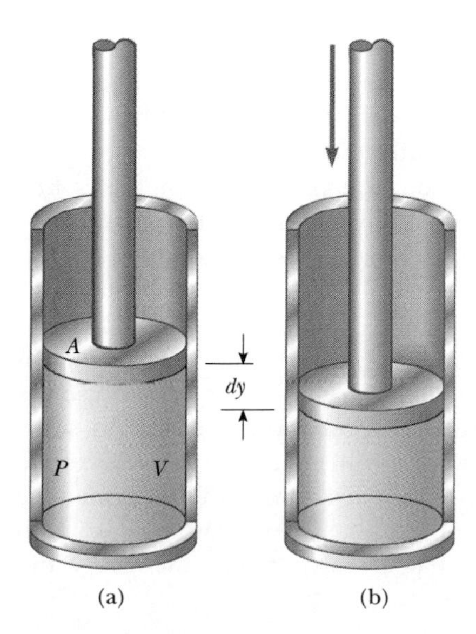

Figure 20.3 Work is done on a gas contained in a cylinder at a pressure P as the piston is pushed downward so that the gas is compressed.

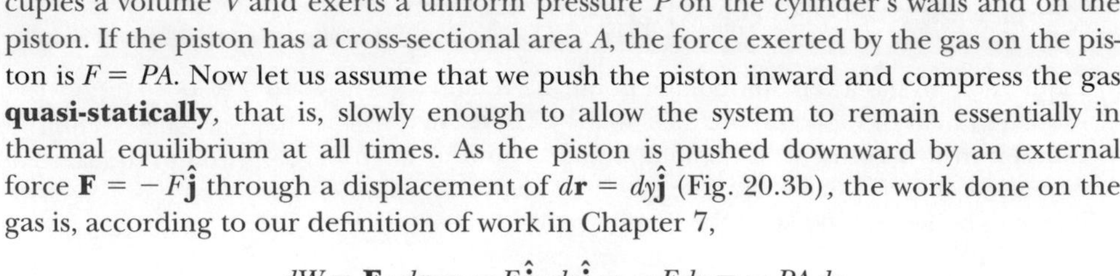

cupies a volume V and exerts a uniform pressure P on the cylinder's walls and on the piston. If the piston has a cross-sectional area A, the force exerted by the gas on the piston is $F = PA$. Now let us assume that we push the piston inward and compress the gas **quasi-statically**, that is, slowly enough to allow the system to remain essentially in thermal equilibrium at all times. As the piston is pushed downward by an external force $\mathbf{F} = -F\hat{\mathbf{j}}$ through a displacement of $d\mathbf{r} = dy\hat{\mathbf{j}}$ (Fig. 20.3b), the work done on the gas is, according to our definition of work in Chapter 7,

$$dW = \mathbf{F} \cdot d\mathbf{r} = -F\hat{\mathbf{j}} \cdot dy\hat{\mathbf{j}} = -F\,dy = -PA\,dy$$

where we have set the magnitude F of the external force equal to PA because the piston is always in equilibrium between the external force and the force from the gas. For this discussion, we assume the mass of the piston is negligible. Because $A\,dy$ is the change in volume of the gas dV, we can express the work done on the gas as

$$dW = -P\,dV \tag{20.7}$$

If the gas is compressed, dV is negative and the work done on the gas is positive. If the gas expands, dV is positive and the work done on the gas is negative. If the volume remains constant, the work done on the gas is zero. The total work done on the gas as its volume changes from V_i to V_f is given by the integral of Equation 20.7:

Work done on a gas

$$W = -\int_{V_i}^{V_f} P\,dV \tag{20.8}$$

To evaluate this integral, one must know how the pressure varies with volume during the process.

In general, the pressure is not constant during a process followed by a gas, but depends on the volume and temperature. If the pressure and volume are known at each step of the process, the state of the gas at each step can be plotted on a graph called a ***PV* diagram**, as in Figure 20.4. This type of diagram allows us to visualize a process through which a gas is progressing. The curve on a PV diagram is called the *path* taken between the initial and final states.

Note that the integral in Equation 20.8 is equal to the area under a curve on a PV diagram. Thus, we can identify an important use for PV diagrams:

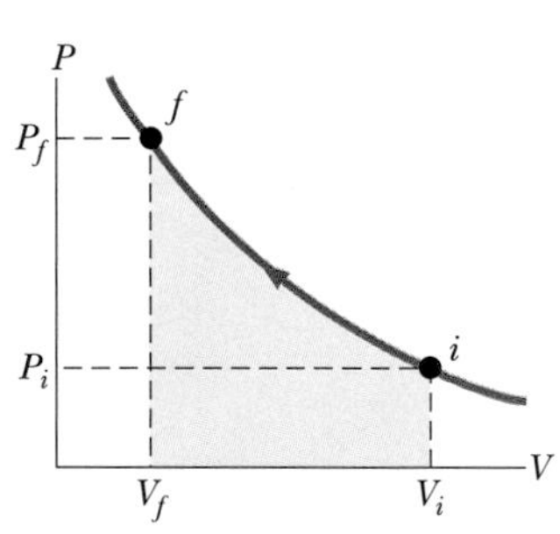

Active Figure 20.4 A gas is compressed quasi-statically (slowly) from state i to state f. The work done on the gas equals the negative of the area under the PV curve.

At the Active Figures link at http://www.pse6.com, you can compress the piston in Figure 20.3 and see the result on the PV diagram in this figure.

> The work done on a gas in a quasi-static process that takes the gas from an initial state to a final state is the negative of the area under the curve on a PV diagram, evaluated between the initial and final states.

As Figure 20.4 suggests, for our process of compressing a gas in the cylinder, the work done depends on the particular path taken between the initial and final states. To illustrate this important point, consider several different paths connecting i and f (Fig. 20.5). In the process depicted in Figure 20.5a, the volume of the gas is first reduced from V_i to V_f at constant pressure P_i and the pressure of the gas then increases from P_i to P_f by heating at constant volume V_f. The work done on the gas along this path is $-P_i(V_f - V_i)$. In Figure 20.5b, the pressure of the gas is increased from P_i to P_f at constant volume V_i and then the volume of the gas is reduced from V_i to V_f at constant pressure P_f. The work done on the gas is $-P_f(V_f - V_i)$, which is greater than that for the process described in Figure 20.5a. It is greater because the piston is moved through the same displacement by a larger force than for the situation in Figure 20.5a. Finally, for the process described in Figure 20.5c, where both P and V change continuously, the work done on the gas has some value intermediate between the values obtained in the first two processes. To evaluate the work in this case, the function $P(V)$ must be known, so that we can evaluate the integral in Equation 20.8.

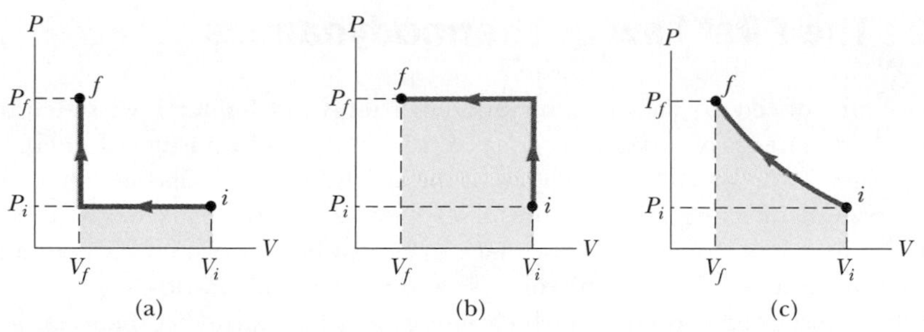

Active Figure 20.5 The work done on a gas as it is taken from an initial state to a final state depends on the path between these states.

At the Active Figures link at http://www.pse6.com, you can choose one of the three paths and see the movement of the piston in Figure 20.3 and of a point on the PV diagram in this figure.

The energy transfer Q into or out of a system by heat also depends on the process. Consider the situations depicted in Figure 20.6. In each case, the gas has the same initial volume, temperature, and pressure, and is assumed to be ideal. In Figure 20.6a, the gas is thermally insulated from its surroundings except at the bottom of the gas-filled region, where it is in thermal contact with an energy reservoir. An *energy reservoir* is a source of energy that is considered to be so great that a finite transfer of energy to or from the reservoir does not change its temperature. The piston is held at its initial position by an external agent—a hand, for instance. When the force holding the piston is reduced slightly, the piston rises very slowly to its final position. Because the piston is moving upward, the gas is doing work on the piston. During this expansion to the final volume V_f, just enough energy is transferred by heat from the reservoir to the gas to maintain a constant temperature T_i.

Now consider the completely thermally insulated system shown in Figure 20.6b. When the membrane is broken, the gas expands rapidly into the vacuum until it occupies a volume V_f and is at a pressure P_f. In this case, the gas does no work because it does not apply a force—no force is required to expand into a vacuum. Furthermore, no energy is transferred by heat through the insulating wall.

The initial and final states of the ideal gas in Figure 20.6a are identical to the initial and final states in Figure 20.6b, but the paths are different. In the first case, the gas does work on the piston, and energy is transferred slowly to the gas by heat. In the second case, no energy is transferred by heat, and the value of the work done is zero. Therefore, we conclude that **energy transfer by heat, like work done, depends on the initial, final, and intermediate states of the system.** In other words, because heat and work depend on the path, neither quantity is determined solely by the end points of a thermodynamic process.

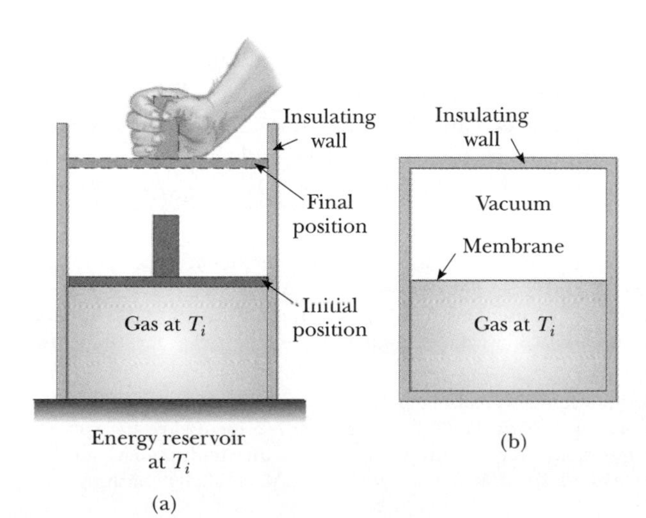

Figure 20.6 (a) A gas at temperature T_i expands slowly while absorbing energy from a reservoir in order to maintain a constant temperature. (b) A gas expands rapidly into an evacuated region after a membrane is broken.

20.5 The First Law of Thermodynamics

When we introduced the law of conservation of energy in Chapter 7, we stated that the change in the energy of a system is equal to the sum of all transfers of energy across the boundary of the system. The first law of thermodynamics is a special case of the law of conservation of energy that encompasses changes in internal energy and energy transfer by heat and work. It is a law that can be applied to many processes and provides a connection between the microscopic and macroscopic worlds.

We have discussed two ways in which energy can be transferred between a system and its surroundings. One is work done on the system, which requires that there be a macroscopic displacement of the point of application of a force. The other is heat, which occurs on a molecular level whenever a temperature difference exists across the boundary of the system. Both mechanisms result in a change in the internal energy of the system and therefore usually result in measurable changes in the macroscopic variables of the system, such as the pressure, temperature, and volume of a gas.

To better understand these ideas on a quantitative basis, suppose that a system undergoes a change from an initial state to a final state. During this change, energy transfer by heat Q to the system occurs, and work W is done on the system. As an example, suppose that the system is a gas in which the pressure and volume change from P_i and V_i to P_f and V_f. If the quantity $Q + W$ is measured for various paths connecting the initial and final equilibrium states, we find that it is the same for all paths connecting the two states. We conclude that the quantity $Q + W$ is determined completely by the initial and final states of the system, and we call this quantity the **change in the internal energy** of the system. Although Q and W both depend on the path, **the quantity $Q + W$ is independent of the path.** If we use the symbol E_{int} to represent the internal energy, then the *change* in internal energy ΔE_{int} can be expressed as[5]

First law of thermodynamics

$$\Delta E_{int} = Q + W \tag{20.9}$$

where all quantities must have the same units of measure for energy. Equation 20.9 is known as the **first law of thermodynamics.** One of the important consequences of the first law of thermodynamics is that there exists a quantity known as internal energy whose value is determined by the state of the system. The internal energy is therefore a state variable like pressure, volume, and temperature.

When a system undergoes an infinitesimal change in state in which a small amount of energy dQ is transferred by heat and a small amount of work dW is done, the internal energy changes by a small amount dE_{int}. Thus, for infinitesimal processes we can express the first law as[6]

$$dE_{int} = dQ + dW$$

The first law of thermodynamics is an energy conservation equation specifying that the only type of energy that changes in the system is the internal energy E_{int}. Let us investigate some special cases in which this condition exists.

First, consider an *isolated system*—that is, one that does not interact with its surroundings. In this case, no energy transfer by heat takes place and the work done on

PITFALL PREVENTION

20.8 Dual Sign Conventions

Some physics and engineering textbooks present the first law as $\Delta E_{int} = Q - W$, with a minus sign between the heat and work. The reason for this is that work is defined in these treatments as the work done *by* the gas rather than *on* the gas, as in our treatment. The equivalent equation to Equation 20.8 in these treatments defines work as $W = \int_{V_i}^{V_f} P\,dV$. Thus, if positive work is done by the gas, energy is leaving the system, leading to the negative sign in the first law.

In your studies in other chemistry or engineering courses, or in your reading of other physics textbooks, be sure to note which sign convention is being used for the first law.

[5] It is an unfortunate accident of history that the traditional symbol for internal energy is U, which is also the traditional symbol for potential energy, as introduced in Chapter 8. To avoid confusion between potential energy and internal energy, we use the symbol E_{int} for internal energy in this book. If you take an advanced course in thermodynamics, however, be prepared to see U used as the symbol for internal energy.

[6] Note that dQ and dW are not true differential quantities because Q and W are not state variables; however, dE_{int} is. Because dQ and dW are *inexact differentials*, they are often represented by the symbols $đQ$ and $đW$. For further details on this point, see an advanced text on thermodynamics, such as R. P. Bauman, *Modern Thermodynamics and Statistical Mechanics*, New York, Macmillan Publishing Co., 1992.

the system is zero; hence, the internal energy remains constant. That is, because $Q = W = 0$, it follows that $\Delta E_{\text{int}} = 0$, and thus $E_{\text{int},\,i} = E_{\text{int},\,f}$. We conclude that **the internal energy E_{int} of an isolated system remains constant.**

Next, consider the case of a system (one not isolated from its surroundings) that is taken through a **cyclic process**—that is, a process that starts and ends at the same state. In this case, the change in the internal energy must again be zero, because E_{int} is a state variable, and therefore the energy Q added to the system must equal the negative of the work W done on the system during the cycle. That is, in a cyclic process,

$$\Delta E_{\text{int}} = 0 \quad \text{and} \quad Q = -W \quad \text{(cyclic process)}$$

On a PV diagram, a cyclic process appears as a closed curve. (The processes described in Figure 20.5 are represented by open curves because the initial and final states differ.) It can be shown that **in a cyclic process, the net work done on the system per cycle equals the area enclosed by the path representing the process on a PV diagram.**

PITFALL PREVENTION

20.9 The First Law

With our approach to energy in this book, the first law of thermodynamics is a special case of Equation 7.17. Some physicists argue that the first law is the general equation for energy conservation, equivalent to Equation 7.17. In this approach, the first law is applied to a closed system (so that there is no matter transfer), heat is interpreted so as to include electromagnetic radiation, and work is interpreted so as to include electrical transmission ("electrical work") and mechanical waves ("molecular work"). Keep this in mind if you run across the first law in your reading of other physics books.

20.6 Some Applications of the First Law of Thermodynamics

The first law of thermodynamics that we discussed in the preceding section relates the changes in internal energy of a system to transfers of energy by work or heat. In this section, we consider applications of the first law to processes through which a gas is taken. As a model, we consider the sample of gas contained in the piston–cylinder apparatus in Figure 20.7. This figure shows work being done on the gas and energy transferring in by heat, so the internal energy of the gas is rising. In the following discussion of various processes, refer back to this figure and mentally alter the directions of the transfer of energy so as to reflect what is happening in the process.

Before we apply the first law of thermodynamics to specific systems, it is useful to first define some idealized thermodynamic processes. An **adiabatic process** is one during which no energy enters or leaves the system by heat—that is, $Q = 0$. An adiabatic process can be achieved either by thermally insulating the walls of the system, such as the cylinder in Figure 20.7, or by performing the process rapidly, so that there is negligible time for energy to transfer by heat. Applying the first law of thermodynamics to an adiabatic process, we see that

$$\Delta E_{\text{int}} = W \quad \text{(adiabatic process)} \tag{20.10}$$

From this result, we see that if a gas is compressed adiabatically such that W is positive, then ΔE_{int} is positive and the temperature of the gas increases. Conversely, the temperature of a gas decreases when the gas expands adiabatically.

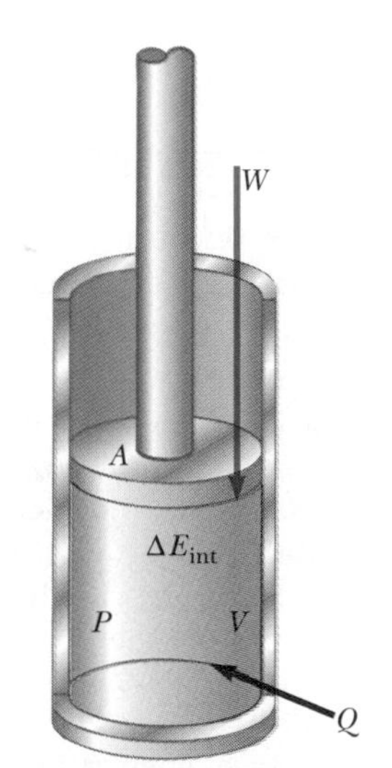

Active Figure 20.7 The first law of thermodynamics equates the change in internal energy E_{int} in a system to the net energy transfer to the system by heat Q and work W. In the situation shown here, the internal energy of the gas increases.

At the Active Figures link at http://www.pse6.com, you can choose one of the four processes for the gas discussed in this section and see the movement of the piston and of a point on a PV diagram.

Adiabatic processes are very important in engineering practice. Some common examples are the expansion of hot gases in an internal combustion engine, the liquefaction of gases in a cooling system, and the compression stroke in a diesel engine.

The process described in Figure 20.6b, called an **adiabatic free expansion,** is unique. The process is adiabatic because it takes place in an insulated container. Because the gas expands into a vacuum, it does not apply a force on a piston as was depicted in Figure 20.6a, so no work is done on or by the gas. Thus, in this adiabatic process, both $Q = 0$ and $W = 0$. As a result, $\Delta E_{\text{int}} = 0$ for this process, as we can see from the first law. That is, **the initial and final internal energies of a gas are equal in an adiabatic free expansion.** As we shall see in the next chapter, the internal energy of an ideal gas depends only on its temperature. Thus, we expect no change in temperature during an adiabatic free expansion. This prediction is in accord with the results of experiments performed at low pressures. (Experiments performed at high pressures for real gases show a slight change in temperature after the expansion. This change is due to intermolecular interactions, which represent a deviation from the model of an ideal gas.)

A process that occurs at constant pressure is called an **isobaric process.** In Figure 20.7, an isobaric process could be established by allowing the piston to move freely so that it is always in equilibrium between the net force from the gas pushing upward and the weight of the piston plus the force due to atmospheric pressure pushing downward. In Figure 20.5, the first process in part (a) and the second process in part (b) are isobaric.

In such a process, the values of the heat and the work are both usually nonzero. The work done on the gas in an isobaric process is simply

Isobaric process

$$W = -P(V_f - V_i) \qquad \text{(isobaric process)} \tag{20.11}$$

where P is the constant pressure.

A process that takes place at constant volume is called an **isovolumetric process.** In Figure 20.7, clamping the piston at a fixed position would ensure an isovolumetric process. In Figure 20.5, the second process in part (a) and the first process in part (b) are isovolumetric.

In such a process, the value of the work done is zero because the volume does not change. Hence, from the first law we see that in an isovolumetric process, because $W = 0$,

Isovolumetric process

$$\Delta E_{\text{int}} = Q \qquad \text{(isovolumetric process)} \tag{20.12}$$

This expression specifies that **if energy is added by heat to a system kept at constant volume, then all of the transferred energy remains in the system as an increase in its internal energy.** For example, when a can of spray paint is thrown into a fire, energy enters the system (the gas in the can) by heat through the metal walls of the can. Consequently, the temperature, and thus the pressure, in the can increases until the can possibly explodes.

Isothermal process

A process that occurs at constant temperature is called an **isothermal process**. In Figure 20.7, this process can be established by immersing the cylinder in Figure 20.7 in an ice-water bath or by putting the cylinder in contact with some other constant-temperature reservoir. A plot of P versus V at constant temperature for an ideal gas yields a hyperbolic curve called an *isotherm.* The internal energy of an ideal gas is a function of temperature only. Hence, in an isothermal process involving an ideal gas, $\Delta E_{\text{int}} = 0$. For an isothermal process, then, we conclude from the first law that the energy transfer Q must be equal to the negative of the work done on the gas—that is, $Q = -W$. Any energy that enters the system by heat is transferred out of the system by work; as a result, no change in the internal energy of the system occurs in an isothermal process.

▲ PITFALL PREVENTION

20.10 $Q \neq 0$ in an Isothermal Process

Do not fall into the common trap of thinking that there must be no transfer of energy by heat if the temperature does not change, as is the case in an isothermal process. Because the cause of temperature change can be either heat *or* work, the temperature can remain constant even if energy enters the gas by heat. This can only happen if the energy entering the gas by heat leaves by work.

Quick Quiz 20.5 In the last three columns of the following table, fill in the boxes with −, +, or 0. For each situation, the system to be considered is identified.

Situation	System	Q	W	ΔE_{int}
(a) Rapidly pumping up a bicycle tire	Air in the pump			
(b) Pan of room-temperature water sitting on a hot stove	Water in the pan			
(c) Air quickly leaking out of a balloon	Air originally in the balloon			

Isothermal Expansion of an Ideal Gas

Suppose that an ideal gas is allowed to expand quasi-statically at constant temperature. This process is described by the PV diagram shown in Figure 20.8. The curve is a hyperbola (see Appendix B, Eq. B.23), and the ideal gas law with T constant indicates that the equation of this curve is $PV =$ constant.

Let us calculate the work done on the gas in the expansion from state i to state f. The work done on the gas is given by Equation 20.8. Because the gas is ideal and the process is quasi-static, we can use the expression $PV = nRT$ for each point on the path. Therefore, we have

$$W = -\int_{V_i}^{V_f} P\,dV = -\int_{V_i}^{V_f} \frac{nRT}{V}\,dV$$

Because T is constant in this case, it can be removed from the integral along with n and R:

$$W = -nRT\int_{V_i}^{V_f} \frac{dV}{V} = -nRT \ln V \Big|_{V_i}^{V_f}$$

To evaluate the integral, we used $\int(dx/x) = \ln x$. Evaluating this at the initial and final volumes, we have

$$W = nRT \ln\left(\frac{V_i}{V_f}\right) \qquad (20.13)$$

Numerically, this work W equals the negative of the shaded area under the PV curve shown in Figure 20.8. Because the gas expands, $V_f > V_i$ and the value for the work done on the gas is negative, as we expect. If the gas is compressed, then $V_f < V_i$ and the work done on the gas is positive.

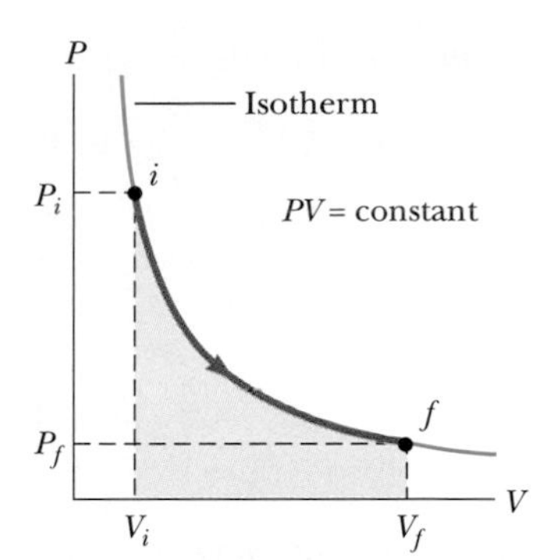

Figure 20.8 The PV diagram for an isothermal expansion of an ideal gas from an initial state to a final state. The curve is a hyperbola.

Quick Quiz 20.6 Characterize the paths in Figure 20.9 as isobaric, isovolumetric, isothermal, or adiabatic. Note that $Q = 0$ for path B.

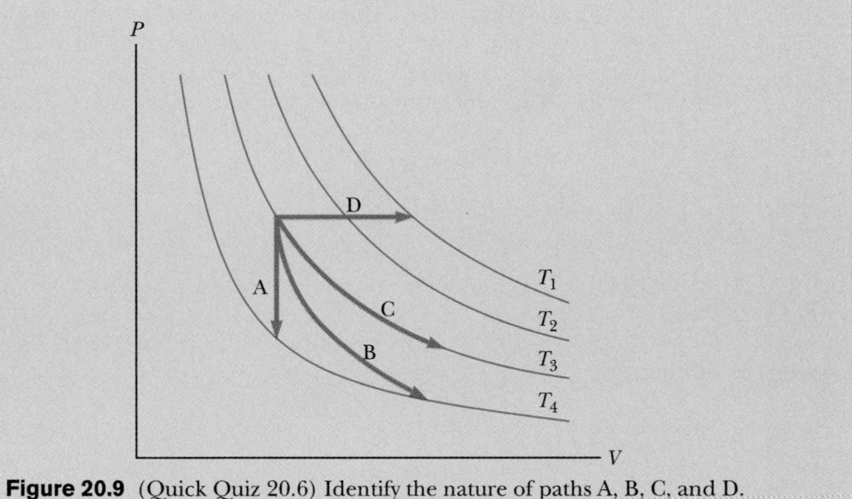

Figure 20.9 (Quick Quiz 20.6) Identify the nature of paths A, B, C, and D.

Example 20.6 An Isothermal Expansion

A 1.0-mol sample of an ideal gas is kept at 0.0°C during an expansion from 3.0 L to 10.0 L.

(A) How much work is done on the gas during the expansion?

Solution Substituting the values into Equation 20.13, we have

$$W = nRT \ln\left(\frac{V_i}{V_f}\right)$$

$$= (1.0 \text{ mol})(8.31 \text{ J/mol}\cdot\text{K})(273 \text{ K}) \ln\left(\frac{3.0 \text{ L}}{10.0 \text{ L}}\right)$$

$$= -2.7 \times 10^3 \text{ J}$$

(B) How much energy transfer by heat occurs with the surroundings in this process?

Solution From the first law, we find that

$$\Delta E_{\text{int}} = Q + W$$
$$0 = Q + W$$
$$Q = -W = \boxed{2.7 \times 10^3 \text{ J}}$$

(C) If the gas is returned to the original volume by means of an isobaric process, how much work is done on the gas?

Solution The work done in an isobaric process is given by Equation 20.11. In this case, the initial volume is 10.0 L and the final volume is 3.0 L, the reverse of the situation in part (A). We are not given the pressure, so we need to incorporate the ideal gas law:

$$W = -P(V_f - V_i) = -\frac{nRT_i}{V_i}(V_f - V_i)$$
$$= -\frac{(1.0 \text{ mol})(8.31 \text{ J/mol} \cdot \text{K})(273 \text{ K})}{10.0 \times 10^{-3} \text{ m}^3} \times (3.0 \times 10^{-3} \text{ m}^3 - 10.0 \times 10^{-3} \text{ m}^3)$$
$$= \boxed{1.6 \times 10^3 \text{ J}}$$

Notice that we use the initial temperature and volume to determine the value of the constant pressure because we do not know the final temperature. The work done on the gas is positive because the gas is being compressed.

Example 20.7 Boiling Water

Suppose 1.00 g of water vaporizes isobarically at atmospheric pressure (1.013×10^5 Pa). Its volume in the liquid state is $V_i = V_{\text{liquid}} = 1.00 \text{ cm}^3$, and its volume in the vapor state is $V_f = V_{\text{vapor}} = 1\,671 \text{ cm}^3$. Find the work done in the expansion and the change in internal energy of the system. Ignore any mixing of the steam and the surrounding air—imagine that the steam simply pushes the surrounding air out of the way.

Solution Because the expansion takes place at constant pressure, the work done on the system (the vaporizing water) as it pushes away the surrounding air is, from Equation 20.11,

$$W = -P(V_f - V_i)$$
$$= -(1.013 \times 10^5 \text{ Pa})(1\,671 \times 10^{-6} \text{ m}^3 - 1.00 \times 10^{-6} \text{ m}^3)$$
$$= \boxed{-169 \text{ J}}$$

To determine the change in internal energy, we must know the energy transfer Q needed to vaporize the water. Using Equation 20.6 and the latent heat of vaporization for water, we have

$$Q = mL_v = (1.00 \times 10^{-3} \text{ kg})(2.26 \times 10^6 \text{ J/kg}) = 2\,260 \text{ J}$$

Hence, from the first law, the change in internal energy is

$$\Delta E_{\text{int}} = Q + W = 2\,260 \text{ J} + (-169 \text{ J}) = \boxed{2.09 \text{ kJ}}$$

The positive value for ΔE_{int} indicates that the internal energy of the system increases. We see that most of the energy (2 090 J/2 260 J = 93%) transferred to the liquid goes into increasing the internal energy of the system. The remaining 7% of the energy transferred leaves the system by work done by the steam on the surrounding atmosphere.

Example 20.8 Heating a Solid

A 1.0-kg bar of copper is heated at atmospheric pressure. If its temperature increases from 20°C to 50°C,

(A) what is the work done on the copper bar by the surrounding atmosphere?

Solution Because the process is isobaric, we can find the work done on the copper bar using Equation 20.11, $W = -P(V_f - V_i)$. We can calculate the change in volume of the copper bar using Equation 19.6. Using the average linear expansion coefficient for copper given in Table 19.1, and remembering that $\beta = 3\alpha$, we obtain

$$\Delta V = \beta V_i \, \Delta T$$
$$= [5.1 \times 10^{-5} \, (°\text{C})^{-1}](50°\text{C} - 20°\text{C}) V_i = 1.5 \times 10^{-3} \, V_i$$

The volume V_i is equal to m/ρ, and Table 14.1 indicates that the density of copper is $8.92 \times 10^3 \text{ kg/m}^3$. Hence,

$$\Delta V = (1.5 \times 10^{-3})\left(\frac{1.0 \text{ kg}}{8.92 \times 10^3 \text{ kg/m}^3}\right)$$
$$= 1.7 \times 10^{-7} \text{ m}^3$$

The work done on the copper bar is

$$W = -P \, \Delta V = -(1.013 \times 10^5 \text{ N/m}^2)(1.7 \times 10^{-7} \text{ m}^3)$$
$$= \boxed{-1.7 \times 10^{-2} \text{ J}}$$

Because this work is negative, work is done *by* the copper bar on the atmosphere.

(B) What quantity of energy is transferred to the copper bar by heat?

Solution Taking the specific heat of copper from Table 20.1 and using Equation 20.4, we find that the energy transferred by heat is

$$Q = mc \, \Delta T = (1.0 \text{ kg})(387 \text{ J/kg} \cdot °\text{C})(30°\text{C})$$
$$= \boxed{1.2 \times 10^4 \text{ J}}$$

(C) What is the increase in internal energy of the copper bar?

Solution From the first law of thermodynamics, we have

$$\Delta E_{\text{int}} = Q + W = 1.2 \times 10^4 \text{ J} + (-1.7 \times 10^{-2} \text{ J})$$

$$= 1.2 \times 10^4 \text{ J}$$

Note that almost all of the energy transferred into the system by heat goes into increasing the internal energy of the copper bar. The fraction of energy used to do work on the surrounding atmosphere is only about 10^{-6}! Hence, when the thermal expansion of a solid or a liquid is analyzed, the small amount of work done on or by the system is usually ignored.

20.7 Energy Transfer Mechanisms

In Chapter 7, we introduced a global approach to energy analysis of physical processes through Equation 7.17, $\Delta E_{\text{system}} = \Sigma T$, where T represents energy transfer. Earlier in this chapter, we discussed two of the terms on the right-hand side of this equation, work and heat. In this section, we explore more details about heat as a means of energy transfer and consider two other energy transfer methods that are often related to temperature changes—convection (a form of matter transfer) and electromagnetic radiation.

Thermal Conduction

The process of energy transfer by heat can also be called **conduction** or **thermal conduction.** In this process, the transfer can be represented on an atomic scale as an exchange of kinetic energy between microscopic particles—molecules, atoms, and free electrons—in which less-energetic particles gain energy in collisions with more energetic particles. For example, if you hold one end of a long metal bar and insert the other end into a flame, you will find that the temperature of the metal in your hand soon increases. The energy reaches your hand by means of conduction. We can understand the process of conduction by examining what is happening to the microscopic particles in the metal. Initially, before the rod is inserted into the flame, the microscopic particles are vibrating about their equilibrium positions. As the flame heats the rod, the particles near the flame begin to vibrate with greater and greater amplitudes. These particles, in turn, collide with their neighbors and transfer some of their energy in the collisions. Slowly, the amplitudes of vibration of metal atoms and electrons farther and farther from the flame increase until, eventually, those in the metal near your hand are affected. This increased vibration is detected by an increase in the temperature of the metal and of your potentially burned hand.

Charles D. Winters

A pan of boiling water sits on a stove burner. Energy enters the water through the bottom of the pan by thermal conduction.

The rate of thermal conduction depends on the properties of the substance being heated. For example, it is possible to hold a piece of asbestos in a flame indefinitely. This implies that very little energy is conducted through the asbestos. In general, metals are good thermal conductors, and materials such as asbestos, cork, paper, and fiberglass are poor conductors. Gases also are poor conductors because the separation distance between the particles is so great. Metals are good thermal conductors because they contain large numbers of electrons that are relatively free to move through the metal and so can transport energy over large distances. Thus, in a good conductor, such as copper, conduction takes place by means of both the vibration of atoms and the motion of free electrons.

Conduction occurs only if there is a difference in temperature between two parts of the conducting medium. Consider a slab of material of thickness Δx and cross-sectional area A. One face of the slab is at a temperature T_c, and the other face is at a temperature $T_h > T_c$ (Fig. 20.10). Experimentally, it is found that the energy Q transfers in a time interval Δt from the hotter face to the colder one. The rate $\mathcal{P} = Q/\Delta t$ at which this energy transfer occurs is found to be proportional to the cross-sectional area and the temperature difference $\Delta T = T_h - T_c$, and inversely proportional to the thickness:

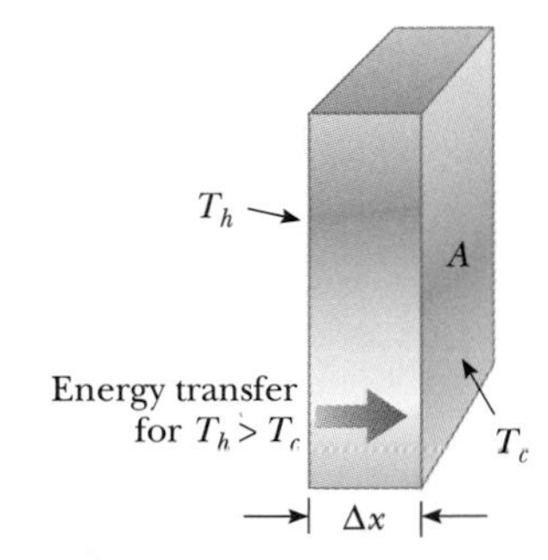

Figure 20.10 Energy transfer through a conducting slab with a cross-sectional area A and a thickness Δx. The opposite faces are at different temperatures T_c and T_h.

$$\mathcal{P} = \frac{Q}{\Delta t} \propto A \frac{\Delta T}{\Delta x}$$

Note that $\mathcal{P}$ has units of watts when Q is in joules and Δt is in seconds. This is not surprising because $\mathcal{P}$ is *power*—the rate of energy transfer by heat. For a slab of infinitesimal thickness dx and temperature difference dT, we can write the **law of thermal conduction** as

Law of thermal conduction

$$\mathcal{P} = kA\left|\frac{dT}{dx}\right| \quad (20.14)$$

where the proportionality constant k is the **thermal conductivity** of the material and $|dT/dx|$ is the **temperature gradient** (the rate at which temperature varies with position).

Suppose that a long, uniform rod of length L is thermally insulated so that energy cannot escape by heat from its surface except at the ends, as shown in Figure 20.11. One end is in thermal contact with an energy reservoir at temperature T_c, and the other end is in thermal contact with a reservoir at temperature $T_h > T_c$. When a steady state has been reached, the temperature at each point along the rod is constant in time. In this case if we assume that k is not a function of temperature, the temperature gradient is the same everywhere along the rod and is

$$\left|\frac{dT}{dx}\right| = \frac{T_h - T_c}{L}$$

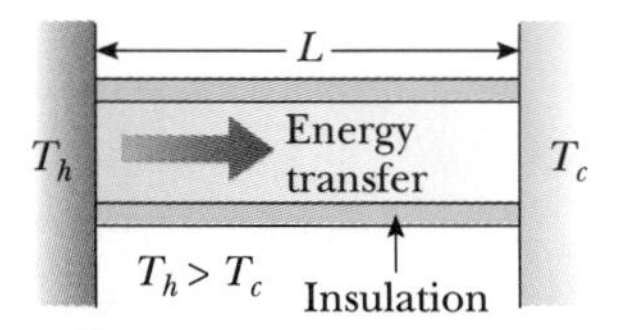

Figure 20.11 Conduction of energy through a uniform, insulated rod of length L. The opposite ends are in thermal contact with energy reservoirs at different temperatures.

Table 20.3

Thermal Conductivities

Substance	Thermal Conductivity (W/m · °C)
Metals (at 25°C)	
Aluminum	238
Copper	397
Gold	314
Iron	79.5
Lead	34.7
Silver	427
Nonmetals (approximate values)	
Asbestos	0.08
Concrete	0.8
Diamond	2 300
Glass	0.8
Ice	2
Rubber	0.2
Water	0.6
Wood	0.08
Gases (at 20°C)	
Air	0.023 4
Helium	0.138
Hydrogen	0.172
Nitrogen	0.023 4
Oxygen	0.023 8

Thus the rate of energy transfer by conduction through the rod is

$$\mathcal{P} = kA\left(\frac{T_h - T_c}{L}\right) \tag{20.15}$$

Substances that are good thermal conductors have large thermal conductivity values, whereas good thermal insulators have low thermal conductivity values. Table 20.3 lists thermal conductivities for various substances. Note that metals are generally better thermal conductors than nonmetals.

For a compound slab containing several materials of thicknesses L_1, L_2, . . . and thermal conductivities k_1, k_2, . . . , the rate of energy transfer through the slab at steady state is

$$\mathcal{P} = \frac{A(T_h - T_c)}{\sum_i (L_i/k_i)} \tag{20.16}$$

where T_c and T_h are the temperatures of the outer surfaces (which are held constant) and the summation is over all slabs. Example 20.9 shows how this equation results from a consideration of two thicknesses of materials.

Example 20.9 Energy Transfer Through Two Slabs

Two slabs of thickness L_1 and L_2 and thermal conductivities k_1 and k_2 are in thermal contact with each other, as shown in Figure 20.12. The temperatures of their outer surfaces are T_c and T_h, respectively, and $T_h > T_c$. Determine the temperature at the interface and the rate of energy transfer by conduction through the slabs in the steady-state condition.

Solution To conceptualize this problem, notice the phrase "in the steady-state condition." We interpret this to mean that energy transfers through the compound slab at the same rate at all points. Otherwise, energy would be building up or disappearing at some point. Furthermore, the temperature will vary with position in the two slabs, most likely at different rates in each part of the compound slab. Thus, there will be some fixed temperature T at the interface when the system is in steady state. We categorize this as a thermal conduction problem and impose the condition that the power is the same in both slabs of material. To analyze the problem, we use Equation 20.15 to express the rate at which energy is transferred through slab 1:

$$(1) \qquad \mathcal{P}_1 = k_1 A\left(\frac{T - T_c}{L_1}\right)$$

The rate at which energy is transferred through slab 2 is

$$(2) \qquad \mathcal{P}_2 = k_2 A\left(\frac{T_h - T}{L_2}\right)$$

When a steady state is reached, these two rates must be equal; hence,

$$k_1 A\left(\frac{T - T_c}{L_1}\right) = k_2 A\left(\frac{T_h - T}{L_2}\right)$$

Solving for T gives

$$(3) \qquad T = \frac{k_1 L_2 T_c + k_2 L_1 T_h}{k_1 L_2 + k_2 L_1}$$

Substituting Equation (3) into either Equation (1) or Equation (2), we obtain

$$(4) \qquad \mathcal{P} = \frac{A(T_h - T_c)}{(L_1/k_1) + (L_2/k_2)}$$

To finalize this problem, note that extension of this procedure to several slabs of materials leads to Equation 20.16.

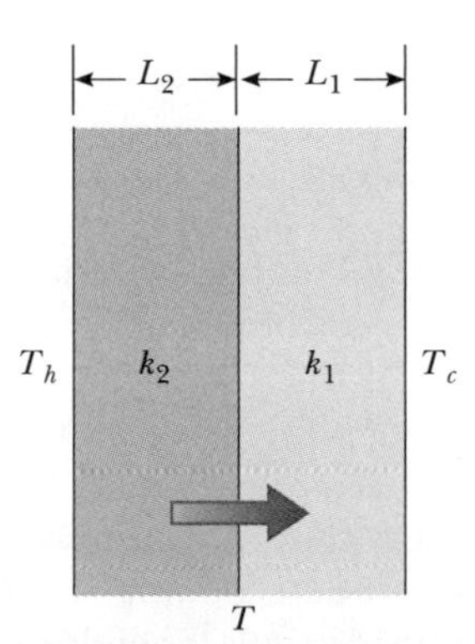

Figure 20.12 (Example 20.9) Energy transfer by conduction through two slabs in thermal contact with each other. At steady state, the rate of energy transfer through slab 1 equals the rate of energy transfer through slab 2.

What If? **Suppose you are building an insulated container with two layers of insulation and the rate of energy transfer determined by Equation (4) turns out to be too high. You have enough room to increase the thickness of one of the two layers by 20%. How would you decide which layer to choose?**

Answer To decrease the power as much as possible, you must increase the denominator in Equation (4) as much as possible. Whichever thickness you choose to increase, L_1 or L_2, you will increase the corresponding term L/k in the denominator by 20%. In order for this percentage change to represent the largest absolute change, you want to take 20% of the larger term. Thus, you should increase the thickness of the layer that has the larger value of L/k.

Quick Quiz 20.7 Will an ice cube wrapped in a wool blanket remain frozen for (a) a shorter length of time (b) the same length of time (c) a longer length of time than an identical ice cube exposed to air at room temperature?

Quick Quiz 20.8 You have two rods of the same length and diameter but they are formed from different materials. The rods will be used to connect two regions of different temperature such that energy will transfer through the rods by heat. They can be connected in series, as in Figure 20.13a, or in parallel, as in Figure 20.13b. In which case is the rate of energy transfer by heat larger? (a) when the rods are in series (b) when the rods are in parallel (c) The rate is the same in both cases.

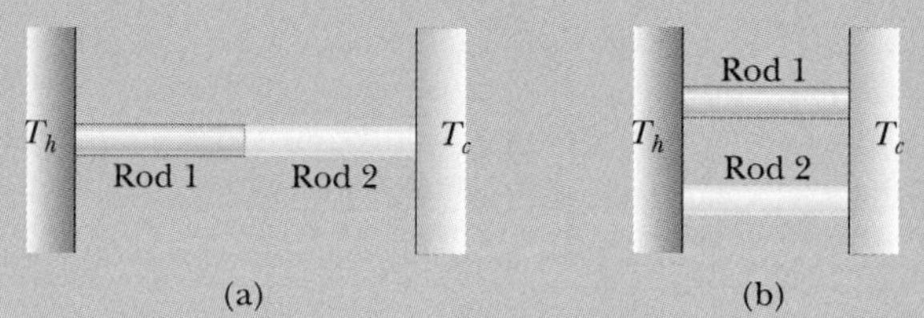

Figure 20.13 (Quick Quiz 20.8) In which case is the rate of energy transfer larger?

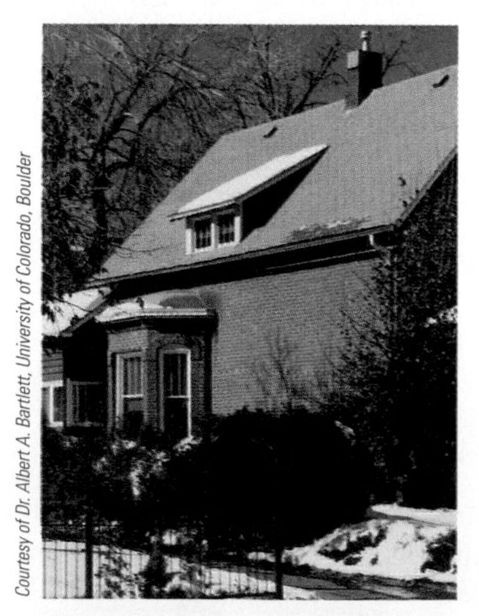

Courtesy of Dr. Albert A. Bartlett, University of Colorado, Boulder

Energy is conducted from the inside to the exterior more rapidly on the part of the roof where the snow has melted. The dormer appears to have been added and insulated. The main roof does not appear to be well insulated.

Home Insulation

In engineering practice, the term L/k for a particular substance is referred to as the ***R* value** of the material. Thus, Equation 20.16 reduces to

$$\mathcal{P} = \frac{A(T_h - T_c)}{\sum_i R_i} \tag{20.17}$$

where $R_i = L_i/k_i$. The R values for a few common building materials are given in Table 20.4. In the United States, the insulating properties of materials used in buildings are usually expressed in U.S. customary units, not SI units. Thus, in Table 20.4, measurements of R values are given as a combination of British thermal units, feet, hours, and degrees Fahrenheit.

At any vertical surface open to the air, a very thin stagnant layer of air adheres to the surface. One must consider this layer when determining the R value for a wall. The thickness of this stagnant layer on an outside wall depends on the speed of the wind. Energy loss from a house on a windy day is greater than the loss on a day when the air is calm. A representative R value for this stagnant layer of air is given in Table 20.4.

Table 20.4

***R* Values for Some Common Building Materials**

Material	*R* value ($ft^2 \cdot °F \cdot h/Btu$)
Hardwood siding (1 in. thick)	0.91
Wood shingles (lapped)	0.87
Brick (4 in. thick)	4.00
Concrete block (filled cores)	1.93
Fiberglass insulation (3.5 in. thick)	10.90
Fiberglass insulation (6 in. thick)	18.80
Fiberglass board (1 in. thick)	4.35
Cellulose fiber (1 in. thick)	3.70
Flat glass (0.125 in. thick)	0.89
Insulating glass (0.25-in. space)	1.54
Air space (3.5 in. thick)	1.01
Stagnant air layer	0.17
Drywall (0.5 in. thick)	0.45
Sheathing (0.5 in. thick)	1.32

Example 20.10 The *R* Value of a Typical Wall Interactive

Calculate the total R value for a wall constructed as shown in Figure 20.14a. Starting outside the house (toward the front in the figure) and moving inward, the wall consists of 4 in. of brick, 0.5 in. of sheathing, an air space 3.5 in. thick, and 0.5 in. of drywall. Do not forget the stagnant air layers inside and outside the house.

Solution Referring to Table 20.4, we find that

$$R_1 \text{ (outside stagnant air layer)} = 0.17 \text{ ft}^2 \cdot °\text{F} \cdot \text{h/Btu}$$
$$R_2 \text{ (brick)} = 4.00 \text{ ft}^2 \cdot °\text{F} \cdot \text{h/Btu}$$
$$R_3 \text{ (sheathing)} = 1.32 \text{ ft}^2 \cdot °\text{F} \cdot \text{h/Btu}$$
$$R_4 \text{ (air space)} = 1.01 \text{ ft}^2 \cdot °\text{F} \cdot \text{h/Btu}$$
$$R_5 \text{ (drywall)} = 0.45 \text{ ft}^2 \cdot °\text{F} \cdot \text{h/Btu}$$
$$R_6 \text{ (inside stagnant air layer)} = 0.17 \text{ ft}^2 \cdot °\text{F} \cdot \text{h/Btu}$$
$$R_{\text{total}} = 7.12 \text{ ft}^2 \cdot °\text{F} \cdot \text{h/Btu}$$

What If? **You are not happy with this total *R* value for the wall. You cannot change the overall structure, but you can fill the air space as in Figure 20.14b. What material should you choose to fill the air space in order to *maximize* the total *R* value?**

Answer Looking at Table 20.4, we see that 3.5 in. of fiberglass insulation is over ten times as effective at insulating the wall as 3.5 in. of air. Thus, we could fill the air space with fiberglass insulation. The result is that we add 10.90 $\text{ft}^2 \cdot °\text{F} \cdot \text{h/Btu}$ of R value and we lose 1.01 $\text{ft}^2 \cdot °\text{F} \cdot \text{h/Btu}$ due to the air space we have replaced, for a total change of 10.90 $\text{ft}^2 \cdot °\text{F} \cdot \text{h/Btu} - 1.01 \text{ ft}^2 \cdot °\text{F} \cdot \text{h/Btu} = 9.89 \text{ ft}^2 \cdot °\text{F} \cdot \text{h/Btu}$. The new total R value is $7.12 \text{ ft}^2 \cdot °\text{F} \cdot \text{h/Btu} + 9.89 \text{ ft}^2 \cdot °\text{F} \cdot \text{h/Btu} = 17.01 \text{ ft}^2 \cdot °\text{F} \cdot \text{h/Btu}$.

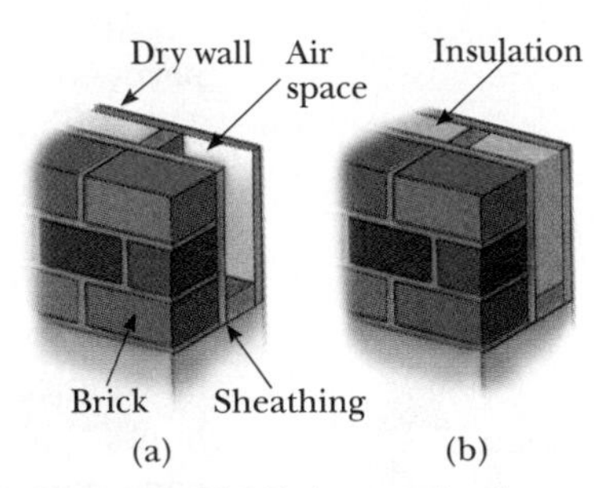

Figure 20.14 (Example 20.10) An exterior house wall containing (a) an air space and (b) insulation.

Study the R values of various types of common building materials at the Interactive Worked Example link at http://www.pse6.com.

Convection

At one time or another, you probably have warmed your hands by holding them over an open flame. In this situation, the air directly above the flame is heated and expands. As a result, the density of this air decreases and the air rises. This hot air warms your

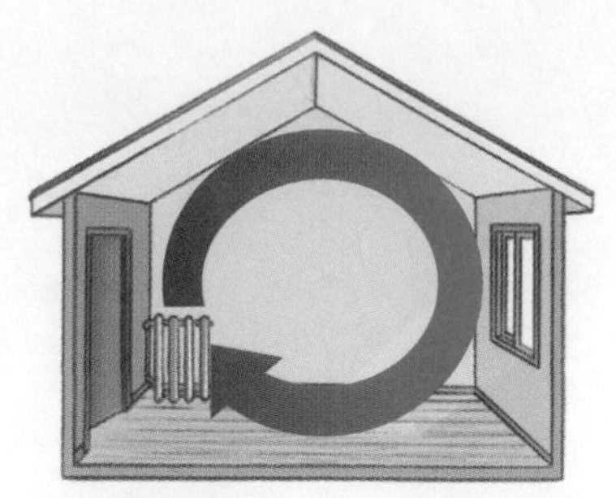

Figure 20.15 Convection currents are set up in a room warmed by a radiator.

hands as it flows by. **Energy transferred by the movement of a warm substance is said to have been transferred by convection.** When the movement results from differences in density, as with air around a fire, it is referred to as *natural convection.* Air flow at a beach is an example of natural convection, as is the mixing that occurs as surface water in a lake cools and sinks (see Section 19.4). When the heated substance is forced to move by a fan or pump, as in some hot-air and hot-water heating systems, the process is called *forced convection.*

If it were not for convection currents, it would be very difficult to boil water. As water is heated in a teakettle, the lower layers are warmed first. This water expands and rises to the top because its density is lowered. At the same time, the denser, cool water at the surface sinks to the bottom of the kettle and is heated.

The same process occurs when a room is heated by a radiator. The hot radiator warms the air in the lower regions of the room. The warm air expands and rises to the ceiling because of its lower density. The denser, cooler air from above sinks, and the continuous air current pattern shown in Figure 20.15 is established.

Radiation

The third means of energy transfer that we shall discuss is **radiation**. All objects radiate energy continuously in the form of electromagnetic waves (see Chapter 34) produced by thermal vibrations of the molecules. You are likely familiar with electromagnetic radiation in the form of the orange glow from an electric stove burner, an electric space heater, or the coils of a toaster.

The rate at which an object radiates energy is proportional to the fourth power of its absolute temperature. This is known as **Stefan's law** and is expressed in equation form as

Stefan's law

$$\mathcal{P} = \sigma A e T^4 \tag{20.18}$$

where $\mathcal{P}$ is the power in watts radiated from the surface of the object, σ is a constant equal to $5.669\,6 \times 10^{-8}\ \mathrm{W/m^2 \cdot K^4}$, A is the surface area of the object in square meters, e is the **emissivity**, and T is the surface temperature in kelvins. The value of e can vary between zero and unity, depending on the properties of the surface of the object. The emissivity is equal to the **absorptivity**, which is the fraction of the incoming radiation that the surface absorbs.

Approximately 1 340 J of electromagnetic radiation from the Sun passes perpendicularly through each 1 $\mathrm{m^2}$ at the top of the Earth's atmosphere every second. This radiation is primarily visible and infrared light accompanied by a significant amount of ultraviolet radiation. We shall study these types of radiation in detail in Chapter 34. Some of this energy is reflected back into space, and some is absorbed by the atmosphere. However, enough energy arrives at the surface of the Earth each day to supply all our energy needs on this planet hundreds of times over—if only it could be captured and used efficiently. The growth in the number of solar energy–powered houses built in this country reflects the increasing efforts being made to use this abundant energy. Radiant energy from the Sun affects our day-to-day existence in a number of ways. For example, it influences the Earth's average temperature, ocean currents, agriculture, and rain patterns.

What happens to the atmospheric temperature at night is another example of the effects of energy transfer by radiation. If there is a cloud cover above the Earth, the water vapor in the clouds absorbs part of the infrared radiation emitted by the Earth and re-emits it back to the surface. Consequently, temperature levels at the surface remain moderate. In the absence of this cloud cover, there is less in the way to prevent this radiation from escaping into space; thus the temperature decreases more on a clear night than on a cloudy one.

As an object radiates energy at a rate given by Equation 20.18, it also absorbs electromagnetic radiation. If the latter process did not occur, an object would eventually radiate all its energy, and its temperature would reach absolute zero. The energy an object absorbs comes from its surroundings, which consist of other objects that radiate energy. If an object is at a temperature T and its surroundings are at an average temperature T_0, then the net rate of energy gained or lost by the object as a result of radiation is

$$\mathcal{P}_{\text{net}} = \sigma A e(T^4 - T_0^{\,4}) \tag{20.19}$$

When an object is in equilibrium with its surroundings, it radiates and absorbs energy at the same rate, and its temperature remains constant. When an object is hotter than its surroundings, it radiates more energy than it absorbs, and its temperature decreases.

An **ideal absorber** is defined as an object that absorbs all the energy incident on it, and for such an object, $e = 1$. An object for which $e = 1$ is often referred to as a **black body.** We shall investigate experimental and theoretical approaches to radiation from a black body in Chapter 40. An ideal absorber is also an ideal radiator of energy. In contrast, an object for which $e = 0$ absorbs none of the energy incident on it. Such an object reflects all the incident energy, and thus is an **ideal reflector.**

The Dewar Flask

The *Dewar flask*[7] is a container designed to minimize energy losses by conduction, convection, and radiation. Such a container is used to store either cold or hot liquids for long periods of time. (A Thermos bottle is a common household equivalent of a Dewar flask.) The standard construction (Fig. 20.16) consists of a double-walled Pyrex glass vessel with silvered walls. The space between the walls is evacuated to minimize energy transfer by conduction and convection. The silvered surfaces minimize energy transfer by radiation because silver is a very good reflector and has very low emissivity. A further reduction in energy loss is obtained by reducing the size of the neck. Dewar flasks are commonly used to store liquid nitrogen (boiling point: 77 K) and liquid oxygen (boiling point: 90 K).

To confine liquid helium (boiling point: 4.2 K), which has a very low heat of vaporization, it is often necessary to use a double Dewar system, in which the Dewar flask containing the liquid is surrounded by a second Dewar flask. The space between the two flasks is filled with liquid nitrogen.

Newer designs of storage containers use "super insulation" that consists of many layers of reflecting material separated by fiberglass. All of this is in a vacuum, and no liquid nitrogen is needed with this design.

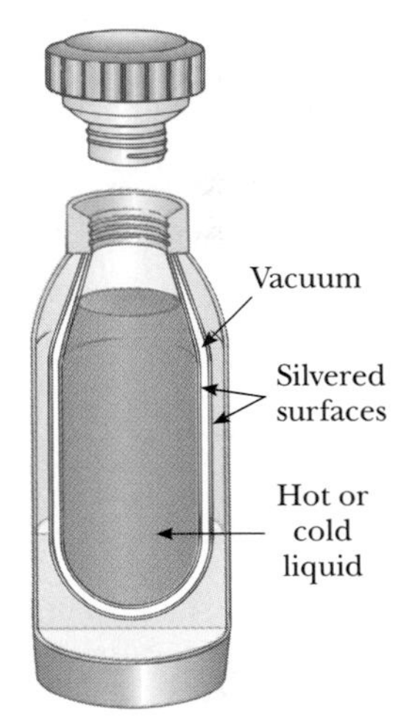

Figure 20.16 A cross-sectional view of a Dewar flask, which is used to store hot or cold substances.

Example 20.11 Who Turned Down the Thermostat?

A student is trying to decide what to wear. The surroundings (his bedroom) are at 20.0°C. If the skin temperature of the unclothed student is 35°C, what is the net energy loss from his body in 10.0 min by radiation? Assume that the emissivity of skin is 0.900 and that the surface area of the student is 1.50 m^2.

Solution Using Equation 20.19, we find that the net rate of energy loss from the skin is

$$\mathcal{P}_{\text{net}} = \sigma A e(T^4 - T_0^{\,4})$$

$$= (5.67 \times 10^{-8}\ \text{W/m}^2\cdot\text{K}^4)(1.50\ \text{m}^2) \times (0.900)[(308\ \text{K})^4 - (293\ \text{K})^4] = 125\ \text{W}$$

At this rate, the total energy lost by the skin in 10 min is

$$Q = \mathcal{P}_{\text{net}}\,\Delta t = (125\ \text{W})(600\ \text{s}) = 7.5 \times 10^4\ \text{J}$$

Note that the energy radiated by the student is roughly equivalent to that produced by two 60-W light bulbs!

[7] Invented by Sir James Dewar (1842–1923).

SUMMARY

Take a practice test for this chapter by clicking on the Practice Test link at http://www.pse6.com.

Internal energy is all of a system's energy that is associated with the system's microscopic components. Internal energy includes kinetic energy of random translation, rotation, and vibration of molecules, potential energy within molecules, and potential energy between molecules.

Heat is the transfer of energy across the boundary of a system resulting from a temperature difference between the system and its surroundings. We use the symbol Q for the amount of energy transferred by this process.

The **calorie** is the amount of energy necessary to raise the temperature of 1 g of water from 14.5°C to 15.5°C. The **mechanical equivalent of heat** is 1 cal = 4.186 J.

The **heat capacity** C of any sample is the amount of energy needed to raise the temperature of the sample by 1°C. The energy Q required to change the temperature of a mass m of a substance by an amount ΔT is

$$Q = mc\,\Delta T \qquad (20.4)$$

where c is the **specific heat** of the substance.

The energy required to change the phase of a pure substance of mass m is

$$Q = \pm\, mL \qquad (20.6)$$

where L is the **latent heat** of the substance and depends on the nature of the phase change and the properties of the substance. The positive sign is used if energy is entering the system, and the negative sign is used if energy is leaving.

The **work done** on a gas as its volume changes from some initial value V_i to some final value V_f is

$$W = -\int_{V_i}^{V_f} P\,dV \qquad (20.8)$$

where P is the pressure, which may vary during the process. In order to evaluate W, the process must be fully specified—that is, P and V must be known during each step. In other words, the work done depends on the path taken between the initial and final states.

The **first law of thermodynamics** states that when a system undergoes a change from one state to another, the change in its internal energy is

$$\Delta E_{\text{int}} = Q + W \qquad (20.9)$$

where Q is the energy transferred into the system by heat and W is the work done on the system. Although Q and W both depend on the path taken from the initial state to the final state, the quantity ΔE_{int} is path-independent.

In a **cyclic process** (one that originates and terminates at the same state), $\Delta E_{\text{int}} = 0$ and, therefore, $Q = -W$. That is, the energy transferred into the system by heat equals the negative of the work done on the system during the process.

In an **adiabatic process**, no energy is transferred by heat between the system and its surroundings ($Q = 0$). In this case, the first law gives $\Delta E_{\text{int}} = W$. That is, the internal energy changes as a consequence of work being done on the system. In the **adiabatic free expansion** of a gas $Q = 0$ and $W = 0$, and so $\Delta E_{\text{int}} = 0$. That is, the internal energy of the gas does not change in such a process.

An **isobaric process** is one that occurs at constant pressure. The work done on a gas in such a process is $W = -P(V_f - V_i)$.

An **isovolumetric process** is one that occurs at constant volume. No work is done in such a process, so $\Delta E_{\text{int}} = Q$.

An **isothermal process** is one that occurs at constant temperature. The work done on an ideal gas during an isothermal process is

$$W = nRT\ln\left(\frac{V_i}{V_f}\right) \qquad (20.13)$$

Energy may be transferred by work, which we addressed in Chapter 7, and by conduction, convection, or radiation. **Conduction** can be viewed as an exchange of kinetic energy between colliding molecules or electrons. The rate of energy transfer by conduction through a slab of area A is

$$\mathcal{P} = kA\left|\frac{dT}{dx}\right| \tag{20.14}$$

where k is the **thermal conductivity** of the material from which the slab is made and $|dT/dx|$ is the **temperature gradient.** This equation can be used in many situations in which the rate of transfer of energy through materials is important.

In **convection,** a warm substance transfers energy from one location to another.

All objects emit **radiation** in the form of electromagnetic waves at the rate

$$\mathcal{P} = \sigma A e T^4 \tag{20.18}$$

An object that is hotter than its surroundings radiates more energy than it absorbs, whereas an object that is cooler than its surroundings absorbs more energy than it radiates.

QUESTIONS

1. Clearly distinguish among temperature, heat, and internal energy.

2. Ethyl alcohol has about half the specific heat of water. If equal-mass samples of alcohol and water in separate beakers are supplied with the same amount of energy, compare the temperature increases of the two liquids.

3. A small metal crucible is taken from a 200°C oven and immersed in a tub full of water at room temperature (this process is often referred to as *quenching*). What is the approximate final equilibrium temperature?

4. What is a major problem that arises in measuring specific heats if a sample with a temperature above 100°C is placed in water?

5. In a daring lecture demonstration, an instructor dips his wetted fingers into molten lead (327°C) and withdraws them quickly, without getting burned. How is this possible? (This is a dangerous experiment, which you should *NOT* attempt.)

6. What is wrong with the following statement? "Given any two objects, the one with the higher temperature contains more heat."

7. Why is a person able to remove a piece of dry aluminum foil from a hot oven with bare fingers, while a burn results if there is moisture on the foil?

8. The air temperature above coastal areas is profoundly influenced by the large specific heat of water. One reason is that the energy released when 1 m^3 of water cools by 1°C will raise the temperature of a much larger volume of air by 1°C. Find this volume of air. The specific heat of air is approximately 1 kJ/kg · °C. Take the density of air to be 1.3 kg/m^3.

9. Concrete has a higher specific heat than soil. Use this fact to explain (partially) why cities have a higher average nighttime temperature than the surrounding countryside. If a city is hotter than the surrounding countryside, would you expect breezes to blow from city to country or from country to city? Explain.

10. Using the first law of thermodynamics, explain why the *total* energy of an isolated system is always constant.

11. When a sealed Thermos bottle full of hot coffee is shaken, what are the changes, if any, in (a) the temperature of the coffee (b) the internal energy of the coffee?

12. Is it possible to convert internal energy to mechanical energy? Explain with examples.

13. The U.S. penny was formerly made mostly of copper and is now made of copper-coated zinc. Can a calorimetric experiment be devised to test for the metal content in a collection of pennies? If so, describe the procedure you would use.

14. Figure Q20.14 shows a pattern formed by snow on the roof of a barn. What causes the alternating pattern of snow-covered and exposed roof?

Courtesy of Dr. Albert A. Bartlett, University of Colorado, Boulder, CO

Figure Q20.14 Alternating patterns on a snow-covered roof.

15. A tile floor in a bathroom may feel uncomfortably cold to your bare feet, but a carpeted floor in an adjoining room at the same temperature will feel warm. Why?

16. Why can potatoes be baked more quickly when a metal skewer has been inserted through them?

17. A piece of paper is wrapped around a rod made half of wood and half of copper. When held over a flame, the paper in contact with the wood burns but the half in contact with the metal does not. Explain.

18. Why do heavy draperies over the windows help keep a home cool in the summer, as well as warm in the winter?

19. If you wish to cook a piece of meat thoroughly on an open fire, why should you not use a high flame? (Note that carbon is a good thermal insulator.)

20. In an experimental house, Styrofoam beads were pumped into the air space between the panes of glass in double windows at night in the winter, and pumped out to holding bins during the day. How would this assist in conserving energy in the house?

21. Pioneers stored fruits and vegetables in underground cellars. Discuss the advantages of this choice for a storage site.

22. The pioneers referred to in the last question found that a large tub of water placed in a storage cellar would prevent their food from freezing on really cold nights. Explain why this is so.

23. When camping in a canyon on a still night, one notices that as soon as the sun strikes the surrounding peaks, a breeze begins to stir. What causes the breeze?

24. Updrafts of air are familiar to all pilots and are used to keep nonmotorized gliders aloft. What causes these currents?

25. If water is a poor thermal conductor, why can its temperature be raised quickly when it is placed over a flame?

26. Why is it more comfortable to hold a cup of hot tea by the handle rather than by wrapping your hands around the cup itself?

27. If you hold water in a paper cup over a flame, you can bring the water to a boil without burning the cup. How is this possible?

28. You need to pick up a very hot cooking pot in your kitchen. You have a pair of hot pads. Should you soak them in cold water or keep them dry, to be able to pick up the pot most comfortably?

29. Suppose you pour hot coffee for your guests, and one of them wants to drink it with cream, several minutes later, and then as warm as possible. In order to have the warmest coffee, should the person add the cream just after the coffee is poured or just before drinking? Explain.

30. Two identical cups both at room temperature are filled with the same amount of hot coffee. One cup contains a metal spoon, while the other does not. If you wait for several minutes, which of the two will have the warmer coffee? Which energy transfer process explains your answer?

31. A warning sign often seen on highways just before a bridge is "Caution—Bridge surface freezes before road surface." Which of the three energy transfer processes discussed in Section 20.7 is most important in causing a bridge surface to freeze before a road surface on very cold days?

32. A professional physics teacher drops one marshmallow into a flask of liquid nitrogen, waits for the most energetic boiling to stop, fishes it out with tongs, shakes it off, pops it into his mouth, chews it up, and swallows it. Clouds of ice crystals issue from his mouth as he crunches noisily and comments on the sweet taste. How can he do this without injury? *Caution:* Liquid nitrogen can be a dangerous substance and you should *not* try this yourself. The teacher might be badly injured if he did not shake it off, if he touched the tongs to a tooth, or if he did not start with a mouthful of saliva.

33. In 1801 Humphry Davy rubbed together pieces of ice inside an ice-house. He took care that nothing in their environment was at a higher temperature than the rubbed pieces. He observed the production of drops of liquid water. Make a table listing this and other experiments or processes, to illustrate each of the following. (a) A system can absorb energy by heat, increase in internal energy, and increase in temperature. (b) A system can absorb energy by heat and increase in internal energy, without an increase in temperature. (c) A system can absorb energy by heat without increasing in temperature or in internal energy. (d) A system can increase in internal energy and in temperature, without absorbing energy by heat. (e) A system can increase in internal energy without absorbing energy by heat or increasing in temperature. (f) **What If?** If a system's temperature increases, is it necessarily true that its internal energy increases?

34. Consider the opening photograph for Part 3 on page 578. Discuss the roles of conduction, convection, and radiation in the operation of the cooling fins on the support posts of the Alaskan oil pipeline.

PROBLEMS

1, 2, 3 = straightforward, intermediate, challenging □ = full solution available in the *Student Solutions Manual and Study Guide*

= coached solution with hints available at http://www.pse6.com = computer useful in solving problem

= paired numerical and symbolic problems

Section 20.1 Heat and Internal Energy

1. On his honeymoon James Joule traveled from England to Switzerland. He attempted to verify his idea of the interconvertibility of mechanical energy and internal energy by measuring the increase in temperature of water that fell in a waterfall. If water at the top of an alpine waterfall has a temperature of 10.0°C and then falls 50.0 m (as at Niagara Falls), what maximum temperature at the bottom of the falls could Joule expect? He did not succeed in measuring the temperature change, partly because evaporation cooled the falling water, and also because his thermometer was not sufficiently sensitive.

2. Consider Joule's apparatus described in Figure 20.1. The mass of each of the two blocks is 1.50 kg, and the insulated tank is filled with 200 g of water. What is the increase in the temperature of the water after the blocks fall through a distance of 3.00 m?

Section 20.2 Specific Heat and Calorimetry

3. The temperature of a silver bar rises by 10.0°C when it absorbs 1.23 kJ of energy by heat. The mass of the bar is 525 g. Determine the specific heat of silver.

4. A 50.0-g sample of copper is at 25.0°C. If 1 200 J of energy is added to it by heat, what is the final temperature of the copper?

5. Systematic use of solar energy can yield a large saving in the cost of winter space heating for a typical house in the north central United States. If the house has good insulation, you may model it as losing energy by heat steadily at the rate 6 000 W on a day in April when the average exterior temperature is 4°C, and when the conventional heating system is not used at all. The passive solar energy collector can consist simply of very large windows in a room facing south. Sunlight shining in during the daytime is absorbed by the floor, interior walls, and objects in the room, raising their temperature to 38°C. As the sun goes down, insulating draperies or shutters are closed over the windows. During the period between 5:00 P.M. and 7:00 A.M. the temperature of the house will drop, and a sufficiently large "thermal mass" is required to keep it from dropping too far. The thermal mass can be a large quantity of stone (with specific heat 850 J/kg · °C) in the floor and the interior walls exposed to sunlight. What mass of stone is required if the temperature is not to drop below 18°C overnight?

6. The *Nova* laser at Lawrence Livermore National Laboratory in California is used in studies of initiating controlled nuclear fusion (Section 45.4). It can deliver a power of 1.60×10^{13} W over a time interval of 2.50 ns. Compare its energy output in one such time interval to the energy required to make a pot of tea by warming 0.800 kg of water from 20.0°C to 100°C.

7. A 1.50-kg iron horseshoe initially at 600°C is dropped into a bucket containing 20.0 kg of water at 25.0°C. What is the final temperature? (Ignore the heat capacity of the container, and assume that a negligible amount of water boils away.)

8. An aluminum cup of mass 200 g contains 800 g of water in thermal equilibrium at 80.0°C. The combination of cup and water is cooled uniformly so that the temperature decreases by 1.50°C per minute. At what rate is energy being removed by heat? Express your answer in watts.

9. An aluminum calorimeter with a mass of 100 g contains 250 g of water. The calorimeter and water are in thermal equilibrium at 10.0°C. Two metallic blocks are placed into the water. One is a 50.0-g piece of copper at 80.0°C. The other block has a mass of 70.0 g and is originally at a temperature of 100°C. The entire system stabilizes at a final temperature of 20.0°C. (a) Determine the specific heat of the unknown sample. (b) Guess the material of the unknown, using the data in Table 20.1.

10. A 3.00-g copper penny at 25.0°C drops 50.0 m to the ground. (a) Assuming that 60.0% of the change in potential energy of the penny–Earth system goes into increasing the internal energy of the penny, determine its final temperature. (b) **What If?** Does the result depend on the mass of the penny? Explain.

11. A combination of 0.250 kg of water at 20.0°C, 0.400 kg of aluminum at 26.0°C, and 0.100 kg of copper at 100°C is mixed in an insulated container and allowed to come to thermal equilibrium. Ignore any energy transfer to or from the container and determine the final temperature of the mixture.

12. If water with a mass m_h at temperature T_h is poured into an aluminum cup of mass m_{Al} containing mass m_c of water at T_c, where $T_h > T_c$, what is the equilibrium temperature of the system?

13. A water heater is operated by solar power. If the solar collector has an area of 6.00 m^2 and the intensity delivered by sunlight is 550 W/m^2, how long does it take to increase the temperature of 1.00 m^3 of water from 20.0°C to 60.0°C?

14. Two thermally insulated vessels are connected by a narrow tube fitted with a valve that is initially closed. One vessel, of volume 16.8 L, contains oxygen at a temperature of 300 K and a pressure of 1.75 atm. The other vessel, of volume 22.4 L, contains oxygen at a temperature of 450 K and a pressure of 2.25 atm. When the valve is opened, the gases in the two vessels mix, and the temperature and pressure become uniform throughout. (a) What is the final temperature? (b) What is the final pressure?

Section 20.3 Latent Heat

15. How much energy is required to change a 40.0-g ice cube from ice at −10.0°C to steam at 110°C?

16. A 50.0-g copper calorimeter contains 250 g of water at 20.0°C. How much steam must be condensed into the water if the final temperature of the system is to reach 50.0°C?

17. A 3.00-g lead bullet at 30.0°C is fired at a speed of 240 m/s into a large block of ice at 0°C, in which it becomes embedded. What quantity of ice melts?

18. Steam at 100°C is added to ice at 0°C. (a) Find the amount of ice melted and the final temperature when the mass of steam is 10.0 g and the mass of ice is 50.0 g. (b) **What If?** Repeat when the mass of steam is 1.00 g and the mass of ice is 50.0 g.

19. A 1.00-kg block of copper at 20.0°C is dropped into a large vessel of liquid nitrogen at 77.3 K. How many kilograms of nitrogen boil away by the time the copper reaches 77.3 K? (The specific heat of copper is 0.092 0 cal/g · °C. The latent heat of vaporization of nitrogen is 48.0 cal/g.)

20. Assume that a hailstone at 0°C falls through air at a uniform temperature of 0°C and lands on a sidewalk also at this temperature. From what initial height must the hailstone fall in order to entirely melt on impact?

21. In an insulated vessel, 250 g of ice at 0°C is added to 600 g of water at 18.0°C. (a) What is the final temperature

of the system? (b) How much ice remains when the system reaches equilibrium?

22. **Review problem.** Two speeding lead bullets, each of mass 5.00 g, and at temperature 20.0°C, collide head-on at speeds of 500 m/s each. Assuming a perfectly inelastic collision and no loss of energy by heat to the atmosphere, describe the final state of the two-bullet system.

Section 20.4 Work and Heat in Thermodynamic Processes

23. A sample of ideal gas is expanded to twice its original volume of 1.00 m^3 in a quasi-static process for which $P = \alpha V^2$, with $\alpha = 5.00\ \text{atm/m}^6$, as shown in Figure P20.23. How much work is done on the expanding gas?

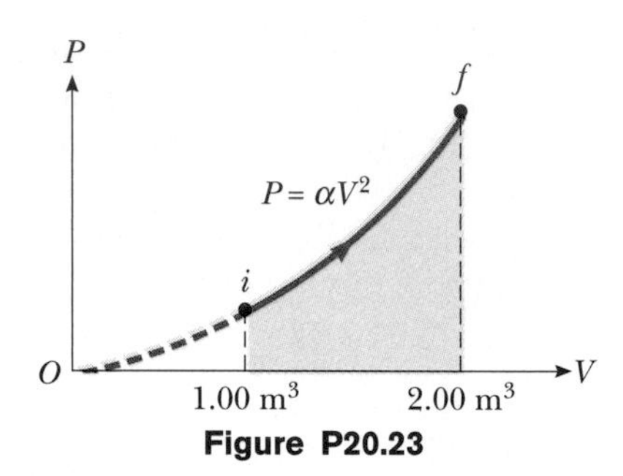

Figure P20.23

24. (a) Determine the work done on a fluid that expands from i to f as indicated in Figure P20.24. (b) **What If?** How much work is performed on the fluid if it is compressed from f to i along the same path?

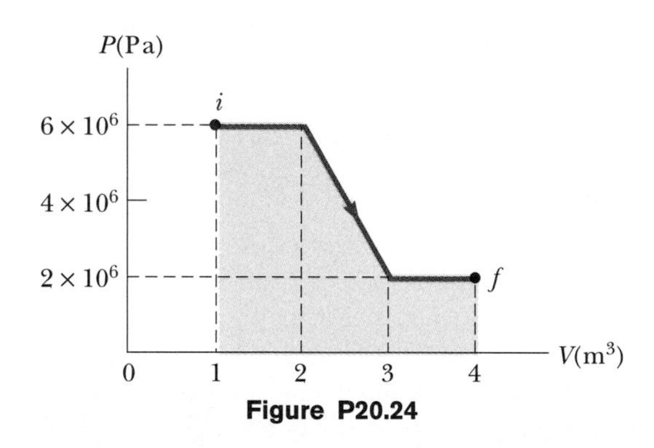

Figure P20.24

25. An ideal gas is enclosed in a cylinder with a movable piston on top of it. The piston has a mass of 8 000 g and an area of 5.00 cm^2 and is free to slide up and down, keeping the pressure of the gas constant. How much work is done on the gas as the temperature of 0.200 mol of the gas is raised from 20.0°C to 300°C?

26. An ideal gas is enclosed in a cylinder that has a movable piston on top. The piston has a mass m and an area A and is free to slide up and down, keeping the pressure of the gas constant. How much work is done on the gas as the temperature of n mol of the gas is raised from T_1 to T_2?

27. One mole of an ideal gas is heated slowly so that it goes from the PV state (P_i, V_i) to $(3P_i, 3V_i)$ in such a way that the pressure is directly proportional to the volume. (a) How much work is done on the gas in the process? (b) How is the temperature of the gas related to its volume during this process?

Section 20.5 The First Law of Thermodynamics

28. A gas is compressed at a constant pressure of 0.800 atm from 9.00 L to 2.00 L. In the process, 400 J of energy leaves the gas by heat. (a) What is the work done on the gas? (b) What is the change in its internal energy?

29. A thermodynamic system undergoes a process in which its internal energy decreases by 500 J. At the same time, 220 J of work is done on the system. Find the energy transferred to or from it by heat.

30. A gas is taken through the cyclic process described in Figure P20.30. (a) Find the net energy transferred to the system by heat during one complete cycle. (b) **What If?** If the cycle is reversed—that is, the process follows the path $ACBA$—what is the net energy input per cycle by heat?

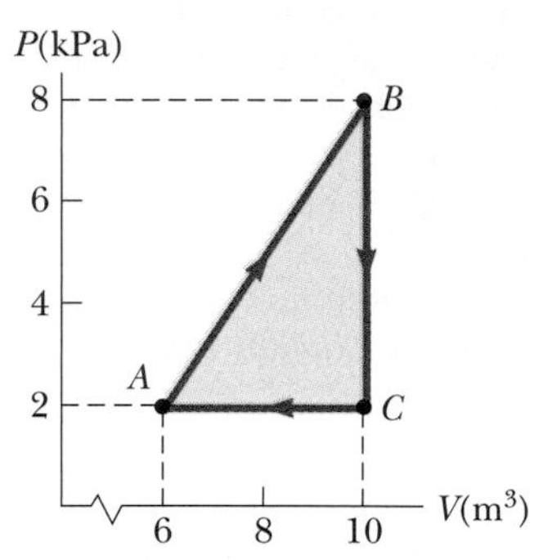

Figure P20.30 Problems 30 and 31.

31. Consider the cyclic process depicted in Figure P20.30. If Q is negative for the process BC and ΔE_{int} is negative for the process CA, what are the signs of Q, W, and ΔE_{int} that are associated with each process?

32. A sample of an ideal gas goes through the process shown in Figure P20.32. From A to B, the process is adiabatic; from B to C, it is isobaric with 100 kJ of energy entering the system by heat. From C to D, the process is isothermal; from D to A, it is isobaric with 150 kJ of energy leaving the system by heat. Determine the difference in internal energy $E_{\text{int},B} - E_{\text{int},A}$.

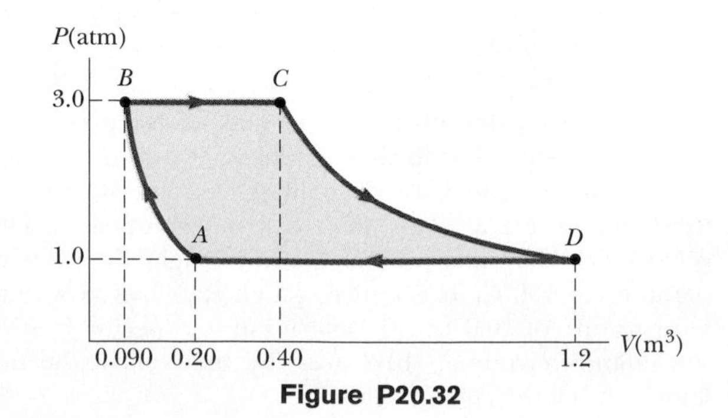

Figure P20.32

33. A sample of an ideal gas is in a vertical cylinder fitted with a piston. As 5.79 kJ of energy is transferred to the gas by heat to raise its temperature, the weight on the piston is adjusted so that the state of the gas changes from point A to point B along the semicircle shown in Figure P20.33. Find the change in internal energy of the gas.

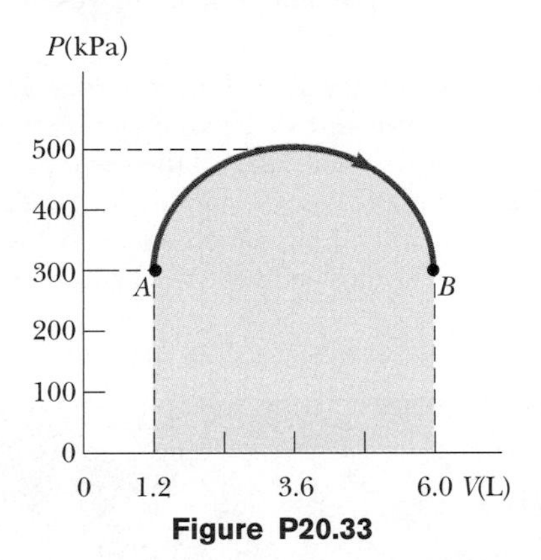

Figure P20.33

Section 20.6 Some Applications of the First Law of Thermodynamics

34. One mole of an ideal gas does 3 000 J of work on its surroundings as it expands isothermally to a final pressure of 1.00 atm and volume of 25.0 L. Determine (a) the initial volume and (b) the temperature of the gas.

35. An ideal gas initially at 300 K undergoes an isobaric expansion at 2.50 kPa. If the volume increases from 1.00 m^3 to 3.00 m^3 and 12.5 kJ is transferred to the gas by heat, what are (a) the change in its internal energy and (b) its final temperature?

36. A 1.00-kg block of aluminum is heated at atmospheric pressure so that its temperature increases from 22.0°C to 40.0°C. Find (a) the work done on the aluminum, (b) the energy added to it by heat, and (c) the change in its internal energy.

37. How much work is done on the steam when 1.00 mol of water at 100°C boils and becomes 1.00 mol of steam at 100°C at 1.00 atm pressure? Assuming the steam to behave as an ideal gas, determine the change in internal energy of the material as it vaporizes.

38. An ideal gas initially at P_i, V_i, and T_i is taken through a cycle as in Figure P20.38. (a) Find the net work done on the gas per cycle. (b) What is the net energy added by heat to the system per cycle? (c) Obtain a numerical value for the net work done per cycle for 1.00 mol of gas initially at 0°C.

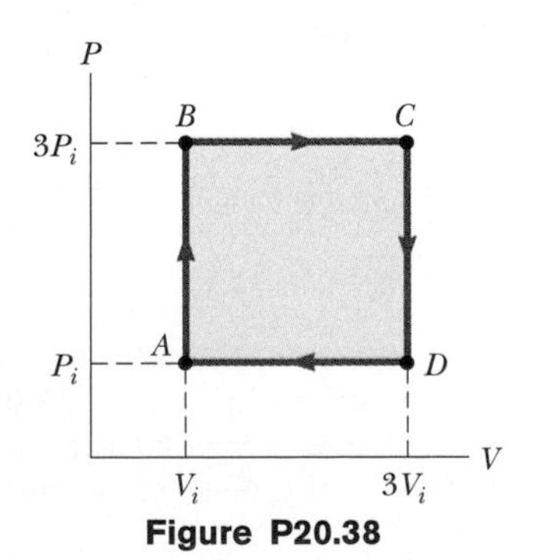

Figure P20.38

39. A 2.00-mol sample of helium gas initially at 300 K and 0.400 atm is compressed isothermally to 1.20 atm. Noting that the helium behaves as an ideal gas, find (a) the final volume of the gas, (b) the work done on the gas, and (c) the energy transferred by heat.

40. In Figure P20.40, the change in internal energy of a gas that is taken from A to C is +800 J. The work done on the gas along path ABC is −500 J. (a) How much energy must be added to the system by heat as it goes from A through B to C? (b) If the pressure at point A is five times that of point C, what is the work done on the system in going from C to D? (c) What is the energy exchanged with the surroundings by heat as the cycle goes from C to A along the green path? (d) If the change in internal energy in going from point D to point A is +500 J, how much energy must be added to the system by heat as it goes from point C to point D?

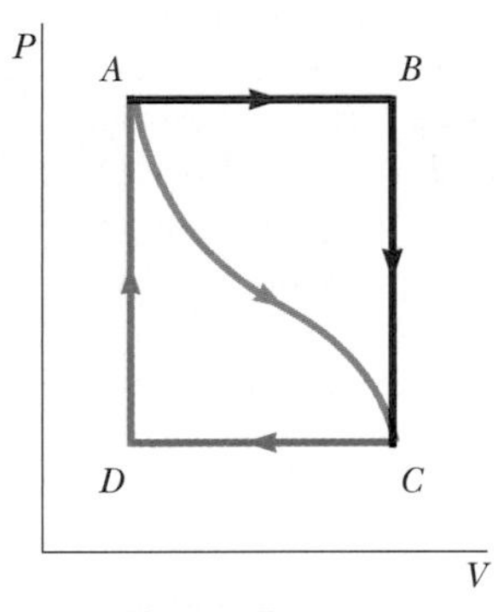

Figure P20.40

Section 20.7 Energy-Transfer Mechanisms

41. A box with a total surface area of 1.20 m^2 and a wall thickness of 4.00 cm is made of an insulating material. A 10.0-W electric heater inside the box maintains the inside temperature at 15.0°C above the outside temperature. Find the thermal conductivity k of the insulating material.

42. A glass window pane has an area of 3.00 m^2 and a thickness of 0.600 cm. If the temperature difference between its faces is 25.0°C, what is the rate of energy transfer by conduction through the window?

43. A bar of gold is in thermal contact with a bar of silver of the same length and area (Fig. P20.43). One end of the compound bar is maintained at 80.0°C while the opposite end is at 30.0°C. When the energy transfer reaches steady state, what is the temperature at the junction?

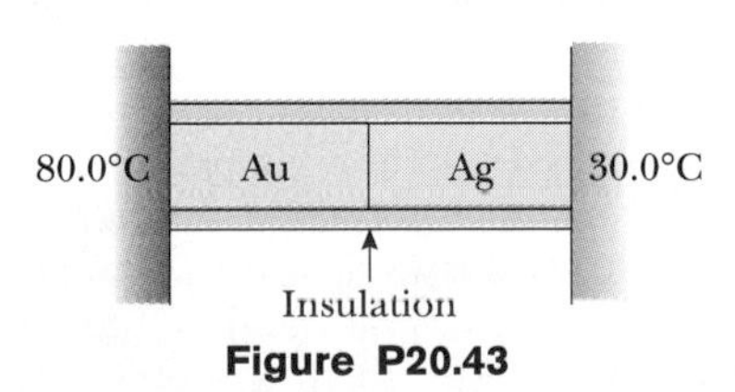

Figure P20.43

44. A thermal window with an area of 6.00 m^2 is constructed of two layers of glass, each 4.00 mm thick, and separated from each other by an air space of 5.00 mm. If the inside surface is at 20.0°C and the outside is at −30.0°C, what is the rate of energy transfer by conduction through the window?

45. A power transistor is a solid-state electronic device. Assume that energy entering the device at the rate of 1.50 W by electrical transmission causes the internal energy of the device to increase. The surface area of the transistor is so small that it tends to overheat. To prevent overheating, the transistor is attached to a larger metal heat sink with fins. The temperature of the heat sink remains constant at 35.0°C under steady-state conditions. The transistor is electrically insulated from the heat sink by a rectangular sheet of mica measuring 8.25 mm by 6.25 mm, and 0.085 2 mm thick. The thermal conductivity of mica is equal to 0.075 3 W/m · °C. What is the operating temperature of the transistor?

46. Calculate the R value of (a) a window made of a single pane of flat glass $\frac{1}{8}$ in. thick, and (b) a thermal window made of two single panes each $\frac{1}{8}$ in. thick and separated by a $\frac{1}{4}$-in. air space. (c) By what factor is the transfer of energy by heat through the window reduced by using the thermal window instead of the single pane window?

47. The surface of the Sun has a temperature of about 5 800 K. The radius of the Sun is 6.96×10^8 m. Calculate the total energy radiated by the Sun each second. Assume that the emissivity of the Sun is 0.965.

48. A large hot pizza floats in outer space. What is the order of magnitude of (a) its rate of energy loss? (b) its rate of temperature change? List the quantities you estimate and the value you estimate for each.

49. The tungsten filament of a certain 100-W light bulb radiates 2.00 W of light. (The other 98 W is carried away by convection and conduction.) The filament has a surface area of 0.250 mm^2 and an emissivity of 0.950. Find the filament's temperature. (The melting point of tungsten is 3 683 K.)

50. At high noon, the Sun delivers 1 000 W to each square meter of a blacktop road. If the hot asphalt loses energy only by radiation, what is its equilibrium temperature?

51. The intensity of solar radiation reaching the top of the Earth's atmosphere is 1 340 W/m^2. The temperature of the Earth is affected by the so-called greenhouse effect of the atmosphere. That effect makes our planet's emissivity for visible light higher than its emissivity for infrared light. For comparison, consider a spherical object with no atmosphere, at the same distance from the Sun as the Earth. Assume that its emissivity is the same for all kinds of electromagnetic waves and that its temperature is uniform over its surface. Identify the projected area over which it absorbs sunlight and the surface area over which it radiates. Compute its equilibrium temperature. Chilly, isn't it? Your calculation applies to (a) the average temperature of the Moon, (b) astronauts in mortal danger aboard the crippled *Apollo 13* spacecraft, and (c) global catastrophe on the Earth if widespread fires should cause a layer of soot to accumulate throughout the upper atmosphere, so that most of the radiation from the Sun were absorbed there rather than at the surface below the atmosphere.

Additional Problems

52. Liquid nitrogen with a mass of 100 g at 77.3 K is stirred into a beaker containing 200 g of 5.00°C water. If the nitrogen leaves the solution as soon as it turns to gas, how much water freezes? (The latent heat of vaporization of nitrogen is 48.0 cal/g, and the latent heat of fusion of water is 79.6 cal/g.)

53. A 75.0-kg cross-country skier moves across the snow (Fig. P20.53). The coefficient of friction between the skis and the snow is 0.200. Assume that all the snow beneath his skis is at 0°C and that all the internal energy generated by friction is added to the snow, which sticks to his skis until it melts. How far would he have to ski to melt 1.00 kg of snow?

Figure P20.53

54. On a cold winter day you buy roasted chestnuts from a street vendor. Into the pocket of your down parka you put the change he gives you—coins constituting 9.00 g of copper at −12.0°C. Your pocket already contains 14.0 g of silver coins at 30.0°C. A short time later the temperature of the copper coins is 4.00°C and is increasing at a rate of 0.500°C/s. At this time, (a) what is the temperature of the silver coins, and (b) at what rate is it changing?

55. An aluminum rod 0.500 m in length and with a cross-sectional area of 2.50 cm^2 is inserted into a thermally insulated vessel containing liquid helium at 4.20 K. The rod is initially at 300 K. (a) If half of the rod is inserted into the helium, how many liters of helium boil off by the time the inserted half cools to 4.20 K? (Assume the upper half does not yet cool.) (b) If the upper end of the rod is maintained at 300 K, what is the approximate boil-off rate of liquid helium after the lower half has reached 4.20 K? (Aluminum has thermal conductivity of 31.0 J/s · cm · K at 4.2 K; ignore its temperature variation. Aluminum has a specific heat of 0.210 cal/g · °C and density of 2.70 g/cm^3. The density of liquid helium is 0.125 g/cm^3.)

56. A copper ring (with mass of 25.0 g, coefficient of linear expansion of 1.70×10^{-5} (°C)$^{-1}$, and specific heat of 9.24×10^{-2} cal/g · °C) has a diameter of 5.00 cm at its temperature of 15.0°C. A spherical aluminum shell (with mass 10.9 g, coefficient of linear expansion 2.40×10^{-5} (°C)$^{-1}$, and specific heat 0.215 cal/g·°C) has a diameter of 5.01 cm at a temperature higher than 15.0°C. The sphere is placed on top of the horizontal ring, and the two are allowed to come to thermal equilibrium without any exchange of energy with the surroundings. As soon as the sphere and ring reach thermal equilibrium, the sphere barely falls through the ring. Find (a) the equilibrium temperature, and (b) the initial temperature of the sphere.

57. A *flow calorimeter* is an apparatus used to measure the specific heat of a liquid. The technique of flow calorimetry involves measuring the temperature difference between the input and output points of a flowing stream of the liquid while energy is added by heat at a known rate. A liquid of density ρ flows through the calorimeter with volume flow rate R. At steady state, a temperature difference ΔT is established between the input and output points when energy is supplied at the rate $\mathcal{P}$. What is the specific heat of the liquid?

58. One mole of an ideal gas is contained in a cylinder with a movable piston. The initial pressure, volume, and temperature are P_i, V_i, and T_i, respectively. Find the work done on the gas for the following processes and show each process on a PV diagram: (a) An isobaric compression in which the final volume is half the initial volume. (b) An isothermal compression in which the final pressure is four times the initial pressure. (c) An isovolumetric process in which the final pressure is three times the initial pressure.

59. One mole of an ideal gas, initially at 300 K, is cooled at constant volume so that the final pressure is one fourth of the initial pressure. Then the gas expands at constant pressure until it reaches the initial temperature. Determine the work done on the gas.

60. **Review problem.** Continue the analysis of Problem 60 in Chapter 19. Following a collision between a large spacecraft and an asteroid, a copper disk of radius 28.0 m and thickness 1.20 m, at a temperature of 850°C, is floating in space, rotating about its axis with an angular speed of 25.0 rad/s. As the disk radiates infrared light, its temperature falls to 20.0°C. No external torque acts on the disk. (a) Find the change in kinetic energy of the disk. (b) Find the change in internal energy of the disk. (b) Find the amount of energy it radiates.

61. **Review problem.** A 670-kg meteorite happens to be composed of aluminum. When it is far from the Earth, its temperature is -15°C and it moves with a speed of 14.0 km/s relative to the Earth. As it crashes into the planet, assume that the resulting additional internal energy is shared equally between the meteor and the planet, and that all of the material of the meteor rises momentarily to the same final temperature. Find this temperature. Assume that the specific heat of liquid and of gaseous aluminum is 1170 J/kg · °C.

62. An iron plate is held against an iron wheel so that a kinetic friction force of 50.0 N acts between the two pieces of metal. The relative speed at which the two surfaces slide over each other is 40.0 m/s. (a) Calculate the rate at which mechanical energy is converted to internal energy. (b) The plate and the wheel each have a mass of 5.00 kg, and each receives 50.0% of the internal energy. If the system is run as described for 10.0 s and each object is then allowed to reach a uniform internal temperature, what is the resultant temperature increase?

63. www A solar cooker consists of a curved reflecting surface that concentrates sunlight onto the object to be warmed (Fig. P20.63). The solar power per unit area reaching the Earth's surface at the location is 600 W/m^2. The cooker faces the Sun and has a diameter of 0.600 m. Assume that 40.0% of the incident energy is transferred to 0.500 L of water in an open container, initially at 20.0°C. How long does it take to completely boil away the water? (Ignore the heat capacity of the container.)

Figure P20.63

64. Water in an electric teakettle is boiling. The power absorbed by the water is 1.00 kW. Assuming that the pressure of vapor in the kettle equals atmospheric pressure, determine the speed of effusion of vapor from the kettle's spout, if the spout has a cross-sectional area of 2.00 cm^2.

65. A cooking vessel on a slow burner contains 10.0 kg of water and an unknown mass of ice in equilibrium at 0°C at time $t = 0$. The temperature of the mixture is measured at various times, and the result is plotted in Figure P20.65. During the first 50.0 min, the mixture remains at 0°C. From 50.0 min to 60.0 min, the temperature increases to 2.00°C. Ignoring the heat capacity of the vessel, determine the initial mass of ice.

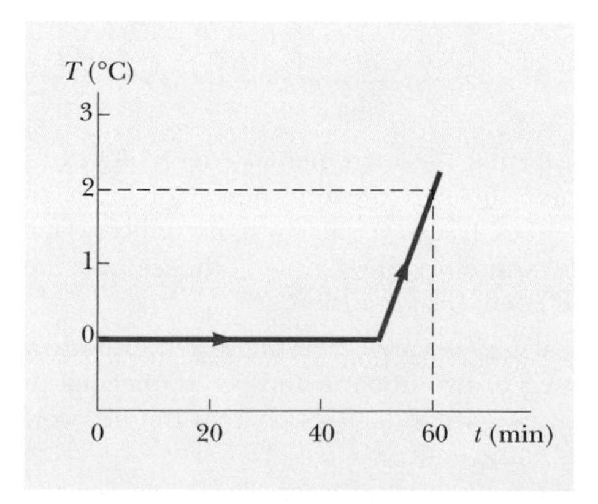

Figure P20.65

66. (a) In air at 0°C, a 1.60-kg copper block at 0°C is set sliding at 2.50 m/s over a sheet of ice at 0°C. Friction brings the block to rest. Find the mass of the ice that melts. To describe the process of slowing down, identify the energy input Q, the work input W, the change in internal energy ΔE_{int}, and the change in mechanical energy ΔK for the block and also for the ice. (b) A 1.60-kg block of ice at 0°C is set sliding at 2.50 m/s over a sheet of copper at 0°C. Friction brings the block to rest. Find the mass of the ice that melts. Identify Q, W, ΔE_{int}, and ΔK for the block and for the metal sheet during the process. (c) A thin 1.60-kg slab of copper at 20°C is set sliding at 2.50 m/s over an identical stationary slab at the same temperature. Friction quickly stops the motion. If no energy is lost to the environment by heat, find the change in temperature of both objects. Identify Q, W, ΔE_{int}, and ΔK for each object during the process.

67. The average thermal conductivity of the walls (including the windows) and roof of the house depicted in Figure P20.67 is 0.480 W/m · °C, and their average thickness is 21.0 cm. The house is heated with natural gas having a heat of combustion (that is, the energy provided per cubic meter of gas burned) of 9 300 kcal/m^3. How many cubic meters of gas must be burned each day to maintain an inside temperature of 25.0°C if the outside temperature is 0.0°C? Disregard radiation and the energy lost by heat through the ground.

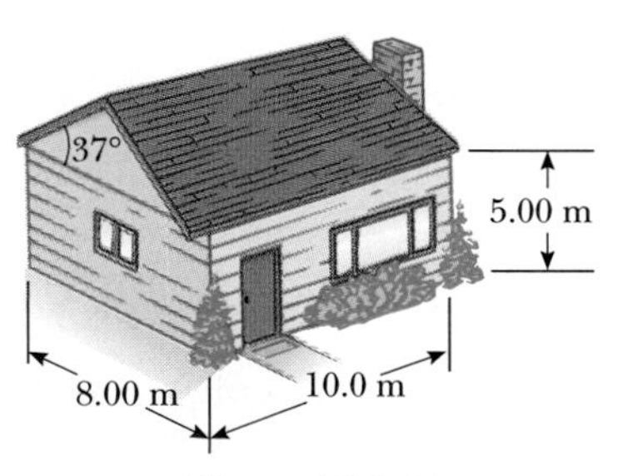

Figure P20.67

68. A pond of water at 0°C is covered with a layer of ice 4.00 cm thick. If the air temperature stays constant at −10.0°C, how long does it take for the ice thickness to increase to 8.00 cm? *Suggestion:* Utilize Equation 20.15 in the form

$$\frac{dQ}{dt} = kA\frac{\Delta T}{x}$$

and note that the incremental energy dQ extracted from the water through the thickness x of ice is the amount required to freeze a thickness dx of ice. That is, $dQ = L\rho A\, dx$, where ρ is the density of the ice, A is the area, and L is the latent heat of fusion.

69. An ideal gas is carried through a thermodynamic cycle consisting of two isobaric and two isothermal processes as shown in Figure P20.69. Show that the net work done on the gas in the entire cycle is given by

$$W_{\text{net}} = -P_1(V_2 - V_1)\ln\frac{P_2}{P_1}$$

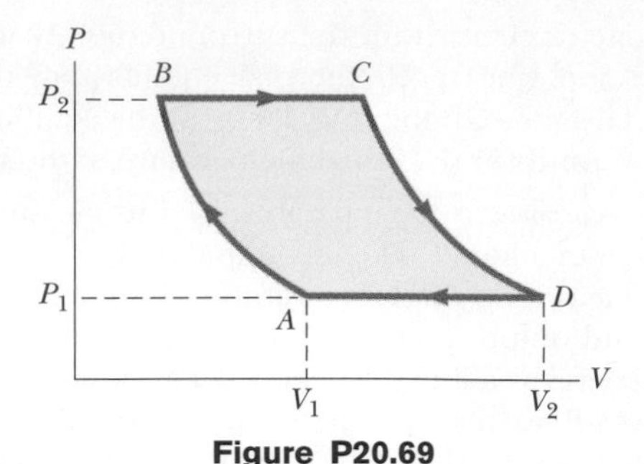

Figure P20.69

70. The inside of a hollow cylinder is maintained at a temperature T_a while the outside is at a lower temperature, T_b (Fig. P20.70). The wall of the cylinder has a thermal conductivity k. Ignoring end effects, show that the rate of energy conduction from the inner to the outer surface in the radial direction is

$$\frac{dQ}{dt} = 2\pi Lk\left[\frac{T_a - T_b}{\ln(b/a)}\right]$$

(*Suggestions:* The temperature gradient is dT/dr. Note that a radial energy current passes through a concentric cylinder of area $2\pi rL$.)

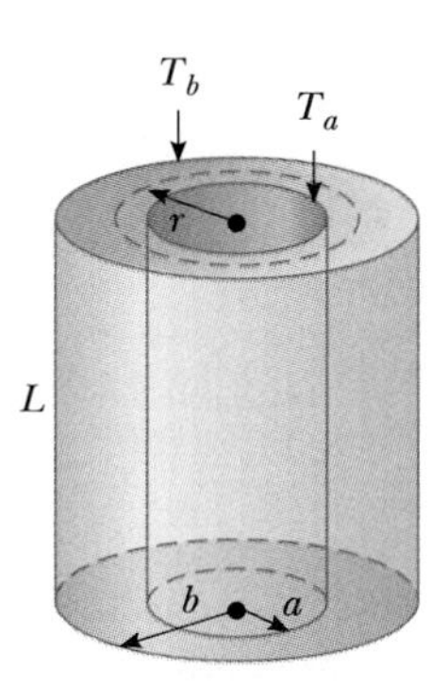

Figure P20.70

71. The passenger section of a jet airliner is in the shape of a cylindrical tube with a length of 35.0 m and an inner radius of 2.50 m. Its walls are lined with an insulating material 6.00 cm in thickness and having a thermal conductivity of 4.00×10^{-5} cal/s · cm · °C. A heater must maintain the interior temperature at 25.0°C while the outside temperature is −35.0°C. What power must be supplied to the heater? (Use the result of Problem 70.)

72. A student obtains the following data in a calorimetry experiment designed to measure the specific heat of aluminum:

Initial temperature of water and calorimeter:	70°C
Mass of water:	0.400 kg
Mass of calorimeter:	0.040 kg
Specific heat of calorimeter:	0.63 kJ/kg · °C
Initial temperature of aluminum:	27°C
Mass of aluminum:	0.200 kg
Final temperature of mixture:	66.3°C

Use these data to determine the specific heat of aluminum. Your result should be within 15% of the value listed in Table 20.1.

73. During periods of high activity, the Sun has more sunspots than usual. Sunspots are cooler than the rest of the luminous layer of the Sun's atmosphere (the photosphere). Paradoxically, the total power output of the active Sun is not lower than average but is the same or slightly higher than average. Work out the details of the following crude model of this phenomenon. Consider a patch of the photosphere with an area of 5.10×10^{14} m^2. Its emissivity is 0.965. (a) Find the power it radiates if its temperature is uniformly 5 800 K, corresponding to the quiet Sun. (b) To represent a sunspot, assume that 10.0% of the area is at 4 800 K and the other 90.0% is at 5 890 K. That is, a section with the surface area of the Earth is 1 000 K cooler than before and a section nine times as large is 90 K warmer. Find the average temperature of the patch. (c) Find the power output of the patch. Compare it with the answer to part (a). (The next sunspot maximum is expected around the year 2012.)

Answers to Quick Quizzes

20.1 Water, glass, iron. Because water has the highest specific heat (4 186 J/kg · °C), it has the smallest change in temperature. Glass is next (837 J/kg · °C), and iron is last (448 J/kg · °C).

20.2 Iron, glass, water. For a given temperature increase, the energy transfer by heat is proportional to the specific heat.

20.3 The figure below shows a graphical representation of the internal energy of the ice in parts A to E as a function of energy added. Notice that this graph looks quite different from Figure 20.2—it doesn't have the flat portions during the phase changes. Regardless of how the temperature is varying in Figure 20.2, the internal energy of the system simply increases linearly with energy input.

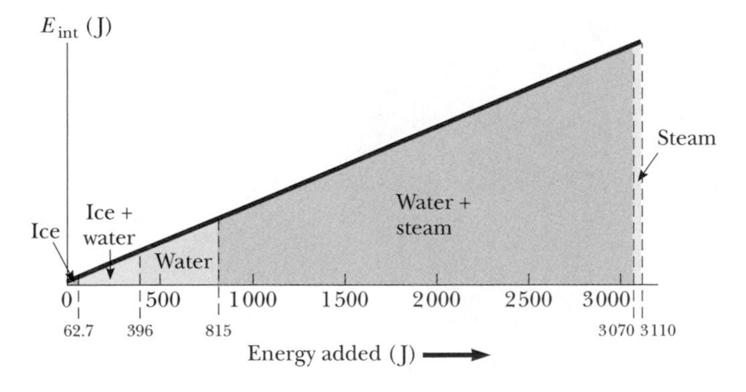

20.4 C, A, E. The slope is the ratio of the temperature change to the amount of energy input. Thus, the slope is proportional to the reciprocal of the specific heat. Water, which has the highest specific heat, has the smallest slope.

20.5

Situation	System	Q	W	ΔE_{int}
(a) Rapidly pumping up a bicycle tire	Air in the pump	0	+	+
(b) Pan of room-temperature water sitting on a hot stove	Water in the pan	+	0	+
(c) Air quickly leaking out of a balloon	Air originally in the balloon	0	−	−

(a) Because the pumping is rapid, no energy enters or leaves the system by heat. Because $W > 0$ when work is done *on* the system, it is positive here. Thus, we see that $\Delta E_{int} = Q + W$ must be positive. The air in the pump is warmer. (b) There is no work done either on or by the system, but energy transfers into the water by heat from the hot burner, making both Q and ΔE_{int} positive. (c) Again no energy transfers into or out of the system by heat, but the air molecules escaping from the balloon do work on the surrounding air molecules as they push them out of the way. Thus W is negative and ΔE_{int} is negative. The decrease in internal energy is evidenced by the fact that the escaping air becomes cooler.

20.6 A is isovolumetric, B is adiabatic, C is isothermal, and D is isobaric.

20.7 (c). The blanket acts as a thermal insulator, slowing the transfer of energy by heat from the air into the cube.

20.8 (b). In parallel, the rods present a larger area through which energy can transfer and a smaller length.

Chapter 21

The Kinetic Theory of Gases

CHAPTER OUTLINE

21.1 Molecular Model of an Ideal Gas

21.2 Molar Specific Heat of an Ideal Gas

21.3 Adiabatic Processes for an Ideal Gas

21.4 The Equipartition of Energy

21.5 The Boltzmann Distribution Law

21.6 Distribution of Molecular Speeds

21.7 Mean Free Path

▲ *Dogs do not have sweat glands like humans. In hot weather, dogs pant to promote evaporation from the tongue. In this chapter, we show that evaporation is a cooling process based on the removal of molecules with high kinetic energy from a liquid. (Frank Oberle/Getty Images)*

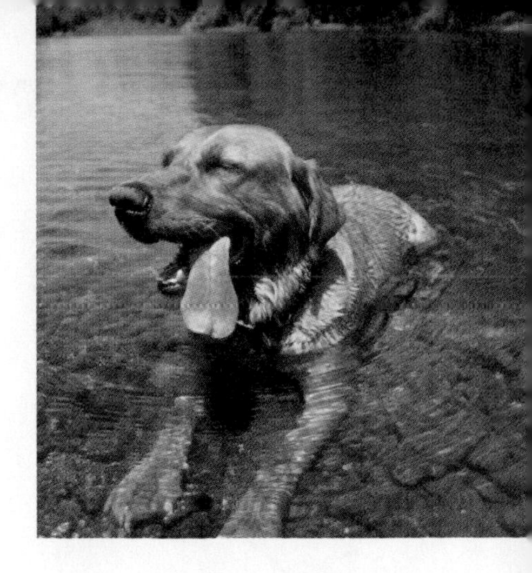

In Chapter 19 we discussed the properties of an ideal gas, using such macroscopic variables as pressure, volume, and temperature. We shall now show that such large-scale properties can be related to a description on a microscopic scale, where matter is treated as a collection of molecules. Newton's laws of motion applied in a statistical manner to a collection of particles provide a reasonable description of thermodynamic processes. To keep the mathematics relatively simple, we shall consider primarily the behavior of gases, because in gases the interactions between molecules are much weaker than they are in liquids or solids. In our model of gas behavior, called **kinetic theory**, gas molecules move about in a random fashion, colliding with the walls of their container and with each other. Kinetic theory provides us with a physical basis for our understanding of the concept of temperature.

21.1 Molecular Model of an Ideal Gas

We begin this chapter by developing a microscopic model of an ideal gas. The model shows that the pressure that a gas exerts on the walls of its container is a consequence of the collisions of the gas molecules with the walls and is consistent with the macroscopic description of Chapter 19. In developing this model, we make the following assumptions:

Assumptions of the molecular model of an ideal gas

1. **The number of molecules in the gas is large, and the average separation between them is large compared with their dimensions.** This means that the molecules occupy a negligible volume in the container. This is consistent with the ideal gas model, in which we imagine the molecules to be point-like.
2. **The molecules obey Newton's laws of motion, but as a whole they move randomly.** By "randomly" we mean that any molecule can move in any direction with any speed. At any given moment, a certain percentage of molecules move at high speeds, and a certain percentage move at low speeds.
3. **The molecules interact only by short-range forces during elastic collisions.** This is consistent with the ideal gas model, in which the molecules exert no long-range forces on each other.
4. **The molecules make elastic collisions with the walls.**
5. **The gas under consideration is a pure substance; that is, all molecules are identical.**

Although we often picture an ideal gas as consisting of single atoms, we can assume that the behavior of molecular gases approximates that of ideal gases rather well at low pressures. Molecular rotations or vibrations have no effect, on the average, on the motions that we consider here.

For our first application of kinetic theory, let us derive an expression for the pressure of N molecules of an ideal gas in a container of volume V in terms of microscopic quantities. The container is a cube with edges of length d (Fig. 21.1). We shall first

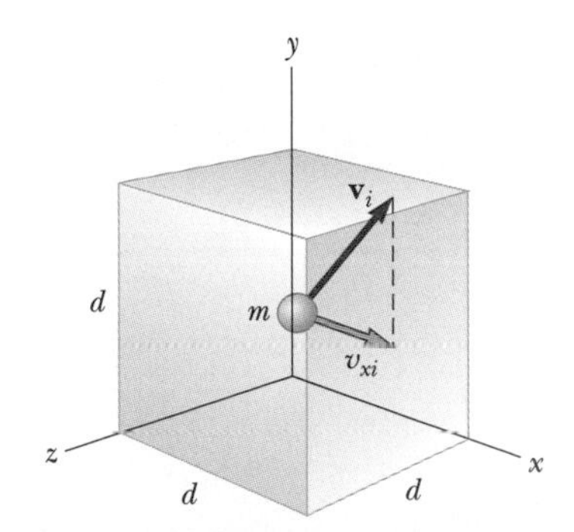

Figure 21.1 A cubical box with sides of length d containing an ideal gas. The molecule shown moves with velocity $\mathbf{v}_i$.

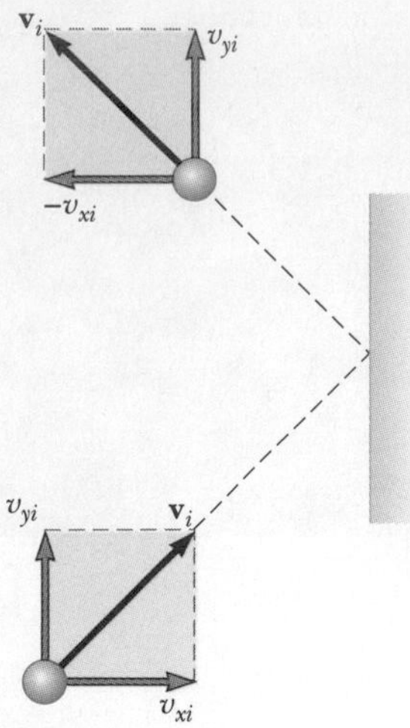

Active Figure 21.2 A molecule makes an elastic collision with the wall of the container. Its x component of momentum is reversed, while its y component remains unchanged. In this construction, we assume that the molecule moves in the xy plane.

At the Active Figures link at http://www.pse6.com, you can observe molecules within a container making collisions with the walls of the container and with each other.

focus our attention on one of these molecules of mass m, and assume that it is moving so that its component of velocity in the x direction is v_{xi} as in Figure 21.2. (The subscript i here refers to the ith molecule, not to an initial value. We will combine the effects of all of the molecules shortly.) As the molecule collides elastically with any wall (assumption 4), its velocity component perpendicular to the wall is reversed because the mass of the wall is far greater than the mass of the molecule. Because the momentum component p_{xi} of the molecule is mv_{xi} before the collision and $-mv_{xi}$ after the collision, the change in the x component of the momentum of the molecule is

$$\Delta p_{xi} = -mv_{xi} - (mv_{xi}) = -2mv_{xi}$$

Because the molecules obey Newton's laws (assumption 2), we can apply the impulse-momentum theorem (Eq. 9.8) to the molecule to give us

$$\overline{F}_{i,\text{on molecule}} \Delta t_{\text{collision}} = \Delta p_{xi} = -2mv_{xi}$$

where $\overline{F}_{i,\text{on molecule}}$ is the x component of the average force that the wall exerts on the molecule during the collision and $\Delta t_{\text{collision}}$ is the duration of the collision. In order for the molecule to make another collision with the same wall after this first collision, it must travel a distance of $2d$ in the x direction (across the container and back). Therefore, the time interval between two collisions with the same wall is

$$\Delta t = \frac{2d}{v_{xi}}$$

The force that causes the change in momentum of the molecule in the collision with the wall occurs only during the collision. However, we can average the force over the time interval for the molecule to move across the cube and back. Sometime during this time interval, the collision occurs, so that the change in momentum for this time interval is the same as that for the short duration of the collision. Thus, we can rewrite the impulse-momentum theorem as

$$\overline{F}_i \Delta t = -2mv_{xi}$$

where $\overline{F}_i$ is the average force component over the time for the molecule to move across the cube and back. Because exactly one collision occurs for each such time interval, this is also the long-term average force on the molecule, over long time intervals containing any number of multiples of Δt.

This equation and the preceding one enable us to express the x component of the long-term average force exerted by the wall on the molecule as

$$\overline{F}_i = \frac{-2mv_{xi}}{\Delta t} = \frac{-2mv_{xi}^2}{2d} = \frac{-mv_{xi}^2}{d}$$

Now, by Newton's third law, the average x component of the force exerted by the molecule on the wall is equal in magnitude and opposite in direction:

$$\overline{F}_{i,\text{on wall}} = -\overline{F}_i = -\left(\frac{-mv_{xi}^2}{d}\right) = \frac{mv_{xi}^2}{d}$$

The total average force $\overline{F}$ exerted by the gas on the wall is found by adding the average forces exerted by the individual molecules. We add terms such as that above for all molecules:

$$\overline{F} = \sum_{i=1}^{N} \frac{mv_{xi}^2}{d} = \frac{m}{d}\sum_{i=1}^{N} v_{xi}^2$$

where we have factored out the length of the box and the mass m, because assumption 5 tells us that all of the molecules are the same. We now impose assumption 1, that the number of molecules is large. For a small number of molecules, the actual force on the wall would vary with time. It would be nonzero during the short interval of a collision of a molecule with the wall and zero when no molecule happens to be hitting the wall. For a very large number of molecules, however, such as Avogadro's number, these variations in force are smoothed out, so that the average force given above is the same over *any* time interval. Thus, the *constant* force F on the wall due to the molecular collisions is

$$F = \frac{m}{d} \sum_{i=1}^{N} v_{xi}^2$$

To proceed further, let us consider how to express the average value of the square of the x component of the velocity for N molecules. The traditional average of a set of values is the sum of the values over the number of values:

$$\overline{v_x^2} = \frac{\sum_{i=1}^{N} v_{xi}^2}{N}$$

The numerator of this expression is contained in the right-hand side of the preceding equation. Thus, combining the two expressions, the total force on the wall can be written

$$F = \frac{m}{d} N\overline{v_x^2} \tag{21.1}$$

Now let us focus again on one molecule with velocity components v_{xi}, v_{yi}, and v_{zi}. The Pythagorean theorem relates the square of the speed of the molecule to the squares of the velocity components:

$$v_i^2 = v_{xi}^2 + v_{yi}^2 + v_{zi}^2$$

Hence, the average value of v^2 for all the molecules in the container is related to the average values of v_x^2, v_y^2, and v_z^2 according to the expression

$$\overline{v^2} = \overline{v_x^2} + \overline{v_y^2} + \overline{v_z^2}$$

Because the motion is completely random (assumption 2), the average values $\overline{v_x^2}$, $\overline{v_y^2}$, and $\overline{v_z^2}$ are equal to each other. Using this fact and the preceding equation, we find that

$$\overline{v^2} = 3\overline{v_x^2}$$

Thus, from Equation 21.1, the total force exerted on the wall is

$$F = \frac{N}{3}\left(\frac{m\overline{v^2}}{d}\right)$$

Using this expression, we can find the total pressure exerted on the wall:

$$P = \frac{F}{A} = \frac{F}{d^2} = \frac{1}{3}\left(\frac{N}{d^3} m\overline{v^2}\right) = \frac{1}{3}\left(\frac{N}{V}\right) m\overline{v^2}$$

$$P = \frac{2}{3}\left(\frac{N}{V}\right)\left(\frac{1}{2}m\overline{v^2}\right) \tag{21.2}$$

Relationship between pressure and molecular kinetic energy

This result indicates that **the pressure of a gas is proportional to the number of molecules per unit volume and to the average translational kinetic energy of the molecules, $\frac{1}{2}m\overline{v^2}$.** In analyzing this simplified model of an ideal gas, we obtain an important result that relates the macroscopic quantity of pressure to a microscopic quantity—the average value of the square of the molecular speed. Thus, we have established a key link between the molecular world and the large-scale world.

You should note that Equation 21.2 verifies some features of pressure with which you are probably familiar. One way to increase the pressure inside a container is to increase the number of molecules per unit volume N/V in the container. This is what you do when you add air to a tire. The pressure in the tire can also be increased by increasing the average translational kinetic energy of the air molecules in the tire.

This can be accomplished by increasing the temperature of that air, as we shall soon show mathematically. This is why the pressure inside a tire increases as the tire warms up during long trips. The continuous flexing of the tire as it moves along the road surface results in work done as parts of the tire distort, causing an increase in internal energy of the rubber. The increased temperature of the rubber results in the transfer of energy by heat into the air inside the tire. This transfer increases the air's temperature, and this increase in temperature in turn produces an increase in pressure.

Molecular Interpretation of Temperature

We can gain some insight into the meaning of temperature by first writing Equation 21.2 in the form

$$PV = \tfrac{2}{3}N(\tfrac{1}{2}m\overline{v^2})$$

Let us now compare this with the equation of state for an ideal gas (Eq. 19.10):

$$PV = Nk_BT$$

Recall that the equation of state is based on experimental facts concerning the macroscopic behavior of gases. Equating the right sides of these expressions, we find that

Temperature is proportional to average kinetic energy

$$T = \frac{2}{3k_B}(\tfrac{1}{2}m\overline{v^2}) \tag{21.3}$$

This result tells us that temperature is a direct measure of average molecular kinetic energy. By rearranging Equation 21.3, we can relate the translational molecular kinetic energy to the temperature:

Average kinetic energy per molecule

$$\tfrac{1}{2}m\overline{v^2} = \tfrac{3}{2}k_BT \tag{21.4}$$

That is, the average translational kinetic energy per molecule is $\frac{3}{2}k_BT$. Because $\overline{v_x^2} = \frac{1}{3}\overline{v^2}$, it follows that

$$\tfrac{1}{2}m\overline{v_x^2} = \tfrac{1}{2}k_BT \tag{21.5}$$

In a similar manner, it follows that the motions in the *y* and *z* directions give us

$$\tfrac{1}{2}m\overline{v_y^2} = \tfrac{1}{2}k_BT \qquad \text{and} \qquad \tfrac{1}{2}m\overline{v_z^2} = \tfrac{1}{2}k_BT$$

Thus, each translational degree of freedom contributes an equal amount of energy, $\frac{1}{2}k_BT$, to the gas. (In general, a "degree of freedom" refers to an independent means by which a molecule can possess energy.) A generalization of this result, known as the **theorem of equipartition of energy**, states that

Theorem of equipartition of energy

each degree of freedom contributes $\frac{1}{2}k_BT$ to the energy of a system, where possible degrees of freedom in addition to those associated with translation arise from rotation and vibration of molecules.

The total translational kinetic energy of *N* molecules of gas is simply *N* times the average energy per molecule, which is given by Equation 21.4:

Total translational kinetic energy of *N* molecules

$$K_{\text{tot trans}} = N(\tfrac{1}{2}m\overline{v^2}) = \tfrac{3}{2}Nk_BT = \tfrac{3}{2}nRT \tag{21.6}$$

where we have used $k_B = R/N_A$ for Boltzmann's constant and $n = N/N_A$ for the number of moles of gas. If we consider a gas in which molecules possess only translational kinetic energy, Equation 21.6 represents the internal energy of the gas. This result implies that **the internal energy of an ideal gas depends only on the temperature.** We will follow up on this point in Section 21.2.

The square root of $\overline{v^2}$ is called the *root-mean-square* (rms) *speed* of the molecules. From Equation 21.4 we find that the rms speed is

$$v_{\text{rms}} = \sqrt{\overline{v^2}} = \sqrt{\frac{3k_BT}{m}} = \sqrt{\frac{3RT}{M}} \quad (21.7)$$

Root-mean-square speed

where M is the molar mass in kilograms per mole and is equal to mN_A. This expression shows that, at a given temperature, lighter molecules move faster, on the average, than do heavier molecules. For example, at a given temperature, hydrogen molecules, whose molar mass is 2.02×10^{-3} kg/mol, have an average speed approximately four times that of oxygen molecules, whose molar mass is 32.0×10^{-3} kg/mol. Table 21.1 lists the rms speeds for various molecules at 20°C.

PITFALL PREVENTION

21.1 The Square Root of the Square?

Notice that taking the square root of $\overline{v^2}$ does not "undo" the square because we have taken an average *between* squaring and taking the square root. While the square root of $(\overline{v})^2$ is $\overline{v}$ because the squaring is done after the averaging, the square root of $\overline{v^2}$ is *not* $\overline{v}$, but rather v_{rms}.

Table 21.1

Some rms Speeds

Gas	Molar mass (g/mol)	v_{rms} at 20°C(m/s)
H_2	2.02	1 902
He	4.00	1 352
H_2O	18.0	637
Ne	20.2	602
N_2 or CO	28.0	511
NO	30.0	494
O_2	32.0	478
CO_2	44.0	408
SO_2	64.1	338

Example 21.1 A Tank of Helium

A tank used for filling helium balloons has a volume of 0.300 m^3 and contains 2.00 mol of helium gas at 20.0°C. Assume that the helium behaves like an ideal gas.

(A) What is the total translational kinetic energy of the gas molecules?

Solution Using Equation 21.6 with $n = 2.00$ mol and $T = 293$ K, we find that

$$K_{\text{tot trans}} = \tfrac{3}{2}nRT = \tfrac{3}{2}(2.00 \text{ mol})(8.31 \text{ J/mol}\cdot\text{K})(293 \text{ K})$$

$$= 7.30 \times 10^3 \text{ J}$$

(B) What is the average kinetic energy per molecule?

Solution Using Equation 21.4, we find that the average kinetic energy per molecule is

$$\tfrac{1}{2}m\overline{v^2} = \tfrac{3}{2}k_BT = \tfrac{3}{2}(1.38 \times 10^{-23} \text{ J/K})(293 \text{ K})$$

$$= 6.07 \times 10^{-21} \text{ J}$$

What If? What if the temperature is raised from 20.0°C to 40.0°C? Because 40.0 is twice as large as 20.0, is the total translational energy of the molecules of the gas twice as large at the higher temperature?

Answer The expression for the total translational energy depends on the temperature, and the value for the temperature must be expressed in kelvins, not in degrees Celsius. Thus, the ratio of 40.0 to 20.0 is *not* the appropriate ratio. Converting the Celsius temperatures to kelvins, 20.0°C is 293 K and 40.0°C is 313 K. Thus, the total translational energy increases by a factor of 313 K/293 K = 1.07.

Quick Quiz 21.1 Two containers hold an ideal gas at the same temperature and pressure. Both containers hold the same type of gas but container B has twice the volume of container A. The average translational kinetic energy per molecule in container B is (a) twice that for container A (b) the same as that for container A (c) half that for container A (d) impossible to determine.

Quick Quiz 21.2 Consider again the situation in Quick Quiz 21.1. The internal energy of the gas in container B is (a) twice that for container A (b) the same as that for container A (c) half that for container A (d) impossible to determine.

Quick Quiz 21.3 Consider again the situation in Quick Quiz 21.1. The rms speed of the gas molecules in container B is (a) twice that for container A (b) the same as that for container A (c) half that for container A (d) impossible to determine.

21.2 Molar Specific Heat of an Ideal Gas

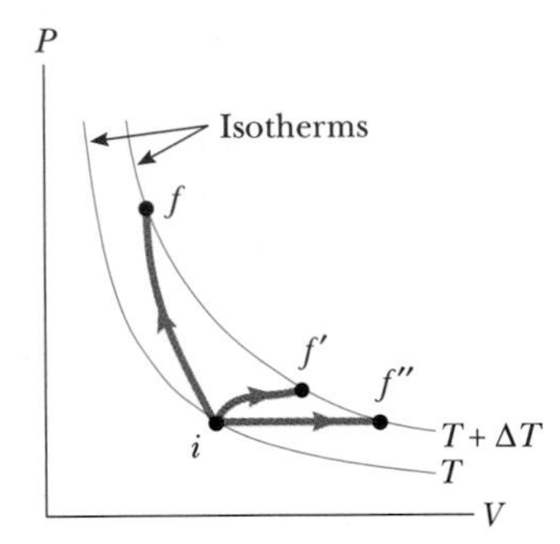

Figure 21.3 An ideal gas is taken from one isotherm at temperature T to another at temperature $T + \Delta T$ along three different paths.

Consider an ideal gas undergoing several processes such that the change in temperature is $\Delta T = T_f - T_i$ for all processes. The temperature change can be achieved by taking a variety of paths from one isotherm to another, as shown in Figure 21.3. Because ΔT is the same for each path, the change in internal energy ΔE_{int} is the same for all paths. However, we know from the first law, $Q = \Delta E_{\text{int}} - W$, that the heat Q is different for each path because W (the negative of the area under the curves) is different for each path. Thus, the heat associated with a given change in temperature does *not* have a unique value.

We can address this difficulty by defining specific heats for two processes that frequently occur: changes at constant volume and changes at constant pressure. Because the number of moles is a convenient measure of the amount of gas, we define the **molar specific heats** associated with these processes with the following equations:

$$Q = nC_V \Delta T \qquad \text{(constant volume)} \tag{21.8}$$

$$Q = nC_P \Delta T \qquad \text{(constant pressure)} \tag{21.9}$$

where C_V is the **molar specific heat at constant volume** and C_P is the **molar specific heat at constant pressure.** When we add energy to a gas by heat at constant pressure, not only does the internal energy of the gas increase, but work is done on the gas because of the change in volume. Therefore, the heat $Q_{\text{constant } P}$ must account for both the increase in internal energy and the transfer of energy out of the system by work. For this reason, $Q_{\text{constant } P}$ is greater than $Q_{\text{constant } V}$ for given values of n and ΔT. Thus, C_P is greater than C_V.

In the previous section, we found that the temperature of a gas is a measure of the average translational kinetic energy of the gas molecules. This kinetic energy is associated with the motion of the center of mass of each molecule. It does not include the energy associated with the internal motion of the molecule—namely, vibrations and rotations about the center of mass. This should not be surprising because the simple kinetic theory model assumes a structureless molecule.

In view of this, let us first consider the simplest case of an ideal monatomic gas, that is, a gas containing one atom per molecule, such as helium, neon, or argon. When energy is added to a monatomic gas in a container of fixed volume, all of the added energy goes into increasing the translational kinetic energy of the atoms. There is no other way to store the energy in a monatomic gas. Therefore, from Equation 21.6, we see that the internal energy E_{int} of N molecules (or n mol) of an ideal monatomic gas is

Internal energy of an ideal monatomic gas

$$E_{\text{int}} = K_{\text{tot trans}} = \tfrac{3}{2}Nk_{\text{B}}T = \tfrac{3}{2}nRT \tag{21.10}$$

Note that for a monatomic ideal gas, E_{int} is a function of T only, and the functional relationship is given by Equation 21.10. In general, the internal energy of an ideal gas is a function of T only, and the exact relationship depends on the type of gas.

If energy is transferred by heat to a system at *constant volume*, then no work is done on the system. That is, $W = -\int P\,dV = 0$ for a constant-volume process. Hence, from the first law of thermodynamics, we see that

$$Q = \Delta E_{\text{int}} \tag{21.11}$$

In other words, all of the energy transferred by heat goes into increasing the internal energy of the system. A constant-volume process from i to f for an ideal gas is described in Figure 21.4, where ΔT is the temperature difference between the two isotherms. Substituting the expression for Q given by Equation 21.8 into Equation 21.11, we obtain

$$\Delta E_{\text{int}} = nC_V\,\Delta T \tag{21.12}$$

If the molar specific heat is constant, we can express the internal energy of a gas as

$$E_{\text{int}} = nC_VT$$

This equation applies to all ideal gases—to gases having more than one atom per molecule as well as to monatomic ideal gases. In the limit of infinitesimal changes, we can use Equation 21.12 to express the molar specific heat at constant volume as

$$C_V = \frac{1}{n}\frac{dE_{\text{int}}}{dT} \tag{21.13}$$

Let us now apply the results of this discussion to the monatomic gas that we have been studying. Substituting the internal energy from Equation 21.10 into Equation 21.13, we find that

$$C_V = \tfrac{3}{2}R \tag{21.14}$$

This expression predicts a value of $C_V = \frac{3}{2}R = 12.5\ \text{J/mol}\cdot\text{K}$ for *all* monatomic gases. This prediction is in excellent agreement with measured values of molar specific heats for such gases as helium, neon, argon, and xenon over a wide range of temperatures (Table 21.2). Small variations in Table 21.2 from the predicted values are due to the

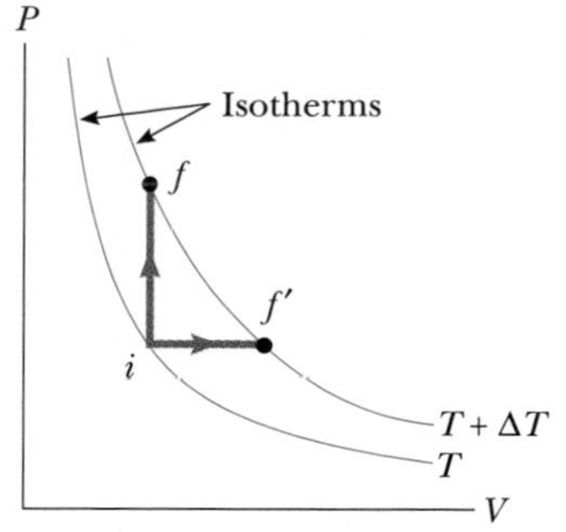

Active Figure 21.4 Energy is transferred by heat to an ideal gas in two ways. For the constant-volume path $i \rightarrow f$, all the energy goes into increasing the internal energy of the gas because no work is done. Along the constant-pressure path $i \rightarrow f'$, part of the energy transferred in by heat is transferred out by work.

At the Active Figures link at http://www.pse6.com, you can choose initial and final temperatures for one mole of an ideal gas undergoing constant-volume and constant pressure processes and measure Q, W, ΔE_{int}, C_V, and C_P.

Table 21.2

Molar Specific Heats of Various Gases

	Molar Specific Heat (J/mol · K)[a]			
Gas	C_P	C_V	$C_P - C_V$	$\gamma = C_P/C_V$
Monatomic Gases				
He	20.8	12.5	8.33	1.67
Ar	20.8	12.5	8.33	1.67
Ne	20.8	12.7	8.12	1.64
Kr	20.8	12.3	8.49	1.69
Diatomic Gases				
H_2	28.8	20.4	8.33	1.41
N_2	29.1	20.8	8.33	1.40
O_2	29.4	21.1	8.33	1.40
CO	29.3	21.0	8.33	1.40
Cl_2	34.7	25.7	8.96	1.35
Polyatomic Gases				
CO_2	37.0	28.5	8.50	1.30
SO_2	40.4	31.4	9.00	1.29
H_2O	35.4	27.0	8.37	1.30
CH_4	35.5	27.1	8.41	1.31

[a] All values except that for water were obtained at 300 K.

fact that real gases are not ideal gases. In real gases, weak intermolecular interactions occur, which are not addressed in our ideal gas model.

Now suppose that the gas is taken along the constant-pressure path $i \to f'$ shown in Figure 21.4. Along this path, the temperature again increases by ΔT. The energy that must be transferred by heat to the gas in this process is $Q = nC_P\,\Delta T$. Because the volume changes in this process, the work done on the gas is $W = -P\,\Delta V$ where P is the constant pressure at which the process occurs. Applying the first law of thermodynamics to this process, we have

$$\Delta E_{\text{int}} = Q + W = nC_P \Delta T + (-P\Delta V) \tag{21.15}$$

In this case, the energy added to the gas by heat is channeled as follows: Part of it leaves the system by work (that is, the gas moves a piston through a displacement), and the remainder appears as an increase in the internal energy of the gas. But the change in internal energy for the process $i \to f'$ is equal to that for the process $i \to f$ because E_{int} depends only on temperature for an ideal gas and because ΔT is the same for both processes. In addition, because $PV = nRT$, we note that for a constant-pressure process, $P\,\Delta V = nR\,\Delta T$. Substituting this value for $P\,\Delta V$ into Equation 21.15 with $\Delta E_{\text{int}} = nC_V\,\Delta T$ (Eq. 21.12) gives

$$nC_V\Delta T = nC_P\Delta T - nR\Delta T$$

$$C_P - C_V = R \tag{21.16}$$

This expression applies to *any* ideal gas. It predicts that the molar specific heat of an ideal gas at constant pressure is greater than the molar specific heat at constant volume by an amount R, the universal gas constant (which has the value $8.31\ \text{J/mol}\cdot\text{K}$). This expression is applicable to real gases, as the data in Table 21.2 show.

Because $C_V = \frac{3}{2}R$ for a monatomic ideal gas, Equation 21.16 predicts a value $C_P = \frac{5}{2}R = 20.8\ \text{J/mol}\cdot\text{K}$ for the molar specific heat of a monatomic gas at constant pressure. The ratio of these molar specific heats is a dimensionless quantity γ (Greek gamma):

Ratio of molar specific heats for a monatomic ideal gas

$$\gamma = \frac{C_P}{C_V} = \frac{5R/2}{3R/2} = \frac{5}{3} = 1.67 \tag{21.17}$$

Theoretical values of C_V, C_P and γ are in excellent agreement with experimental values obtained for monatomic gases, but they are in serious disagreement with the values for the more complex gases (see Table 21.2). This is not surprising because the value $C_V = \frac{3}{2}R$ was derived for a monatomic ideal gas and we expect some additional contribution to the molar specific heat from the internal structure of the more complex molecules. In Section 21.4, we describe the effect of molecular structure on the molar specific heat of a gas. The internal energy—and, hence, the molar specific heat—of a complex gas must include contributions from the rotational and the vibrational motions of the molecule.

In the case of solids and liquids heated at constant pressure, very little work is done because the thermal expansion is small. Consequently, C_P and C_V are approximately equal for solids and liquids.

Quick Quiz 21.4 How does the internal energy of an ideal gas change as it follows path $i \to f$ in Figure 21.4? (a) E_{int} increases. (b) E_{int} decreases. (c) E_{int} stays the same. (d) There is not enough information to determine how E_{int} changes.

Quick Quiz 21.5 How does the internal energy of an ideal gas change as it follows path $f \to f'$ along the isotherm labeled $T + \Delta T$ in Figure 21.4? (a) E_{int} increases. (b) E_{int} decreases. (c) E_{int} stays the same. (d) There is not enough information to determine how E_{int} changes.

Example 21.2 Heating a Cylinder of Helium

A cylinder contains 3.00 mol of helium gas at a temperature of 300 K.

(A) If the gas is heated at constant volume, how much energy must be transferred by heat to the gas for its temperature to increase to 500 K?

Solution For the constant-volume process, we have

$$Q_1 = nC_V \, \Delta T$$

Because $C_V = 12.5 \text{ J/mol} \cdot \text{K}$ for helium and $\Delta T = 200$ K, we obtain

$$Q_1 = (3.00 \text{ mol})(12.5 \text{ J/mol} \cdot \text{K})(200 \text{ K})$$
$$= 7.50 \times 10^3 \text{ J}$$

(B) How much energy must be transferred by heat to the gas at constant pressure to raise the temperature to 500 K?

Solution Making use of Table 21.2, we obtain

$$Q_2 = nC_P \, \Delta T$$
$$= (3.00 \text{ mol})(20.8 \text{ J/mol} \cdot \text{K})(200 \text{ K})$$
$$= 12.5 \times 10^3 \text{ J}$$

Note that this is larger than Q_1, due to the transfer of energy out of the gas by work in the constant pressure process.

21.3 Adiabatic Processes for an Ideal Gas

As we noted in Section 20.6, an **adiabatic process** is one in which no energy is transferred by heat between a system and its surroundings. For example, if a gas is compressed (or expanded) very rapidly, very little energy is transferred out of (or into) the system by heat, and so the process is nearly adiabatic. Such processes occur in the cycle of a gasoline engine, which we discuss in detail in the next chapter. Another example of an adiabatic process is the very slow expansion of a gas that is thermally insulated from its surroundings.

Suppose that an ideal gas undergoes an adiabatic expansion. At any time during the process, we assume that the gas is in an equilibrium state, so that the equation of state $PV = nRT$ is valid. As we show below, the pressure and volume of an ideal gas at any time during an adiabatic process are related by the expression

$$PV^\gamma = \text{constant} \tag{21.18}$$

Relationship between *P* and *V* for an adiabatic process involving an ideal gas

where $\gamma = C_P/C_V$ is assumed to be constant during the process. Thus, we see that all three variables in the ideal gas law—P, V, and T—change during an adiabatic process.

Proof That PV^γ = Constant for an Adiabatic Process

When a gas is compressed adiabatically in a thermally insulated cylinder, no energy is transferred by heat between the gas and its surroundings; thus, $Q = 0$. Let us imagine an infinitesimal change in volume dV and an accompanying infinitesimal change in temperature dT. The work done on the gas is $-P\,dV$. Because the internal energy of an ideal gas depends only on temperature, the change in the internal energy in an adiabatic process is the same as that for an isovolumetric process between the same temperatures, $dE_{\text{int}} = nC_V\,dT$ (Eq. 21.12). Hence, the first law of thermodynamics, $\Delta E_{\text{int}} = Q + W$, with $Q = 0$ becomes

$$dE_{\text{int}} = nC_V\,dT = -P\,dV$$

Taking the total differential of the equation of state of an ideal gas, $PV = nRT$, we see that

$$P\,dV + V\,dP = nR\,dT$$

Eliminating dT from these two equations, we find that

$$P\,dV + V\,dP = -\frac{R}{C_V}\,P\,dV$$

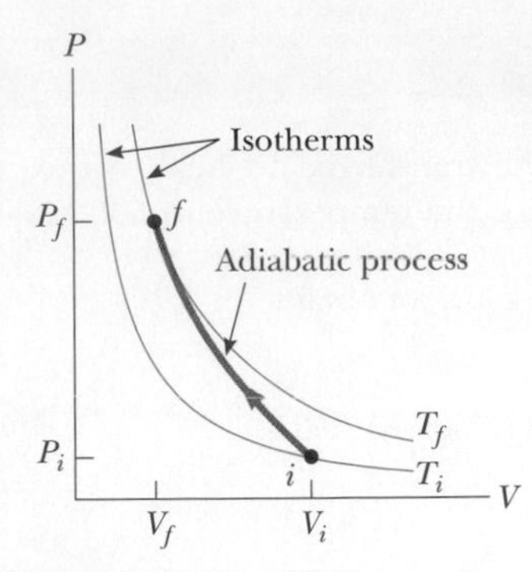

Figure 21.5 The *PV* diagram for an adiabatic compression. Note that $T_f > T_i$ in this process, so the temperature of the gas increases.

Substituting $R = C_P - C_V$ and dividing by *PV*, we obtain

$$\frac{dV}{V} + \frac{dP}{P} = -\left(\frac{C_P - C_V}{C_V}\right)\frac{dV}{V} = (1 - \gamma)\frac{dV}{V}$$

$$\frac{dP}{P} + \gamma\frac{dV}{V} = 0$$

Integrating this expression, we have

$$\ln P + \gamma \ln V = \text{constant}$$

which is equivalent to Equation 21.18:

$$PV^{\gamma} = \text{constant}$$

The *PV* diagram for an adiabatic compression is shown in Figure 21.5. Because $\gamma > 1$, the *PV* curve is steeper than it would be for an isothermal compression. By the definition of an adiabatic process, no energy is transferred by heat into or out of the system. Hence, from the first law, we see that ΔE_{int} is positive (work is done on the gas, so its internal energy increases) and so ΔT also is positive. Thus, the temperature of the gas increases ($T_f > T_i$) during an adiabatic compression. Conversely, the temperature decreases if the gas expands adiabatically.[1] Applying Equation 21.18 to the initial and final states, we see that

$$P_i V_i^{\gamma} = P_f V_f^{\gamma} \tag{21.19}$$

Relationship between *T* and *V* for an adiabatic process involving an ideal gas

Using the ideal gas law, we can express Equation 21.19 as

$$T_i V_i^{\gamma - 1} = T_f V_f^{\gamma - 1} \tag{21.20}$$

Example 21.3 A Diesel Engine Cylinder

Air at 20.0°C in the cylinder of a diesel engine is compressed from an initial pressure of 1.00 atm and volume of 800.0 cm^3 to a volume of 60.0 cm^3. Assume that air behaves as an ideal gas with $\gamma = 1.40$ and that the compression is adiabatic. Find the final pressure and temperature of the air.

Solution Conceptualize by imagining what happens if we compress a gas into a smaller volume. Our discussion above and Figure 21.5 tell us that the pressure and temperature both increase. We categorize this as a problem involving an adiabatic compression. To analyze the problem, we use Equation 21.19 to find the final pressure:

$$P_f = P_i\left(\frac{V_i}{V_f}\right)^{\gamma} = (1.00\text{ atm})\left(\frac{800.0\text{ cm}^3}{60.0\text{ cm}^3}\right)^{1.40}$$

$$= 37.6\text{ atm}$$

Because $PV = nRT$ is valid throughout an ideal gas process and because no gas escapes from the cylinder,

$$\frac{P_i V_i}{T_i} = \frac{P_f V_f}{T_f}$$

$$T_f = \frac{P_f V_f}{P_i V_i} T_i = \frac{(37.6\text{ atm})(60.0\text{ cm}^3)}{(1.00\text{ atm})(800.0\text{ cm}^3)}(293\text{ K})$$

$$= 826\text{ K} = 553°\text{C}$$

To finalize the problem, note that the temperature of the gas has increased by a factor of 2.82. The high compression in a diesel engine raises the temperature of the fuel enough to cause its combustion without the use of spark plugs.

21.4 The Equipartition of Energy

We have found that predictions based on our model for molar specific heat agree quite well with the behavior of monatomic gases but not with the behavior of complex gases (see Table 21.2). The value predicted by the model for the quantity $C_P - C_V = R$, however, is the same for all gases. This is not surprising because this difference is the result of the work done on the gas, which is independent of its molecular structure.

[1] In the adiabatic free expansion discussed in Section 20.6, the temperature remains constant. This is a special process in which no work is done because the gas expands into a vacuum. In general, the temperature decreases in an adiabatic expansion in which work is done.

To clarify the variations in C_V and C_P in gases more complex than monatomic gases, let us explore further the origin of molar specific heat. So far, we have assumed that the sole contribution to the internal energy of a gas is the translational kinetic energy of the molecules. However, the internal energy of a gas includes contributions from the translational, vibrational, and rotational motion of the molecules. The rotational and vibrational motions of molecules can be activated by collisions and therefore are "coupled" to the translational motion of the molecules. The branch of physics known as *statistical mechanics* has shown that, for a large number of particles obeying the laws of Newtonian mechanics, the available energy is, on the average, shared equally by each independent degree of freedom. Recall from Section 21.1 that the equipartition theorem states that, at equilibrium, each degree of freedom contributes $\frac{1}{2}k_BT$ of energy per molecule.

Let us consider a diatomic gas whose molecules have the shape of a dumbbell (Fig. 21.6). In this model, the center of mass of the molecule can translate in the x, y, and z directions (Fig. 21.6a). In addition, the molecule can rotate about three mutually perpendicular axes (Fig. 21.6b). We can neglect the rotation about the y axis because the molecule's moment of inertia I_y and its rotational energy $\frac{1}{2}I_y\omega^2$ about this axis are negligible compared with those associated with the x and z axes. (If the two atoms are taken to be point masses, then I_y is identically zero.) Thus, there are five degrees of freedom for translation and rotation: three associated with the translational motion and two associated with the rotational motion. Because each degree of freedom contributes, on the average, $\frac{1}{2}k_BT$ of energy per molecule, the internal energy for a system of N molecules, ignoring vibration for now, is

$$E_{\text{int}} = 3N(\tfrac{1}{2}k_BT) + 2N(\tfrac{1}{2}k_BT) = \tfrac{5}{2}Nk_BT = \tfrac{5}{2}nRT$$

We can use this result and Equation 21.13 to find the molar specific heat at constant volume:

$$C_V = \frac{1}{n}\frac{dE_{\text{int}}}{dT} = \frac{1}{n}\frac{d}{dT}(\tfrac{5}{2}nRT) = \tfrac{5}{2}R \qquad (21.21)$$

From Equations 21.16 and 21.17, we find that

$$C_P = C_V + R = \tfrac{7}{2}R$$

$$\gamma = \frac{C_P}{C_V} = \frac{\frac{7}{2}R}{\frac{5}{2}R} = \frac{7}{5} = 1.40$$

These results agree quite well with most of the data for diatomic molecules given in Table 21.2. This is rather surprising because we have not yet accounted for the possible vibrations of the molecule.

In the model for vibration, the two atoms are joined by an imaginary spring (see Fig. 21.6c). The vibrational motion adds two more degrees of freedom, which correspond to the kinetic energy and the potential energy associated with vibrations along the length of the molecule. Hence, classical physics and the equipartition theorem in a model that includes all three types of motion predict a total internal energy of

$$E_{\text{int}} = 3N(\tfrac{1}{2}k_BT) + 2N(\tfrac{1}{2}k_BT) + 2N(\tfrac{1}{2}k_BT) = \tfrac{7}{2}Nk_BT = \tfrac{7}{2}nRT$$

and a molar specific heat at constant volume of

$$C_V = \frac{1}{n}\frac{dE_{\text{int}}}{dT} = \frac{1}{n}\frac{d}{dT}(\tfrac{7}{2}nRT) = \tfrac{7}{2}R \qquad (21.22)$$

This value is inconsistent with experimental data for molecules such as H_2 and N_2 (see Table 21.2) and suggests a breakdown of our model based on classical physics.

It might seem that our model is a failure for predicting molar specific heats for diatomic gases. We can claim some success for our model, however, if measurements of molar specific heat are made over a wide temperature range, rather than at the

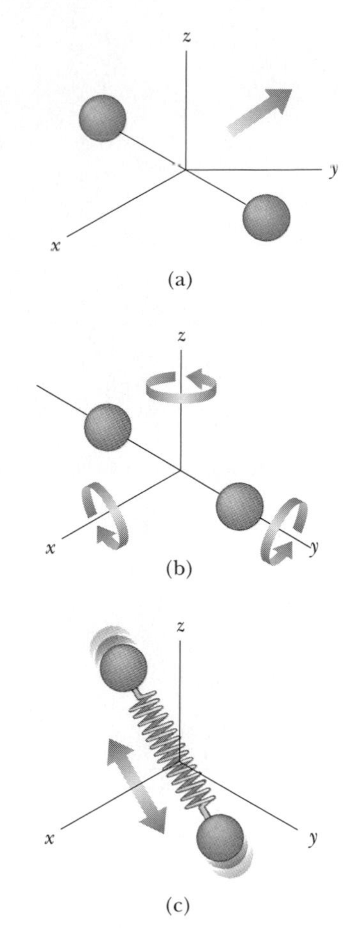

Figure 21.6 Possible motions of a diatomic molecule: (a) translational motion of the center of mass, (b) rotational motion about the various axes, and (c) vibrational motion along the molecular axis.

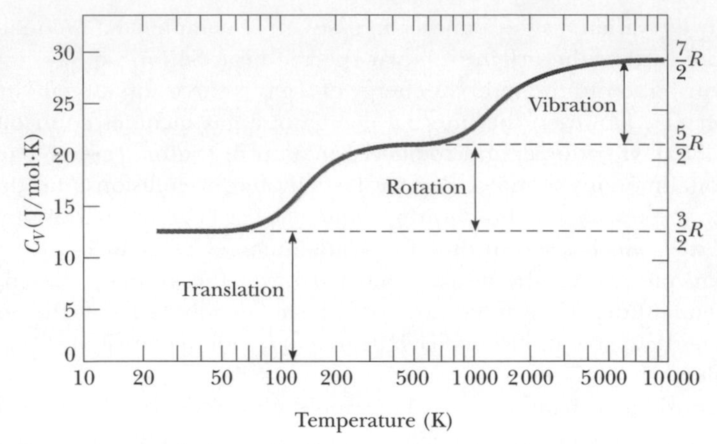

Figure 21.7 The molar specific heat of hydrogen as a function of temperature. The horizontal scale is logarithmic. Note that hydrogen liquefies at 20 K.

single temperature that gives us the values in Table 21.2. Figure 21.7 shows the molar specific heat of hydrogen as a function of temperature. There are three plateaus in the curve. The remarkable feature of these plateaus is that they are at the values of the molar specific heat predicted by Equations 21.14, 21.21, and 21.22! For low temperatures, the diatomic hydrogen gas behaves like a monatomic gas. As the temperature rises to room temperature, its molar specific heat rises to a value for a diatomic gas, consistent with the inclusion of rotation but not vibration. For high temperatures, the molar specific heat is consistent with a model including all types of motion.

Before addressing the reason for this mysterious behavior, let us make a brief remark about polyatomic gases. For molecules with more than two atoms, the vibrations are more complex than for diatomic molecules and the number of degrees of freedom is even larger. This results in an even higher predicted molar specific heat, which is in qualitative agreement with experiment. For the polyatomic gases shown in Table 21.2 we see that the molar specific heats are higher than those for diatomic gases. The more degrees of freedom available to a molecule, the more "ways" there are to store energy, resulting in a higher molar specific heat.

A Hint of Energy Quantization

Our model for molar specific heats has been based so far on purely classical notions. It predicts a value of the specific heat for a diatomic gas that, according to Figure 21.7, only agrees with experimental measurements made at high temperatures. In order to explain why this value is only true at high temperatures and why the plateaus exist in Figure 21.7, we must go beyond classical physics and introduce some quantum physics into the model. In Chapter 18, we discussed quantization of frequency for vibrating strings and air columns. This is a natural result whenever waves are subject to boundary conditions.

Quantum physics (Chapters 40 to 43) shows that atoms and molecules can be described by the physics of waves under boundary conditions. Consequently, these waves have quantized frequencies. Furthermore, in quantum physics, the energy of a system is proportional to the frequency of the wave representing the system. Hence, **the energies of atoms and molecules are quantized.**

For a molecule, quantum physics tells us that the rotational and vibrational energies are quantized. Figure 21.8 shows an **energy-level diagram** for the rotational and vibrational quantum states of a diatomic molecule. The lowest allowed state is called the **ground state**. Notice that vibrational states are separated by larger energy gaps than are rotational states.

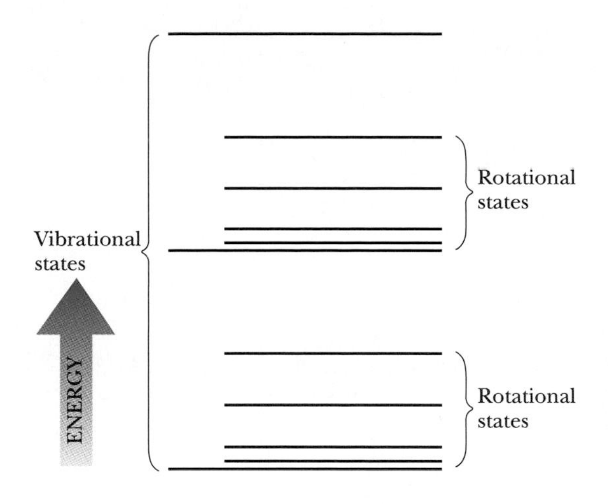

Figure 21.8 An energy-level diagram for vibrational and rotational states of a diatomic molecule. Note that the rotational states lie closer together in energy than the vibrational states.

At low temperatures, the energy that a molecule gains in collisions with its neighbors is generally not large enough to raise it to the first excited state of either rotation or vibration. Thus, even though rotation and vibration are classically allowed, they do not occur at low temperatures. All molecules are in the ground state for rotation and vibration. Thus, the only contribution to the molecules' average energy is from translation, and the specific heat is that predicted by Equation 21.14.

As the temperature is raised, the average energy of the molecules increases. In some collisions, a molecule may have enough energy transferred to it from another molecule to excite the first rotational state. As the temperature is raised further, more molecules can be excited to this state. The result is that rotation begins to contribute to the internal energy and the molar specific heat rises. At about room temperature in Figure 21.7, the second plateau has been reached and rotation contributes fully to the molar specific heat. The molar specific heat is now equal to the value predicted by Equation 21.21.

There is no contribution at room temperature from vibration, because the molecules are still in the ground vibrational state. The temperature must be raised even further to excite the first vibrational state. This happens in Figure 21.7 between 1 000 K and 10 000 K. At 10 000 K on the right side of the figure, vibration is contributing fully to the internal energy and the molar specific heat has the value predicted by Equation 21.22.

The predictions of this model are supportive of the theorem of equipartition of energy. In addition, the inclusion in the model of energy quantization from quantum physics allows a full understanding of Figure 21.7.

Quick Quiz 21.6 The molar specific heat of a diatomic gas is measured at constant volume and found to be 29.1 J/mol · K. The types of energy that are contributing to the molar specific heat are (a) translation only (b) translation and rotation only (c) translation and vibration only (d) translation, rotation, and vibration.

Quick Quiz 21.7 The molar specific heat of a gas is measured at constant volume and found to be $11R/2$. The gas is most likely to be (a) monatomic (b) diatomic (c) polyatomic.

The Molar Specific Heat of Solids

The molar specific heats of solids also demonstrate a marked temperature dependence. Solids have molar specific heats that generally decrease in a nonlinear manner with decreasing temperature and approach zero as the temperature approaches

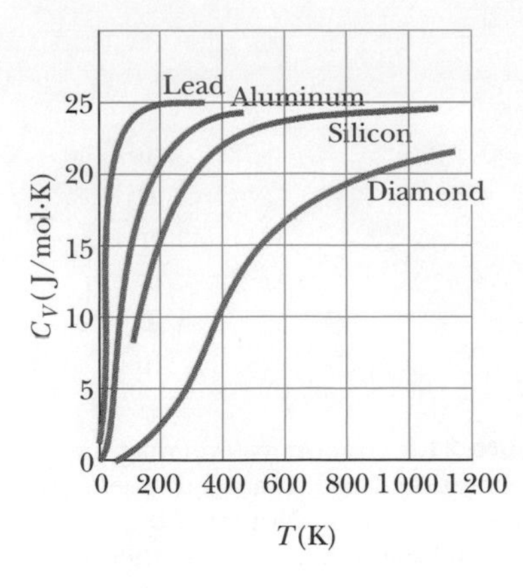

Figure 21.9 Molar specific heat of four solids. As T approaches zero, the molar specific heat also approaches zero.

absolute zero. At high temperatures (usually above 300 K), the molar specific heats approach the value of $3R \approx 25$ J/mol · K, a result known as the *DuLong–Petit law.* The typical data shown in Figure 21.9 demonstrate the temperature dependence of the molar specific heats for several solids.

We can explain the molar specific heat of a solid at high temperatures using the equipartition theorem. For small displacements of an atom from its equilibrium position, each atom executes simple harmonic motion in the x, y, and z directions. The energy associated with vibrational motion in the x direction is

$$E = \tfrac{1}{2}mv_x^{\,2} + \tfrac{1}{2}kx^2$$

The expressions for vibrational motions in the y and z directions are analogous. Therefore, each atom of the solid has six degrees of freedom. According to the equipartition theorem, this corresponds to an average vibrational energy of $6(\frac{1}{2}k_BT) = 3k_BT$ per atom. Therefore, the internal energy of a solid consisting of N atoms is

Total internal energy of a solid

$$E_{\text{int}} = 3Nk_BT = 3nRT \tag{21.23}$$

From this result, we find that the molar specific heat of a solid at constant volume is

Molar specific heat of a solid at constant volume

$$C_V = \frac{1}{n}\frac{dE_{\text{int}}}{dT} = 3R \tag{21.24}$$

This result is in agreement with the empirical DuLong–Petit law. The discrepancies between this model and the experimental data at low temperatures are again due to the inadequacy of classical physics in describing the world at the atomic level.

21.5 The Boltzmann Distribution Law

PITFALL PREVENTION

21.2 The Distribution Function

Notice that the distribution function $n_V(E)$ is defined in terms of the number of molecules with energy in the range E to $E + dE$ rather than in terms of the number of molecules with energy E. Because the number of molecules is finite and the number of possible values of the energy is infinite, the number of molecules with an *exact* energy E may be zero.

Thus far we have considered only average values of the energies of molecules in a gas and have not addressed the distribution of energies among molecules. In reality, the motion of the molecules is extremely chaotic. Any individual molecule is colliding with others at an enormous rate—typically, a billion times per second. Each collision results in a change in the speed and direction of motion of each of the participant molecules. Equation 21.7 shows that rms molecular speeds increase with increasing temperature. What is the relative number of molecules that possess some characteristic, such as energy within a certain range?

We shall address this question by considering the **number density** $n_V(E)$. This quantity, called a *distribution function,* is defined so that $n_V(E)\,dE$ is the number of molecules per unit volume with energy between E and $E + dE$. (Note that the ratio of the number of molecules that have the desired characteristic to the total number of molecules is the probability that a particular molecule has that characteristic.) In general, the number density is found from statistical mechanics to be

$$n_V(E) = n_0 e^{-E/k_B T} \quad (21.25)$$

Boltzmann distribution law

where n_0 is defined such that $n_0\,dE$ is the number of molecules per unit volume having energy between $E = 0$ and $E = dE$. This equation, known as the **Boltzmann distribution law,** is important in describing the statistical mechanics of a large number of molecules. It states that **the probability of finding the molecules in a particular energy state varies exponentially as the negative of the energy divided by $k_B T$.** All the molecules would fall into the lowest energy level if the thermal agitation at a temperature T did not excite the molecules to higher energy levels.

Example 21.4 Thermal Excitation of Atomic Energy Levels **Interactive**

As we discussed in Section 21.4, atoms can occupy only certain discrete energy levels. Consider a gas at a temperature of 2 500 K whose atoms can occupy only two energy levels separated by 1.50 eV, where 1 eV (electron volt) is an energy unit equal to 1.60×10^{-19} J (Fig. 21.10). Determine the ratio of the number of atoms in the higher energy level to the number in the lower energy level.

Solution Equation 21.25 gives the relative number of atoms in a given energy level. In this case, the atom has two possible energies, E_1 and E_2, where E_1 is the lower energy level. Hence, the ratio of the number of atoms in the higher energy level to the number in the lower energy level is

$$(1) \qquad \frac{n_V(E_2)}{n_V(E_1)} = \frac{n_0 e^{-E_2/k_B T}}{n_0 e^{-E_1/k_B T}} = e^{-(E_2 - E_1)/k_B T}$$

In this problem, $E_2 - E_1 = 1.50$ eV, and the denominator of the exponent is

$$k_B T = (1.38 \times 10^{-23}\ \text{J/K})(2\,500\ \text{K})\left(\frac{1\ \text{eV}}{1.60 \times 10^{-19}\ \text{J}}\right)$$

$$= 0.216\ \text{eV}$$

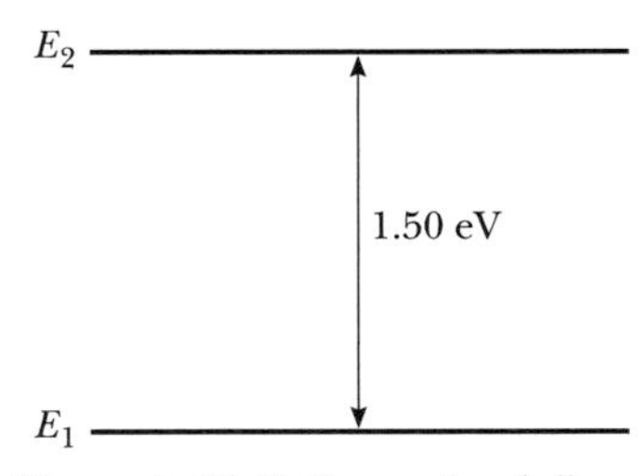

Figure 21.10 (Example 21.4) Energy-level diagram for a gas whose atoms can occupy two energy states.

Therefore, the required ratio is

$$\frac{n_V(E_2)}{n_V(E_1)} = e^{-1.50\ \text{eV}/0.216\ \text{eV}} = e^{-6.94}$$

$$= 9.64 \times 10^{-4}$$

This result indicates that at $T = 2\,500$ K, only a small fraction of the atoms are in the higher energy level. In fact, for every atom in the higher energy level, there are about 1 000 atoms in the lower level. The number of atoms in the higher level increases at even higher temperatures, but the distribution law specifies that at equilibrium there are always more atoms in the lower level than in the higher level.

What If? What if the energy levels in Figure 21.10 were closer together in energy? Would this increase or decrease the fraction of the atoms in the upper energy level?

Answer If the excited level is lower in energy than that in Figure 21.10, it would be easier for thermal agitation to excite atoms to this level, and the fraction of atoms in this energy level would be larger. Let us see this mathematically by expressing Equation (1) as

$$r_2 = e^{-(E_2 - E_1)/k_B T}$$

where r_2 is the ratio of atoms having energy E_2 to those with energy E_1. Differentiating with respect to E_2, we find

$$\frac{dr_2}{dE_2} = \frac{d}{dE_2}\left(e^{-(E_2 - E_1)/k_B T}\right) = -\frac{1}{k_B T} e^{-(E_2 - E_1)/k_B T} < 0$$

Because the derivative has a negative value, we see that as E_2 decreases, r_2 increases.

At the Interactive Worked Example link at **http://www.pse6.com,** *you can investigate the effects of changing the temperature and the energy difference between the states.*

21.6 Distribution of Molecular Speeds

In 1860 James Clerk Maxwell (1831–1879) derived an expression that describes the distribution of molecular speeds in a very definite manner. His work and subsequent developments by other scientists were highly controversial because direct detection of molecules could not be achieved experimentally at that time. However, about 60 years later, experiments were devised that confirmed Maxwell's predictions.

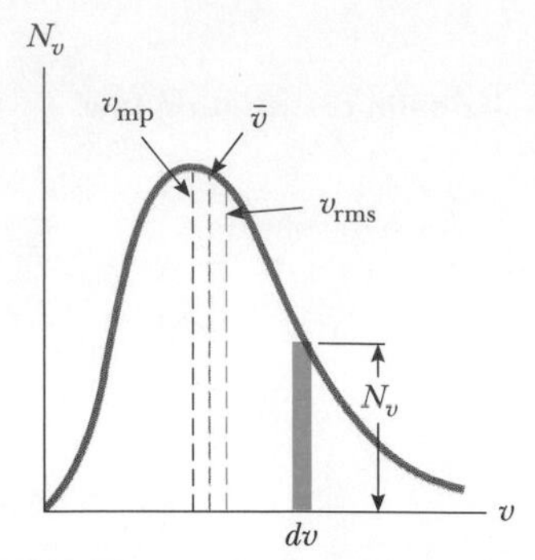

Active Figure 21.11 The speed distribution of gas molecules at some temperature. The number of molecules having speeds in the range v to $v + dv$ is equal to the area of the shaded rectangle, $N_v dv$. The function N_v approaches zero as v approaches infinity.

At the Active Figures link at http://www.pse6.com, you can move the blue triangle and measure the number of molecules with speeds within a small range.

Let us consider a container of gas whose molecules have some distribution of speeds. Suppose we want to determine how many gas molecules have a speed in the range from, for example, 400 to 410 m/s. Intuitively, we expect that the speed distribution depends on temperature. Furthermore, we expect that the distribution peaks in the vicinity of v_{rms}. That is, few molecules are expected to have speeds much less than or much greater than v_{rms} because these extreme speeds result only from an unlikely chain of collisions.

The observed speed distribution of gas molecules in thermal equilibrium is shown in Figure 21.11. The quantity N_v, called the **Maxwell–Boltzmann speed distribution function,** is defined as follows. If N is the total number of molecules, then the number of molecules with speeds between v and $v + dv$ is $dN = N_v\,dv$. This number is also equal to the area of the shaded rectangle in Figure 21.11. Furthermore, the fraction of molecules with speeds between v and $v + dv$ is $(N_v\,dv)/N$. This fraction is also equal to the probability that a molecule has a speed in the range v to $v + dv$.

The fundamental expression that describes the distribution of speeds of N gas molecules is

$$N_v = 4\pi N\left(\frac{m}{2\pi k_B T}\right)^{3/2} v^2 e^{-mv^2/2k_B T} \tag{21.26}$$

where m is the mass of a gas molecule, k_B is Boltzmann's constant, and T is the absolute temperature.[2] Observe the appearance of the Boltzmann factor $e^{-E/k_B T}$ with $E = \frac{1}{2}mv^2$.

As indicated in Figure 21.11, the average speed is somewhat lower than the rms speed. The *most probable speed* v_{mp} is the speed at which the distribution curve reaches a peak. Using Equation 21.26, one finds that

$$v_{rms} = \sqrt{\overline{v^2}} = \sqrt{\frac{3k_B T}{m}} = 1.73\sqrt{\frac{k_B T}{m}} \tag{21.27}$$

$$\bar{v} = \sqrt{\frac{8k_B T}{\pi m}} = 1.60\sqrt{\frac{k_B T}{m}} \tag{21.28}$$

$$v_{mp} = \sqrt{\frac{2k_B T}{m}} = 1.41\sqrt{\frac{k_B T}{m}} \tag{21.29}$$

Equation 21.27 has previously appeared as Equation 21.7. The details of the derivations of these equations from Equation 21.26 are left for the student (see Problems 39 and 65). From these equations, we see that

$$v_{rms} > \bar{v} > v_{mp}$$

Figure 21.12 represents speed distribution curves for nitrogen, N_2. The curves were obtained by using Equation 21.26 to evaluate the distribution function at various speeds and at two temperatures. Note that the peak in the curve shifts to the right as T increases, indicating that the average speed increases with increasing temperature, as expected. The asymmetric shape of the curves is due to the fact that the lowest speed possible is zero while the upper classical limit of the speed is infinity. (In Chapter 39, we will show that the actual upper limit is the speed of light.)

Equation 21.26 shows that the distribution of molecular speeds in a gas depends both on mass and on temperature. At a given temperature, the fraction of molecules with speeds exceeding a fixed value increases as the mass decreases. This explains why lighter molecules, such as H_2 and He, escape more readily from the Earth's atmosphere than do heavier molecules, such as N_2 and O_2. (See the discussion of escape speed in Chapter 13. Gas molecules escape even more readily from the Moon's surface than from the Earth's because the escape speed on the Moon is lower than that on the Earth.)

The speed distribution curves for molecules in a liquid are similar to those shown in Figure 21.12. We can understand the phenomenon of evaporation of a liquid from this distribution in speeds, using the fact that some molecules in the liquid are more

Ludwig Boltzmann

Austrian physicist (1844–1906)

Boltzmann made many important contributions to the development of the kinetic theory of gases, electromagnetism, and thermodynamics. His pioneering work in the field of kinetic theory led to the branch of physics known as statistical mechanics. *(Courtesy of AIP Niels Bohr Library, Lande Collection)*

[2] For the derivation of this expression, see an advanced textbook on thermodynamics, such as that by R. P. Bauman, *Modern Thermodynamics with Statistical Mechanics*, New York, Macmillan Publishing Co., 1992.

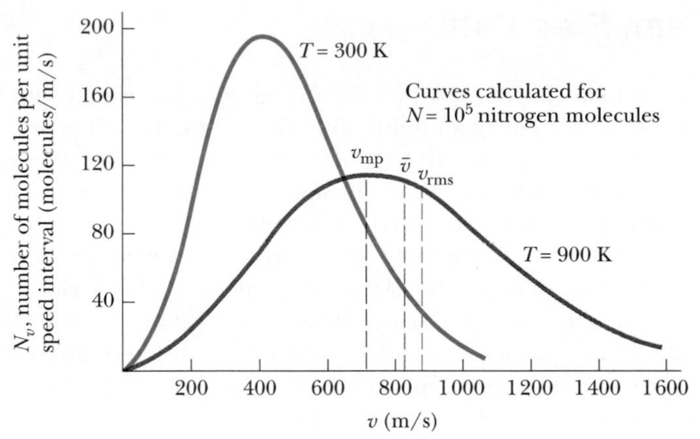

Active Figure 21.12 The speed distribution function for 10^5 nitrogen molecules at 300 K and 900 K. The total area under either curve is equal to the total number of molecules, which in this case equals 10^5. Note that $v_{rms} > \bar{v} > v_{mp}$.

At the Active Figures link at http://www.pse6.com, you can set the desired temperature and see the effect on the distribution curve.

energetic than others. Some of the faster-moving molecules in the liquid penetrate the surface and leave the liquid even at temperatures well below the boiling point. The molecules that escape the liquid by evaporation are those that have sufficient energy to overcome the attractive forces of the molecules in the liquid phase. Consequently, the molecules left behind in the liquid phase have a lower average kinetic energy; as a result, the temperature of the liquid decreases. Hence, evaporation is a cooling process. For example, an alcohol-soaked cloth often is placed on a feverish head to cool and comfort a patient.

Quick Quiz 21.8 Consider the qualitative shapes of the two curves in Figure 21.12, without regard for the numerical values or labels in the graph. Suppose you have two containers of gas *at the same temperature.* Container A has 10^5 nitrogen molecules and container B has 10^5 hydrogen molecules. The correct qualitative matching between the containers and the two curves in Figure 21.12 is (a) container A corresponds to the blue curve and container B to the brown curve (b) container B corresponds to the blue curve and container A to the brown curve (c) both containers correspond to the same curve.

Example 21.5 A System of Nine Particles

Nine particles have speeds of 5.00, 8.00, 12.0, 12.0, 12.0, 14.0, 14.0, 17.0, and 20.0 m/s.

(A) Find the particles' average speed.

Solution The average speed of the particles is the sum of the speeds divided by the total number of particles:

$$\bar{v} = \frac{(5.00 + 8.00 + 12.0 + 12.0 + 12.0 + 14.0 + 14.0 + 17.0 + 20.0)\ \text{m/s}}{9}$$

$$= 12.7\ \text{m/s}$$

(B) What is the rms speed of the particles?

Solution The average value of the square of the speed is

$$\overline{v^2} = \frac{(5.00^2 + 8.00^2 + 12.0^2 + 12.0^2 + 12.0^2 + 14.0^2 + 14.0^2 + 17.0^2 + 20.0^2)\ \text{m}^2/\text{s}^2}{9}$$

$$= 178\ \text{m}^2/\text{s}^2$$

Hence, the rms speed of the particles is

$$v_{rms} = \sqrt{\overline{v^2}} = \sqrt{178\ \text{m}^2/\text{s}^2} = 13.3\ \text{m/s}$$

(C) What is the most probable speed of the particles?

Solution Three of the particles have a speed of 12.0 m/s, two have a speed of 14.0 m/s, and the remaining have different speeds. Hence, we see that the most probable speed v_{mp} is

12.0 m/s.

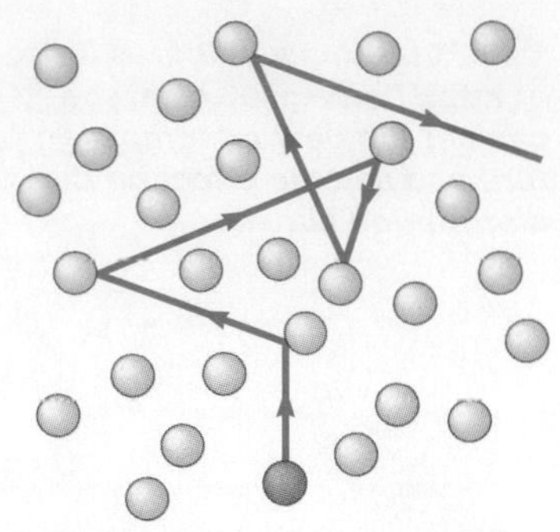

Figure 21.13 A molecule moving through a gas collides with other molecules in a random fashion. This behavior is sometimes referred to as a *random-walk process.* The mean free path increases as the number of molecules per unit volume decreases. Note that the motion is not limited to the plane of the paper.

21.7 Mean Free Path

Most of us are familiar with the fact that the strong odor associated with a gas such as ammonia may take a fraction of a minute to diffuse throughout a room. However, because average molecular speeds are typically several hundred meters per second at room temperature, we might expect a diffusion time of much less than one second. The reason for this difference is that molecules collide with one other because they are not geometrical points. Therefore, they do not travel from one side of a room to the other in a straight line. Between collisions, the molecules move with constant speed along straight lines. The average distance between collisions is called the **mean free path**. The path of an individual molecule is random and resembles that shown in Figure 21.13. As we would expect from this description, the mean free path is related to the diameter of the molecules and the density of the gas.

We now describe how to estimate the mean free path for a gas molecule. For this calculation, we assume that the molecules are spheres of diameter d. We see from Figure 21.14a that no two molecules collide unless their paths, assumed perpendicular to the page in Figure 21.14a are less than a distance d apart as the molecules approach each other. An equivalent way to describe the collisions is to imagine that one of the molecules has a diameter $2d$ and that the rest are geometrical points (Fig. 21.14b). Let us choose the large molecule to be one moving with the average speed $\overline{v}$. In a time interval Δt, this molecule travels a distance $\overline{v}\,\Delta t$. In this time interval, the molecule sweeps out a cylinder having a cross-sectional area πd^2 and a length $\overline{v}\,\Delta t$ (Fig. 21.15). Hence, the volume of the cylinder is $\pi d^2\overline{v}\,\Delta t$. If n_V is the number of molecules per unit volume, then the number of point-size molecules in the cylinder is $(\pi d^2\overline{v}\,\Delta t)\,n_V$. The molecule of equivalent diameter $2d$ collides with every molecule in this cylinder in the time interval Δt. Hence, the number of collisions in the time interval Δt is equal to the number of molecules in the cylinder, $(\pi d^2\overline{v}\,\Delta t)\,n_V$.

The mean free path ℓ equals the average distance $\overline{v}\,\Delta t$ traveled in a time interval Δt divided by the number of collisions that occur in that time interval:

$$\ell = \frac{\overline{v}\,\Delta t}{(\pi d^2\overline{v}\,\Delta t)\,n_V} = \frac{1}{\pi d^2 n_V}$$

Because the number of collisions in a time interval Δt is $(\pi d^2\overline{v}\,\Delta t)\,n_V$, the number of collisions per unit time interval, or **collision frequency *f*,** is

$$f = \pi d^2\overline{v}n_V$$

The inverse of the collision frequency is the average time interval between collisions, known as the **mean free time.**

Our analysis has assumed that molecules in the cylinder are stationary. When the motion of these molecules is included in the calculation, the correct results are

Mean free path

$$\ell = \frac{1}{\sqrt{2}\,\pi d^2 n_V} \tag{21.30}$$

Collision frequency

$$f = \sqrt{2}\,\pi d^2\overline{v}n_V = \frac{\overline{v}}{\ell} \tag{21.31}$$

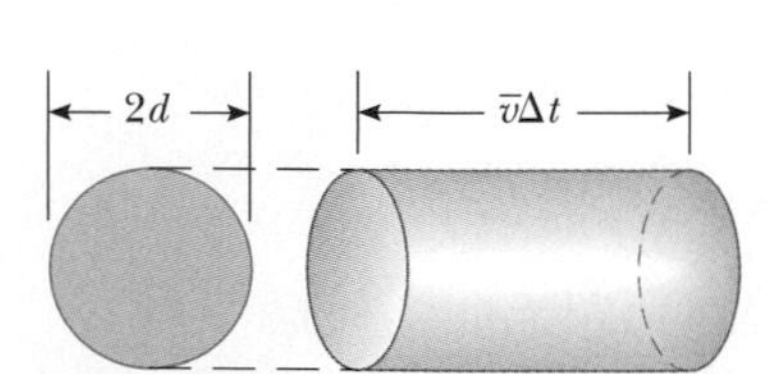

Figure 21.15 In a time interval Δt, a molecule of effective diameter $2d$ and moving to the right sweeps out a cylinder of length $\overline{v}\Delta t$ where $\overline{v}$ is its average speed. In this time interval, it collides with every point molecule within this cylinder.

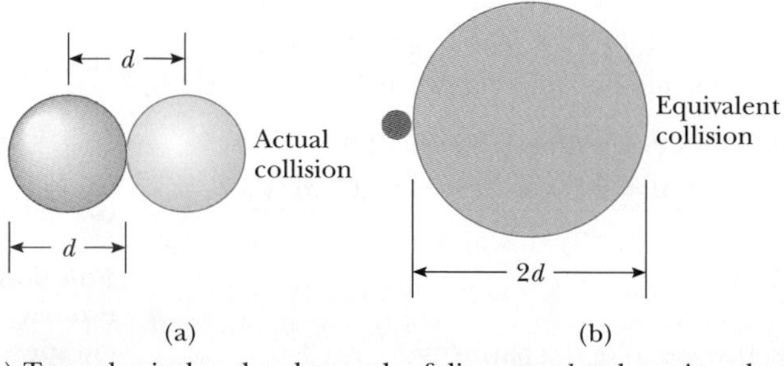

Figure 21.14 (a) Two spherical molecules, each of diameter d and moving along paths perpendicular to the page, collide if their paths are within a distance d of each other. (b) The collision between the two molecules is equivalent to a point molecule colliding with a molecule having an effective diameter of $2d$.

Example 21.6 Bouncing Around in the Air

Approximate the air around you as a collection of nitrogen molecules, each having a diameter of 2.00×10^{-10} m.

(A) How far does a typical molecule move before it collides with another molecule?

Solution Assuming that the gas is ideal, we can use the equation $PV = Nk_BT$ to obtain the number of molecules per unit volume under typical room conditions:

$$n_V = \frac{N}{V} = \frac{P}{k_BT} = \frac{1.01 \times 10^5\ \text{N/m}^2}{(1.38 \times 10^{-23}\ \text{J/K})(293\ \text{K})}$$

$$= 2.50 \times 10^{25}\ \text{molecules/m}^3$$

Hence, the mean free path is

$$\ell = \frac{1}{\sqrt{2}\pi d^2 n_V}$$

$$= \frac{1}{\sqrt{2}\pi(2.00 \times 10^{-10}\ \text{m})^2(2.50 \times 10^{25}\ \text{molecules/m}^3)}$$

$$= \boxed{2.25 \times 10^{-7}\ \text{m}}$$

This value is about 10^3 times greater than the molecular diameter.

(B) On average, how frequently does one molecule collide with another?

Solution Because the rms speed of a nitrogen molecule at 20.0°C is 511 m/s (see Table 21.1), we know from Equations 21.27 and 21.28 that $\overline{v} = (1.60/1.73)(511\ \text{m/s}) = 473\ \text{m/s}$. Therefore, the collision frequency is

$$f = \frac{\overline{v}}{\ell} = \frac{473\ \text{m/s}}{2.25 \times 10^{-7}\ \text{m}} = \boxed{2.10 \times 10^9/\text{s}}$$

The molecule collides with other molecules at the average rate of about two billion times each second!

The mean free path ℓ is *not* the same as the average separation between particles. In fact, the average separation d between particles is approximately $n_V^{-1/3}$. In this example, the average molecular separation is

$$d = \frac{1}{n_V^{1/3}} = \frac{1}{(2.5 \times 10^{25})^{1/3}} = 3.4 \times 10^{-9}\ \text{m}$$

SUMMARY

Take a practice test for this chapter by clicking on the Practice Test link at http://www.pse6.com.

The pressure of N molecules of an ideal gas contained in a volume V is

$$P = \frac{2}{3}\frac{N}{V}\left(\tfrac{1}{2}m\overline{v^2}\right) \qquad (21.2)$$

The average translational kinetic energy per molecule of a gas, $\frac{1}{2}m\overline{v^2}$, is related to the temperature T of the gas through the expression

$$\tfrac{1}{2}m\overline{v^2} = \tfrac{3}{2}k_BT \qquad (21.4)$$

where k_B is Boltzmann's constant. Each translational degree of freedom (x, y, or z) has $\frac{1}{2}k_BT$ of energy associated with it.

The **theorem of equipartition of energy** states that the energy of a system in thermal equilibrium is equally divided among all degrees of freedom.

The internal energy of N molecules (or n mol) of an ideal monatomic gas is

$$E_{\text{int}} = \tfrac{3}{2}Nk_BT = \tfrac{3}{2}nRT \qquad (21.10)$$

The change in internal energy for n mol of any ideal gas that undergoes a change in temperature ΔT is

$$\Delta E_{\text{int}} = nC_V\Delta T \qquad (21.12)$$

where C_V is the molar specific heat at constant volume.

The molar specific heat of an ideal monatomic gas at constant volume is $C_V = \frac{3}{2}R$; the molar specific heat at constant pressure is $C_P = \frac{5}{2}R$. The ratio of specific heats is given by $\gamma = C_P/C_V = \frac{5}{3}$.

If an ideal gas undergoes an adiabatic expansion or compression, the first law of thermodynamics, together with the equation of state, shows that

$$PV^{\gamma} = \text{constant} \qquad (21.18)$$

The **Boltzmann distribution law** describes the distribution of particles among available energy states. The relative number of particles having energy between E and $E + dE$ is $n_V(E)\,dE$, where

$$n_V(E) = n_0 e^{-E/k_B T} \tag{21.25}$$

The **Maxwell–Boltzmann speed distribution function** describes the distribution of speeds of molecules in a gas:

$$N_v = 4\pi N \left(\frac{m}{2\pi k_B T}\right)^{3/2} v^2 e^{-mv^2/2k_B T} \tag{21.26}$$

This expression enables us to calculate the **root-mean-square speed,** the **average speed,** and **the most probable speed:**

$$v_{\text{rms}} = \sqrt{\overline{v^2}} = \sqrt{\frac{3k_B T}{m}} = 1.73\sqrt{\frac{k_B T}{m}} \tag{21.27}$$

$$\overline{v} = \sqrt{\frac{8k_B T}{\pi m}} = 1.60\sqrt{\frac{k_B T}{m}} \tag{21.28}$$

$$v_{\text{mp}} = \sqrt{\frac{2k_B T}{m}} = 1.41\sqrt{\frac{k_B T}{m}} \tag{21.29}$$

QUESTIONS

1. Dalton's law of partial pressures states that the total pressure of a mixture of gases is equal to the sum of the partial pressures of gases making up the mixture. Give a convincing argument for this law based on the kinetic theory of gases.
2. One container is filled with helium gas and another with argon gas. If both containers are at the same temperature, which molecules have the higher rms speed? Explain.
3. A gas consists of a mixture of He and N_2 molecules. Do the lighter He molecules travel faster than the N_2 molecules? Explain.
4. Although the average speed of gas molecules in thermal equilibrium at some temperature is greater than zero, the average velocity is zero. Explain why this statement must be true.
5. When alcohol is rubbed on your body, it lowers your skin temperature. Explain this effect.
6. A liquid partially fills a container. Explain why the temperature of the liquid decreases if the container is then partially evacuated. (Using this technique, it is possible to freeze water at temperatures above 0°C.)
7. A vessel containing a fixed volume of gas is cooled. Does the mean free path of the molecules increase, decrease, or remain constant in the cooling process? What about the collision frequency?
8. A gas is compressed at a constant temperature. What happens to the mean free path of the molecules in this process?
9. If a helium-filled balloon initially at room temperature is placed in a freezer, will its volume increase, decrease, or remain the same?
10. Which is denser, dry air or air saturated with water vapor? Explain.
11. What happens to a helium-filled balloon released into the air? Will it expand or contract? Will it stop rising at some height?
12. Why does a diatomic gas have a greater energy content per mole than a monatomic gas at the same temperature?
13. An ideal gas is contained in a vessel at 300 K. If the temperature is increased to 900 K, by what factor does each one of the following change? (a) The average kinetic energy of the molecules. (b) The rms molecular speed. (c) The average momentum change of one molecule in a collision with a wall. (d) The rate of collisions of molecules with walls. (e) The pressure of the gas.
14. A vessel is filled with gas at some equilibrium pressure and temperature. Can all gas molecules in the vessel have the same speed?
15. In our model of the kinetic theory of gases, molecules were viewed as hard spheres colliding elastically with the walls of the container. Is this model realistic?
16. In view of the fact that hot air rises, why does it generally become cooler as you climb a mountain? (Note that air has low thermal conductivity.)
17. Inspecting the magnitudes of C_V and C_P for the diatomic and polyatomic gases in Table 21.2, we find that the values increase with increasing molecular mass. Give a qualitative explanation of this observation.

PROBLEMS

1, 2, 3 = straightforward, intermediate, challenging □ = full solution available in the *Student Solutions Manual and Study Guide*

= coached solution with hints available at http://www.pse6.com = computer useful in solving problem

= paired numerical and symbolic problems

Section 21.1 Molecular Model of an Ideal Gas

1. In a 30.0-s interval, 500 hailstones strike a glass window of area $0.600\ m^2$ at an angle of 45.0° to the window surface. Each hailstone has a mass of 5.00 g and moves with a speed of 8.00 m/s. Assuming the collisions are elastic, find the average force and pressure on the window.

2. In a period of 1.00 s, 5.00×10^{23} nitrogen molecules strike a wall with an area of $8.00\ cm^2$. If the molecules move with a speed of 300 m/s and strike the wall head-on in elastic collisions, what is the pressure exerted on the wall? (The mass of one N_2 molecule is 4.68×10^{-26} kg.)

3. A sealed cubical container 20.0 cm on a side contains three times Avogadro's number of molecules at a temperature of 20.0°C. Find the force exerted by the gas on one of the walls of the container.

4. A 2.00-mol sample of oxygen gas is confined to a 5.00-L vessel at a pressure of 8.00 atm. Find the average translational kinetic energy of an oxygen molecule under these conditions.

5. A spherical balloon of volume $4\,000\ cm^3$ contains helium at an (inside) pressure of 1.20×10^5 Pa. How many moles of helium are in the balloon if the average kinetic energy of the helium atoms is 3.60×10^{-22} J?

6. Use the definition of Avogadro's number to find the mass of a helium atom.

7. (a) How many atoms of helium gas fill a balloon having a diameter of 30.0 cm at 20.0°C and 1.00 atm? (b) What is the average kinetic energy of the helium atoms? (c) What is the root-mean-square speed of the helium atoms?

8. Given that the rms speed of a helium atom at a certain temperature is 1 350 m/s, find by proportion the rms speed of an oxygen (O_2) molecule at this temperature. The molar mass of O_2 is 32.0 g/mol, and the molar mass of He is 4.00 g/mol.

9. A cylinder contains a mixture of helium and argon gas in equilibrium at 150°C. (a) What is the average kinetic energy for each type of gas molecule? (b) What is the root-mean-square speed of each type of molecule?

10. A 5.00-L vessel contains nitrogen gas at 27.0°C and a pressure of 3.00 atm. Find (a) the total translational kinetic energy of the gas molecules and (b) the average kinetic energy per molecule.

11. (a) Show that 1 Pa = 1 J/m³. (b) Show that the density in space of the translational kinetic energy of an ideal gas is $3P/2$.

Section 21.2 Molar Specific Heat of an Ideal Gas

Note: You may use data in Table 21.2 about particular gases. Here we define a "monatomic ideal gas" to have molar specific heats $C_V = 3R/2$ and $C_P = 5R/2$, and a "diatomic ideal gas" to have $C_V = 5R/2$ and $C_P = 7R/2$.

12. Calculate the change in internal energy of 3.00 mol of helium gas when its temperature is increased by 2.00 K.

13. A 1.00-mol sample of hydrogen gas is heated at constant pressure from 300 K to 420 K. Calculate (a) the energy transferred to the gas by heat, (b) the increase in its internal energy, and (c) the work done on the gas.

14. A 1.00-mol sample of air (a diatomic ideal gas) at 300 K, confined in a cylinder under a heavy piston, occupies a volume of 5.00 L. Determine the final volume of the gas after 4.40 kJ of energy is transferred to the air by heat.

15. In a constant-volume process, 209 J of energy is transferred by heat to 1.00 mol of an ideal monatomic gas initially at 300 K. Find (a) the increase in internal energy of the gas, (b) the work done on it, and (c) its final temperature.

16. A house has well-insulated walls. It contains a volume of $100\ m^3$ of air at 300 K. (a) Calculate the energy required to increase the temperature of this diatomic ideal gas by 1.00°C. (b) **What If?** If this energy could be used to lift an object of mass m through a height of 2.00 m, what is the value of m?

17. An incandescent lightbulb contains a volume V of argon at pressure P_i. The bulb is switched on and constant power $\mathcal{P}$ is transferred to the argon for a time interval Δt. (a) Show that the pressure P_f in the bulb at the end of this process is $P_f = P_i[1 + (\mathcal{P}\Delta t R)/(P_i V C_V)]$. (b) Find the pressure in a spherical light bulb 10.0 cm in diameter 4.00 s after it is switched on, given that it has initial pressure 1.00 atm and that 3.60 W of power is transferred to the gas.

18. A vertical cylinder with a heavy piston contains air at a temperature of 300 K. The initial pressure is 200 kPa, and the initial volume is $0.350\ m^3$. Take the molar mass of air as 28.9 g/mol and assume that $C_V = 5R/2$. (a) Find the specific heat of air at constant volume in units of J/kg · °C. (b) Calculate the mass of the air in the cylinder. (c) Suppose the piston is held fixed. Find the energy input required to raise the temperature of the air to 700 K. (d) **What If?** Assume again the conditions of the initial state and that the heavy piston is free to move. Find the energy input required to raise the temperature to 700 K.

19. A 1-L Thermos bottle is full of tea at 90°C. You pour out one cup and immediately screw the stopper back on. Make an order-of-magnitude estimate of the change in temperature of the tea remaining in the flask that results from the admission of air at room temperature. State the quantities you take as data and the values you measure or estimate for them.

20. A 1.00-mol sample of a diatomic ideal gas has pressure P and volume V. When the gas is heated, its pressure triples and its volume doubles. This heating process includes two steps, the first at constant pressure and the second at constant volume. Determine the amount of energy transferred to the gas by heat.

21. A 1.00-mol sample of an ideal monatomic gas is at an initial temperature of 300 K. The gas undergoes an isovolumetric process acquiring 500 J of energy by heat. It then undergoes an isobaric process losing this same amount of energy by heat. Determine (a) the new temperature of the gas and (b) the work done on the gas.

22. A vertical cylinder with a movable piston contains 1.00 mol of a diatomic ideal gas. The volume of the gas is V_i, and its temperature is T_i. Then the cylinder is set on a stove and additional weights are piled onto the piston as it moves up, in such a way that the pressure is proportional to the volume and the final volume is $2V_i$. (a) What is the final temperature? (b) How much energy is transferred to the gas by heat?

23. A container has a mixture of two gases: n_1 mol of gas 1 having molar specific heat C_1 and n_2 mol of gas 2 of molar specific heat C_2. (a) Find the molar specific heat of the mixture. (b) **What If?** What is the molar specific heat if the mixture has m gases in the amounts $n_1, n_2, n_3, \ldots, n_m$, with molar specific heats $C_1, C_2, C_3, \ldots, C_m$, respectively?

Section 21.3 Adiabatic Processes for an Ideal Gas

24. During the compression stroke of a certain gasoline engine, the pressure increases from 1.00 atm to 20.0 atm. If the process is adiabatic and the fuel–air mixture behaves as a diatomic ideal gas, (a) by what factor does the volume change and (b) by what factor does the temperature change? (c) Assuming that the compression starts with 0.016 0 mol of gas at 27.0°C, find the values of Q, W, and ΔE_{int} that characterize the process.

25. A 2.00-mol sample of a diatomic ideal gas expands slowly and adiabatically from a pressure of 5.00 atm and a volume of 12.0 L to a final volume of 30.0 L. (a) What is the final pressure of the gas? (b) What are the initial and final temperatures? (c) Find Q, W, and ΔE_{int}.

26. Air (a diatomic ideal gas) at 27.0°C and atmospheric pressure is drawn into a bicycle pump that has a cylinder with an inner diameter of 2.50 cm and length 50.0 cm. The down stroke adiabatically compresses the air, which reaches a gauge pressure of 800 kPa before entering the tire (Fig. P21.26). Determine (a) the volume of the compressed air and (b) the temperature of the compressed air. (c) **What If?** The pump is made of steel and has an inner wall that is 2.00 mm thick. Assume that 4.00 cm of the cylinder's length is allowed to come to thermal equilibrium with the air. What will be the increase in wall temperature?

27. Air in a thundercloud expands as it rises. If its initial temperature is 300 K and no energy is lost by thermal conduction on expansion, what is its temperature when the initial volume has doubled?

28. The largest bottle ever made by blowing glass has a volume of about 0.720 m^3. Imagine that this bottle is filled with air that behaves as an ideal diatomic gas. The bottle is held with its opening at the bottom and rapidly submerged into the ocean. No air escapes or mixes with the water. No energy is exchanged with the ocean by heat. (a) If the final volume of the air is 0.240 m^3, by what factor does the internal energy of the air increase? (b) If the bottle is submerged so that the air temperature doubles, how much volume is occupied by air?

George Sample

Figure P21.26

29. A 4.00-L sample of a diatomic ideal gas with specific heat ratio 1.40, confined to a cylinder, is carried through a closed cycle. The gas is initially at 1.00 atm and at 300 K. First, its pressure is tripled under constant volume. Then, it expands adiabatically to its original pressure. Finally, the gas is compressed isobarically to its original volume. (a) Draw a PV diagram of this cycle. (b) Determine the volume of the gas at the end of the adiabatic expansion. (c) Find the temperature of the gas at the start of the adiabatic expansion. (d) Find the temperature at the end of the cycle. (e) What was the net work done on the gas for this cycle?

30. A diatomic ideal gas ($\gamma = 1.40$) confined to a cylinder is put through a closed cycle. Initially the gas is at P_i, V_i, and T_i. First, its pressure is tripled under constant volume. It then expands adiabatically to its original pressure and finally is compressed isobarically to its original volume. (a) Draw a PV diagram of this cycle. (b) Determine the volume at the end of the adiabatic expansion. Find (c) the temperature of the gas at the start of the adiabatic expansion and (d) the temperature at the end of the cycle. (e) What was the net work done on the gas for this cycle?

31. How much work is required to compress 5.00 mol of air at 20.0°C and 1.00 atm to one tenth of the original volume (a) by an isothermal process? (b) by an adiabatic process? (c) What is the final pressure in each of these two cases?

32. During the power stroke in a four-stroke automobile engine, the piston is forced down as the mixture of combustion products and air undergoes an adiabatic expansion (Fig. P21.32). Assume that (1) the engine is running at 2 500 cycles/min, (2) the gauge pressure right before the expansion is 20.0 atm, (3) the volumes of the mixture right before and after the expansion are 50.0 and 400 cm^3, respectively, (4) the time involved in the expansion is one-fourth that of the total cycle, and (5) the mixture behaves like an ideal gas with specific heat ratio 1.40. Find the average power generated during the expansion.

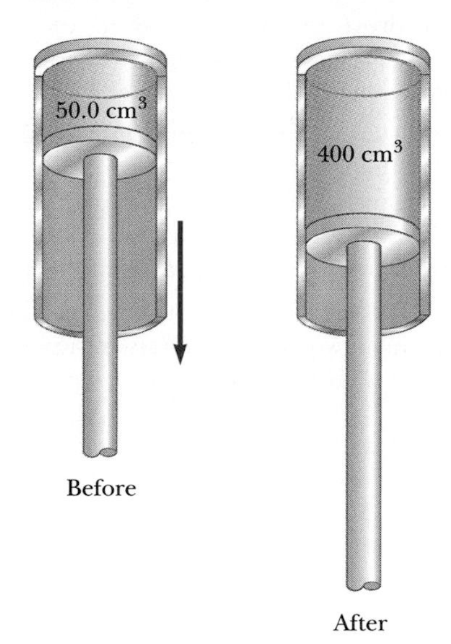

Figure P21.32

Section 21.4 The Equipartition of Energy

33. Consider 2.00 mol of an ideal diatomic gas. (a) Find the total heat capacity of the gas at constant volume and at constant pressure assuming the molecules rotate but do not vibrate. (b) **What If?** Repeat, assuming the molecules both rotate and vibrate.

34. A certain molecule has f degrees of freedom. Show that an ideal gas consisting of such molecules has the following properties: (1) its total internal energy is $fnRT/2$; (2) its molar specific heat at constant volume is $fR/2$; (3) its molar specific heat at constant pressure is $(f + 2)R/2$; (4) its specific heat ratio is $\gamma = C_P/C_V = (f + 2)/f$.

35. In a crude model (Fig. P21.35) of a rotating diatomic molecule of chlorine (Cl_2), the two Cl atoms are 2.00×10^{-10} m apart and rotate about their center of mass with angular speed $\omega = 2.00 \times 10^{12}$ rad/s. What is the rotational kinetic energy of one molecule of Cl_2, which has a molar mass of 70.0 g/mol?

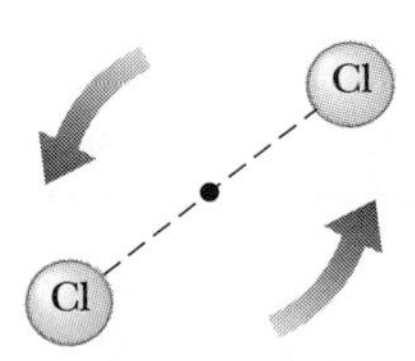

Figure P21.35

Section 21.5 The Boltzmann Distribution Law

Section 21.6 Distribution of Molecular Speeds

36. One cubic meter of atomic hydrogen at 0°C and atmospheric pressure contains approximately 2.70×10^{25} atoms. The first excited state of the hydrogen atom has an energy of 10.2 eV above the lowest energy level, called the ground state. Use the Boltzmann factor to find the number of atoms in the first excited state at 0°C and at 10 000°C.

37. Fifteen identical particles have various speeds: one has a speed of 2.00 m/s; two have speeds of 3.00 m/s; three have speeds of 5.00 m/s; four have speeds of 7.00 m/s; three have speeds of 9.00 m/s; and two have speeds of 12.0 m/s. Find (a) the average speed, (b) the rms speed, and (c) the most probable speed of these particles.

38. Two gases in a mixture diffuse through a filter at rates proportional to the gases' rms speeds. (a) Find the ratio of speeds for the two isotopes of chlorine, ^{35}Cl and ^{37}Cl, as they diffuse through the air. (b) Which isotope moves faster?

39. From the Maxwell–Boltzmann speed distribution, show that the most probable speed of a gas molecule is given by Equation 21.29. Note that the most probable speed corresponds to the point at which the slope of the speed distribution curve dN_v/dv is zero.

40. Helium gas is in thermal equilibrium with liquid helium at 4.20 K. Even though it is on the point of condensation, model the gas as ideal and determine the most probable speed of a helium atom (mass $= 6.64 \times 10^{-27}$ kg) in it.

41. **Review problem.** At what temperature would the average speed of helium atoms equal (a) the escape speed from Earth, 1.12×10^4 m/s and (b) the escape speed from the Moon, 2.37×10^3 m/s? (See Chapter 13 for a discussion of escape speed, and note that the mass of a helium atom is 6.64×10^{-27} kg.)

42. A gas is at 0°C. If we wish to double the rms speed of its molecules, to what temperature must the gas be brought?

43. Assume that the Earth's atmosphere has a uniform temperature of 20°C and uniform composition, with an effective molar mass of 28.9 g/mol. (a) Show that the number density of molecules depends on height according to

$$n_V(y) = n_0 e^{-mgy/k_B T}$$

where n_0 is the number density at sea level, where $y = 0$. This result is called the *law of atmospheres.* (b) Commercial jetliners typically cruise at an altitude of 11.0 km. Find the ratio of the atmospheric density there to the density at sea level.

44. *If you can't walk to outer space, can you at least walk halfway?* Using the law of atmospheres from Problem 43, we find that the average height of a molecule in the Earth's atmosphere is given by

$$\bar{y} = \frac{\int_0^\infty y n_V(y)\, dy}{\int_0^\infty n_V(y)\, dy} = \frac{\int_0^\infty y e^{-mgy/k_B T} dy}{\int_0^\infty e^{-mgy/k_B T} dy}$$

(a) Prove that this average height is equal to $k_B T/mg$. (b) Evaluate the average height, assuming the temperature is 10°C and the molecular mass is 28.9 u.

Section 21.7 Mean Free Path

45. In an ultra-high-vacuum system, the pressure is measured to be 1.00×10^{-10} torr (where 1 torr = 133 Pa). Assuming the molecular diameter is 3.00×10^{-10} m, the average molecular speed is 500 m/s, and the temperature is 300 K, find (a) the number of molecules in a volume of 1.00 m^3, (b) the mean free path of the molecules, and (c) the collision frequency.

46. In deep space the number density of particles can be one particle per cubic meter. Using the average temperature of 3.00 K and assuming the particle is H_2 with a diameter of 0.200 nm, (a) determine the mean free path of the particle and the average time between collisions. (b) **What If?** Repeat part (a) assuming a density of one particle per cubic centimeter.

47. Show that the mean free path for the molecules of an ideal gas is

$$\ell = \frac{k_B T}{\sqrt{2}\pi d^2 P}$$

where d is the molecular diameter.

48. In a tank full of oxygen, how many molecular diameters d (on average) does an oxygen molecule travel (at 1.00 atm and 20.0°C) before colliding with another O_2 molecule? (The diameter of the O_2 molecule is approximately 3.60×10^{-10} m.)

49. Argon gas at atmospheric pressure and 20.0°C is confined in a 1.00-m^3 vessel. The effective hard-sphere diameter of the argon atom is 3.10×10^{-10} m. (a) Determine the mean free path ℓ. (b) Find the pressure when $\ell = 1.00$ m. (c) Find the pressure when $\ell = 3.10 \times 10^{-10}$ m.

Additional Problems

50. The dimensions of a room are 4.20 m × 3.00 m × 2.50 m. (a) Find the number of molecules of air in the room at atmospheric pressure and 20.0°C. (b) Find the mass of this air, assuming that the air consists of diatomic molecules with molar mass 28.9 g/mol. (c) Find the average kinetic energy of one molecule. (d) Find the root-mean-square molecular speed. (e) On the assumption that the molar specific heat is a constant independent of temperature, we have $E_{int} = 5nRT/2$. Find the internal energy in the air. (f) **What If?** Find the internal energy of the air in the room at 25.0°C.

51. The function $E_{int} = 3.50nRT$ describes the internal energy of a certain ideal gas. A sample comprising 2.00 mol of the gas always starts at pressure 100 kPa and temperature 300 K. For each one of the following processes, determine the final pressure, volume, and temperature; the change in internal energy of the gas; the energy added to the gas by heat; and the work done on the gas. (a) The gas is heated at constant pressure to 400 K. (b) The gas is heated at constant volume to 400 K. (c) The gas is compressed at constant temperature to 120 kPa. (d) The gas is compressed adiabatically to 120 kPa.

52. Twenty particles, each of mass m and confined to a volume V, have various speeds: two have speed v; three have speed $2v$; five have speed $3v$; four have speed $4v$; three have speed $5v$; two have speed $6v$; one has speed $7v$. Find (a) the average speed, (b) the rms speed, (c) the most probable speed, (d) the pressure the particles exert on the walls of the vessel, and (e) the average kinetic energy per particle.

53. A cylinder containing n mol of an ideal gas undergoes an adiabatic process. (a) Starting with the expression $W = -\int P dV$ and using the condition PV^γ = constant, show that the work done on the gas is

$$W = \left(\frac{1}{\gamma - 1}\right)(P_f V_f - P_i V_i)$$

(b) Starting with the first law of thermodynamics in differential form, prove that the work done on the gas is also equal to $nC_V(T_f - T_i)$. Show that this result is consistent with the equation in part (a).

54. As a 1.00-mol sample of a monatomic ideal gas expands adiabatically, the work done on it is −2 500 J. The initial temperature and pressure of the gas are 500 K and 3.60 atm. Calculate (a) the final temperature and (b) the final pressure. You may use the result of Problem 53.

55. A cylinder is closed at both ends and has insulating walls. It is divided into two compartments by a perfectly insulating partition that is perpendicular to the axis of the cylinder. Each compartment contains 1.00 mol of oxygen, which behaves as an ideal gas with $\gamma = 7/5$. Initially the two compartments have equal volumes, and their temperatures are 550 K and 250 K. The partition is then allowed to move slowly until the pressures on its two sides are equal. Find the final temperatures in the two compartments. You may use the result of Problem 53.

56. An air rifle shoots a lead pellet by allowing high-pressure air to expand, propelling the pellet down the rifle barrel. Because this process happens very quickly, no appreciable thermal conduction occurs, and the expansion is essentially adiabatic. Suppose that the rifle starts by admitting to the barrel 12.0 cm^3 of compressed air, which behaves as an ideal gas with $\gamma = 1.40$. The air expands behind a 1.10-g pellet and pushes on it as a piston with cross-sectional area 0.030 0 cm^2, as the pellet moves 50.0 cm along the gun barrel. The pellet emerges with muzzle speed 120 m/s. Use the result of problem 53 to find the initial pressure required.

57. **Review problem.** Oxygen at pressures much greater than 1 atm is toxic to lung cells. Assume that a deep-sea diver breathes a mixture of oxygen (O_2) and helium (He). By weight, what ratio of helium to oxygen must be used if the diver is at an ocean depth of 50.0 m?

58. A vessel contains 1.00×10^4 oxygen molecules at 500 K. (a) Make an accurate graph of the Maxwell–Boltzmann speed distribution function versus speed with points at speed intervals of 100 m/s. (b) Determine the most probable speed from this graph. (c) Calculate the average and rms speeds for the molecules and label these points on your graph. (d) From the graph, estimate the fraction of molecules with speeds in the range 300 m/s to 600 m/s.

59. The compressibility κ of a substance is defined as the fractional change in volume of that substance for a given change in pressure:

$$\kappa = -\frac{1}{V}\frac{dV}{dP}$$

(a) Explain why the negative sign in this expression ensures that κ is always positive. (b) Show that if an ideal gas is compressed isothermally, its compressibility is given by $\kappa_1 = 1/P$. (c) **What If?** Show that if an ideal gas is compressed adiabatically, its compressibility is given by $\kappa_2 = 1/\gamma P$. (d) Determine values for κ_1 and κ_2 for a monatomic ideal gas at a pressure of 2.00 atm.

60. **Review problem.** (a) Show that the speed of sound in an ideal gas is

$$v = \sqrt{\frac{\gamma RT}{M}}$$

where M is the molar mass. Use the general expression for the speed of sound in a fluid from Section 17.1, the definition of the bulk modulus from Section 12.4, and the result of Problem 59 in this chapter. As a sound wave passes through a gas, the compressions are either so rapid or so far apart that thermal conduction is prevented by a negligible time interval or by effective thickness of insulation. The compressions and rarefactions are adiabatic. (b) Compute the theoretical speed of sound in air at 20°C and compare it with the value in Table 17.1. Take $M = 28.9$ g/mol. (c) Show that the speed of sound in an ideal gas is

$$v = \sqrt{\frac{\gamma k_B T}{m}}$$

where m is the mass of one molecule. Compare it with the most probable, average, and rms molecular speeds.

61. Model air as a diatomic ideal gas with $M = 28.9$ g/mol. A cylinder with a piston contains 1.20 kg of air at 25.0°C and 200 kPa. Energy is transferred by heat into the system as it is allowed to expand, with the pressure rising to 400 kPa. Throughout the expansion, the relationship between pressure and volume is given by

$$P = CV^{1/2}$$

where C is a constant. (a) Find the initial volume. (b) Find the final volume. (c) Find the final temperature. (d) Find the work done on the air. (e) Find the energy transferred by heat.

62. *Smokin'!* A pitcher throws a 0.142-kg baseball at 47.2 m/s (Fig. P21.62). As it travels 19.4 m, the ball slows to a speed of 42.5 m/s because of air resistance. Find the change in temperature of the air through which it passes. To find the greatest possible temperature change, you may make the following assumptions: Air has a molar specific heat of $C_P = 7R/2$ and an equivalent molar mass of 28.9 g/mol. The process is so rapid that the cover of the baseball acts as thermal insulation, and the temperature of the ball itself does not change. A change in temperature happens initially only for the air in a cylinder 19.4 m in length and 3.70 cm in radius. This air is initially at 20.0°C.

63. For a Maxwellian gas, use a computer or programmable calculator to find the numerical value of the ratio $N_v(v)/N_v(v_{mp})$ for the following values of v: $v = (v_{mp}/50)$, $(v_{mp}/10)$, $(v_{mp}/2)$, v_{mp}, $2v_{mp}$, $10v_{mp}$, and $50v_{mp}$. Give your results to three significant figures.

64. Consider the particles in a gas centrifuge, a device used to separate particles of different mass by whirling them in a circular path of radius r at angular speed ω. The force acting toward the center of the circular path on a given particle is $m\omega^2 r$. (a) Discuss how a gas centrifuge can be used to separate particles of different mass. (b) Show that the density of the particles as a function of r is

$$n(r) = n_0 e^{mr^2\omega^2/2k_B T}$$

Figure P21.62 John Lackey, the first rookie to win a World Series game 7 in 93 years, pitches for the Anaheim Angels during the final game of the 2002 World Series.

65. Verify Equations 21.27 and 21.28 for the rms and average speed of the molecules of a gas at a temperature T. Note that the average value of v^n is

$$\overline{v^n} = \frac{1}{N}\int_0^\infty v^n N_v\,dv$$

Use the table of definite integrals in Appendix B (Table B.6).

66. On the PV diagram for an ideal gas, one isothermal curve and one adiabatic curve pass through each point. Prove that the slope of the adiabat is steeper than the slope of the isotherm by the factor γ.

67. A sample of monatomic ideal gas occupies 5.00 L at atmospheric pressure and 300 K (point A in Figure P21.67). It is heated at constant volume to 3.00 atm (point B). Then it is allowed to expand isothermally to 1.00 atm (point C) and at last compressed isobarically to its original state. (a) Find the number of moles in the sample. (b) Find the temperature at points B and C and the volume at point C. (c) Assuming that the molar specific heat does not depend on temperature, so that $E_{int} = 3nRT/2$, find the internal energy at points A, B, and C. (d) Tabulate P, V, T, and E_{int} for the states at points A, B, and C. (e) Now consider the processes $A \rightarrow B$, $B \rightarrow C$, and $C \rightarrow A$. Describe just how to carry out each process experimentally. (f) Find Q, W, and ΔE_{int} for each of the processes. (g) For the whole cycle $A \rightarrow B \rightarrow C \rightarrow A$ find Q, W, and ΔE_{int}.

68. This problem can help you to think about the size of molecules. In the city of Beijing a restaurant keeps a pot of chicken broth simmering continuously. Every morning it is topped up to contain 10.0 L of water, along with a fresh chicken, vegetables, and spices. The soup is thoroughly stirred. The molar mass of water is 18.0 g/mol. (a) Find

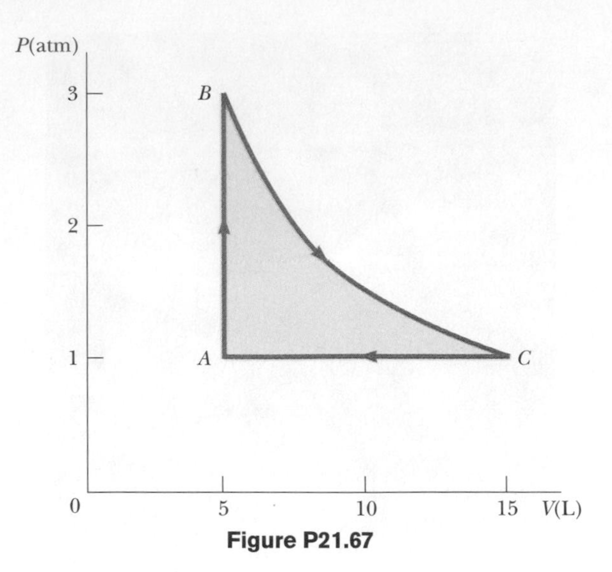

Figure P21.67

the number of molecules of water in the pot. (b) During a certain month, 90.0% of the broth was served each day to people who then emigrated immediately. Of the water molecules in the pot on the first day of the month, when was the last one likely to have been ladled out of the pot? (c) The broth has been simmering for centuries, through wars, earthquakes, and stove repairs. Suppose the water that was in the pot long ago has thoroughly mixed into the Earth's hydrosphere, of mass 1.32×10^{21} kg. How many of the water molecules originally in the pot are likely to be present in it again today?

69. **Review problem.** (a) If it has enough kinetic energy, a molecule at the surface of the Earth can "escape the Earth's gravitation," in the sense that it can continue to move away from the Earth forever, as discussed in Section 13.7. Using the principle of conservation of energy, show that the minimum kinetic energy needed for "escape" is mgR_E, where m is the mass of the molecule, g is the free-fall acceleration at the surface, and R_E is the radius of the Earth. (b) Calculate the temperature for which the minimum escape kinetic energy is ten times the average kinetic energy of an oxygen molecule.

70. Using multiple laser beams, physicists have been able to cool and trap sodium atoms in a small region. In one experiment the temperature of the atoms was reduced to 0.240 mK. (a) Determine the rms speed of the sodium atoms at this temperature. The atoms can be trapped for about 1.00 s. The trap has a linear dimension of roughly 1.00 cm. (b) Approximately how long would it take an atom to wander out of the trap region if there were no trapping action?

Answers to Quick Quizzes

21.1 (b). The average translational kinetic energy per molecule is a function only of temperature.

21.2 (a). Because there are twice as many molecules and the temperature of both containers is the same, the total energy in B is twice that in A.

21.3 (b). Because both containers hold the same type of gas, the rms speed is a function only of temperature.

21.4 (a). According to Equation 21.10, E_{int} is a function of temperature only. Because the temperature increases, the internal energy increases.

21.5 (c). Along an isotherm, T is constant by definition. Therefore, the internal energy of the gas does not change.

21.6 (d). The value of 29.1 J/mol·K is $7R/2$. According to Figure 21.7, this suggests that all three types of motion are occurring.

21.7 (c). The highest possible value of C_V for a diatomic gas is $7R/2$, so the gas must be polyatomic.

21.8 (a). Because the hydrogen atoms are lighter than the nitrogen molecules, they move with a higher average speed and the distribution curve is stretched out more along the horizontal axis. See Equation 21.26 for a mathematical statement of the dependence of N_v on m.

B.C. **By John Hart**

Chapter 22

Heat Engines, Entropy, and the Second Law of Thermodynamics

▲ *This cutaway image of an automobile engine shows two pistons that have work done on them by an explosive mixture of air and fuel, ultimately leading to the motion of the automobile. This apparatus can be modeled as a heat engine, which we study in this chapter. (Courtesy of Ford Motor Company)*

CHAPTER OUTLINE

22.1 Heat Engines and the Second Law of Thermodynamics

22.2 Heat Pumps and Refrigerators

22.3 Reversible and Irreversible Processes

22.4 The Carnot Engine

22.5 Gasoline and Diesel Engines

22.6 Entropy

22.7 Entropy Changes in Irreversible Processes

22.8 Entropy on a Microscopic Scale

The first law of thermodynamics, which we studied in Chapter 20, is a statement of conservation of energy. This law states that a change in internal energy in a system can occur as a result of energy transfer by heat or by work, or by both. As was stated in Chapter 20, the law makes no distinction between the results of heat and the results of work—either heat or work can cause a change in internal energy. However, there is an important distinction between heat and work that is not evident from the first law. One manifestation of this distinction is that it is impossible to design a device that, operating in a cyclic fashion, takes in energy by heat and expels an *equal* amount of energy by work. A cyclic device that takes in energy by heat and expels a *fraction* of this energy by work is possible and is called a *heat engine.*

Lord Kelvin

British physicist and mathematician (1824–1907)

Born William Thomson in Belfast, Kelvin was the first to propose the use of an absolute scale of temperature. The Kelvin temperature scale is named in his honor. Kelvin's work in thermodynamics led to the idea that energy cannot pass spontaneously from a colder object to a hotter object.
(J. L. Charmet/SPL/Photo Researchers, Inc.)

Although the first law of thermodynamics is very important, it makes no distinction between processes that occur spontaneously and those that do not. However, only certain types of energy-conversion and energy-transfer processes actually take place in nature. The *second law of thermodynamics,* the major topic in this chapter, establishes which processes do and which do not occur. The following are examples of processes that do not violate the principle of conservation of energy if they proceed in either direction, but are observed to proceed in only one direction, governed by the second law:

- When two objects at different temperatures are placed in thermal contact with each other, the net transfer of energy by heat is always from the warmer object to the cooler object, never from the cooler to the warmer.
- A rubber ball dropped to the ground bounces several times and eventually comes to rest, but a ball lying on the ground never gathers internal energy from the ground and begins bouncing on its own.
- An oscillating pendulum eventually comes to rest because of collisions with air molecules and friction at the point of suspension. The mechanical energy of the system is converted to internal energy in the air, the pendulum, and the suspension; the reverse conversion of energy never occurs.

All these processes are *irreversible*—that is, they are processes that occur naturally in one direction only. No irreversible process has ever been observed to run backward—if it were to do so, it would violate the second law of thermodynamics.[1]

From an engineering standpoint, perhaps the most important implication of the second law is the limited efficiency of heat engines. The second law states that a machine that operates in a cycle, taking in energy by heat and expelling an equal amount of energy by work, cannot be constructed.

[1] Although we have never *observed* a process occurring in the time-reversed sense, it is *possible* for it to occur. As we shall see later in the chapter, however, the probability of such a process occurring is infinitesimally small. From this viewpoint, we say that processes occur with a vastly greater probability in one direction than in the opposite direction.

22.1 Heat Engines and the Second Law of Thermodynamics

A **heat engine** is a device that takes in energy by heat[2] and, operating in a cyclic process, expels a fraction of that energy by means of work. For instance, in a typical process by which a power plant produces electricity, coal or some other fuel is burned, and the high-temperature gases produced are used to convert liquid water to steam. This steam is directed at the blades of a turbine, setting it into rotation. The mechanical energy associated with this rotation is used to drive an electric generator. Another device that can be modeled as a heat engine—the internal combustion engine in an automobile—uses energy from a burning fuel to perform work on pistons that results in the motion of the automobile.

A heat engine carries some working substance through a cyclic process during which (1) the working substance absorbs energy by heat from a high-temperature energy reservoir, (2) work is done by the engine, and (3) energy is expelled by heat to a lower-temperature reservoir. As an example, consider the operation of a steam engine (Fig. 22.1), which uses water as the working substance. The water in a boiler absorbs energy from burning fuel and evaporates to steam, which then does work by expanding against a piston. After the steam cools and condenses, the liquid water produced returns to the boiler and the cycle repeats.

It is useful to represent a heat engine schematically as in Figure 22.2. The engine absorbs a quantity of energy $|Q_h|$ from the hot reservoir. For this discussion of heat engines, we will use absolute values to make all energy transfers positive and will indicate the direction of transfer with an explicit positive or negative sign. The engine does work W_{eng} (so that *negative* work $W = -W_{\text{eng}}$ is done *on* the engine), and then gives up a quantity of energy $|Q_c|$ to the cold reservoir. Because the working substance goes

Figure 22.1 This steam-driven locomotive runs from Durango to Silverton, Colorado. It obtains its energy by burning wood or coal. The generated energy vaporizes water into steam, which powers the locomotive. (This locomotive must take on water from tanks located along the route to replace steam lost through the funnel.) Modern locomotives use diesel fuel instead of wood or coal. Whether old-fashioned or modern, such locomotives can be modeled as heat engines, which extract energy from a burning fuel and convert a fraction of it to mechanical energy.

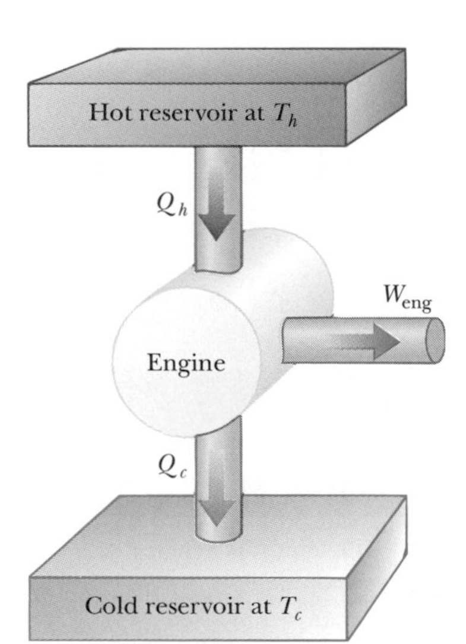

Active Figure 22.2 Schematic representation of a heat engine. The engine does work W_{eng}. The arrow at the top represents energy $Q_h > 0$ entering the engine. At the bottom, $Q_c < 0$ represents energy leaving the engine.

At the Active Figures link at http://www.pse6.com, you can select the efficiency of the engine and observe the transfer of energy.

[2] We will use heat as our model for energy transfer into a heat engine. Other methods of energy transfer are also possible in the model of a heat engine, however. For example, the Earth's atmosphere can be modeled as a heat engine, in which the input energy transfer is by means of electromagnetic radiation from the Sun. The output of the atmospheric heat engine causes the wind structure in the atmosphere.

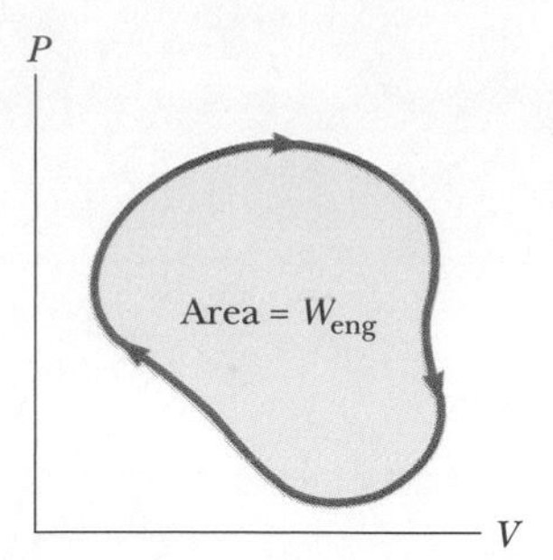

Figure 22.3 *PV* diagram for an arbitrary cyclic process taking place in an engine. The value of the net work done by the engine in one cycle equals the area enclosed by the curve.

through a cycle, its initial and final internal energies are equal, and so $\Delta E_{int} = 0$. Hence, from the first law of thermodynamics, $\Delta E_{int} = Q + W = Q - W_{eng}$, and with no change in internal energy, **the net work W_{eng} done by a heat engine is equal to the net energy Q_{net} transferred to it.** As we can see from Figure 22.2, $Q_{net} = |Q_h| - |Q_c|$; therefore,

$$W_{eng} = |Q_h| - |Q_c| \tag{22.1}$$

If the working substance is a gas, **the net work done in a cyclic process is the area enclosed by the curve representing the process on a *PV* diagram.** This is shown for an arbitrary cyclic process in Figure 22.3.

The **thermal efficiency** e of a heat engine is defined as the ratio of the net work done by the engine during one cycle to the energy input at the higher temperature during the cycle:

Thermal efficiency of a heat engine

$$e = \frac{W_{eng}}{|Q_h|} = \frac{|Q_h| - |Q_c|}{|Q_h|} = 1 - \frac{|Q_c|}{|Q_h|} \tag{22.2}$$

We can think of the efficiency as the ratio of what you gain (work) to what you give (energy transfer at the higher temperature). In practice, all heat engines expel only a fraction of the input energy Q_h by mechanical work and consequently their efficiency is always less than 100%. For example, a good automobile engine has an efficiency of about 20%, and diesel engines have efficiencies ranging from 35% to 40%.

Equation 22.2 shows that a heat engine has 100% efficiency ($e = 1$) only if $|Q_c| = 0$—that is, if no energy is expelled to the cold reservoir. In other words, a heat engine with perfect efficiency would have to expel all of the input energy by work. On the basis of the fact that efficiencies of real engines are well below 100%, the **Kelvin–Planck form of the second law of thermodynamics** states the following:

> It is impossible to construct a heat engine that, operating in a cycle, produces no effect other than the input of energy by heat from a reservoir and the performance of an equal amount of work.

This statement of the second law means that, during the operation of a heat engine, W_{eng} can never be equal to $|Q_h|$, or, alternatively, that some energy $|Q_c|$ must be rejected to the environment. Figure 22.4 is a schematic diagram of the impossible "perfect" heat engine.

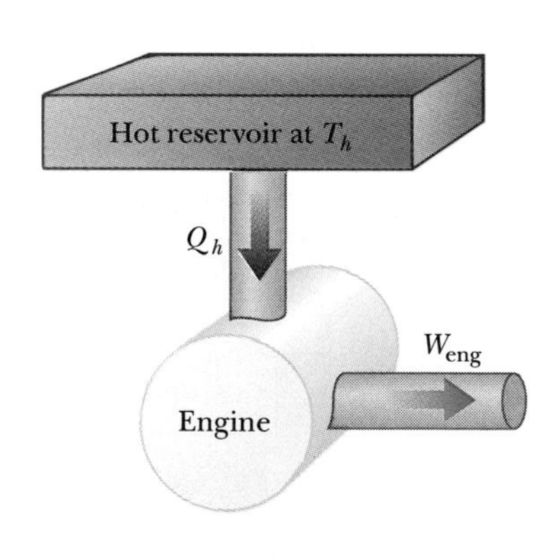

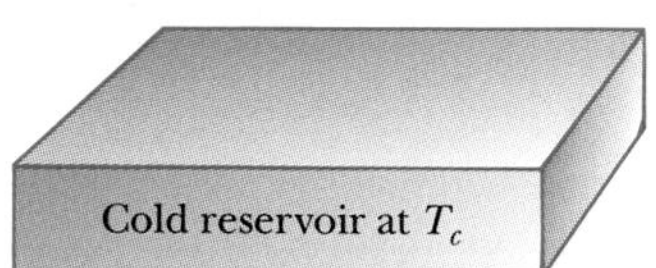

The impossible engine

Figure 22.4 Schematic diagram of a heat engine that takes in energy from a hot reservoir and does an equivalent amount of work. It is impossible to construct such a perfect engine.

Quick Quiz 22.1 The energy input to an engine is 3.00 times greater than the work it performs. What is its thermal efficiency? (a) 3.00 (b) 1.00 (c) 0.333 (d) impossible to determine

Quick Quiz 22.2 For the engine of Quick Quiz 22.1, what fraction of the energy input is expelled to the cold reservoir? (a) 0.333 (b) 0.667 (c) 1.00 (d) impossible to determine

Example 22.1 The Efficiency of an Engine

An engine transfers 2.00×10^3 J of energy from a hot reservoir during a cycle and transfers 1.50×10^3 J as exhaust to a cold reservoir.

(A) Find the efficiency of the engine.

Solution The efficiency of the engine is given by Equation 22.2 as

$$e = 1 - \frac{|Q_c|}{|Q_h|} = 1 - \frac{1.50 \times 10^3 \text{ J}}{2.00 \times 10^3 \text{ J}} = 0.250, \text{ or } 25.0\%$$

(B) How much work does this engine do in one cycle?

Solution The work done is the difference between the input and output energies:

$$W_{\text{eng}} = |Q_h| - |Q_c| = 2.00 \times 10^3\text{ J} - 1.50 \times 10^3\text{ J}$$

$$= \boxed{5.0 \times 10^2\text{ J}}$$

What If? Suppose you were asked for the power output of this engine? Do you have sufficient information to answer this question?

Answer No, you do not have enough information. The power of an engine is the *rate* at which work is done by the engine. You know how much work is done per cycle but you have no information about the time interval associated with one cycle. However, if you were told that the engine operates at 2 000 rpm (revolutions per minute), you could relate this rate to the period of rotation T of the mechanism of the engine. If we assume that there is one thermodynamic cycle per revolution, then the power is

$$\mathcal{P} = \frac{W_{\text{eng}}}{T} = \frac{5.0 \times 10^2\text{ J}}{\left(\frac{1}{2000}\text{ min}\right)}\left(\frac{1\text{ min}}{60\text{ s}}\right) = 1.7 \times 10^4\text{ W}$$

22.2 Heat Pumps and Refrigerators

In a heat engine, the direction of energy transfer is from the hot reservoir to the cold reservoir, which is the natural direction. The role of the heat engine is to process the energy from the hot reservoir so as to do useful work. What if we wanted to transfer energy from the cold reservoir to the hot reservoir? Because this is not the natural direction of energy transfer, we must put some energy into a device in order to accomplish this. Devices that perform this task are called **heat pumps** or **refrigerators.** For example, we cool homes in summer using heat pumps called *air conditioners.* The air conditioner transfers energy from the cool room in the home to the warm air outside.

In a refrigerator or heat pump, the engine takes in energy $|Q_c|$ from a cold reservoir and expels energy $|Q_h|$ to a hot reservoir (Fig. 22.5). This can be accomplished only if work is done *on* the engine. From the first law, we know that the energy given up to the hot reservoir must equal the sum of the work done and the energy taken in from the cold reservoir. Therefore, the refrigerator or heat pump transfers energy from a colder body (for example, the contents of a kitchen refrigerator or the winter air outside a building) to a hotter body (the air in the kitchen or a room in the building). In practice, it is desirable to carry out this process with a minimum of work. If it could be accomplished without doing any work, then the refrigerator or heat pump would be "perfect" (Fig. 22.6). Again, the existence of such a device would be in violation of the second law of thermodynamics, which in the form of the **Clausius statement**[3] states:

PITFALL PREVENTION

22.1 The First and Second Laws

Notice the distinction between the first and second laws of thermodynamics. If a gas undergoes a *one-time isothermal process* $\Delta E_{\text{int}} = Q + W = 0$. Therefore, the first law allows *all* energy input by heat to be expelled by work. In a heat engine, however, in which a substance undergoes a cyclic process, only a *portion* of the energy input by heat can be expelled by work according to the second law.

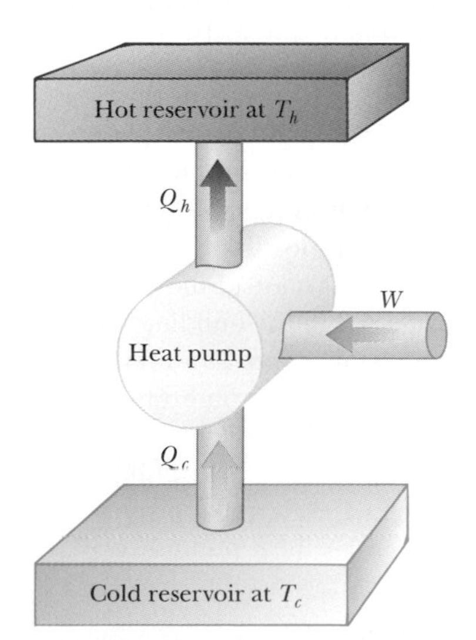

Active Figure 22.5 Schematic diagram of a heat pump, which takes in energy $Q_c > 0$ from a cold reservoir and expels energy $Q_h < 0$ to a hot reservoir. Work W is done *on* the heat pump. A refrigerator works the same way.

At the Active Figures link at http://www.pse6.com, you can select the COP of the heat pump and observe the transfer of energy.

[3] First expressed by Rudolf Clausius (1822–1888).

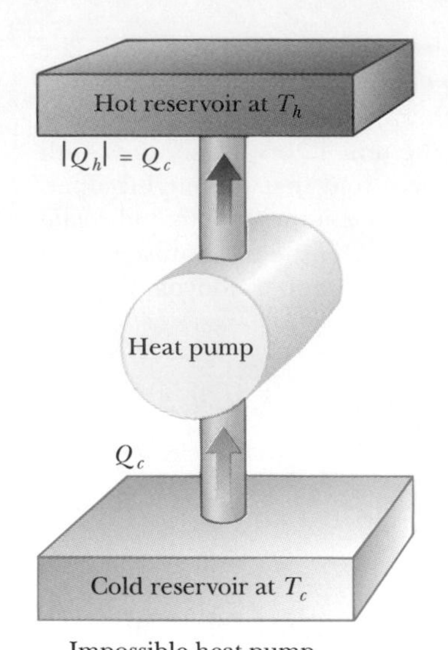

Impossible heat pump

Figure 22.6 Schematic diagram of an impossible heat pump or refrigerator—that is, one that takes in energy from a cold reservoir and expels an equivalent amount of energy to a hot reservoir without the input of energy by work.

It is impossible to construct a cyclical machine whose sole effect is to transfer energy continuously by heat from one object to another object at a higher temperature without the input of energy by work.

In simpler terms, **energy does not transfer spontaneously by heat from a cold object to a hot object.** This direction of energy transfer requires an input of energy to a heat pump, which is often supplied by means of electricity.

The Clausius and Kelvin–Planck statements of the second law of thermodynamics appear, at first sight, to be unrelated, but in fact they are equivalent in all respects. Although we do not prove so here, if either statement is false, then so is the other.[4]

Heat pumps have long been used for cooling homes and buildings, and they are now becoming increasingly popular for heating them as well. The heat pump contains two sets of metal coils that can exchange energy by heat with the surroundings: one set on the outside of the building, in contact with the air or buried in the ground, and the other set in the interior of the building. In the heating mode, a circulating fluid flowing through the coils absorbs energy from the outside and releases it to the interior of the building from the interior coils. The fluid is cold and at low pressure when it is in the external coils, where it absorbs energy by heat from either the air or the ground. The resulting warm fluid is then compressed and enters the interior coils as a hot, high-pressure fluid, where it releases its stored energy to the interior air.

An air conditioner is simply a heat pump with its exterior and interior coils interchanged, so that it operates in the cooling mode. Energy is absorbed into the circulating fluid in the interior coils; then, after the fluid is compressed, energy leaves the fluid through the external coils. The air conditioner must have a way to release energy to the outside. Otherwise, the work done on the air conditioner would represent energy added to the air inside the house, and the temperature would increase. In the same manner, a refrigerator cannot cool the kitchen if the refrigerator door is left open. The amount of energy leaving the external coils (Fig. 22.7) behind or underneath the refrigerator is greater than the amount of energy removed from the food. The difference between the energy out and the energy in is the work done by the electricity supplied to the refrigerator.

Figure 22.7 The coils on the back of a refrigerator transfer energy by heat to the air. The second law of thermodynamics states that this amount of energy must be greater than the amount of energy removed from the contents of the refrigerator, due to the input of energy by work.

The effectiveness of a heat pump is described in terms of a number called the **coefficient of performance** (COP). In the heating mode, the COP is defined as the ratio of the energy transferred to the hot reservoir to the work required to transfer that energy:

$$\text{COP (heating mode)} \equiv \frac{\text{energy transferred at high temperature}}{\text{work done by heat pump}} = \frac{|Q_h|}{W} \qquad (22.3)$$

Note that the COP is similar to the thermal efficiency for a heat engine in that it is a ratio of what you gain (energy delivered to the interior of the building) to what you give (work input). Because $|Q_h|$ is generally greater than W, typical values for the COP are greater than unity. It is desirable for the COP to be as high as possible, just as it is desirable for the thermal efficiency of an engine to be as high as possible.

If the outside temperature is 25°F (−4°C) or higher, a typical value of the COP for a heat pump is about 4. That is, the amount of energy transferred to the building is about four times greater than the work done by the motor in the heat pump. However, as the outside temperature decreases, it becomes more difficult for the heat pump to extract sufficient energy from the air, and so the COP decreases. In fact, the COP can fall below unity for temperatures below about 15°F (−9°C). Thus, the use of heat pumps that extract energy from the air, while satisfactory in moderate climates, is not appropriate in areas where winter temperatures are very low. It is possible to use heat pumps in colder

[4] See, for example, R. P. Bauman, *Modern Thermodynamics and Statistical Mechanics*, New York, Macmillan Publishing Co., 1992.

areas by burying the external coils deep in the ground. In this case, the energy is extracted from the ground, which tends to be warmer than the air in the winter.

For a heat pump operating in the cooling mode, "what you gain" is energy removed from the cold reservoir. The most effective refrigerator or air conditioner is one that removes the greatest amount of energy from the cold reservoir in exchange for the least amount of work. Thus, for these devices we define the COP in terms of $|Q_c|$:

$$\text{COP (cooling mode)} = \frac{|Q_c|}{W} \tag{22.4}$$

A good refrigerator should have a high COP, typically 5 or 6.

Quick Quiz 22.3 The energy entering an electric heater by electrical transmission can be converted to internal energy with an efficiency of 100%. By what factor does the cost of heating your home change when you replace your electric heating system with an electric heat pump that has a COP of 4.00? Assume that the motor running the heat pump is 100% efficient. (a) 4.00 (b) 2.00 (c) 0.500 (d) 0.250

Example 22.2 Freezing Water

A certain refrigerator has a COP of 5.00. When the refrigerator is running, its power input is 500 W. A sample of water of mass 500 g and temperature 20.0°C is placed in the freezer compartment. How long does it take to freeze the water to ice at 0°C? Assume that all other parts of the refrigerator stay at the same temperature and there is no leakage of energy from the exterior, so that the operation of the refrigerator results only in energy being extracted from the water.

Solution Conceptualize this problem by realizing that energy leaves the water, reducing its temperature and then freezing it into ice. The time interval required for this entire process is related to the rate at which energy is withdrawn from the water, which, in turn is related to the power input of the refrigerator. We categorize this problem as one in which we will need to combine our understanding of temperature changes and phase changes from Chapter 20 with our understanding of heat pumps from the current chapter. To analyze the problem, we first find the amount of energy that we must extract from 500 g of water at 20°C to turn it into ice at 0°C. Using Equations 20.4 and 20.6,

$$\begin{aligned} |Q_c| &= |mc\,\Delta T + mL_f| = m\,|c\,\Delta T + L_f| \\ &= (0.500\text{ kg})[(4\,186\text{ J/kg}\cdot{}^\circ\text{C})(20.0^\circ\text{C}) + 3.33\times10^5\text{ J/kg}] \\ &= 2.08\times10^5\text{ J} \end{aligned}$$

Now we use Equation 22.4 to find out how much energy we need to provide to the refrigerator to extract this much energy from the water:

$$\text{COP} = \frac{|Q_c|}{W} \longrightarrow W = \frac{|Q_c|}{\text{COP}} = \frac{2.08\times10^5\text{ J}}{5.00}$$

$$W = 4.17\times10^4\text{ J}$$

Using the power rating of the refrigerator, we find out the time interval required for the freezing process to occur:

$$\mathcal{P} = \frac{W}{\Delta t} \longrightarrow \Delta t = \frac{W}{\mathcal{P}} = \frac{4.17\times10^4\text{ J}}{500\text{ W}} = 83.3\text{ s}$$

To finalize this problem, note that this time interval is very different from that of our everyday experience; this suggests the difficulties with our assumptions. Only a small part of the energy extracted from the refrigerator interior in a given time interval will come from the water. Energy must also be extracted from the container in which the water is placed, and energy that continuously leaks into the interior from the exterior must be continuously extracted. In reality, the time interval for the water to freeze is much longer than 83.3 s.

22.3 Reversible and Irreversible Processes

In the next section we discuss a theoretical heat engine that is the most efficient possible. To understand its nature, we must first examine the meaning of reversible and irreversible processes. In a **reversible** process, the system undergoing the process can be

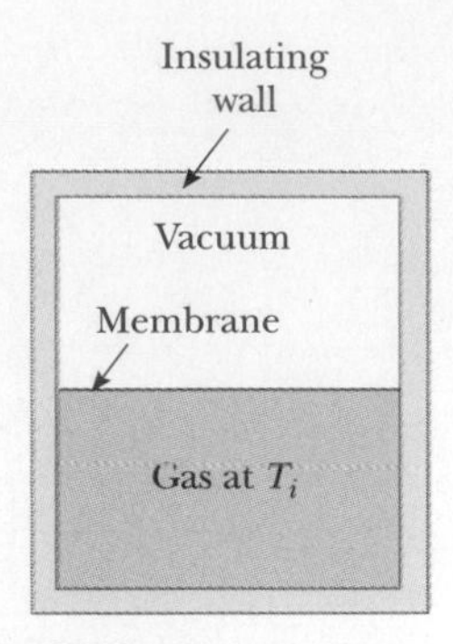

Figure 22.8 Adiabatic free expansion of a gas.

returned to its initial conditions along the same path on a *PV* diagram, and every point along this path is an equilibrium state. A process that does not satisfy these requirements is **irreversible.**

All natural processes are known to be irreversible. From the endless number of examples that could be selected, let us examine the adiabatic free expansion of a gas, which was already discussed in Section 20.6, and show that it cannot be reversible. Consider a gas in a thermally insulated container, as shown in Figure 22.8. A membrane separates the gas from a vacuum. When the membrane is punctured, the gas expands freely into the vacuum. As a result of the puncture, the system has changed because it occupies a greater volume after the expansion. Because the gas does not exert a force through a displacement, it does no work on the surroundings as it expands. In addition, no energy is transferred to or from the gas by heat because the container is insulated from its surroundings. Thus, in this adiabatic process, the system has changed but the surroundings have not.

For this process to be reversible, we need to be able to return the gas to its original volume and temperature without changing the surroundings. Imagine that we try to reverse the process by compressing the gas to its original volume. To do so, we fit the container with a piston and use an engine to force the piston inward. During this process, the surroundings change because work is being done by an outside agent on the system. In addition, the system changes because the compression increases the temperature of the gas. We can lower the temperature of the gas by allowing it to come into contact with an external energy reservoir. Although this step returns the gas to its original conditions, the surroundings are again affected because energy is being added to the surroundings from the gas. If this energy could somehow be used to drive the engine that compressed the gas, then the net energy transfer to the surroundings would be zero. In this way, the system and its surroundings could be returned to their initial conditions, and we could identify the process as reversible. However, the Kelvin–Planck statement of the second law specifies that the energy removed from the gas to return the temperature to its original value cannot be completely converted to mechanical energy in the form of the work done by the engine in compressing the gas. Thus, we must conclude that the process is irreversible.

PITFALL PREVENTION

22.2 All Real Processes Are Irreversible

The reversible process is an idealization—all real processes on Earth are irreversible.

We could also argue that the adiabatic free expansion is irreversible by relying on the portion of the definition of a reversible process that refers to equilibrium states. For example, during the expansion, significant variations in pressure occur throughout the gas. Thus, there is no well-defined value of the pressure for the entire system at any time between the initial and final states. In fact, the process cannot even be represented as a path on a *PV* diagram. The *PV* diagram for an adiabatic free expansion would show the initial and final conditions as points, but these points would not be connected by a path. Thus, because the intermediate conditions between the initial and final states are not equilibrium states, the process is irreversible.

Although all real processes are irreversible, some are almost reversible. If a real process occurs very slowly such that the system is always very nearly in an equilibrium state, then the process can be approximated as being reversible. Suppose that a gas is compressed isothermally in a piston–cylinder arrangement in which the gas is in thermal contact with an energy reservoir, and we continuously transfer just enough energy from the gas to the reservoir during the process to keep the temperature constant. For example, imagine that the gas is compressed very slowly by dropping grains of sand onto a frictionless piston, as shown in Figure 22.9. As each grain lands on the piston and compresses the gas a bit, the system deviates from an equilibrium state, but is so close to one that it achieves a new equilibrium state in a relatively short time interval. Each grain added represents a change to a new equilibrium state but the differences between states are so small that we can approximate the entire process as occurring through continuous equilibrium states. We can reverse the process by slowly removing grains from the piston.

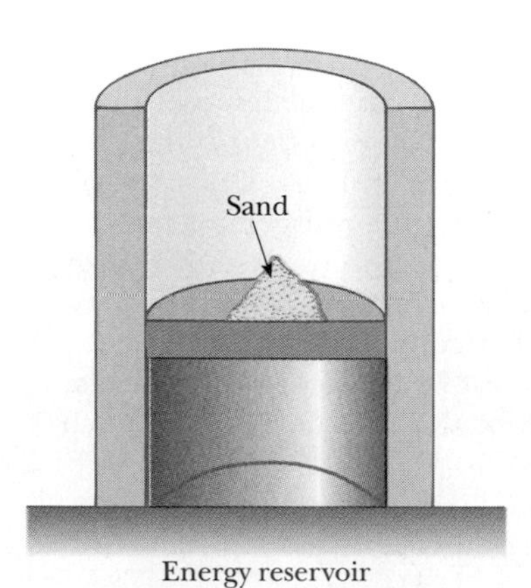

Figure 22.9 A gas in thermal contact with an energy reservoir is compressed slowly as individual grains of sand drop onto the piston. The compression is isothermal and reversible.

A general characteristic of a reversible process is that no dissipative effects (such as turbulence or friction) that convert mechanical energy to internal energy can be present. Such effects can be impossible to eliminate completely. Hence, it is not surprising that real processes in nature are irreversible.

22.4 The Carnot Engine

In 1824 a French engineer named Sadi Carnot described a theoretical engine, now called a **Carnot engine,** which is of great importance from both practical and theoretical viewpoints. He showed that a heat engine operating in an ideal, reversible cycle—called a **Carnot cycle**—between two energy reservoirs is the most efficient engine possible. Such an ideal engine establishes an upper limit on the efficiencies of all other engines. That is, the net work done by a working substance taken through the Carnot cycle is the greatest amount of work possible for a given amount of energy supplied to the substance at the higher temperature. **Carnot's theorem** can be stated as follows:

> No real heat engine operating between two energy reservoirs can be more efficient than a Carnot engine operating between the same two reservoirs.

Sadi Carnot
French engineer (1796–1832)

Carnot was the first to show the quantitative relationship between work and heat. In 1824 he published his only work–*Reflections on the Motive Power of Heat*–which reviewed the industrial, political, and economic importance of the steam engine. In it, he defined work as "weight lifted through a height." *(J.-L. Charmet/Science Photo Library/Photo Researchers, Inc.)*

To argue the validity of this theorem, imagine two heat engines operating between the *same* energy reservoirs. One is a Carnot engine with efficiency e_C, and the other is an engine with efficiency e, where we assume that $e > e_C$. The more efficient engine is used to drive the Carnot engine as a Carnot refrigerator. The output by work of the more efficient engine is matched to the input by work of the Carnot refrigerator. For the *combination* of the engine and refrigerator, no exchange by work with the surroundings occurs. Because we have assumed that the engine is more efficient than the refrigerator, the net result of the combination is a transfer of energy from the cold to the hot reservoir without work being done on the combination. According to the Clausius statement of the second law, this is impossible. Hence, the assumption that $e > e_C$ must be false. **All real engines are less efficient than the Carnot engine because they do not operate through a reversible cycle.** The efficiency of a real engine is further reduced by such practical difficulties as friction and energy losses by conduction.

To describe the Carnot cycle taking place between temperatures T_c and T_h, we assume that the working substance is an ideal gas contained in a cylinder fitted with a movable piston at one end. The cylinder's walls and the piston are thermally nonconducting. Four stages of the Carnot cycle are shown in Figure 22.10, and the *PV* diagram for the cycle is shown in Figure 22.11. The Carnot cycle consists of two adiabatic processes and two isothermal processes, all reversible:

1. Process $A \rightarrow B$ (Fig. 22.10a) is an isothermal expansion at temperature T_h. The gas is placed in thermal contact with an energy reservoir at temperature T_h. During the expansion, the gas absorbs energy $|Q_h|$ from the reservoir through the base of the cylinder and does work W_{AB} in raising the piston.
2. In process $B \rightarrow C$ (Fig. 22.10b), the base of the cylinder is replaced by a thermally nonconducting wall, and the gas expands adiabatically—that is, no energy enters or leaves the system by heat. During the expansion, the temperature of the gas decreases from T_h to T_c and the gas does work W_{BC} in raising the piston.
3. In process $C \rightarrow D$ (Fig. 22.10c), the gas is placed in thermal contact with an energy reservoir at temperature T_c and is compressed isothermally at temperature T_c. During this time, the gas expels energy $|Q_c|$ to the reservoir, and the work done by the piston on the gas is W_{CD}.
4. In the final process $D \rightarrow A$ (Fig. 22.10d), the base of the cylinder is replaced by a nonconducting wall, and the gas is compressed adiabatically. The temperature of the gas increases to T_h, and the work done by the piston on the gas is W_{DA}.

PITFALL PREVENTION

22.3 Don't Shop for a Carnot Engine

The Carnot engine is an idealization—do not expect a Carnot engine to be developed for commercial use. We explore the Carnot engine only for theoretical considerations.

At the Active Figures link at http://www.pse6.com, you can observe the motion of the piston in the Carnot cycle while you also observe the cycle on the PV diagram of Figure 22.11.

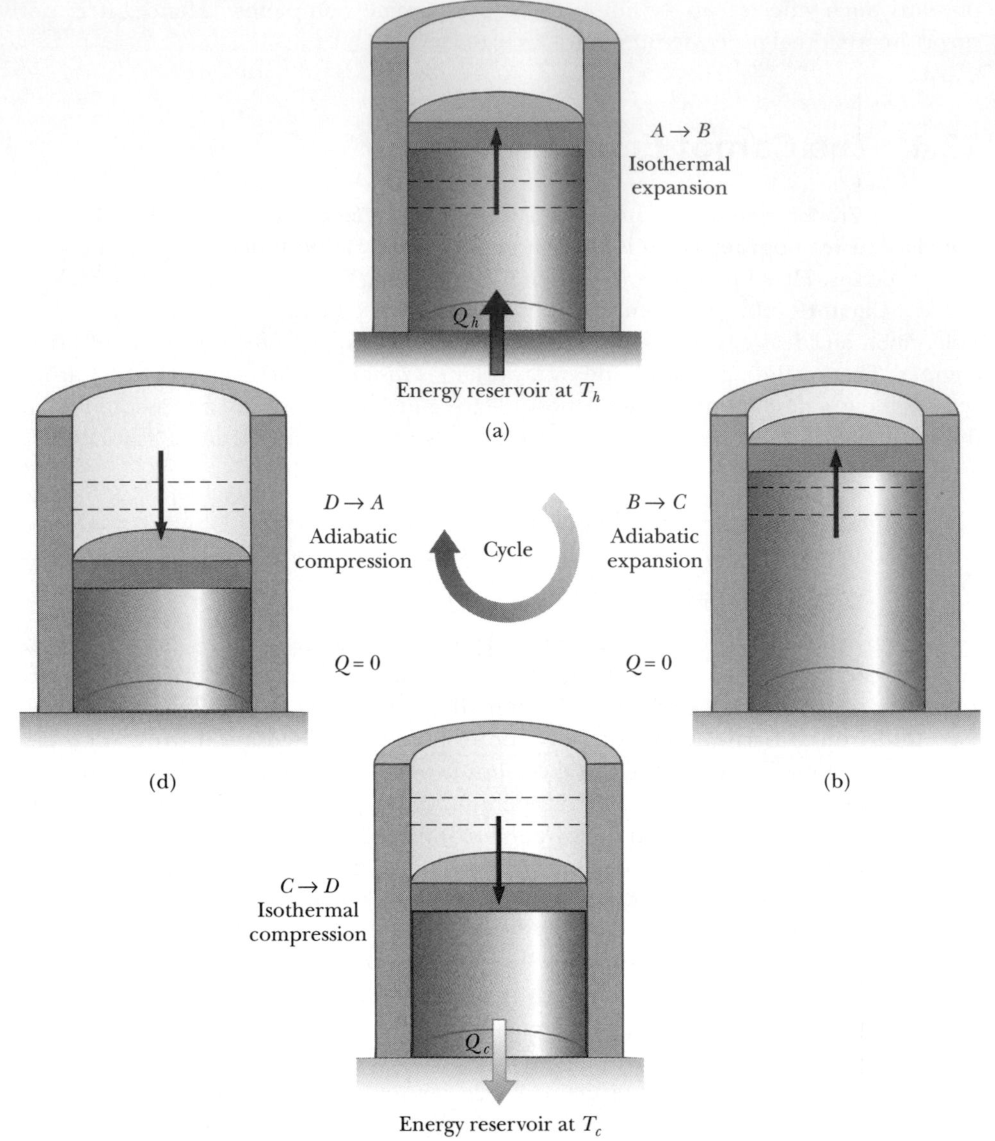

Active Figure 22.10 The Carnot cycle. (a) In process $A \rightarrow B$, the gas expands isothermally while in contact with a reservoir at T_h. (b) In process $B \rightarrow C$, the gas expands adiabatically ($Q = 0$). (c) In process $C \rightarrow D$, the gas is compressed isothermally while in contact with a reservoir at $T_c < T_h$. (d) In process $D \rightarrow A$, the gas is compressed adiabatically. The arrows on the piston indicate the direction of its motion during each process.

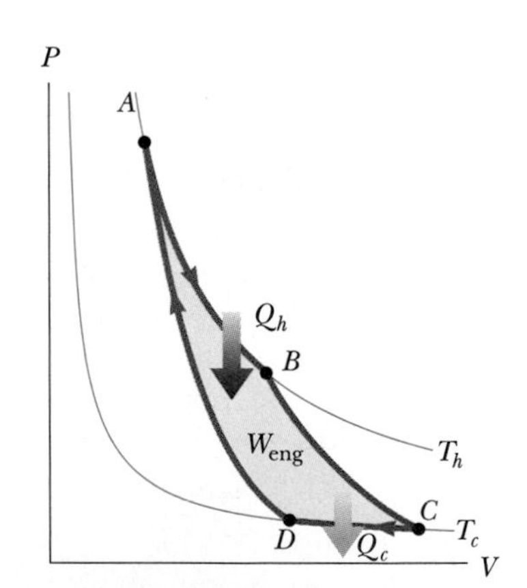

Active Figure 22.11 *PV* diagram for the Carnot cycle. The net work done W_{eng} equals the net energy transferred into the Carnot engine in one cycle, $|Q_h| - |Q_c|$. Note that $\Delta E_{int} = 0$ for the cycle.

At the Active Figures link at http://www.pse6.com, you can observe the Carnot cycle on the PV diagram while you also observe the motion of the piston in Figure 22.10.

The net work done in this reversible, cyclic process is equal to the area enclosed by the path *ABCDA* in Figure 22.11. As we demonstrated in Section 22.1, because the change in internal energy is zero, the net work W_{eng} done by the gas in one cycle equals the net energy transferred into the system, $|Q_h| - |Q_c|$. The thermal efficiency of the engine is given by Equation 22.2:

$$e = \frac{W_{eng}}{|Q_h|} = \frac{|Q_h| - |Q_c|}{|Q_h|} = 1 - \frac{|Q_c|}{|Q_h|}$$

In Example 22.3, we show that for a Carnot cycle

$$\frac{|Q_c|}{|Q_h|} = \frac{T_c}{T_h} \tag{22.5}$$

Hence, the thermal efficiency of a Carnot engine is

$$e_C = 1 - \frac{T_c}{T_h} \tag{22.6}$$

Efficiency of a Carnot engine

This result indicates that **all Carnot engines operating between the same two temperatures have the same efficiency.**[5]

Equation 22.6 can be applied to any working substance operating in a Carnot cycle between two energy reservoirs. According to this equation, the efficiency is zero if $T_c = T_h$, as one would expect. The efficiency increases as T_c is lowered and as T_h is raised. However, the efficiency can be unity (100%) only if $T_c = 0$ K. Such reservoirs are not available; thus, the maximum efficiency is always less than 100%. In most practical cases, T_c is near room temperature, which is about 300 K. Therefore, one usually strives to increase the efficiency by raising T_h. Theoretically, a Carnot-cycle heat engine run in reverse constitutes the most effective heat pump possible, and it determines the maximum COP for a given combination of hot and cold reservoir temperatures. Using Equations 22.1 and 22.3, we see that the maximum COP for a heat pump in its heating mode is

$$\text{COP}_C \text{ (heating mode)} = \frac{|Q_h|}{W}$$

$$= \frac{|Q_h|}{|Q_h| - |Q_c|} = \frac{1}{1 - \dfrac{|Q_c|}{|Q_h|}} = \frac{1}{1 - \dfrac{T_c}{T_h}} = \frac{T_h}{T_h - T_c}$$

The Carnot COP for a heat pump in the cooling mode is

$$\text{COP}_C \text{ (cooling mode)} = \frac{T_c}{T_h - T_c}$$

As the difference between the temperatures of the two reservoirs approaches zero in this expression, the theoretical COP approaches infinity. In practice, the low temperature of the cooling coils and the high temperature at the compressor limit the COP to values below 10.

Quick Quiz 22.4 Three engines operate between reservoirs separated in temperature by 300 K. The reservoir temperatures are as follows: Engine A: $T_h = 1\,000$ K, $T_c = 700$ K; Engine B: $T_h = 800$ K, $T_c = 500$ K; Engine C: $T_h = 600$ K, $T_c = 300$ K. Rank the engines in order of theoretically possible efficiency, from highest to lowest.

[5] In order for the processes in the Carnot cycle to be reversible, they must be carried out infinitesimally slowly. Thus, although the Carnot engine is the most efficient engine possible, it has zero power output, because it takes an infinite time interval to complete one cycle! For a real engine, the short time interval for each cycle results in the working substance reaching a high temperature lower than that of the hot reservoir and a low temperature higher than that of the cold reservoir. An engine undergoing a Carnot cycle between this narrower temperature range was analyzed by Curzon and Ahlborn (*Am. J. Phys.*, **43**(1), 22, 1975), who found that the efficiency at maximum power output depends only on the reservoir temperatures T_c and T_h, and is given by $e_{C\text{-}A} = 1 - (T_c/T_h)^{1/2}$. The Curzon–Ahlborn efficiency $e_{C\text{-}A}$ provides a closer approximation to the efficiencies of real engines than does the Carnot efficiency.

Example 22.3 Efficiency of the Carnot Engine

Show that the efficiency of a heat engine operating in a Carnot cycle using an ideal gas is given by Equation 22.6.

Solution During the isothermal expansion (process $A \rightarrow B$ in Fig. 22.10), the temperature of the gas does not change. Thus, its internal energy remains constant. The work done on a gas during an isothermal process is given by Equation 20.13. According to the first law,

$$|Q_h| = |-W_{AB}| = nRT_h \ln \frac{V_B}{V_A}$$

In a similar manner, the energy transferred to the cold reservoir during the isothermal compression $C \rightarrow D$ is

$$|Q_c| = |-W_{CD}| = nRT_c \ln \frac{V_C}{V_D}$$

Dividing the second expression by the first, we find that

$$(1) \qquad \frac{|Q_c|}{|Q_h|} = \frac{T_c}{T_h} \frac{\ln(V_C/V_D)}{\ln(V_B/V_A)}$$

We now show that the ratio of the logarithmic quantities is unity by establishing a relationship between the ratio of volumes. For any quasi-static, adiabatic process, the temperature and volume are related by Equation 21.20:

$$T_i V_i^{\gamma-1} = T_f V_f^{\gamma-1}$$

Applying this result to the adiabatic processes $B \rightarrow C$ and $D \rightarrow A$, we obtain

$$T_h V_B^{\gamma-1} = T_c V_C^{\gamma-1}$$

$$T_h V_A^{\gamma-1} = T_c V_D^{\gamma-1}$$

Dividing the first equation by the second, we obtain

$$(V_B/V_A)^{\gamma-1} = (V_C/V_D)^{\gamma-1}$$

$$(2) \qquad \frac{V_B}{V_A} = \frac{V_C}{V_D}$$

Substituting Equation (2) into Equation (1), we find that the logarithmic terms cancel, and we obtain the relationship

$$\frac{|Q_c|}{|Q_h|} = \frac{T_c}{T_h}$$

Using this result and Equation 22.2, we see that the thermal efficiency of the Carnot engine is

$$e_C = 1 - \frac{|Q_c|}{|Q_h|} = 1 - \frac{T_c}{T_h}$$

which is Equation 22.6, the one we set out to prove.

Example 22.4 The Steam Engine

A steam engine has a boiler that operates at 500 K. The energy from the burning fuel changes water to steam, and this steam then drives a piston. The cold reservoir's temperature is that of the outside air, approximately 300 K. What is the maximum thermal efficiency of this steam engine?

Solution Using Equation 22.6, we find that the maximum thermal efficiency for any engine operating between these temperatures is

$$e_C = 1 - \frac{T_c}{T_h} = 1 - \frac{300 \text{ K}}{500 \text{ K}} = 0.400 \quad \text{or} \quad 40.0\%$$

You should note that this is the highest *theoretical* efficiency of the engine. In practice, the efficiency is considerably lower.

What If? Suppose we wished to increase the theoretical efficiency of this engine and we could do so by increasing T_h by ΔT or by decreasing T_c by the same ΔT. Which would be more effective?

Answer A given ΔT would have a larger fractional effect on a smaller temperature, so we would expect a larger change in efficiency if we alter T_c by ΔT. Let us test this numerically. Increasing T_h by 50 K, corresponding to $T_h = 550$ K, would give a maximum efficiency of

$$e_C = 1 - \frac{T_c}{T_h} = 1 - \frac{300 \text{ K}}{550 \text{ K}} = 0.455$$

Decreasing T_c by 50 K, corresponding to $T_c = 250$ K, would give a maximum efficiency of

$$e_C = 1 - \frac{T_c}{T_h} = 1 - \frac{250 \text{ K}}{500 \text{ K}} = 0.500$$

While changing T_c is *mathematically* more effective, often changing T_h is *practically* more feasible.

Example 22.5 The Carnot Efficiency

The highest theoretical efficiency of a certain engine is 30.0%. If this engine uses the atmosphere, which has a temperature of 300 K, as its cold reservoir, what is the temperature of its hot reservoir?

Solution We use the Carnot efficiency to find T_h:

$$e_C = 1 - \frac{T_c}{T_h}$$

$$T_h = \frac{T_c}{1 - e_C} = \frac{300 \text{ K}}{1 - 0.300} = 429 \text{ K}$$

22.5 Gasoline and Diesel Engines

In a gasoline engine, six processes occur in each cycle; five of these are illustrated in Figure 22.12. In this discussion, we consider the interior of the cylinder above the piston to be the system that is taken through repeated cycles in the operation of the engine. For a given cycle, the piston moves up and down twice. This represents a four-stroke cycle consisting of two upstrokes and two downstrokes. The processes in the cycle can be approximated by the **Otto cycle**, shown in the *PV* diagram in Figure 22.13. In the following discussion, refer to Figure 22.12 for the pictorial representation of the strokes and to Figure 22.13 for the significance on the *PV* diagram of the letter designations below:

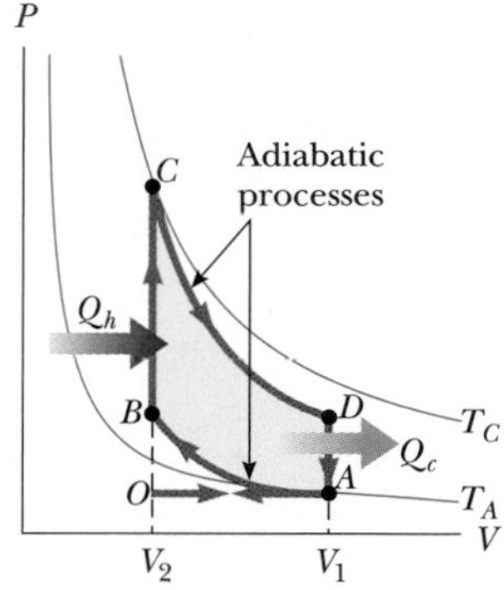

Active Figure 22.13 *PV* diagram for the Otto cycle, which approximately represents the processes occurring in an internal combustion engine.

At the Active Figures link at http://www.pse6.com, you can observe the Otto cycle on the PV diagram while you observe the motion of the piston and crankshaft in Figure 22.12.

1. During the *intake stroke* $O \rightarrow A$ (Fig. 22.12a), the piston moves downward, and a gaseous mixture of air and fuel is drawn into the cylinder at atmospheric pressure. In this process, the volume increases from V_2 to V_1. This is the energy input part of the cycle—energy enters the system (the interior of the cylinder) as potential energy stored in the fuel.
2. During the *compression stroke* $A \rightarrow B$ (Fig. 22.12b), the piston moves upward, the air–fuel mixture is compressed adiabatically from volume V_1 to volume V_2, and the temperature increases from T_A to T_B. The work done on the gas is positive, and its value is equal to the negative of the area under the curve *AB* in Figure 22.13.
3. In process $B \rightarrow C$, combustion occurs when the spark plug fires (Fig. 22.12c). This is not one of the strokes of the cycle because it occurs in a very short period of time while the piston is at its highest position. The combustion represents a rapid transformation from potential energy stored in chemical bonds in the fuel to internal energy associated with molecular motion, which is related to temperature. During this time, the pressure and temperature in the cylinder increase rapidly, with the temperature rising from T_B to T_C. The volume, however, remains approximately constant because of the short time interval. As a result, approximately no work is done on or by the gas. We can model this process in the *PV* diagram (Fig. 22.13) as

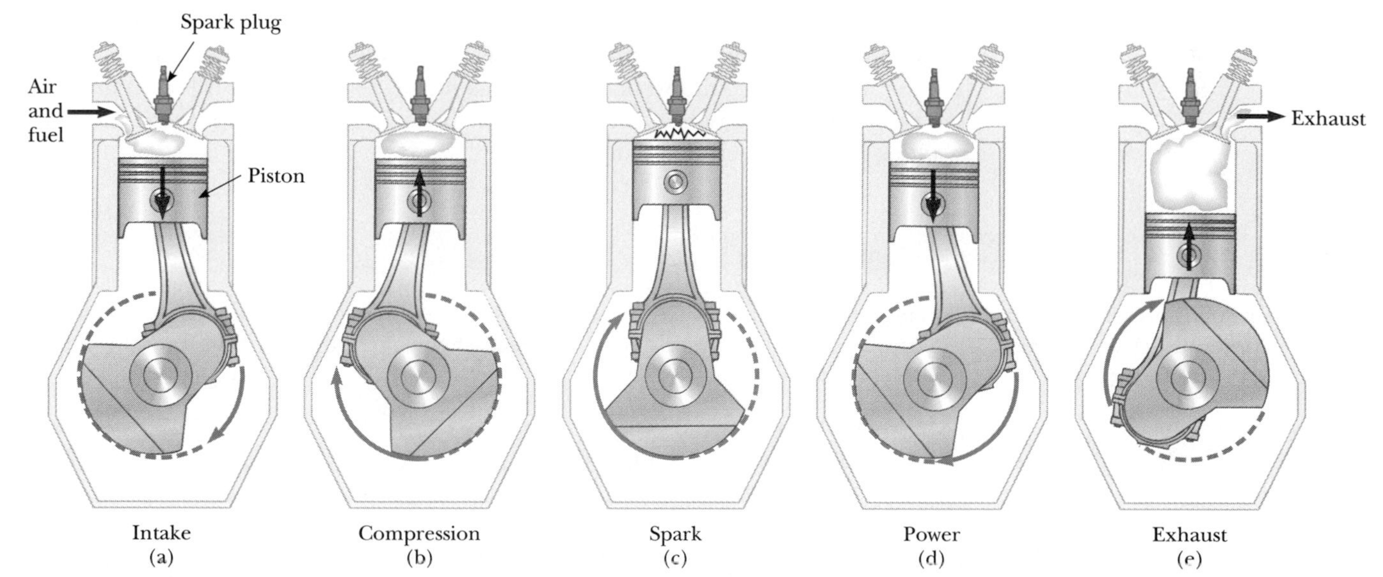

Active Figure 22.12 The four-stroke cycle of a conventional gasoline engine. The arrows on the piston indicate the direction of its motion during each process. (a) In the intake stroke, air and fuel enter the cylinder. (b) The intake valve is then closed, and the air–fuel mixture is compressed by the piston. (c) The mixture is ignited by the spark plug, with the result that the temperature of the mixture increases at essentially constant volume. (d) In the power stroke, the gas expands against the piston. (e) Finally, the residual gases are expelled, and the cycle repeats.

At the Active Figures link at http://www.pse6.com, you can observe the motion of the piston and crankshaft while you also observe the cycle on the PV diagram of Figure 22.13.

that process in which the energy $|Q_h|$ enters the system. (However, in reality this process is a *conversion* of energy already in the cylinder from process $O \rightarrow A$.)

4. In the *power stroke* $C \rightarrow D$ (Fig. 22.12d), the gas expands adiabatically from V_2 to V_1. This expansion causes the temperature to drop from T_C to T_D. Work is done by the gas in pushing the piston downward, and the value of this work is equal to the area under the curve CD.
5. In the process $D \rightarrow A$ (not shown in Fig. 22.12), an exhaust valve is opened as the piston reaches the bottom of its travel, and the pressure suddenly drops for a short time interval. During this interval, the piston is almost stationary and the volume is approximately constant. Energy is expelled from the interior of the cylinder and continues to be expelled during the next process.
6. In the final process, the *exhaust stroke* $A \rightarrow O$ (Fig. 22.12e), the piston moves upward while the exhaust valve remains open. Residual gases are exhausted at atmospheric pressure, and the volume decreases from V_1 to V_2. The cycle then repeats.

If the air–fuel mixture is assumed to be an ideal gas, then the efficiency of the Otto cycle is

$$e = 1 - \frac{1}{(V_1/V_2)^{\gamma-1}} \qquad \text{(Otto cycle)} \qquad (22.7)$$

where γ is the ratio of the molar specific heats C_P/C_V for the fuel–air mixture and V_1/V_2 is the **compression ratio.** Equation 22.7, which we derive in Example 22.6, shows that the efficiency increases as the compression ratio increases. For a typical compression ratio of 8 and with $\gamma = 1.4$, we predict a theoretical efficiency of 56% for an engine operating in the idealized Otto cycle. This value is much greater than that achieved in real engines (15% to 20%) because of such effects as friction, energy transfer by conduction through the cylinder walls, and incomplete combustion of the air–fuel mixture.

Diesel engines operate on a cycle similar to the Otto cycle but do not employ a spark plug. The compression ratio for a diesel engine is much greater than that for a gasoline engine. Air in the cylinder is compressed to a very small volume, and, as a consequence, the cylinder temperature at the end of the compression stroke is very high. At this point, fuel is injected into the cylinder. The temperature is high enough for the fuel–air mixture to ignite without the assistance of a spark plug. Diesel engines are more efficient than gasoline engines because of their greater compression ratios and resulting higher combustion temperatures.

Example 22.6 Efficiency of the Otto Cycle

Show that the thermal efficiency of an engine operating in an idealized Otto cycle (see Figs. 22.12 and 22.13) is given by Equation 22.7. Treat the working substance as an ideal gas.

Solution First, let us calculate the work done on the gas during each cycle. No work is done during processes $B \rightarrow C$ and $D \rightarrow A$. The work done on the gas during the adiabatic compression $A \rightarrow B$ is positive, and the work done on the gas during the adiabatic expansion $C \rightarrow D$ is negative. The value of the net work done equals the area of the shaded region bounded by the closed curve in Figure 22.13. Because the change in internal energy for one cycle is zero, we see from the first law that the net work done during one cycle equals the net energy transfer to the system:

$$W_{\text{eng}} = |Q_h| - |Q_c|$$

Because processes $B \rightarrow C$ and $D \rightarrow A$ take place at constant volume, and because the gas is ideal, we find from the definition of molar specific heat (Eq. 21.8) that

$$|Q_h| = nC_V(T_C - T_B) \qquad \text{and} \qquad |Q_c| = nC_V(T_D - T_A)$$

Using these expressions together with Equation 22.2, we obtain for the thermal efficiency

$$(1) \qquad e = \frac{W_{\text{eng}}}{|Q_h|} = 1 - \frac{|Q_c|}{|Q_h|} = 1 - \frac{T_D - T_A}{T_C - T_B}$$

We can simplify this expression by noting that processes $A \rightarrow B$ and $C \rightarrow D$ are adiabatic and hence obey Equation 21.20. For the two adiabatic processes, then,

$$A \rightarrow B: \qquad T_A V_A^{\gamma-1} = T_B V_B^{\gamma-1}$$

$$C \rightarrow D: \qquad T_C V_C^{\gamma-1} = T_D V_D^{\gamma-1}$$

Using these equations and relying on the fact that $V_A = V_D = V_1$ and $V_B = V_C = V_2$, we find that

$$T_A V_1^{\gamma-1} = T_B V_2^{\gamma-1}$$

$$(2) \qquad T_A = T_B\left(\frac{V_2}{V_1}\right)^{\gamma-1}$$

$$T_D V_1^{\gamma-1} = T_C V_2^{\gamma-1}$$

$$(3) \qquad T_D = T_C\left(\frac{V_2}{V_1}\right)^{\gamma-1}$$

Subtracting Equation (2) from Equation (3) and rearranging, we find that

$$(4) \qquad \frac{T_D - T_A}{T_C - T_B} = \left(\frac{V_2}{V_1}\right)^{\gamma-1}$$

Substituting Equation (4) into Equation (1), we obtain for the thermal efficiency

$$(5) \qquad e = 1 - \frac{1}{(V_1/V_2)^{\gamma-1}}$$

which is Equation 22.7.

We can also express this efficiency in terms of temperatures by noting from Equations (2) and (3) that

$$\left(\frac{V_2}{V_1}\right)^{\gamma-1} = \frac{T_A}{T_B} = \frac{T_D}{T_C}$$

Therefore, Equation (5) becomes

$$(6) \qquad e = 1 - \frac{T_A}{T_B} = 1 - \frac{T_D}{T_C}$$

During the Otto cycle, the lowest temperature is T_A and the highest temperature is T_C. Therefore, the efficiency of a Carnot engine operating between reservoirs at these two temperatures, which is given by the expression $e_C = 1 - (T_A/T_C)$, is *greater* than the efficiency of the Otto cycle given by Equation (6), as expected.

Application Models of Gasoline and Diesel Engines

We can use the thermodynamic principles discussed in this and earlier chapters to model the performance of gasoline and diesel engines. In both types of engine, a gas is first compressed in the cylinders of the engine and then the fuel–air mixture is ignited. Work is done on the gas during compression, but significantly more work is done on the piston by the mixture as the products of combustion expand in the cylinder. The power of the engine is transferred from the piston to the crankshaft by the connecting rod.

Two important quantities of either engine are the **displacement volume,** which is the volume displaced by the piston as it moves from the bottom to the top of the cylinder, and the compression ratio r, which is the ratio of the maximum and minimum volumes of the cylinder, as discussed earlier. Most gasoline and diesel engines operate with a four-stroke cycle (intake, compression, power, exhaust), in which the net work of the intake and exhaust strokes can be considered negligible. Therefore, power is developed only once for every two revolutions of the crankshaft (see Fig. 22.12).

In a diesel engine, only air (and no fuel) is present in the cylinder at the beginning of the compression. In the idealized diesel cycle of Figure 22.14, air in the cylinder undergoes an adiabatic compression from A to B. Starting at B, fuel is injected into the cylinder. The high temperature of the mixture causes combustion of the fuel–air mixture. Fuel continues to be injected in such a way that during the time interval while the fuel is being injected, the fuel–air mixture undergoes a constant-pressure expansion to an intermediate volume V_C ($B \to C$). At C, the fuel injection is cut off and the power stroke is an adiabatic expansion back to $V_D = V_A$ ($C \to D$). The exhaust valve is opened, and a constant-volume output of energy occurs ($D \to A$) as the cylinder empties.

To simplify our calculations, we assume that the mixture in the cylinder is air modeled as an ideal gas. We use specific heats c instead of molar specific heats C and assume constant values for air at 300 K. We express the specific heats and the universal gas constant in terms of unit masses rather than moles. Thus, $c_V = 0.718$ kJ/kg·K, $c_P = 1.005$ kJ/kg·K, $\gamma = c_P/c_V = 1.40$, and $R = c_P - c_V = 0.287$ kJ/kg·K $= 0.287$ kPa·m^3/kg·K.

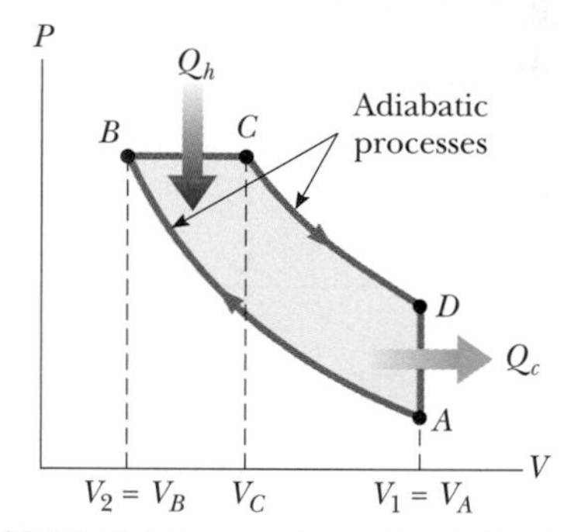

Figure 22.14 *PV* diagram for an ideal diesel engine.

A 3.00-L Gasoline Engine

Let us calculate the power delivered by a six-cylinder gasoline engine that has a displacement volume of 3.00 L operating at 4 000 rpm and having a compression ratio of $r = 9.50$. The air–fuel mixture enters a cylinder at atmospheric pressure and an ambient temperature of 27°C. During combustion, the mixture reaches a temperature of 1 350°C.

First, let us calculate the work done in an individual cylinder. Using the initial pressure $P_A = 100$ kPa, and the initial temperature $T_A = 300$ K, we calculate the initial volume and the mass of the air–fuel mixture. We know that the ratio of the initial and final volumes is the compression ratio,

$$\frac{V_A}{V_B} = r = 9.50$$

We also know that the difference in volumes is the displacement volume. The 3.00-L rating of the engine is the total displacement volume for all six cylinders. Thus, for one cylinder,

$$V_A - V_B = \frac{3.00 \text{ L}}{6} = 0.500 \times 10^{-3} \text{ m}^3$$

Solving these two equations simultaneously, we find the initial and final volumes:

$$V_A = 0.559 \times 10^{-3} \text{ m}^3 \qquad V_B = 0.588 \times 10^{-4} \text{ m}^3$$

Using the ideal gas law (in the form $PV = mRT$, because we are using the universal gas constant in terms of mass rather than moles), we can find the mass of the air–fuel mixture:

$$m = \frac{P_A V_A}{RT_A} = \frac{(100 \text{ kPa})(0.559 \times 10^{-3} \text{ m}^3)}{(0.287 \text{ kPa}\cdot\text{m}^3/\text{kg}\cdot\text{K})(300 \text{ K})}$$
$$= 6.49 \times 10^{-4} \text{ kg}$$

Process $A \rightarrow B$ (see Fig. 22.13) is an adiabatic compression, and this means that $PV^\gamma =$ constant; hence,

$$P_B V_B{}^\gamma = P_A V_A{}^\gamma$$

$$P_B = P_A\left(\frac{V_A}{V_B}\right)^\gamma = P_A(r)^\gamma = (100 \text{ kPa})(9.50)^{1.40}$$
$$= 2.34 \times 10^3 \text{ kPa}$$

Using the ideal gas law, we find that the temperature after the compression is

$$T_B = \frac{P_B V_B}{mR} = \frac{(2.34 \times 10^3 \text{ kPa})(0.588 \times 10^{-4} \text{ m}^3)}{(6.49 \times 10^{-4} \text{ kg})(0.287 \text{ kPa}\cdot\text{m}^3/\text{kg}\cdot\text{K})}$$
$$= 739 \text{ K}$$

In process $B \rightarrow C$, the combustion that transforms the potential energy in chemical bonds into internal energy of molecular motion occurs at constant volume; thus, $V_C = V_B$. Combustion causes the temperature to increase to $T_C =$ 1 350°C = 1 623 K. Using this value and the ideal gas law, we can calculate P_C:

$$P_C = \frac{mRT_C}{V_C}$$
$$= \frac{(6.49 \times 10^{-4} \text{ kg})(0.287 \text{ kPa}\cdot\text{m}^3/\text{kg}\cdot\text{K})(1\,623 \text{ K})}{(0.588 \times 10^{-4} \text{ m}^3)}$$
$$= 5.14 \times 10^3 \text{ kPa}$$

Process $C \rightarrow D$ is an adiabatic expansion; the pressure after the expansion is

$$P_D = P_C\left(\frac{V_C}{V_D}\right)^\gamma = P_C\left(\frac{V_B}{V_A}\right)^\gamma = P_C\left(\frac{1}{r}\right)^\gamma$$
$$= (5.14 \times 10^3 \text{ kPa})\left(\frac{1}{9.50}\right)^{1.40} = 220 \text{ kPa}$$

Using the ideal gas law again, we find the final temperature:

$$T_D = \frac{P_D V_D}{mR} = \frac{(220 \text{ kPa})(0.559 \times 10^{-3} \text{ m}^3)}{(6.49 \times 10^{-4} \text{ kg})(0.287 \text{ kPa}\cdot\text{m}^3/\text{kg}\cdot\text{K})}$$
$$= 660 \text{ K}$$

Now that we have the temperatures at the beginning and end of each process of the cycle, we can calculate the net energy transfer and net work done in each cylinder every two cycles:

$$|Q_h| = |Q_{\text{in}}| = mc_V(T_C - T_B)$$
$$= (6.49 \times 10^{-4} \text{ kg})(0.718 \text{ kJ/kg}\cdot\text{K})(1\,623 - 739 \text{ K})$$
$$= 0.412 \text{ kJ}$$
$$|Q_c| = |Q_{\text{out}}| = mc_V(T_D - T_A)$$
$$= (6.49 \times 10^{-4} \text{ kg})(0.718 \text{ kJ/kg}\cdot\text{K})(660 \text{ K} - 300 \text{ K})$$
$$= 0.168 \text{ kJ}$$
$$W_{\text{net}} = |Q_{\text{in}}| - |Q_{\text{out}}| = 0.244 \text{ kJ}$$

From Equation 22.2, the efficiency is $e = W_{\text{net}}/|Q_{\text{in}}| = 59\%$. (We can also use Equation 22.7 to calculate the efficiency directly from the compression ratio.)

Recalling that power is delivered every other revolution of the crankshaft, we find that the net power for the six-cylinder engine operating at 4 000 rpm is

$$\mathcal{P}_{\text{net}} = 6(\tfrac{1}{2} \text{ rev})[(4\,000 \text{ rev/min})(1 \text{ min}/60 \text{ s})](0.244 \text{ kJ})$$
$$= 48.8 \text{ kW} = 65 \text{ hp}$$

A 2.00-L Diesel Engine

Let us calculate the power delivered by a four-cylinder diesel engine that has a displacement volume of 2.00 L and is operating at 3 000 rpm. The compression ratio is $r = V_A/V_B = 22.0$, and the **cutoff ratio,** which is the ratio of the volume change during the constant-pressure process $B \rightarrow C$ in Figure 22.14, is $r_c = V_C/V_B = 2.00$. The air enters each cylinder at the beginning of the compression cycle at atmospheric pressure and at an ambient temperature of 27°C.

Our model of the diesel engine is similar to our model of the gasoline engine except that now the fuel is injected at point B and the mixture self-ignites near the end of the compression cycle $A \rightarrow B$, when the temperature reaches the ignition temperature. We assume that the energy input occurs in the constant-pressure process $B \rightarrow C$, and that the expansion process continues from C to D with no further energy transfer by heat.

Let us calculate the work done in an individual cylinder that has an initial volume of $V_A = (2.00 \times 10^{-3} \text{ m}^3)/4 = 0.500 \times 10^{-3} \text{ m}^3$. Because the compression ratio is quite high, we approximate the maximum cylinder volume to be the displacement volume. Using the initial pressure $P_A = 100$ kPa and initial temperature $T_A = 300$ K, we can calculate the mass of the air in the cylinder using the ideal gas law:

$$m = \frac{P_A V_A}{RT_A} = \frac{(100 \text{ kPa})(0.500 \times 10^{-3} \text{ m}^3)}{(0.287 \text{ kPa}\cdot\text{m}^3/\text{kg}\cdot\text{K})(300 \text{ K})}$$
$$= 5.81 \times 10^{-4} \text{ kg}$$

Process $A \rightarrow B$ is an adiabatic compression, so $PV^\gamma =$ constant; thus,

$$P_B V_B{}^\gamma = P_A V_A{}^\gamma$$

$$P_B = P_A\left(\frac{V_A}{V_B}\right)^\gamma = (100 \text{ kPa})(22.0)^{1.40} = 7.58 \times 10^3 \text{ kPa}$$

Using the ideal gas law, we find that the temperature of the air after the compression is

$$T_B = \frac{P_B V_B}{mR} = \frac{(7.58 \times 10^3 \text{ kPa})(0.500 \times 10^{-3} \text{ m}^3)(1/22.0)}{(5.81 \times 10^{-4} \text{ kg})(0.287 \text{ kPa}\cdot\text{m}^3/\text{kg}\cdot\text{K})}$$
$$= 1.03 \times 10^3 \text{ K}$$

Process $B \rightarrow C$ is a constant-pressure expansion; thus, $P_C = P_B$. We know from the cutoff ratio of 2.00 that the volume doubles in this process. According to the ideal gas law, a doubling of volume in an isobaric process results in a doubling of the temperature, so

$$T_C = 2T_B = 2.06 \times 10^3 \text{ K}$$

Process $C \rightarrow D$ is an adiabatic expansion; therefore,

$$P_D = P_C\left(\frac{V_C}{V_D}\right)^{\gamma} = P_C\left(\frac{V_C}{V_B}\frac{V_B}{V_D}\right)^{\gamma} = P_C\left(r_c\frac{1}{r}\right)^{\gamma}$$

$$= (7.57 \times 10^3 \text{ kPa})\left(\frac{2.00}{22.0}\right)^{1.40}$$

$$= 264 \text{ kPa}$$

We find the temperature at D from the ideal gas law:

$$T_D = \frac{P_D V_D}{mR} = \frac{(264 \text{ kPa})(0.500 \times 10^{-3} \text{ m}^3)}{(5.81 \times 10^{-4} \text{ kg})(0.287 \text{ kPa}\cdot\text{m}^3/\text{kg}\cdot\text{K})}$$

$$= 792 \text{ K}$$

Now that we have the temperatures at the beginning and the end of each process, we can calculate the net energy transfer by heat and the net work done in each cylinder every two cycles:

$$|Q_h| = |Q_{\text{in}}| = mc_P(T_C - T_B) = 0.601 \text{ kJ}$$

$$|Q_c| = |Q_{\text{out}}| = mc_V(T_D - T_A) = 0.205 \text{ kJ}$$

$$W_{\text{net}} = |Q_{\text{in}}| - |Q_{\text{out}}| = 0.396 \text{ kJ}$$

The efficiency is $e = W_{\text{net}}/|Q_{\text{in}}| = 66\%$.

The net power for the four-cylinder engine operating at 3 000 rpm is

$$\mathcal{P}_{\text{net}} = 4(\tfrac{1}{2}\text{ rev})[(3\,000 \text{ rev/min})(1 \text{ min}/60 \text{ s})](0.396 \text{ kJ})$$

$$= 39.6 \text{ kW} = 53 \text{ hp}$$

Modern engine design goes beyond this very simple thermodynamic treatment, which uses idealized cycles.

22.6 Entropy

The zeroth law of thermodynamics involves the concept of temperature, and the first law involves the concept of internal energy. Temperature and internal energy are both state variables—that is, they can be used to describe the thermodynamic state of a system. Another state variable—this one related to the second law of thermodynamics—is **entropy** S. In this section we define entropy on a macroscopic scale as it was first expressed by Clausius in 1865.

Entropy was originally formulated as a useful concept in thermodynamics; however, its importance grew as the field of statistical mechanics developed because the analytical techniques of statistical mechanics provide an alternative means of interpreting entropy and a more global significance to the concept. In statistical mechanics, the behavior of a substance is described in terms of the statistical behavior of its atoms and molecules. One of the main results of this treatment is that **isolated systems tend toward disorder and that entropy is a measure of this disorder.** For example, consider the molecules of a gas in the air in your room. If half of the gas molecules had velocity vectors of equal magnitude directed toward the left and the other half had velocity vectors of the same magnitude directed toward the right, the situation would be very ordered. However, such a situation is extremely unlikely. If you could actually view the molecules, you would see that they move haphazardly in all directions, bumping into one another, changing speed upon collision, some going fast and others going slowly. This situation is highly disordered.

The cause of the tendency of an isolated system toward disorder is easily explained. To do so, we distinguish between *microstates* and *macrostates* of a system. A **microstate** is a particular configuration of the individual constituents of the system. For example, the description of the ordered velocity vectors of the air molecules in your room refers to a particular microstate, and the more likely haphazard motion is another microstate—one that represents disorder. A **macrostate** is a description of the conditions of the system from a macroscopic point of view and makes use of macroscopic variables such as pressure, density, and temperature for gases.

For any given macrostate of the system, a number of microstates are possible. For example, the macrostate of a four on a pair of dice can be formed from the possible microstates 1-3, 2-2, and 3-1. It is assumed that all microstates are equally probable. However, when all possible macrostates are examined, it is found that macrostates

PITFALL PREVENTION

22.4 Entropy Is Abstract

Entropy is one of the most abstract notions in physics, so follow the discussion in this and the subsequent sections very carefully. Do not confuse energy with entropy—even though the names sound similar, they are very different concepts.

associated with disorder have far more possible microstates than those associated with order. For example, there is only one microstate associated with the macrostate of a royal flush in a poker hand of five spades, laid out in order from ten to ace (Fig. 22.15a). This is a highly ordered hand. However, there are many microstates (the set of five individual cards in a poker hand) associated with a worthless hand in poker (Fig. 22.15b).

The probability of being dealt the royal flush in spades is exactly the same as the probability of being dealt any *particular* worthless hand. Because there are so many worthless hands, however, the probability of a macrostate of a worthless hand is far larger than the probability of a macrostate of a royal flush in spades.

(a)

(b)

Figure 22.15 (a) A royal flush is a highly ordered poker hand with low probability of occurring. (b) A disordered and worthless poker hand. The probability of this *particular* hand occurring is the same as that of the royal flush. There are so many worthless hands, however, that the probability of being dealt a worthless hand is much higher than that of a royal flush.

Quick Quiz 22.5 Suppose that you select four cards at random from a standard deck of playing cards and end up with a macrostate of four deuces. How many microstates are associated with this macrostate?

Quick Quiz 22.6 Suppose you pick up two cards at random from a standard deck of playing cards and end up with a macrostate of two aces. How many microstates are associated with this macrostate?

We can also imagine ordered macrostates and disordered macrostates in physical processes, not just in games of dice and poker. The probability of a system moving in time from an ordered macrostate to a disordered macrostate is far greater than the probability of the reverse, because there are more microstates in a disordered macrostate.

If we consider a system and its surroundings to include the entire Universe, then the Universe is always moving toward a macrostate corresponding to greater disorder. Because entropy is a measure of disorder, an alternative way of stating this is **the entropy of the Universe increases in all real processes.** This is yet another statement of the second law of thermodynamics that can be shown to be equivalent to the Kelvin–Planck and Clausius statements.

The original formulation of entropy in thermodynamics involves the transfer of energy by heat during a reversible process. Consider any infinitesimal process in which a system changes from one equilibrium state to another. If dQ_r is the amount of energy transferred by heat when the system follows a reversible path between the states, then the change in entropy dS is equal to this amount of energy for the reversible process divided by the absolute temperature of the system:

$$dS = \frac{dQ_r}{T} \tag{22.8}$$

We have assumed that the temperature is constant because the process is infinitesimal. Because we have claimed that entropy is a state variable, **the change in entropy during a process depends only on the end points and therefore is independent of the actual path followed. Consequently, the entropy change for an irreversible process can be determined by calculating the entropy change for a reversible process that connects the same initial and final states.**

The subscript r on the quantity dQ_r is a reminder that the transferred energy is to be measured along a reversible path, even though the system may actually have followed some irreversible path. When energy is absorbed by the system, dQ_r is positive and the entropy of the system increases. When energy is expelled by the system, dQ_r is negative and the entropy of the system decreases. Note that Equation 22.8 defines not entropy but rather the *change* in entropy. Hence, the meaningful quantity in describing a process is the *change* in entropy.

To calculate the change in entropy for a *finite* process, we must recognize that T is generally not constant. If dQ_r is the energy transferred by heat when the system follows an arbitrary reversible process between the same initial and final states as the irreversible process, then

$$\Delta S = \int_i^f dS = \int_i^f \frac{dQ_r}{T} \tag{22.9}$$

Change in entropy for a finite process

As with an infinitesimal process, the change in entropy ΔS of a system going from one state to another has the same value for *all* paths connecting the two states. That is, the finite change in entropy ΔS of a system depends only on the properties of the initial and final equilibrium states. Thus, we are free to choose a particular reversible path over which to evaluate the entropy in place of the actual path, as long as the initial and final states are the same for both paths. This point is explored further in Section 22.7.

Quick Quiz 22.7 Which of the following is true for the entropy change of a system that undergoes a reversible, adiabatic process? (a) $\Delta S < 0$ (b) $\Delta S = 0$ (c) $\Delta S > 0$

Quick Quiz 22.8 An ideal gas is taken from an initial temperature T_i to a higher final temperature T_f along two different reversible paths: Path A is at constant pressure; Path B is at constant volume. The relation between the entropy changes of the gas for these paths is (a) $\Delta S_A > \Delta S_B$ (b) $\Delta S_A = \Delta S_B$ (c) $\Delta S_A < \Delta S_B$.

Let us consider the changes in entropy that occur in a Carnot heat engine that operates between the temperatures T_c and T_h. In one cycle, the engine takes in energy Q_h from the hot reservoir and expels energy Q_c to the cold reservoir. These energy transfers occur only during the isothermal portions of the Carnot cycle; thus, the constant temperature can be brought out in front of the integral sign in Equation 22.9. The integral then simply has the value of the total amount of energy transferred by heat. Thus, the total change in entropy for one cycle is

$$\Delta S = \frac{|Q_h|}{T_h} - \frac{|Q_c|}{T_c}$$

where the negative sign represents the fact that $|Q_c|$ is positive, but this term must represent energy leaving the engine. In Example 22.3 we showed that, for a Carnot engine,

$$\frac{|Q_c|}{|Q_h|} = \frac{T_c}{T_h}$$

Using this result in the previous expression for ΔS, we find that the total change in entropy for a Carnot engine operating in a cycle is *zero*:

$$\Delta S = 0$$

Now consider a system taken through an arbitrary (non-Carnot) reversible cycle. Because entropy is a state variable—and hence depends only on the properties of a given equilibrium state—we conclude that $\Delta S = 0$ for *any* reversible cycle. In general, we can write this condition in the mathematical form

$$\oint \frac{dQ_r}{T} = 0 \tag{22.10}$$

where the symbol $\oint$ indicates that the integration is over a closed path.

Quasi-Static, Reversible Process for an Ideal Gas

Suppose that an ideal gas undergoes a quasi-static, reversible process from an initial state having temperature T_i and volume V_i to a final state described by T_f and V_f. Let us calculate the change in entropy of the gas for this process.

Writing the first law of thermodynamics in differential form and rearranging the terms, we have $dQ_r = dE_{\text{int}} - dW$, where $dW = -P\,dV$. For an ideal gas, recall that $dE_{\text{int}} = nC_V\,dT$ (Eq. 21.12), and from the ideal gas law, we have $P = nRT/V$. Therefore, we can express the energy transferred by heat in the process as

$$dQ_r = dE_{\text{int}} + P\,dV = nC_V\,dT + nRT\frac{dV}{V}$$

We cannot integrate this expression as it stands because the last term contains two variables, T and V. However, if we divide all terms by T, each of the terms on the right-hand side depends on only one variable:

$$\frac{dQ_r}{T} = nC_V\frac{dT}{T} + nR\frac{dV}{V} \qquad (22.11)$$

Assuming that C_V is constant over the process, and integrating Equation 22.11 from the initial state to the final state, we obtain

$$\Delta S = \int_i^f \frac{dQ_r}{T} = nC_V \ln\frac{T_f}{T_i} + nR\ln\frac{V_f}{V_i} \qquad (22.12)$$

This expression demonstrates mathematically what we argued earlier—ΔS depends only on the initial and final states and is independent of the path between the states. We can claim this because we have not specified the path taken between the initial and final states. We have only required that the path be reversible. Also, note in Equation 22.12 that ΔS can be positive or negative, depending on the values of the initial and final volumes and temperatures. Finally, for a cyclic process ($T_i = T_f$ and $V_i = V_f$), we see from Equation 22.12 that $\Delta S = 0$. This is further evidence that entropy is a state variable.

Example 22.7 Change in Entropy–Melting

A solid that has a latent heat of fusion L_f melts at a temperature T_m.

(A) Calculate the change in entropy of this substance when a mass m of the substance melts.

Solution Let us assume that the melting occurs so slowly that it can be considered a reversible process. In this case the temperature can be regarded as constant and equal to T_m. Making use of Equations 22.9 and that for the latent heat of fusion $Q = mL_f$ (Eq. 20.6, choosing the positive sign because energy is entering the ice), we find that

$$\Delta S = \int \frac{dQ_r}{T} = \frac{1}{T_m}\int dQ = \frac{Q}{T_m} = \frac{mL_f}{T_m}$$

Note that we are able to remove T_m from the integral because the process is modeled as isothermal. Note also that ΔS is positive.

(B) Estimate the value of the change in entropy of an ice cube when it melts.

Solution Let us assume an ice tray makes cubes that are about 3 cm on a side. The volume per cube is then (very roughly) 30 cm^3. This much liquid water has a mass of 30 g. From Table 20.2 we find that the latent heat of fusion of ice is 3.33×10^5 J/kg. Substituting these values into our answer for part (A), we find that

$$\Delta S = \frac{mL_f}{T_m} = \frac{(0.03\text{ kg})(3.33 \times 10^5\text{ J/kg})}{273\text{ K}} = 4 \times 10^1\text{ J/K}$$

We retain only one significant figure, in keeping with the nature of our estimations.

What If? Suppose you did not have Equation 22.9 available so that you could not calculate an entropy change. How could you argue from the statistical description of entropy that the changes in entropy for parts (A) and (B) should be positive?

Answer When a solid melts, its entropy increases because the molecules are much more disordered in the liquid state than they are in the solid state. The positive value for ΔS also means that the substance in its liquid state does not spontaneously transfer energy from itself to the surroundings and freeze because to do so would involve a spontaneous increase in order and a decrease in entropy.

22.7 Entropy Changes in Irreversible Processes

By definition, a calculation of the change in entropy for a system requires information about a reversible path connecting the initial and final equilibrium states. To calculate changes in entropy for real (irreversible) processes, we must remember that entropy (like internal energy) depends only on the *state* of the system. That is, entropy is a state variable. Hence, the change in entropy when a system moves between any two equilibrium states depends only on the initial and final states.

We can calculate the entropy change in some irreversible process between two equilibrium states by devising a reversible process (or series of reversible processes) between the same two states and computing $\Delta S = \int dQ_r/T$ for the reversible process. In irreversible processes, it is critically important that we distinguish between Q, the actual energy transfer in the process, and Q_r, the energy that would have been transferred by heat along a reversible path. Only Q_r is the correct value to be used in calculating the entropy change.

As we show in the following examples, the change in entropy for a system and its surroundings is always positive for an irreversible process. In general, the total entropy—and therefore the disorder—always increases in an irreversible process. Keeping these considerations in mind, we can state the second law of thermodynamics as follows:

The total entropy of an isolated system that undergoes a change cannot decrease.

Furthermore, **if the process is irreversible, then the total entropy of an isolated system always increases. In a reversible process, the total entropy of an isolated system remains constant.**

When dealing with a system that is not isolated from its surroundings, remember that the increase in entropy described in the second law is that of the system *and* its surroundings. When a system and its surroundings interact in an irreversible process, the increase in entropy of one is greater than the decrease in entropy of the other. Hence, we conclude that **the change in entropy of the Universe must be greater than zero for an irreversible process and equal to zero for a reversible process.** Ultimately, the entropy of the Universe should reach a maximum value. At this value, the Universe will be in a state of uniform temperature and density. All physical, chemical, and biological processes will cease because a state of perfect disorder implies that no energy is available for doing work. This gloomy state of affairs is sometimes referred to as the heat death of the Universe.

Quick Quiz 22.9 True or false: The entropy change in an adiabatic process must be zero because $Q = 0$.

Entropy Change in Thermal Conduction

Let us now consider a system consisting of a hot reservoir and a cold reservoir that are in thermal contact with each other and isolated from the rest of the Universe. A process occurs during which energy Q is transferred by heat from the hot reservoir at temperature T_h to the cold reservoir at temperature T_c. The process as described is irreversible, and so we must find an equivalent reversible process. Let us assume that the objects are connected by a poor thermal conductor whose temperature spans the range from T_c to T_h. This conductor transfers energy slowly, and its state does not change during the process. Under this assumption, the energy transfer to or from each object is reversible, and we may set $Q = Q_r$.

Because the cold reservoir absorbs energy Q, its entropy increases by Q/T_c. At the same time, the hot reservoir loses energy Q, and so its entropy change is $-Q/T_h$. Because $T_h > T_c$, the increase in entropy of the cold reservoir is greater than the decrease in entropy of the hot reservoir. Therefore, the change in entropy of the system (and of the Universe) is greater than zero:

$$\Delta S_U = \frac{Q}{T_c} + \frac{-Q}{T_h} > 0$$

Example 22.8 Which Way Does the Energy Go?

A large, cold object is at 273 K, and a second large, hot object is at 373 K. Show that it is impossible for a small amount of energy—for example, 8.00 J—to be transferred spontaneously by heat from the cold object to the hot one without a decrease in the entropy of the Universe and therefore a violation of the second law.

Solution We assume that, during the energy transfer, the two objects do not undergo a temperature change. This is not a necessary assumption; we make it only to avoid complicating the situation by having to use integral calculus in our calculations. The entropy change of the hot object is

$$\Delta S_h = \frac{Q_r}{T_h} = \frac{8.00\ \text{J}}{373\ \text{K}} = 0.021\,4\ \text{J/K}$$

The cold object loses energy, and its entropy change is

$$\Delta S_c = \frac{Q_r}{T_c} = \frac{-8.00\ \text{J}}{273\ \text{K}} = -0.029\,3\ \text{J/K}$$

We consider the two objects to be isolated from the rest of the Universe. Thus, the entropy change of the Universe is just that of our two-object system, which is

$$\Delta S_U = \Delta S_c + \Delta S_h = -0.007\,9\ \text{J/K}$$

This decrease in entropy of the Universe is in violation of the second law. That is, **the spontaneous transfer of energy by heat from a cold to a hot object cannot occur.**

Suppose energy were to continue to transfer spontaneously from a cold object to a hot object, in violation of the second law. We can describe this impossible energy transfer in terms of disorder. Before the transfer, a certain degree of order is associated with the different temperatures of the objects. The hot object's molecules have a higher average energy than the cold object's molecules. If energy spontaneously transfers from the cold object to the hot object, then, over a period of time, the cold object will become colder and the hot object will become hotter. The difference in average molecular energy will become even greater; this would represent an increase in order for the system and a violation of the second law.

In comparison, the process that does occur naturally is the transfer of energy from the hot object to the cold object. In this process, the difference in average molecular energy decreases; this represents a more random distribution of energy and an increase in disorder.

Entropy Change in a Free Expansion

Let us again consider the adiabatic free expansion of a gas occupying an initial volume V_i (Fig. 22.16). In this situation, a membrane separating the gas from an evacuated region is broken, and the gas expands (irreversibly) to a volume V_f. What are the changes in entropy of the gas and of the Universe during this process?

The process is neither reversible nor quasi-static. The work done by the gas against the vacuum is zero, and because the walls are insulating, no energy is transferred by heat during the expansion. That is, $W = 0$ and $Q = 0$. Using the first law, we see that the change in internal energy is zero. Because the gas is ideal, E_{int} depends on temperature only, and we conclude that $\Delta T = 0$ or $T_i = T_f$.

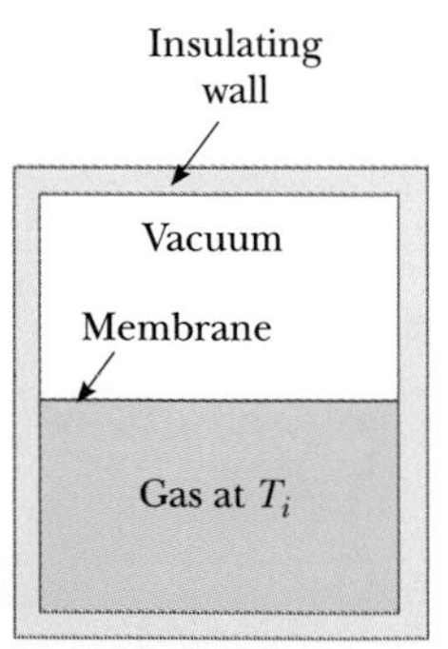

Figure 22.16 Adiabatic free expansion of a gas. When the membrane separating the gas from the evacuated region is ruptured, the gas expands freely and irreversibly. As a result, it occupies a greater final volume. The container is thermally insulated from its surroundings; thus, $Q = 0$.

To apply Equation 22.9, we cannot use $Q = 0$, the value for the irreversible process, but must instead find Q_r; that is, we must find an equivalent reversible path that shares the same initial and final states. A simple choice is an isothermal, reversible expansion in which the gas pushes slowly against a piston while energy enters the gas by heat from a reservoir to hold the temperature constant. Because T is constant in this process, Equation 22.9 gives

$$\Delta S = \int_i^f \frac{dQ_r}{T} = \frac{1}{T}\int_i^f dQ_r$$

For an isothermal process, the first law of thermodynamics specifies that $\int_i^f dQ_r$ is equal to the negative of the work done on the gas during the expansion from V_i to V_f, which is given by Equation 20.13. Using this result, we find that the entropy change for the gas is

$$\Delta S = nR\ln\frac{V_f}{V_i} \tag{22.13}$$

Because $V_f > V_i$, we conclude that ΔS is positive. This positive result indicates that both the entropy and the disorder of the gas *increase* as a result of the irreversible, adiabatic expansion.

It is easy to see that the gas is more disordered after the expansion. Instead of being concentrated in a relatively small space, the molecules are scattered over a larger region.

Because the free expansion takes place in an insulated container, no energy is transferred by heat from the surroundings. (Remember that the isothermal, reversible expansion is only a *replacement* process that we use to calculate the entropy change for the gas; it is not the *actual* process.) Thus, the free expansion has no effect on the surroundings, and the entropy change of the surroundings is zero. Thus, the entropy change for the Universe is positive; this is consistent with the second law.

Entropy Change in Calorimetric Processes

A substance of mass m_1, specific heat c_1, and initial temperature T_c is placed in thermal contact with a second substance of mass m_2, specific heat c_2, and initial temperature $T_h > T_c$. The two substances are contained in a calorimeter so that no energy is lost to the surroundings. The system of the two substances is allowed to reach thermal equilibrium. What is the total entropy change for the system?

First, let us calculate the final equilibrium temperature T_f. Using the techniques of Section 20.2—namely, Equation 20.5, $Q_{\text{cold}} = -Q_{\text{hot}}$, and Equation 20.4, $Q = mc\,\Delta T$, we obtain

$$m_1c_1\,\Delta T_c = -m_2c_2\,\Delta T_h$$

$$m_1c_1\,(T_f - T_c) = -m_2c_2\,(T_f - T_h)$$

Solving for T_f, we have

$$T_f = \frac{m_1c_1T_c + m_2c_2T_h}{m_1c_1 + m_2c_2} \tag{22.14}$$

The process is irreversible because the system goes through a series of nonequilibrium states. During such a transformation, the temperature of the system at any time is not well defined because different parts of the system have different temperatures. However, we can imagine that the hot substance at the initial temperature T_h is slowly cooled to the temperature T_f as it comes into contact with a series of reservoirs differing infinitesimally in temperature, the first reservoir being at T_h and the last being at T_f. Such a series of very small changes in temperature would approximate a reversible process. We imagine doing the same thing for the cold substance. Applying Equation 22.9 and noting that $dQ = mc\,dT$ for an infinitesimal change, we have

$$\Delta S = \int_1 \frac{dQ_{\text{cold}}}{T} + \int_2 \frac{dQ_{\text{hot}}}{T} = m_1c_1\int_{T_c}^{T_f}\frac{dT}{T} + m_2c_2\int_{T_h}^{T_f}\frac{dT}{T}$$

where we have assumed that the specific heats remain constant. Integrating, we find that

$$\Delta S = m_1c_1\ln\frac{T_f}{T_c} + m_2c_2\ln\frac{T_f}{T_h} \tag{22.15}$$

where T_f is given by Equation 22.14. If Equation 22.14 is substituted into Equation 22.15, we can show that one of the terms in Equation 22.15 is always positive and the other is always negative. (You may want to verify this for yourself.) The positive term is always greater than the negative term, and this results in a positive value for ΔS. Thus, we conclude that the entropy of the Universe increases in this irreversible process.

Finally, you should note that Equation 22.15 is valid only when no mixing of different substances occurs, because a further entropy increase is associated with the increase in disorder during the mixing. If the substances are liquids or gases and mixing occurs, the result applies only if the two fluids are identical, as in the following example.

Example 22.9 Calculating ΔS for a Calorimetric Process

Suppose that 1.00 kg of water at 0.00°C is mixed with an equal mass of water at 100°C. After equilibrium is reached, the mixture has a uniform temperature of 50.0°C. What is the change in entropy of the system?

Solution We can calculate the change in entropy from Equation 22.15 using the given values $m_1 = m_2 = 1.00$ kg, $c_1 = c_2 = 4\,186\,\text{J/kg}\cdot\text{K}$, $T_1 = 273$ K, $T_2 = 373$ K, and $T_f = 323$ K:

$$\Delta S = m_1 c_1 \ln \frac{T_f}{T_1} + m_2 c_2 \ln \frac{T_f}{T_2}$$

$$\Delta S = (1.00\text{ kg})(4\,186\text{ J/kg}\cdot\text{K}) \ln\left(\frac{323\text{ K}}{273\text{ K}}\right)$$

$$+ (1.00\text{ kg})(4\,186\text{ J/kg}\cdot\text{K}) \ln\left(\frac{323\text{ K}}{373\text{ K}}\right)$$

$$= 704\text{ J/K} - 602\text{ J/K} = 102\text{ J/K}$$

That is, as a result of this irreversible process, the increase in entropy of the cold water is greater than the decrease in entropy of the warm water. Consequently, the increase in entropy of the system is 102 J/K.

22.8 Entropy on a Microscopic Scale[6]

As we have seen, we can approach entropy by relying on macroscopic concepts. We can also treat entropy from a microscopic viewpoint through statistical analysis of molecular motions. We now use a microscopic model to investigate once again the free expansion of an ideal gas, which was discussed from a macroscopic point of view in the preceding section.

In the kinetic theory of gases, gas molecules are represented as particles moving randomly. Let us suppose that the gas is initially confined to a volume V_i, as shown in Figure 22.17a. When the partition separating V_i from a larger container is removed, the molecules eventually are distributed throughout the greater volume V_f (Fig. 22.17b). For a given uniform distribution of gas in the volume, there are a large number of equivalent microstates, and we can relate the entropy of the gas to the number of microstates corresponding to a given macrostate.

We count the number of microstates by considering the variety of molecular locations involved in the free expansion. The instant after the partition is removed (and before the molecules have had a chance to rush into the other half of the container), all the molecules are in the initial volume. We assume that each molecule occupies some microscopic volume V_m. The total number of possible locations of a single molecule in a macroscopic initial volume V_i is the ratio $w_i = V_i/V_m$, which is a huge number. We use w_i here to represent the number of *ways* that the molecule can be placed in the volume, or the number of microstates, which is equivalent to the number of available locations. We assume that the probabilities of a molecule occupying any of these locations are equal.

V_i Vacuum

(a)

As more molecules are added to the system, the number of possible ways that the molecules can be positioned in the volume multiplies. For example, if we consider two molecules, for every possible placement of the first, all possible placements of the second are available. Thus, there are w_1 ways of locating the first molecule, and for each of these, there are w_2 ways of locating the second molecule. The total number of ways of locating the two molecules is $w_1 w_2$.

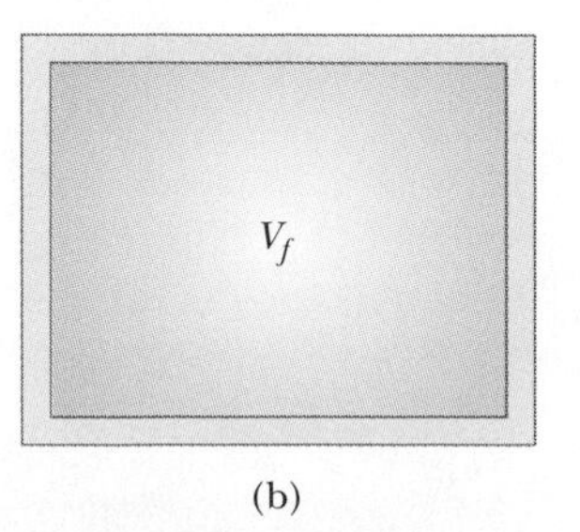

(b)

Figure 22.17 In a free expansion, the gas is allowed to expand into a region that was previously evacuated.

Neglecting the very small probability of having two molecules occupy the same location, each molecule may go into any of the V_i/V_m locations, and so the number of ways of locating N molecules in the volume becomes $W_i = w_i^N = (V_i/V_m)^N$. ($W_i$ is not to be confused with work.) Similarly, when the volume is increased to V_f, the number of ways of locating N molecules increases to $W_f = w_f^N = (V_f/V_m)^N$. The ratio of the number of ways of placing the molecules in the volume for the initial and final configurations is

[6] This section was adapted from A. Hudson and R. Nelson, *University Physics*, Philadelphia, Saunders College Publishing, 1990.

$$\frac{W_f}{W_i} = \frac{(V_f/V_m)^N}{(V_i/V_m)^N} = \left(\frac{V_f}{V_i}\right)^N$$

If we now take the natural logarithm of this equation and multiply by Boltzmann's constant, we find that

$$k_B \ln\left(\frac{W_f}{W_i}\right) = nN_A k_B \ln\left(\frac{V_f}{V_i}\right)$$

where we have used the equality $N = nN_A$. We know from Equation 19.11 that $N_A k_B$ is the universal gas constant R; thus, we can write this equation as

$$k_B \ln W_f - k_B \ln W_i = nR \ln\left(\frac{V_f}{V_i}\right) \qquad (22.16)$$

From Equation 22.13 we know that when n mol of a gas undergoes a free expansion from V_i to V_f, the change in entropy is

$$S_f - S_i = nR \ln\left(\frac{V_f}{V_i}\right) \qquad (22.17)$$

Note that the right-hand sides of Equations 22.16 and 22.17 are identical. Thus, from the left-hand sides, we make the following important connection between entropy and the number of microstates for a given macrostate:

$$S \equiv k_B \ln W \qquad (22.18)$$

Entropy (microscopic definition)

The more microstates there are that correspond to a given macrostate, the greater is the entropy of that macrostate. As we have discussed previously, there are many more microstates associated with disordered macrostates than with ordered macrostates. Thus, Equation 22.18 indicates mathematically that **entropy is a measure of disorder.** Although in our discussion we used the specific example of the free expansion of an ideal gas, a more rigorous development of the statistical interpretation of entropy would lead us to the same conclusion.

We have stated that individual microstates are equally probable. However, because there are far more microstates associated with a disordered macrostate than with an ordered microstate, a disordered macrostate is much more probable than an ordered one.

Figure 22.18 shows a real-world example of this concept. There are two possible macrostates for the carnival game—winning a goldfish and winning a black fish. Because only one jar in the array of jars contains a black fish, only one possible microstate corresponds to the macrostate of winning a black fish. A large number of microstates are described by the coin's falling into a jar containing a goldfish. Thus, for the macrostate of winning a goldfish, there are many equivalent microstates. As a result, the probability of winning a goldfish is much greater than the probability of winning a black fish. If there are 24 goldfish and 1 black fish, the probability of winning the black fish is 1 in 25. This assumes that all microstates have the same probability, a situation

Figure 22.18 By tossing a coin into a jar, the carnival-goer can win the fish in the jar. It is more likely that the coin will land in a jar containing a goldfish than in the one containing the black fish.

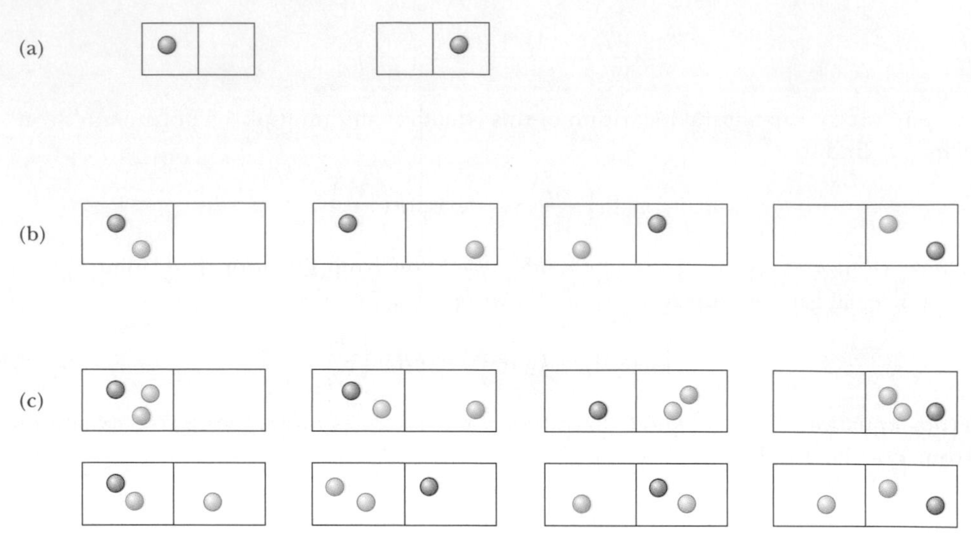

At the Active Figures link at http://www.pse6.com, you can choose the number of molecules to put in the container and measure the probability of all of them being in the left hand side.

Active Figure 22.19 (a) One molecule in a two-sided container has a 1-in-2 chance of being on the left side. (b) Two molecules have a 1-in-4 chance of being on the left side at the same time. (c) Three molecules have a 1-in-8 chance of being on the left side at the same time.

that may not be quite true for the situation shown in Figure 22.18. For example, if you are an accurate coin tosser and you are aiming for the edge of the array of jars, then the probability of the coin's landing in a jar near the edge is likely to be greater than the probability of its landing in a jar near the center.

Let us consider a similar type of probability problem for 100 molecules in a container. At any given moment, the probability of one molecule being in the left part of the container shown in Figure 22.19a as a result of random motion is $\frac{1}{2}$. If there are two molecules, as shown in Figure 22.19b, the probability of both being in the left part is $\left(\frac{1}{2}\right)^2$ or 1 in 4. If there are three molecules (Fig. 22.19c), the probability of all of them being in the left portion at the same moment is $\left(\frac{1}{2}\right)^3$, or 1 in 8. For 100 independently moving molecules, the probability that the 50 fastest ones will be found in the left part at any moment is $\left(\frac{1}{2}\right)^{50}$. Likewise, the probability that the remaining 50 slower molecules will be found in the right part at any moment is $\left(\frac{1}{2}\right)^{50}$. Therefore, the probability of finding this fast-slow separation as a result of random motion is the product $\left(\frac{1}{2}\right)^{50}\left(\frac{1}{2}\right)^{50} = \left(\frac{1}{2}\right)^{100}$, which corresponds to about 1 in 10^{30}. When this calculation is extrapolated from 100 molecules to the number in 1 mol of gas (6.02×10^{23}), the ordered arrangement is found to be *extremely* improbable!

Conceptual Example 22.10 Let's Play Marbles! **Interactive**

Suppose you have a bag of 100 marbles. Fifty of the marbles are red, and 50 are green. You are allowed to draw four marbles from the bag according to the following rules. Draw one marble, record its color, and return it to the bag. Shake the bag and then draw another marble. Continue this process until you have drawn and returned four marbles. What are the possible macrostates for this set of events? What is the most likely macrostate? What is the least likely macrostate?

Solution Because each marble is returned to the bag before the next one is drawn, and the bag is shaken, the probability of drawing a red marble is always the same as the probability of drawing a green one. All the possible microstates and macrostates are shown in Table 22.1. As this table indicates, there is only one way to draw a macrostate of four red marbles, and so there is only one microstate for that macrostate. However, there are four possible microstates that correspond to the macrostate of one green marble and three red marbles; six microstates that correspond to two green marbles and two red marbles; four microstates that correspond to three green marbles and one red marble; and one microstate that corresponds to four green marbles. The most likely, and most disordered, macrostate—two red marbles and two green marbles—corresponds to the largest number of microstates. The least likely, most ordered macrostates—four red marbles or four green marbles—correspond to the smallest number of microstates.

Table 22.1

Possible Results of Drawing Four Marbles from a Bag

Macrostate	Possible Microstates	Total Number of Microstates
All R	RRRR	1
1G, 3R	RRRG, RRGR, RGRR, GRRR	4
2G, 2R	RRGG, RGRG, GRRG, RGGR, GRGR, GGRR	6
3G, 1R	GGGR, GGRG, GRGG, RGGG	4
All G	GGGG	1

Explore the generation of microstates and macrostates at the Interactive Worked Example link at **http://www.pse6.com.**

Example 22.11 Adiabatic Free Expansion–One Last Time

Let us verify that the macroscopic and microscopic approaches to the calculation of entropy lead to the same conclusion for the adiabatic free expansion of an ideal gas. Suppose that an ideal gas expands to four times its initial volume. As we have seen for this process, the initial and final temperatures are the same.

(A) Using a macroscopic approach, calculate the entropy change for the gas.

(B) Using statistical considerations, calculate the change in entropy for the gas and show that it agrees with the answer you obtained in part (A).

Solution

(A) Using Equation 22.13, we have

$$\Delta S = nR\ln\left(\frac{V_f}{V_i}\right) = nR\ln\left(\frac{4V_i}{V_i}\right) = nR\ln 4$$

(B) The number of microstates available to a single molecule in the initial volume V_i is $w_i = V_i/V_m$. For N molecules, the number of available microstates is

$$W_i = w_i^N = \left(\frac{V_i}{V_m}\right)^N$$

The number of microstates for all N molecules in the final volume $V_f = 4V_i$ is

$$W_f = \left(\frac{V_f}{V_m}\right)^N = \left(\frac{4V_i}{V_m}\right)^N$$

Thus, the ratio of the number of final microstates to initial microstates is

$$\frac{W_f}{W_i} = 4^N$$

Using Equation 22.18, we obtain

$$\Delta S = k_B\ln W_f - k_B\ln W_i = k_B\ln\left(\frac{W_f}{W_i}\right)$$

$$= k_B\ln(4^N) = Nk_B\ln 4 = nR\ln 4$$

The answer is the same as that for part (A), which dealt with macroscopic parameters.

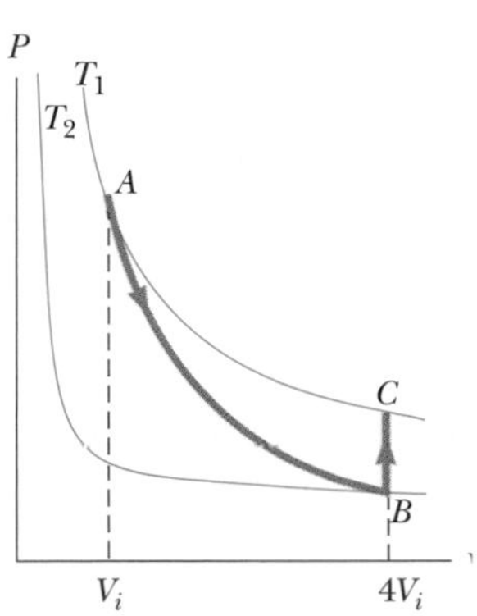

Figure 22.20 (Example 22.11) A gas expands to four times its initial volume and back to the initial temperature by means of a two-step process.

What If? **In part (A) we used Equation 22.13, which was based on a reversible isothermal process connecting the initial and final states. What if we were to choose a different reversible process? Would we arrive at the same result?**

Answer We *must* arrive at the same result because entropy is a state variable. For example, consider the two-step process in Figure 22.20—a reversible adiabatic expansion from V_i to $4V_i$, ($A \rightarrow B$) during which the temperature drops from T_1 to T_2, and a reversible isovolumetric process ($B \rightarrow C$) that takes the gas back to the initial temperature T_1.

During the reversible adiabatic process, $\Delta S = 0$ because $Q_r = 0$. During the reversible isovolumetric process ($B \rightarrow C$), we have from Equation 22.9,

$$\Delta S = \int_i^f \frac{dQ_r}{T} = \int_B^C \frac{nC_V\,dT}{T} = nC_V \ln\left(\frac{T_1}{T_2}\right)$$

Now, we can find the relationship of temperature T_2 to T_1 from Equation 21.20 for the adiabatic process:

$$\frac{T_1}{T_2} = \left(\frac{4V_i}{V_i}\right)^{\gamma-1} = (4)^{\gamma-1}$$

Thus,

$$\Delta S = nC_V \ln(4)^{\gamma-1} = nC_V(\gamma - 1)\ln 4$$

$$= nC_V\left(\frac{C_P}{C_V} - 1\right)\ln 4 = n(C_P - C_V)\ln 4 = nR\ln 4$$

and we do indeed obtain the exact same result for the entropy change.

SUMMARY

Take a practice test for this chapter by clicking on the Practice Test link at http://www.pse6.com.

A **heat engine** is a device that takes in energy by heat and, operating in a cyclic process, expels a fraction of that energy by means of work. The net work done by a heat engine in carrying a working substance through a cyclic process ($\Delta E_{\text{int}} = 0$) is

$$W_{\text{eng}} = |Q_h| - |Q_c| \tag{22.1}$$

where $|Q_h|$ is the energy taken in from a hot reservoir and $|Q_c|$ is the energy expelled to a cold reservoir.

The **thermal efficiency** e of a heat engine is

$$e = \frac{W_{\text{eng}}}{|Q_h|} = 1 - \frac{|Q_c|}{|Q_h|} \tag{22.2}$$

The **second law of thermodynamics** can be stated in the following two ways:

- It is impossible to construct a heat engine that, operating in a cycle, produces no effect other than the input of energy by heat from a reservoir and the performance of an equal amount of work (the Kelvin–Planck statement).
- It is impossible to construct a cyclical machine whose sole effect is to transfer energy continuously by heat from one object to another object at a higher temperature without the input of energy by work (the Clausius statement).

In a **reversible** process, the system can be returned to its initial conditions along the same path on a *PV* diagram, and every point along this path is an equilibrium state. A process that does not satisfy these requirements is **irreversible. Carnot's theorem** states that no real heat engine operating (irreversibly) between the temperatures T_c and T_h can be more efficient than an engine operating reversibly in a Carnot cycle between the same two temperatures.

The **thermal efficiency** of a heat engine operating in the Carnot cycle is

$$e_C = 1 - \frac{T_c}{T_h} \tag{22.6}$$

The second law of thermodynamics states that when real (irreversible) processes occur, the degree of disorder in the system plus the surroundings increases. When a process occurs in an isolated system, the state of the system becomes more disordered. The measure of disorder in a system is called **entropy** S. Thus, another way in which the second law can be stated is

- The entropy of the Universe increases in all real processes.

The **change in entropy** dS of a system during a process between two infinitesimally separated equilibrium states is

$$dS = \frac{dQ_r}{T} \tag{22.8}$$

where dQ_r is the energy transfer by heat for a reversible process that connects the initial and final states. The change in entropy of a system during an arbitrary process between an initial state and a final state is

$$\Delta S = \int_i^f \frac{dQ_r}{T} \tag{22.9}$$

The value of ΔS for the system is the same for all paths connecting the initial and final states. The change in entropy for a system undergoing any reversible, cyclic process is zero, and when such a process occurs, the entropy of the Universe remains constant.

From a microscopic viewpoint, the entropy of a given macrostate is defined as

$$S \equiv k_B \ln W \tag{22.18}$$

where k_B is Boltzmann's constant and W is the number of microstates of the system corresponding to the macrostate.

QUESTIONS

1. What are some factors that affect the efficiency of automobile engines?
2. In practical heat engines, which are we better able to control: the temperature of the hot reservoir or the temperature of the cold reservoir? Explain.
3. A steam-driven turbine is one major component of an electric power plant. Why is it advantageous to have the temperature of the steam as high as possible?
4. Is it possible to construct a heat engine that creates no thermal pollution? What does this tell us about environmental considerations for an industrialized society?
5. Does the second law of thermodynamics contradict or correct the first law? Argue for your answer.
6. "The first law of thermodynamics says you can't really win, and the second law says you can't even break even." Explain how this statement applies to a particular device or process; alternatively, argue against the statement.
7. In solar ponds constructed in Israel, the Sun's energy is concentrated near the bottom of a salty pond. With the proper layering of salt in the water, convection is prevented, and temperatures of 100°C may be reached. Can you estimate the maximum efficiency with which useful energy can be extracted from the pond?
8. Can a heat pump have a coefficient of performance less than unity? Explain.
9. Give various examples of irreversible processes that occur in nature. Give an example of a process in nature that is nearly reversible.
10. A heat pump is to be installed in a region where the average outdoor temperature in the winter months is -20°C. In view of this, why would it be advisable to place the outdoor compressor unit deep in the ground? Why are heat pumps not commonly used for heating in cold climates?
11. The device shown in Figure Q22.11, called a thermoelectric converter, uses a series of semiconductor cells to convert internal energy to electric potential energy, which we will study in Chapter 25. In the photograph at the left,

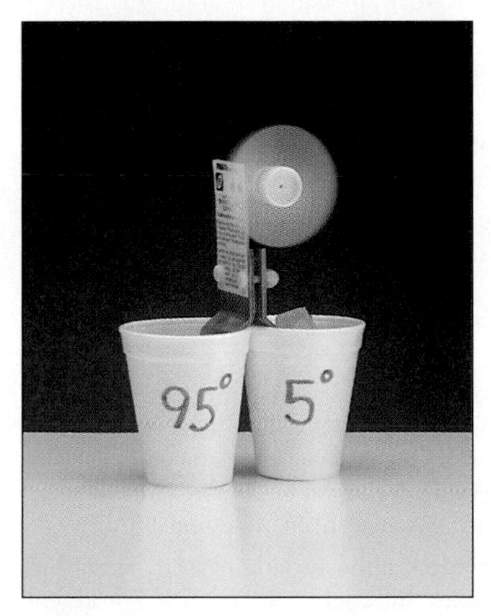

Courtesy of PASCO Scientific Company

Figure Q22.11

both legs of the device are at the same temperature, and no electric potential energy is produced. However, when one leg is at a higher temperature than the other, as in the photograph on the right, electric potential energy is produced as the device extracts energy from the hot reservoir and drives a small electric motor. (a) Why does the temperature differential produce electric potential energy in this demonstration? (b) In what sense does this intriguing experiment demonstrate the second law of thermodynamics?

12. Discuss three common examples of natural processes that involve an increase in entropy. Be sure to account for all parts of each system under consideration.

13. Discuss the change in entropy of a gas that expands (a) at constant temperature and (b) adiabatically.

14. A thermodynamic process occurs in which the entropy of a system changes by -8.0 J/K. According to the second law of thermodynamics, what can you conclude about the entropy change of the environment?

15. If a supersaturated sugar solution is allowed to evaporate slowly, sugar crystals form in the container. Hence, sugar molecules go from a disordered form (in solution) to a highly ordered crystalline form. Does this process violate the second law of thermodynamics? Explain.

16. How could you increase the entropy of 1 mol of a metal that is at room temperature? How could you decrease its entropy?

17. Suppose your roommate is "Mr. Clean" and tidies up your messy room after a big party. Because your roommate is creating more order, does this represent a violation of the second law of thermodynamics?

18. Discuss the entropy changes that occur when you (a) bake a loaf of bread and (b) consume the bread.

19. "Energy is the mistress of the Universe and entropy is her shadow." Writing for an audience of general readers, argue for this statement with examples. Alternatively, argue for the view that entropy is like a decisive hands-on executive instantly determining what will happen, while energy is like a wretched back-office bookkeeper telling us how little we can afford.

20. A classmate tells you that it is just as likely for all the air molecules in the room you are both in to be concentrated in one corner (with the rest of the room being a vacuum) as it is for the air molecules to be distributed uniformly about the room in their current state. Is this a true statement? Why doesn't the situation he describes actually happen?

21. If you shake a jar full of jellybeans of different sizes, the larger beans tend to appear near the top, and the smaller ones tend to fall to the bottom. Why? Does this process violate the second law of thermodynamics?

PROBLEMS

1, 2, 3 = straightforward, intermediate, challenging □ = full solution available in the *Student Solutions Manual and Study Guide*

= coached solution with hints available at http://www.pse6.com = computer useful in solving problem

= paired numerical and symbolic problems

Section 22.1 Heat Engines and the Second Law of Thermodynamics

1. A heat engine takes in 360 J of energy from a hot reservoir and performs 25.0 J of work in each cycle. Find (a) the efficiency of the engine and (b) the energy expelled to the cold reservoir in each cycle.

2. A heat engine performs 200 J of work in each cycle and has an efficiency of 30.0%. For each cycle, how much energy is (a) taken in and (b) expelled by heat?

3. A particular heat engine has a useful power output of 5.00 kW and an efficiency of 25.0%. The engine expels 8 000 J of exhaust energy in each cycle. Find (a) the energy taken in during each cycle and (b) the time interval for each cycle.

4. Heat engine *X* takes in four times more energy by heat from the hot reservoir than heat engine *Y*. Engine *X* delivers two times more work, and it rejects seven times more energy by heat to the cold reservoir than heat engine *Y*. Find the efficiency of (a) heat engine *X* and (b) heat engine *Y*.

5. A multicylinder gasoline engine in an airplane, operating at 2 500 rev/min, takes in energy 7.89×10^3 J and exhausts 4.58×10^3 J for each revolution of the crankshaft. (a) How many liters of fuel does it consume in 1.00 h of operation if the heat of combustion is 4.03×10^7 J/L? (b) What is the mechanical power output of the engine? Ignore friction and express the answer in horsepower. (c) What is the torque exerted by the crankshaft on the load? (d) What power must the exhaust and cooling system transfer out of the engine?

6. Suppose a heat engine is connected to two energy reservoirs, one a pool of molten aluminum (660°C) and the other a block of solid mercury (−38.9°C). The engine runs by freezing 1.00 g of aluminum and melting 15.0 g of mercury during each cycle. The heat of fusion of aluminum is 3.97×10^5 J/kg; the heat of fusion of mercury is 1.18×10^4 J/kg. What is the efficiency of this engine?

Section 22.2 Heat Pumps and Refrigerators

7. A refrigerator has a coefficient of performance equal to 5.00. The refrigerator takes in 120 J of energy from a cold reservoir in each cycle. Find (a) the work required in each cycle and (b) the energy expelled to the hot reservoir.

8. A refrigerator has a coefficient of performance of 3.00. The ice tray compartment is at −20.0°C, and the room temperature is 22.0°C. The refrigerator can convert 30.0 g of water at 22.0°C to 30.0 g of ice at −20.0°C each minute. What input power is required? Give your answer in watts.

9. In 1993 the federal government instituted a requirement that all room air conditioners sold in the United States must have an energy efficiency ratio (EER) of 10 or higher. The EER is defined as the ratio of the cooling capacity of the air conditioner, measured in Btu/h, to its electrical power requirement in watts. (a) Convert the EER of 10.0 to dimensionless form, using the conversion 1 Btu = 1 055 J. (b) What is the appropriate name for this dimensionless quantity? (c) In the 1970s it was common to find room air conditioners with EERs of 5 or lower. Compare the operating costs for 10 000-Btu/h air conditioners with EERs of 5.00 and 10.0. Assume that each air conditioner operates for 1 500 h during the summer in a city where electricity costs 10.0¢ per kWh.

Section 22.3 Reversible and Irreversible Processes

Section 22.4 The Carnot Engine

10. A Carnot engine has a power output of 150 kW. The engine operates between two reservoirs at 20.0°C and 500°C. (a) How much energy does it take in per hour? (b) How much energy is lost per hour in its exhaust?

11. One of the most efficient heat engines ever built is a steam turbine in the Ohio valley, operating between 430°C and 1 870°C on energy from West Virginia coal to produce electricity for the Midwest. (a) What is its maximum theoretical efficiency? (b) The actual efficiency of the engine is 42.0%. How much useful power does the engine deliver if it takes in 1.40×10^5 J of energy each second from its hot reservoir?

12. A heat engine operating between 200°C and 80.0°C achieves 20.0% of the maximum possible efficiency. What energy input will enable the engine to perform 10.0 kJ of work?

13. An ideal gas is taken through a Carnot cycle. The isothermal expansion occurs at 250°C, and the isothermal compression takes place at 50.0°C. The gas takes in 1 200 J of energy from the hot reservoir during the isothermal expansion. Find (a) the energy expelled to the cold reservoir in each cycle and (b) the net work done by the gas in each cycle.

14. The exhaust temperature of a Carnot heat engine is 300°C. What is the intake temperature if the efficiency of the engine is 30.0%?

15. A Carnot heat engine uses a steam boiler at 100°C as the high-temperature reservoir. The low-temperature reservoir is the outside environment at 20.0°C. Energy is exhausted to the low-temperature reservoir at the rate of 15.4 W. (a) Determine the useful power output of the heat engine. (b) How much steam will it cause to condense in the high-temperature reservoir in 1.00 h?

16. A power plant operates at a 32.0% efficiency during the summer when the sea water used for cooling is at 20.0°C. The plant uses 350°C steam to drive turbines. If the plant's efficiency changes in the same proportion as the ideal efficiency, what would be the plant's efficiency in the winter, when the sea water is 10.0°C?

17. Argon enters a turbine at a rate of 80.0 kg/min, a temperature of 800°C and a pressure of 1.50 MPa. It expands adiabatically as it pushes on the turbine blades and exits at pressure 300 kPa. (a) Calculate its temperature at exit. (b) Calculate the (maximum) power output of the turning turbine. (c) The turbine is one component of a model closed-cycle gas turbine engine. Calculate the maximum efficiency of the engine.

18. An electric power plant that would make use of the temperature gradient in the ocean has been proposed. The system is to operate between 20.0°C (surface water temperature) and 5.00°C (water temperature at a depth of about 1 km). (a) What is the maximum efficiency of such a system? (b) If the useful power output of the plant is 75.0 MW, how much energy is taken in from the warm reservoir per hour? (c) In view of your answer to part (a), do you think such a system is worthwhile? Note that the "fuel" is free.

19. Here is a clever idea. Suppose you build a two-engine device such that the exhaust energy output from one heat engine is the input energy for a second heat engine. We say that the two engines are running *in series*. Let e_1 and e_2 represent the efficiencies of the two engines. (a) The overall efficiency of the two-engine device is defined as the total work output divided by the energy put into the first engine by heat. Show that the overall efficiency is given by

$$e = e_1 + e_2 - e_1 e_2$$

(b) **What If?** Assume the two engines are Carnot engines. Engine 1 operates between temperatures T_h and T_i. The gas in engine 2 varies in temperature between T_i and T_c. In terms of the temperatures, what is the efficiency of the combination engine? (c) What value of the intermediate temperature T_i will result in equal work being done by each of the two engines in series? (d) What value of T_i will result in each of the two engines in series having the same efficiency?

20. A 20.0%-efficient real engine is used to speed up a train from rest to 5.00 m/s. It is known that an ideal (Carnot) engine using the same cold and hot reservoirs would accelerate the same train from rest to a speed of 6.50 m/s using the same amount of fuel. The engines use air at 300 K as a cold reservoir. Find the temperature of the steam serving as the hot reservoir.

21. A firebox is at 750 K, and the ambient temperature is 300 K. The efficiency of a Carnot engine doing 150 J of work as it transports energy between these constant-temperature baths is 60.0%. The Carnot engine must take in energy 150 J/0.600 = 250 J from the hot reservoir and must put out 100 J of energy by heat into the environment. To follow Carnot's reasoning, suppose that some other heat engine S could have efficiency 70.0%. (a) Find the energy input and wasted energy output of engine S as it does 150 J of work. (b) Let engine S operate as in part (a) and run the Carnot engine in reverse. Find the total energy the firebox puts out as both engines operate together, and the total energy transferred to the environment. Show that the Clausius statement of the second law of thermodynamics is violated. (c) Find the energy input and work output of engine S as it puts out exhaust energy of 100 J. (d) Let engine S operate as in (c) and contribute 150 J of its work output to running the Carnot engine in reverse. Find the total energy the firebox puts out as both engines operate together, the total work output, and the total energy transferred to the environment. Show that the Kelvin–Planck statement of the second law is violated. Thus our assumption about the efficiency of engine S must be false. (e) Let the engines operate together through one cycle as in part (d). Find the change in entropy of the Universe. Show that the entropy statement of the second law is violated.

22. At point A in a Carnot cycle, 2.34 mol of a monatomic ideal gas has a pressure of 1 400 kPa, a volume of 10.0 L, and a temperature of 720 K. It expands isothermally to point B, and then expands adiabatically to point C where its volume is 24.0 L. An isothermal compression brings it to point D, where its volume is 15.0 L. An adiabatic process returns the gas to point A. (a) Determine all the unknown pressures, volumes and temperatures as you fill in the following table:

	P	V	T
A	1 400 kPa	10.0 L	720 K
B			
C		24.0 L	
D		15.0 L	

(b) Find the energy added by heat, the work done by the engine, and the change in internal energy for each of the steps $A \rightarrow B$, $B \rightarrow C$, $C \rightarrow D$, and $D \rightarrow A$. (c) Calculate the efficiency W_{net}/Q_h. Show that it is equal to $1 - T_C/T_A$, the Carnot efficiency.

23. What is the coefficient of performance of a refrigerator that operates with Carnot efficiency between temperatures $-3.00°\text{C}$ and $+27.0°\text{C}$?

24. What is the maximum possible coefficient of performance of a heat pump that brings energy from outdoors at $-3.00°\text{C}$ into a 22.0°C house? Note that the work done to run the heat pump is also available to warm up the house.

25. An ideal refrigerator or ideal heat pump is equivalent to a Carnot engine running in reverse. That is, energy Q_c is taken in from a cold reservoir and energy Q_h is rejected to a hot reservoir. (a) Show that the work that must be supplied to run the refrigerator or heat pump is

$$W = \frac{T_h - T_c}{T_c} Q_c$$

(b) Show that the coefficient of performance of the ideal refrigerator is

$$\text{COP} = \frac{T_c}{T_h - T_c}$$

26. A heat pump, shown in Figure P22.26, is essentially an air conditioner installed backward. It extracts energy from colder air outside and deposits it in a warmer room. Suppose that the ratio of the actual energy entering the room to the work done by the device's motor is 10.0% of the theoretical maximum ratio. Determine the energy entering the room per joule of work done by the motor, given that the inside temperature is 20.0°C and the outside temperature is $-5.00°\text{C}$.

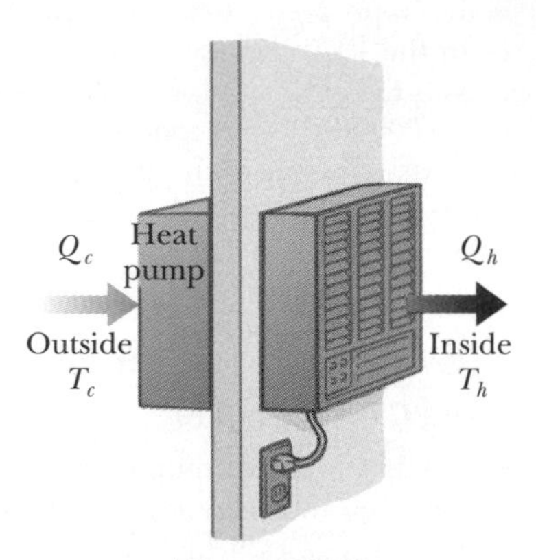

Figure P22.26

27. How much work does an ideal Carnot refrigerator require to remove 1.00 J of energy from helium at 4.00 K and reject this energy to a room-temperature (293-K) environment?

28. A refrigerator maintains a temperature of 0°C in the cold compartment with a room temperature of 25.0°C. It removes energy from the cold compartment at the rate of 8 000 kJ/h. (a) What minimum power is required to operate the refrigerator? (b) The refrigerator exhausts energy into the room at what rate?

29. If a 35.0%-efficient Carnot heat engine (Fig. 22.2) is run in reverse so as to form a refrigerator (Fig. 22.5), what would be this refrigerator's coefficient of performance?

30. Two Carnot engines have the same efficiency. One engine runs in reverse as a heat pump, and the other runs in reverse as a refrigerator. The coefficient of performance of the heat pump is 1.50 times the coefficient of performance of the refrigerator. Find (a) the coefficient of performance of the refrigerator, (b) the coefficient of performance of the heat pump, and (c) the efficiency of each heat engine.

Section 22.5 Gasoline and Diesel Engines

31. In a cylinder of an automobile engine, just after combustion, the gas is confined to a volume of 50.0 cm^3 and has an initial pressure of 3.00×10^6 Pa. The piston moves outward to a final volume of 300 cm^3, and the gas expands without energy loss by heat. (a) If $\gamma = 1.40$ for the gas, what is the final pressure? (b) How much work is done by the gas in expanding?

32. A gasoline engine has a compression ratio of 6.00 and uses a gas for which $\gamma = 1.40$. (a) What is the efficiency of the engine if it operates in an idealized Otto cycle? (b) **What If?** If the actual efficiency is 15.0%, what fraction of the fuel is wasted as a result of friction and energy losses by heat that could by avoided in a reversible engine? (Assume complete combustion of the air–fuel mixture.)

33. A 1.60-L gasoline engine with a compression ratio of 6.20 has a useful power output of 102 hp. Assuming the engine operates in an idealized Otto cycle, find the energy taken in and the energy exhausted each second. Assume the fuel–air mixture behaves like an ideal gas with $\gamma = 1.40$.

34. The compression ratio of an Otto cycle, as shown in Figure 22.13, is $V_A/V_B = 8.00$. At the beginning A of the compression process, 500 cm^3 of gas is at 100 kPa and 20.0°C. At the beginning of the adiabatic expansion the temperature is $T_C = 750$°C. Model the working fluid as an ideal gas with $E_{\text{int}} = nC_VT = 2.50nRT$ and $\gamma = 1.40$. (a) Fill in the table below to follow the states of the gas:

	T (K)	P (kPa)	V (cm^3)	E_{int}
A	293	100	500	
B				
C	1 023			
D				
A				

(b) Fill in the table below to follow the processes:

	Q (input)	W (output)	ΔE_{int}
$A \rightarrow B$			
$B \rightarrow C$			
$C \rightarrow D$			
$D \rightarrow A$			
$ABCDA$			

(c) Identify the energy input Q_h, the energy exhaust Q_c, and the net output work W_{eng}. (d) Calculate the thermal efficiency. (e) Find the number of crankshaft revolutions per minute required for a one-cylinder engine to have an output power of 1.00 kW = 1.34 hp. Note that the thermodynamic cycle involves four piston strokes.

Section 22.6 Entropy

35. An ice tray contains 500 g of liquid water at 0°C. Calculate the change in entropy of the water as it freezes slowly and completely at 0°C.

36. At a pressure of 1 atm, liquid helium boils at 4.20 K. The latent heat of vaporization is 20.5 kJ/kg. Determine the entropy change (per kilogram) of the helium resulting from vaporization.

37. Calculate the change in entropy of 250 g of water heated slowly from 20.0°C to 80.0°C. (*Suggestion:* Note that $dQ = mc\,dT$.)

38. In making raspberry jelly, 900 g of raspberry juice is combined with 930 g of sugar. The mixture starts at room temperature, 23.0°C, and is slowly heated on a stove until it reaches 220°F. It is then poured into heated jars and allowed to cool. Assume that the juice has the same specific heat as water. The specific heat of sucrose is 0.299 cal/g · °C. Consider the heating process. (a) Which of the following terms describe(s) this process: adiabatic, isobaric, isothermal, isovolumetric, cyclic, reversible, isentropic? (b) How much energy does the mixture absorb? (c) What is the minimum change in entropy of the jelly while it is heated?

39. What change in entropy occurs when a 27.9-g ice cube at − 12°C is transformed into steam at 115°C?

Section 22.7 Entropy Changes in Irreversible Processes

40. The temperature at the surface of the Sun is approximately 5 700 K, and the temperature at the surface of the Earth is approximately 290 K. What entropy change occurs when 1 000 J of energy is transferred by radiation from the Sun to the Earth?

41. A 1 500-kg car is moving at 20.0 m/s. The driver brakes to a stop. The brakes cool off to the temperature of the surrounding air, which is nearly constant at 20.0°C. What is the total entropy change?

42. A 1.00-kg iron horseshoe is taken from a forge at 900°C and dropped into 4.00 kg of water at 10.0°C. Assuming that no energy is lost by heat to the surroundings, determine the total entropy change of the horseshoe-plus-water system.

43. How fast are you personally making the entropy of the Universe increase right now? Compute an order-of-magnitude estimate, stating what quantities you take as data and the values you measure or estimate for them.

44. A rigid tank of small mass contains 40.0 g of argon, initially at 200°C and 100 kPa. The tank is placed into a reservoir at 0°C and allowed to cool to thermal equilibrium. (a) Calculate the volume of the tank. (b) Calculate the change in internal energy of the argon. (c) Calculate the energy transferred by heat. (d) Calculate the change in entropy of the argon. (e) Calculate the change in entropy of the constant-temperature bath.

45. A 1.00-mol sample of H_2 gas is contained in the left-hand side of the container shown in Figure P22.45, which has equal volumes left and right. The right-hand side is evacuated. When the valve is opened, the gas streams into the right-hand side. What is the final entropy change of the gas? Does the temperature of the gas change?

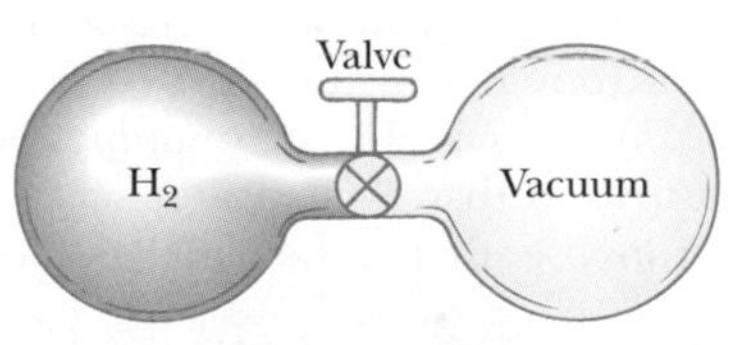

Figure P22.45

46. A 2.00-L container has a center partition that divides it into two equal parts, as shown in Figure P22.46. The left side contains H_2 gas, and the right side contains O_2 gas. Both gases are at room temperature and at atmospheric pressure. The partition is removed, and the gases are allowed to mix. What is the entropy increase of the system?

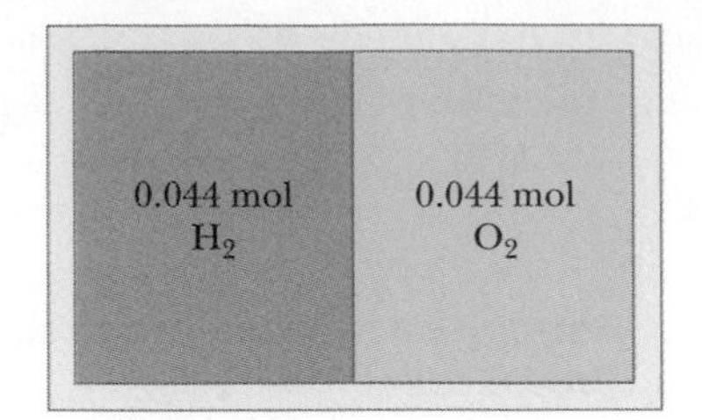

Figure P22.46

47. A 1.00-mol sample of an ideal monatomic gas, initially at a pressure of 1.00 atm and a volume of 0.025 0 m^3, is heated to a final state with a pressure of 2.00 atm and a volume of 0.040 0 m^3. Determine the change in entropy of the gas in this process.

48. A 1.00-mol sample of a diatomic ideal gas, initially having pressure P and volume V, expands so as to have pressure $2P$ and volume $2V$. Determine the entropy change of the gas in the process.

Section 22.8 Entropy on a Microscopic Scale

49. If you toss two dice, what is the total number of ways in which you can obtain (a) a 12 and (b) a 7?

50. Prepare a table like Table 22.1 for the following occurrence. You toss four coins into the air simultaneously and then record the results of your tosses in terms of the numbers of heads and tails that result. For example, HHTH and HTHH are two possible ways in which three heads and one tail can be achieved. (a) On the basis of your table, what is the most probable result of a toss? In terms of entropy, (b) what is the most ordered state and (c) what is the most disordered state?

51. Repeat the procedure used to construct Table 22.1 (a) for the case in which you draw three marbles from your bag rather than four and (b) for the case in which you draw five rather than four.

Additional Problems

52. Every second at Niagara Falls (Fig. P22.52), some 5 000 m^3 of water falls a distance of 50.0 m. What is the increase in entropy per second due to the falling water? Assume that the mass of the surroundings is so great that its temperature and that of the water stay nearly constant at 20.0°C. Suppose that a negligible amount of water evaporates.

CORBIS/Stock Market

Figure P22.52 Niagara Falls, a popular tourist attraction.

53. A house loses energy through the exterior walls and roof at a rate of 5 000 J/s = 5.00 kW when the interior temperature is 22.0°C and the outside temperature is −5.00°C. Calculate the electric power required to maintain the interior temperature at 22.0°C for the following two cases. (a) The electric power is used in electric resistance heaters (which convert all of the energy transferred in by electrical transmission into internal energy). (b) **What If?** The electric power is used to drive an electric motor that operates the compressor of a heat pump, which has a coefficient of performance equal to 60.0% of the Carnot-cycle value.

54. How much work is required, using an ideal Carnot refrigerator, to change 0.500 kg of tap water at 10.0°C into ice at −20.0°C? Assume the temperature of the freezer compartment is held at −20.0°C and the refrigerator exhausts energy into a room at 20.0°C.

55. A heat engine operates between two reservoirs at T_2 = 600 K and T_1 = 350 K. It takes in 1 000 J of energy from the higher-temperature reservoir and performs 250 J of work. Find (a) the entropy change of the Universe ΔS_U for this process and (b) the work W that could have been done by an ideal Carnot engine operating between these two reservoirs. (c) Show that the difference between the amounts of work done in parts (a) and (b) is $T_1 \Delta S_U$.

56. Two identically constructed objects, surrounded by thermal insulation, are used as energy reservoirs for a Carnot engine. The finite reservoirs both have mass m and specific heat c. They start out at temperatures T_h and T_c, where $T_h > T_c$. (a) Show that the engine will stop working when the final temperature of each object is $(T_h\ T_c)^{1/2}$. (b) Show that the total work done by the

Carnot engine is

$$W_{\text{eng}} = mc(T_h^{1/2} - T_c^{1/2})^2$$

57. In 1816 Robert Stirling, a Scottish clergyman, patented the *Stirling engine*, which has found a wide variety of applications ever since. Fuel is burned externally to warm one of the engine's two cylinders. A fixed quantity of inert gas moves cyclically between the cylinders, expanding in the hot one and contracting in the cold one. Figure P22.57 represents a model for its thermodynamic cycle. Consider n mol of an ideal monatomic gas being taken once through the cycle, consisting of two isothermal processes at temperatures $3T_i$ and T_i and two constant-volume processes. Determine, in terms of n, R, and T_i, (a) the net energy transferred by heat to the gas and (b) the efficiency of the engine. A Stirling engine is easier to manufacture than an internal combustion engine or a turbine. It can run on burning garbage. It can run on the energy of sunlight and produce no material exhaust.

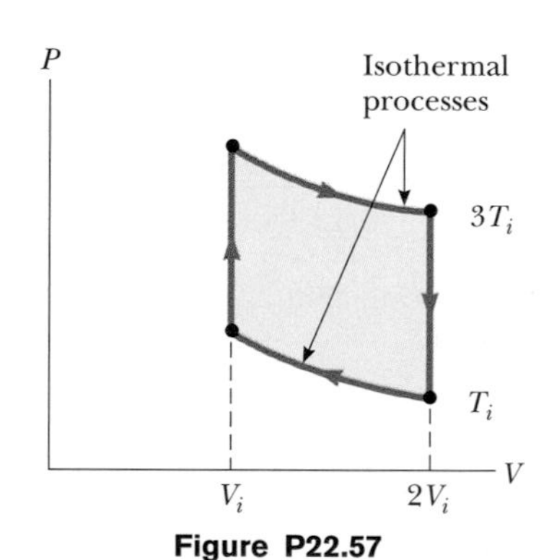

Figure P22.57

58. An electric power plant has an overall efficiency of 15.0%. The plant is to deliver 150 MW of power to a city, and its turbines use coal as the fuel. The burning coal produces steam that drives the turbines. This steam is then condensed to water at 25.0°C by passing it through cooling coils in contact with river water. (a) How many metric tons of coal does the plant consume each day (1 metric ton $= 10^3$ kg)? (b) What is the total cost of the fuel per year if the delivered price is \$8.00/metric ton? (c) If the river water is delivered at 20.0°C, at what minimum rate must it flow over the cooling coils in order that its temperature not exceed 25.0°C? (*Note:* The heat of combustion of coal is 33.0 kJ/g.)

59. A power plant, having a Carnot efficiency, produces 1 000 MW of electrical power from turbines that take in steam at 500 K and reject water at 300 K into a flowing river. The water downstream is 6.00 K warmer due to the output of the power plant. Determine the flow rate of the river.

60. A power plant, having a Carnot efficiency, produces electric power $\mathcal{P}$ from turbines that take in energy from steam at temperature T_h and discharge energy at temperature T_c through a heat exchanger into a flowing river. The water downstream is warmer by ΔT due to the output of the power plant. Determine the flow rate of the river.

61. An athlete whose mass is 70.0 kg drinks 16 oz (453.6 g) of refrigerated water. The water is at a temperature of 35.0°F. (a) Ignoring the temperature change of the body that results from the water intake (so that the body is regarded as a reservoir always at 98.6°F), find the entropy increase of the entire system. (b) **What If?** Assume that the entire body is cooled by the drink and that the average specific heat of a person is equal to the specific heat of liquid water. Ignoring any other energy transfers by heat and any metabolic energy release, find the athlete's temperature after she drinks the cold water, given an initial body temperature of 98.6°F. Under these assumptions, what is the entropy increase of the entire system? Compare this result with the one you obtained in part (a).

62. A 1.00-mol sample of an ideal monatomic gas is taken through the cycle shown in Figure P22.62. The process $A \rightarrow B$ is a reversible isothermal expansion. Calculate (a) the net work done by the gas, (b) the energy added to the gas by heat, (c) the energy exhausted from the gas by heat, and (d) the efficiency of the cycle.

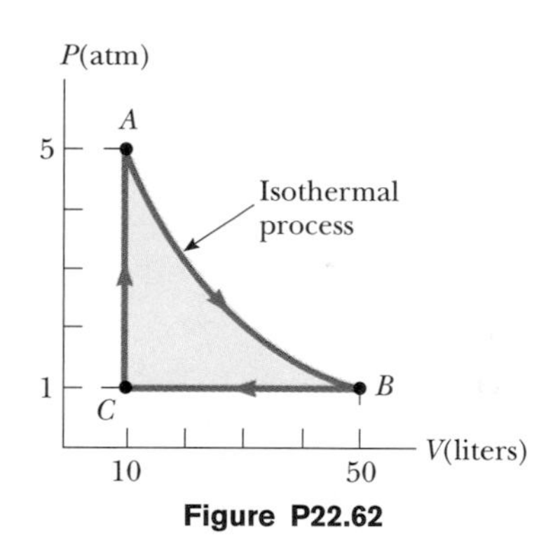

Figure P22.62

63. A biology laboratory is maintained at a constant temperature of 7.00°C by an air conditioner, which is vented to the air outside. On a typical hot summer day the outside temperature is 27.0°C and the air conditioning unit emits energy to the outside at a rate of 10.0 kW. Model the unit as having a coefficient of performance equal to 40.0% of the coefficient of performance of an ideal Carnot device. (a) At what rate does the air conditioner remove energy from the laboratory? (b) Calculate the power required for the work input. (c) Find the change in entropy produced by the air conditioner in 1.00 h. (d) **What If?** The outside temperature increases to 32.0°C. Find the fractional change in the coefficient of performance of the air conditioner.

64. A 1.00-mol sample of an ideal gas expands isothermally, doubling in volume. (a) Show that the work it does in ex-

panding is $W = RT \ln 2$. (b) Because the internal energy E_{int} of an ideal gas depends solely on its temperature, the change in internal energy is zero during the expansion. It follows from the first law that the energy input to the gas by heat during the expansion is equal to the energy output by work. Why does this conversion *not* violate the second law?

65. A 1.00-mol sample of a monatomic ideal gas is taken through the cycle shown in Figure P22.65. At point *A*, the pressure, volume, and temperature are P_i, V_i, and T_i, respectively. In terms of R and T_i, find (a) the total energy entering the system by heat per cycle, (b) the total energy leaving the system by heat per cycle, (c) the efficiency of an engine operating in this cycle, and (d) the efficiency of an engine operating in a Carnot cycle between the same temperature extremes.

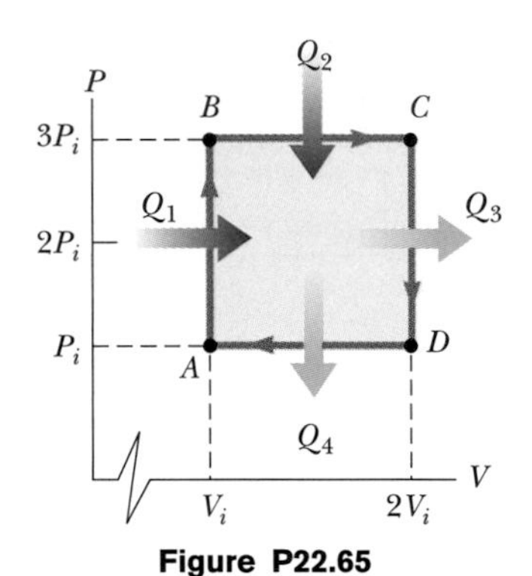

Figure P22.65

66. A sample consisting of n mol of an ideal gas undergoes a reversible isobaric expansion from volume V_i to volume $3V_i$. Find the change in entropy of the gas by calculating $\int_i^f dQ/T$ where $dQ = nC_P\,dT$.

67. A system consisting of n mol of an ideal gas undergoes two reversible processes. It starts with pressure P_i and volume V_i, expands isothermally, and then contracts adiabatically to reach a final state with pressure P_i and volume $3V_i$. (a) Find its change in entropy in the isothermal process. The entropy does not change in the adiabatic process. (b) **What If?** Explain why the answer to part (a) must be the same as the answer to Problem 66.

68. Suppose you are working in a patent office, and an inventor comes to you with the claim that her heat engine, which employs water as a working substance, has a thermodynamic efficiency of 0.61. She explains that it operates between energy reservoirs at 4°C and 0°C. It is a very complicated device, with many pistons, gears, and pulleys, and the cycle involves freezing and melting. Does her claim that $e = 0.61$ warrant serious consideration? Explain.

69. An idealized diesel engine operates in a cycle known as the *air-standard diesel cycle,* shown in Figure 22.14. Fuel is sprayed into the cylinder at the point of maximum compression, *B*. Combustion occurs during the expansion $B \rightarrow C$, which is modeled as an isobaric process. Show that the efficiency of an engine operating in this idealized diesel cycle is

$$e = 1 - \frac{1}{\gamma}\left(\frac{T_D - T_A}{T_C - T_B}\right)$$

70. A 1.00-mol sample of an ideal gas ($\gamma = 1.40$) is carried through the Carnot cycle described in Figure 22.11. At point *A*, the pressure is 25.0 atm and the temperature is 600 K. At point *C*, the pressure is 1.00 atm and the temperature is 400 K. (a) Determine the pressures and volumes at points *A*, *B*, *C*, and *D*. (b) Calculate the net work done per cycle. (c) Determine the efficiency of an engine operating in this cycle.

71. Suppose 1.00 kg of water at 10.0°C is mixed with 1.00 kg of water at 30.0°C at constant pressure. When the mixture has reached equilibrium, (a) what is the final temperature? (b) Take $c_P = 4.19$ kJ/kg·K for water and show that the entropy of the system increases by

$$\Delta S = 4.19 \ln\left[\left(\frac{293}{283}\right)\left(\frac{293}{303}\right)\right] \text{kJ/K}$$

(c) Verify numerically that $\Delta S > 0$. (d) Is the mixing an irreversible process?

Answers to Quick Quizzes

22.1 (c). Equation 22.2 gives this result directly.

22.2 (b). The work represents one third of the input energy. The remainder, two thirds, must be expelled to the cold reservoir.

22.3 (d). The COP of 4.00 for the heat pump means that you are receiving four times as much energy as the energy entering by electrical transmission. With four times as much energy per unit of energy from electricity, you need only one fourth as much electricity.

22.4 C, B, A. Although all three engines operate over a 300-K temperature difference, the efficiency depends on the ratio of temperatures, not the difference.

22.5 One microstate—all four deuces.

22.6 Six microstates—club–diamond, club–heart, club–spade, diamond–heart, diamond–spade, heart–spade. The macrostate of two aces is more probable than that of four deuces in Quick Quiz 22.5 because there are six times as many microstates for this particular macrostate compared to the macrostate of four deuces. Thus, in a hand of poker, two of a kind is less valuable than four of a kind.

22.7 (b). Because the process is reversible and adiabatic, $Q_r = 0$; therefore, $\Delta S = 0$.

22.8 (a). From the first law of thermodynamics, for these two reversible processes, $Q_r = \Delta E_{\text{int}} - W$. During the constant-volume process, $W = 0$, while the work W is nonzero and negative during the constant-pressure expansion. Thus, Q_r is larger for the constant pressure process, leading to a larger value for the change in entropy. In terms of entropy as disorder, during the constant-pressure process, the gas must expand. The increase in volume results in more ways of locating the molecules of the gas in a container, resulting in a larger increase in entropy.

22.9 False. The determining factor for the entropy change is Q_r, not Q. If the adiabatic process is not reversible, the entropy change is not necessarily zero because a reversible path between the same initial and final states may involve energy transfer by heat.

Some Physical Constants[a]

Quantity	Symbol	Value[b]
Atomic mass unit	u	$1.660\,538\,73\,(13) \times 10^{-27}$ kg $931.494\,013\,(37)$ MeV/c^2
Avogadro's number	N_A	$6.022\,141\,99\,(47) \times 10^{23}$ particles/mol
Bohr magneton	$\mu_B = \frac{e\hbar}{2m_e}$	$9.274\,008\,99\,(37) \times 10^{-24}$ J/T
Bohr radius	$a_0 = \frac{\hbar^2}{m_e e^2 k_e}$	$5.291\,772\,083\,(19) \times 10^{-11}$ m
Boltzmann's constant	$k_B = \frac{R}{N_A}$	$1.380\,650\,3\,(24) \times 10^{-23}$ J/K
Compton wavelength	$\lambda_C = \frac{h}{m_e c}$	$2.426\,310\,215\,(18) \times 10^{-12}$ m
Coulomb constant	$k_e = \frac{1}{4\pi\epsilon_0}$	$8.987\,551\,788\ldots \times 10^9$ N·m²/C² (exact)
Deuteron mass	m_d	$3.343\,583\,09\,(26) \times 10^{-27}$ kg $2.013\,553\,212\,71\,(35)$ u
Electron mass	m_e	$9.109\,381\,88\,(72) \times 10^{-31}$ kg $5.485\,799\,110\,(12) \times 10^{-4}$ u $0.510\,998\,902\,(21)$ MeV/c^2
Electron volt	eV	$1.602\,176\,462\,(63) \times 10^{-19}$ J
Elementary charge	e	$1.602\,176\,462\,(63) \times 10^{-19}$ C
Gas constant	R	$8.314\,472\,(15)$ J/K·mol
Gravitational constant	G	$6.673\,(10) \times 10^{-11}$ N·m²/kg²
Josephson frequency–voltage ratio	$\frac{2e}{h}$	$4.835\,978\,98\,(19) \times 10^{14}$ Hz/V
Magnetic flux quantum	$\Phi_0 = \frac{h}{2e}$	$2.067\,833\,636\,(81) \times 10^{-15}$ T·m²
Neutron mass	m_n	$1.674\,927\,16\,(13) \times 10^{-27}$ kg $1.008\,664\,915\,78\,(55)$ u $939.565\,330\,(38)$ MeV/c^2
Nuclear magneton	$\mu_n = \frac{e\hbar}{2m_p}$	$5.050\,783\,17\,(20) \times 10^{-27}$ J/T
Permeability of free space	μ_0	$4\pi \times 10^{-7}$ T·m/A (exact)
Permittivity of free space	$\epsilon_0 = \frac{1}{\mu_0 c^2}$	$8.854\,187\,817\ldots \times 10^{-12}$ C²/N·m² (exact)
Planck's constant	h	$6.626\,068\,76\,(52) \times 10^{-34}$ J·s
	$\hbar = \frac{h}{2\pi}$	$1.054\,571\,596\,(82) \times 10^{-34}$ J·s
Proton mass	m_p	$1.672\,621\,58\,(13) \times 10^{-27}$ kg $1.007\,276\,466\,88\,(13)$ u $938.271\,998\,(38)$ MeV/c^2
Rydberg constant	R_H	$1.097\,373\,156\,854\,9\,(83) \times 10^7$ m^{-1}
Speed of light in vacuum	c	$2.997\,924\,58 \times 10^8$ m/s (exact)

[a] These constants are the values recommended in 1998 by CODATA, based on a least-squares adjustment of data from different measurements. For a more complete list, see P. J. Mohr and B. N. Taylor, *Rev. Mod. Phys.* 72:351, 2000.

[b] The numbers in parentheses for the values above represent the uncertainties of the last two digits.

Solar System Data

Body	Mass (kg)	Mean Radius (m)	Period (s)	Distance from the Sun (m)
Mercury	3.18×10^{23}	2.43×10^{6}	7.60×10^{6}	5.79×10^{10}
Venus	4.88×10^{24}	6.06×10^{6}	1.94×10^{7}	1.08×10^{11}
Earth	5.98×10^{24}	6.37×10^{6}	3.156×10^{7}	1.496×10^{11}
Mars	6.42×10^{23}	3.37×10^{6}	5.94×10^{7}	2.28×10^{11}
Jupiter	1.90×10^{27}	6.99×10^{7}	3.74×10^{8}	7.78×10^{11}
Saturn	5.68×10^{26}	5.85×10^{7}	9.35×10^{8}	1.43×10^{12}
Uranus	8.68×10^{25}	2.33×10^{7}	2.64×10^{9}	2.87×10^{12}
Neptune	1.03×10^{26}	2.21×10^{7}	5.22×10^{9}	4.50×10^{12}
Pluto	$\approx 1.4 \times 10^{22}$	$\approx 1.5 \times 10^{6}$	7.82×10^{9}	5.91×10^{12}
Moon	7.36×10^{22}	1.74×10^{6}	—	—
Sun	1.991×10^{30}	6.96×10^{8}	—	—

Physical Data Often Used[a]

Average Earth–Moon distance	3.84×10^{8} m
Average Earth–Sun distance	1.496×10^{11} m
Average radius of the Earth	6.37×10^{6} m
Density of air (20°C and 1 atm)	$1.20\ \text{kg/m}^3$
Density of water (20°C and 1 atm)	$1.00 \times 10^{3}\ \text{kg/m}^3$
Free-fall acceleration	$9.80\ \text{m/s}^2$
Mass of the Earth	5.98×10^{24} kg
Mass of the Moon	7.36×10^{22} kg
Mass of the Sun	1.99×10^{30} kg
Standard atmospheric pressure	1.013×10^{5} Pa

[a] These are the values of the constants as used in the text.

Some Prefixes for Powers of Ten

Power	Prefix	Abbreviation	Power	Prefix	Abbreviation
10^{-24}	yocto	y	10^{1}	deka	da
10^{-21}	zepto	z	10^{2}	hecto	h
10^{-18}	atto	a	10^{3}	kilo	k
10^{-15}	femto	f	10^{6}	mega	M
10^{-12}	pico	p	10^{9}	giga	G
10^{-9}	nano	n	10^{12}	tera	T
10^{-6}	micro	μ	10^{15}	peta	P
10^{-3}	milli	m	10^{18}	exa	E
10^{-2}	centi	c	10^{21}	zetta	Z
10^{-1}	deci	d	10^{24}	yotta	Y

Standard Abbreviations and Symbols for Units

Symbol	Unit	Symbol	Unit
A	ampere	K	kelvin
u	atomic mass unit	kg	kilogram
atm	atmosphere	kmol	kilomole
Btu	British thermal unit	L	liter
C	coulomb	lb	pound
°C	degree Celsius	ly	lightyear
cal	calorie	m	meter
d	day	min	minute
eV	electron volt	mol	mole
°F	degree Fahrenheit	N	newton
F	farad	Pa	pascal
ft	foot	rad	radian
G	gauss	rev	revolution
g	gram	s	second
H	henry	T	tesla
h	hour	V	volt
hp	horsepower	W	watt
Hz	hertz	Wb	weber
in.	inch	yr	year
J	joule	Ω	ohm

Mathematical Symbols Used in the Text and Their Meaning

Symbol	Meaning
$=$	is equal to
$\equiv$	is defined as
$\neq$	is not equal to
$\propto$	is proportional to
$\sim$	is on the order of
$>$	is greater than
$<$	is less than
$\gg (\ll)$	is much greater (less) than
$\approx$	is approximately equal to
Δx	the change in x
$\sum_{i=1}^{N} x_i$	the sum of all quantities x_i from $i = 1$ to $i = N$
$\lvert x \rvert$	the magnitude of x (always a nonnegative quantity)
$\Delta x \rightarrow 0$	Δx approaches zero
$\frac{dx}{dt}$	the derivative of x with respect to t
$\frac{\partial x}{\partial t}$	the partial derivative of x with respect to t
$\int$	integral

Conversions[a]

Length

1 in. = 2.54 cm (exact)
1 m = 39.37 in. = 3.281 ft
1 ft = 0.304 8 m
12 in. = 1 ft
3 ft = 1 yd
1 yd = 0.914 4 m
1 km = 0.621 mi
1 mi = 1.609 km
1 mi = 5 280 ft
1 μm = 10^{-6} m = 10^3 nm
1 lightyear = 9.461×10^{15} m

Area

1 m^2 = 10^4 cm^2 = 10.76 ft^2
1 ft^2 = 0.092 9 m^2 = 144 $in.^2$
1 $in.^2$ = 6.452 cm^2

Volume

1 m^3 = 10^6 cm^3 = 6.102×10^4 $in.^3$
1 ft^3 = 1 728 $in.^3$ = 2.83×10^{-2} m^3
1 L = 1 000 cm^3 = 1.057 6 qt = 0.035 3 ft^3
1 ft^3 = 7.481 gal = 28.32 L = 2.832×10^{-2} m^3
1 gal = 3.786 L = 231 $in.^3$

Mass

1 000 kg = 1 t (metric ton)
1 slug = 14.59 kg
1 u = 1.66×10^{-27} kg = 931.5 MeV/c^2

Force

1 N = 0.224 8 lb
1 lb = 4.448 N

Velocity

1 mi/h = 1.47 ft/s = 0.447 m/s = 1.61 km/h
1 m/s = 100 cm/s = 3.281 ft/s
1 mi/min = 60 mi/h = 88 ft/s

Acceleration

1 m/s^2 = 3.28 ft/s^2 = 100 cm/s^2
1 ft/s^2 = 0.304 8 m/s^2 = 30.48 cm/s^2

Pressure

1 bar = 10^5 N/m^2 = 14.50 lb/$in.^2$
1 atm = 760 mm Hg = 76.0 cm Hg
1 atm = 14.7 lb/$in.^2$ = 1.013×10^5 N/m^2
1 Pa = 1 N/m^2 = 1.45×10^{-4} lb/$in.^2$

Time

1 yr = 365 days = 3.16×10^7 s
1 day = 24 h = 1.44×10^3 min = 8.64×10^4 s

Energy

1 J = 0.738 ft·lb
1 cal = 4.186 J
1 Btu = 252 cal = 1.054×10^3 J
1 eV = 1.6×10^{-19} J
1 kWh = 3.60×10^6 J

Power

1 hp = 550 ft·lb/s = 0.746 kW
1 W = 1 J/s = 0.738 ft·lb/s
1 Btu/h = 0.293 W

Some Approximations Useful for Estimation Problems

1 m ≈ 1 yd
1 kg ≈ 2 lb
1 N ≈ $\frac{1}{4}$ lb
1 L ≈ $\frac{1}{4}$ gal
1 m/s ≈ 2 mi/h
1 yr ≈ $\pi \times 10^7$ s
60 mi/h ≈ 100 ft/s
1 km ≈ $\frac{1}{2}$ mi

[a] See Table A.1 of Appendix A for a more complete list.

The Greek Alphabet

Alpha	A	α	Iota	I	ι	Rho	P	ρ
Beta	B	β	Kappa	K	κ	Sigma	Σ	σ
Gamma	Γ	γ	Lambda	Λ	λ	Tau	T	τ
Delta	Δ	δ	Mu	M	μ	Upsilon	Υ	υ
Epsilon	E	ϵ	Nu	N	ν	Phi	Φ	ϕ
Zeta	Z	ζ	Xi	Ξ	ξ	Chi	X	χ
Eta	H	η	Omicron	O	o	Psi	Ψ	ψ
Theta	Θ	θ	Pi	Π	π	Omega	Ω	ω